DYNAMICAL SYMMETRY

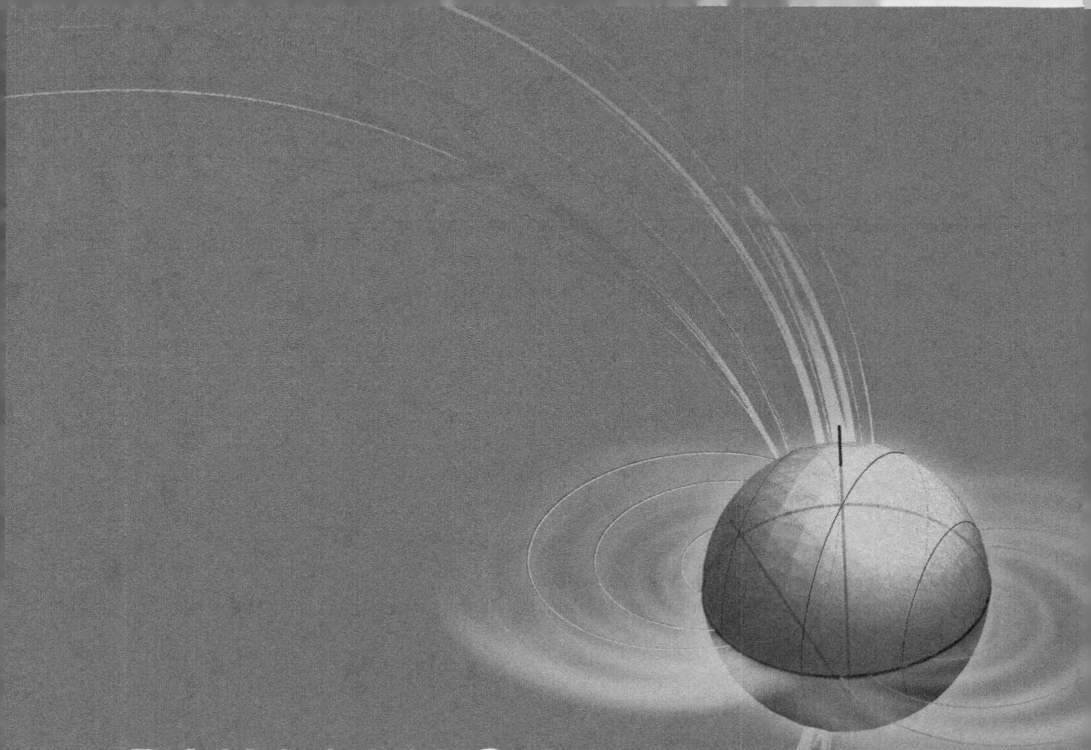

DYNAMICAL SYMMETRY

Carl E Wulfman
University of the Pacific, USA

World Scientific

NEW JERSEY • LONDON • SINGAPORE • BEIJING • SHANGHAI • HONG KONG • TAIPEI • CHENNAI

Published by

World Scientific Publishing Co. Pte. Ltd.
5 Toh Tuck Link, Singapore 596224
USA office: 27 Warren Street, Suite 401-402, Hackensack, NJ 07601
UK office: 57 Shelton Street, Covent Garden, London WC2H 9HE

British Library Cataloguing-in-Publication Data
A catalogue record for this book is available from the British Library.

First published 2011 (Hardcover)
Reprinted 2016 (in paperback edition)
ISBN 978-981-3203-62-4

DYNAMICAL SYMMETRY

ISBN-13 978-981-4291-36-1
ISBN-10 981-4291-36-6

Typeset by Stallion Press
Email: enquiries@stallionpress.com

Printed in Singapore

Dedication

To my fellow scientists who paused to encourage as they explored new vistas — Leonello Paoloni, John Platt, Marcos Moshinsky, Brian Judd, Brian Wybourne, Raphy Levine, and John Avery.

Preface

In his *Harmonices Mundi*, published in 1619, Johannes Kepler exclaims, "Finally I have brought to light and verified beyond all my hopes and expectations that the whole Nature of Harmonies permeates to the fullest extent, and in all its details, the motion of heavenly bodies; not, it is true, in the manner in which I had earlier thought, but in a totally different, altogether complete way."

This book deals with concepts of symmetry which, nearly four centuries after Kepler, continue to provide ever deepening, and often surprising, understandings of interrelations between phenomena associated with moving bodies.

Kepler's religious convictions and his intuition led him to believe that geometric symmetries and musical harmonies are expressed in natural phenomena. The development of the calculus and Newtonian physics in the latter half of the seventeenth century made it possible to deduce and extend Kepler's three laws of planetary motion, but they were seldom seen as expressions of symmetries and harmonies in the natural world. Then, two and a half centuries after we were given Harmonices Mundi, our concepts of symmetry were profoundly deepened by Sophus Lie. Lie saw that any system governed by differential equations exhibits symmetries in a heretofore unrecognized sense. Because so many laws of nature can be formulated as differential equations, Lie's concept of symmetry makes it possible to establish a host of relationships between natural phenomena. By mid twentieth century, physicists had come to realize that relationships expressing Lie symmetries, though easily overlooked, are surprisingly useful. Now, much modern physics has become a search, first for symmetries, then, secondly, for the dynamics that produces them.

Readers of Lie's works cannot help but realize that much of his mathematics was guided by geometric intuitions. The extraordinary fertility of his thought processes, beyond a shadow of doubt, arose in part from the extraordinary visual capabilities of the human mind. Just as a picture can be "worth a thousand words", so also can a geometric illustration, even a non-Euclidean one, suggest a host of mathematical and physical relationships. For this reason, the first three chapters of this book use interconnected geometrical, analytic, and physical concepts of symmetry to stimulate questions as well as insights in the mind of the reader.

A more formal treatment of the subject begins in Chapter 4. Among other things, Chapters 4 and 6 establish the manner in which Lie symmetries are defined by differential equations. Chapter 5 explains how one determines the invariance transformations of functions, functionals, and equations. Readers may wish to shift back and forth between the first three chapters, and these chapters.

The remaining chapters of the book deal with physical and chemical consequences of the symmetries implied in the dynamical equations of mechanics, quantum mechanics, and electromagnetism. Invariance transformations of partial differential equations are discussed in Chapter 9, the first chapter dealing with Schrödinger equations.

A number of monographs currently in print deal with the invariance properties of differential equations, and several recent and very fine ones deal with the utilization of these properties in applied mathematics and engineering. A list of these will be found in the Bibliography at the end of this Preface. Helpful, but out of print, monographs are also listed.

The approach taken in this book contrasts with those to be found in the books of the Bibliography: here, geometric concepts play a greater role. It is hoped that the reader's natural geometric understanding of ordinary symmetry, will, like Lie's, be extended into a partly intuitive, geometric as well as analytic, understanding of symmetries not seen in the space of our common experience. A good deal of emphasis is placed upon dynamical symmetries of classical and quantum mechanical systems of interest to both chemists and physicists. Readers interested in engineering or applied mathematics may also find that the book provides viewpoints and analyses which can enrich their studies.

References

[1] G. W. Bluman, S. Kumei, *Symmetries and Differential Equations* (Springer-Verlag, N.Y., 1989).

[2] J. E. Campbell, *Continuous Groups*; reprint of the 1903 edition of *Introductory Treatise on Lie's Theory of Finite Continuous Transformation Groups* (Chelsea, Bronx, N.Y, 1966).

[3] B. J. Cantwell, *Introduction to Symmetry Analysis* (Cambridge University Press, Cambridge, 2002).

[4] A. Cohen, *The Lie Theory of One-Parameter Groups*; reprint of the 1911 edition (Stechert, N.Y., 1931).

[5] N. H. Ibragimov, *CRC Handbook of Lie Group Analysis of Differential Equations, Vol. 1, 1994–6* (CRC Press, N.Y. 1996).

[6] S. Lie, *Theorie der Transformation Gruppen, I*; translated by M. Ackerman, entitled *Sophus Lie's 1880 Transformation Group Paper*; comments by R. Hermann (Math Sci Press, Brookline, Mass, 1975).

[7] Peter J. Olver, *Applications of Lie Groups to Differential Equations* (Springer-Verlag, N.Y., 1986).

[8] J. M. Page, *Ordinary differential Equations with an Introduction to Lie's Theory of the Group of One Parameter* (Macmillan, London, 1897).

[9] H. Stephani, *Differential Equations: Their Solution Using Symmetries* (Cambridge University Press, Cambridge, 1989).

[10] P. Hydon, *Symmetry Methods for Differential Equations* (Cambridge University Press, Cambridge, 2000).

AUDIENCE: Primarily physicists, physical chemists, applied mathematicians and other natural scientists with a mathematical bent.

BACKGROUND ASSUMED: Differential and integral calculus, introduction to ordinary differential equations. The familiarity with classical mechanics and/or quantum mechanics attained by undergraduate students of physics by their junior or senior year, or by first-year graduate students of physical chemistry.

Acknowledgments

It is a great pleasure to be able to acknowledge my indebtedness to those who have provided so much of the mental stimulation leading to these pages: to my parents Eugene and Jean, my wife Constance, and my brother, David; to some very special teachers, Christian Rondestvedt, George Uhlenbeck, and Michael Dewar; and to my students Fred Yarnell, Tai-ichi Shibuya, Sukeyuki Kumei, Yutaka Kitigawara, Yuuzi Takahata, and Gerald Hyatt.

Jerry Blakefield, Rochelle Wolber, and Jack Bench provided invaluable help checking and proofreading the manuscript, and Eve Wallis greatly improved many of the figures.

Generous grants from the Research Corporation, the Lindbergh Foundation, the University of the Pacific, and the Fulbright Commission, as well as grants from the National Science Foundation, supported research that was instrumental to the development of this book.

Acknowledgments

Contents

CHAPTER 1

Introduction

Methinks that all of nature and the graceful sky
are set into symbols in geometrium.[1]

1.1 On Geometric Symmetry and Invariance in the Sciences

The concept of symmetry that is commonly used in the physical and natural sciences is that of geometric symmetry as usually conceived. A benzene molecule is said to have the symmetry of a regular hexagon, a salt crystal has a lattice with cubic symmetry, mammals have approximate bilateral symmetry, and an ellipsoid has lower symmetry than a sphere.

The geometric symmetry of an object expresses an invariance of the relation between an observer and the object — a relation that persists after some particular alteration of the coordinates which determine this relationship. The symmetry of a sphere provides an extreme example, since the relationship between a sphere and an observer is unchanged by *any* rotation of the sphere, or the observer, about the center of the sphere. The symmetries of isolated atoms and molecules as well as of crystal lattices are all defined by the particular rotations, translations, reflections, and inversions that leave invariant the relationship between these physical objects and the coordinate system of an observer. Associated with each geometric object is the set of *symmetry operations* that leave invariant the relation between the object and a coordinate system.

These conceptions of symmetry and operations of symmetry have their basis in Euclidean geometry. Euclid defined two triangles as congruent if they can be superposed by rotations and translations. All two- or three-dimensional figures, polyhedra, surfaces, or solids that can be interconverted by rotations and translations are congruent. In Euclidean geometry, objects that can be superposed with the aid of translations, rotations, or uniform dilatations, have the same symmetry.

At the beginning of the Renaissance, Euclid's *Elements* appeared to correctly and elegantly describe geometric relations here on earth. It was natural to hope that geometric simplicity would find further expression in the celestial works of God. Visualizing the sun at the center of the solar system, Copernicus believed he had uncovered a wonderful role played by the symmetry and simplicity of regular circular motion. Kepler thought to extend this "Divine Harmony" by embedding the circular orbits of the planets between concentric polyhedra. But intellectual honesty and a careful analysis of the accuracy of Tyco Brahe's observations forced Kepler to recognize that the orbit of Mars is not circular. Thereafter, years of struggle led him to the happy realization that nature expresses harmonies more profound than those he, a mortal, at first anticipated. Kepler's three laws, and much of the intellectual journey that led to them, are all to be found in his *Harmonices Mundi*.[2]

"Standing on the shoulders of giants," Newton concluded that material objects move in the space of Euclidean geometry, and that in the absence of material objects, no point in the space is physically distinguished from any other point. To maintain this translational invariance, he assumed that real mass points are subject to a universal law of gravitation. He also assumed that no point in time is fundamentally distinguished from any other point in time: a time-line in his mechanics has a translational invariance analogous to that of the infinite straight line. Newton's laws of motion consequently remain the same when the motion is viewed from coordinate systems with different origins and timed by clocks with different initial settings. They are also the same when objects are viewed by observers moving with constant relative velocity, V. If Newton's equations of motion are obeyed by an object in an isolated system, the Euclidean symmetry of the object is the same for all observers moving at constant velocities with respect to the object.

The development of Lie's theory of groups in the latter half of the nineteenth century was followed by Noether's realization that the conservation laws of physics were consequences of the invariance properties of space and time presupposed by Newton.[3] The translational invariance of Newton's laws of motion implies the law of conservation of momentum, their rotational invariance implies the law of conservation of angular momentum, and their invariance as time evolves implies the law of conservation of mechanical energy. (A simple proof of these statements can be found in Chapter 7.)

The invariance properties of Newton's equations have many further consequences. An outstanding early example of the exploitation of the rotational and translational invariance inherent in Newtonian mechanics is provided by Maxwell's 1859 derivation of the Maxwell–Boltzmann distribution law of statistical mechanics.[4]

In his early studies of electromagnetism, Maxwell visualized Faraday's lines of force as tubes of flow associated with the motion of incompressible inviscid fluids.[5a] Such motions can conserve energy, momentum, and angular momentum. To preserve these conservation laws of Newtonian mechanics, Maxwell imagined that a displacement current flowed between the plates of charging or discharging capacitors. This allowed him to consistently suppose that Faraday's electric and magnetic fields carry, but neither create nor destroy, energy, momentum, and angular momentum. Preserving these conservation laws, Maxwell derived his equations governing electromagnetic fields.[5b] Maxwell's discovery that light is a self-propagating electromagnetic field, and the continual experimental verifications of his theory, naturally led to the expectation that the theory was essentially correct. In applying Maxwell's theory, it was expected that electromagnetic waves moved relative to the medium, *aether*, supposed to carry them. When the experiments of Michelson and Morley showed that motion with respect to such a medium is unobservable, it appeared that, as in Newtonian mechanics, only relative motions of physical objects were physically significant. Galilean invariance survived, but only briefly, until the Fitzgerald contraction of meter sticks and time dilation of clocks dealt it a mortal blow.

Then, in 1905, Albert Einstein published his profound analysis of how light signals are used in comparing time, distance, and mass measurements made by observers in relative motion.[6] Einstein concluded that physical space is not the Euclidean space assumed by Newton. Realizing that the invariance properties of physical measurements require the merging of space and time into spacetime, Einstein replaced the separate concept of distance between two points in space, and the concept of time interval between two instants, by a single concept: the spacetime interval between two events. As the spacetime interval is not a four-dimensional generalization of the Euclidean distance in ordinary space, geometric relations between events in spacetime are *non-Euclidean.*

These discoveries have, of course, required major modifications in the laws and equations of physics — modifications that continue to occupy physicists to this day. Here we wish only to call attention to a conceptual

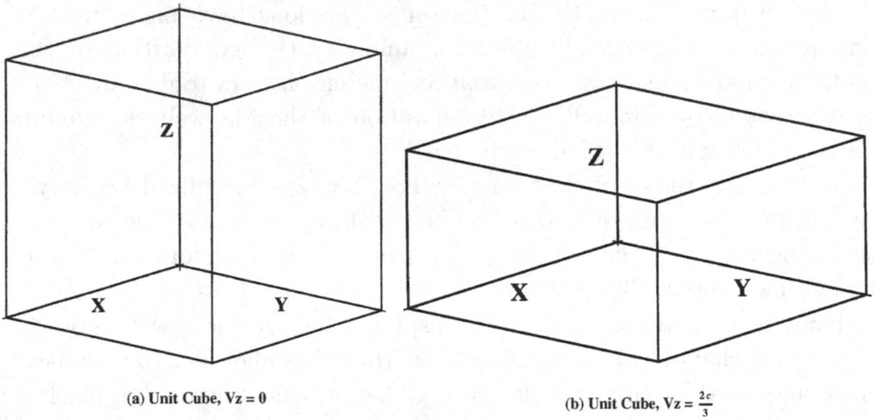

Fig. 1.1 A unit cube viewed by an observer in its rest frame and by an oberserver moving with two thirds the velocity of light in its rest frame.

problem raised by the experimentally established validity of the theory of special relativity. (We reach ahead into Sec. 2.10 to do so.) **Figure 1.1a** depicts a cubic crystal in the spatial coordinate system of an observer at rest with respect to the crystal. **Figure 1.1b** depicts the same crystal as seen by a second observer moving relative to the first with a velocity 87% of that of light. If both observers are to agree on the symmetry of the crystal, their concepts of physical symmetry must be altered: standard concepts of physical symmetry are incompatible with the standard interpretations of relativistic effects.

But that is not all.

Eulenbeck and Goudsmit's discovery of electron spin in 1925 provided a hint that, even with the modifications required by relativity, accepted concepts of symmetry might be inadequate for an understanding of physical phenomena.[7] From a physical standpoint, the conceptual problems were dealt with by allowing the magnetic moment vector of individual electrons to express a symmetry not previously recognized in nature. As explained in Sec. 2.5 this symmetry cannot exist everywhere in ordinary three-dimensional Euclidean space.

In the decade following the invention of quantum mechanics in 1926, there appeared a further hint that accepted concepts of physical symmetry might require further modification. Shortly after the inception of quantum theory it was realized that quantum numbers could have a dual interpretation: they were found to fix both the eigenvalues of observable quantities and the symmetry properties of wavefunctions. The quantum number m, of atomic spectroscopy, fixes both the component of an atom's angular momentum, $m(\mathrm{h}/2\pi)$, on some chosen axis, and the rotational symmetry of its wavefunction about that axis. Knowing the azimuthal quantum number l, one knows both the square of the total angular momentum of the atom, $l\ (l+1)(\mathrm{h}/2\pi)^2$, and the allowed values of m. Symmetry-based arguments require that m can only take on the values $-l, -l+1, \ldots, l-1, l$. For a given value of l there are thus $2l+1$ wavefunctions with different rotational symmetries about the chosen axis, and different angular momentum components, m, along this axis. The connection between symmetries and observables in quantum mechanics is so unique and so direct that the connection is a major theme in Herman Weyl's 1931 monograph, *Group Theory and Quantum Mechanics.*[8]

By 1931, physicists had begun to thoroughly develop the consequences of the fact that the potential energy of interaction of two charged particles depends only upon the distance between them. In a one-electron atom the electrostatic potential energy is the same as that of an electron in the field of a nucleus, so it depends only on the radial distance between the electron and nucleus. In a frame of reference centered on the nucleus, the electrostatic environment in which the electron moves is invariant under rotations. As the kinetic energy of the electron is also invariant under rotations, all states having the same radial distribution for the electron must have the same energy, provided one neglects the small energies associated with spin–orbit coupling. Observable states with the same energy and radial distribution need not have definite angular momentum, but they must be interconverted by rotations. As noted above, arguments of symmetry require that the number of these states is $2l+1$, where the azimuthal quantum number l can have values $0, 1, 2 \ldots$. Spherical symmetry requires that the energy levels of hydrogen, and other one-electron atoms, be $(2l+1)$-fold degenerate.

They are not.

As Pauli, Schrödinger, and Heisenberg found, Bohr's expression $\mathrm{E}_n = -\frac{(\mathrm{Z}e/n)^2}{2a_0}$ for the energy of one-electron atoms with nuclear charge Z_e applies to all n^2 eigenstates of principal quantum number n. The levels with energy E_n are n^2-fold degenerate, and n is always greater than l.

By the early 1930's it was realized that the spatial symmetry of a system often completely accounted for any degeneracy of its energy levels. In cases where spatial symmetry did not account for the degeneracy, it became customary to speak of *accidental degeneracy.* The equality of the energy of a $2s$ state with the energies of the $2p$ states of the hydrogen atom was *an accident* as was the degeneracy of the 3*s*, 3*p* and 3*d* states (and so on). It is natural to suspect that all these accidents afflicting every hydrogen atom in the universe might have a cause.

1.2 Fock's Discovery

In 1935 Vladimir Fock uncovered the cause: *the electron in a one-electron atom moves as though it were in an environment with the symmetry of a hypersphere in four-space.*[9] Fock found a set of transformations which convert wave functions of all bound states of the hydrogen atom into (hyper)-spherical harmonics Y_{nlm} in a four-dimensional Euclidean space. States of the same n, but differing l, can be interconverted by rotations in this four-dimensional space. The effective symmetry of the atom is that of an object, a hypersphere, described by an equation such as

$$\mathrm{y}_1^2 + \mathrm{y}_2^2 + \mathrm{y}_3^2 + \mathrm{y}_4^2 = \mathrm{R}^2, \tag{1.2.1}$$

in which R is a real constant, and the y's are Cartesian coordinates of a point. It is this hyperspherical symmetry which causes all states of the same principal quantum number n to be degenerate *and* requires them to be n^2 in number.

In Sec. 1.4 below we will identify the space in which Fock's hypersphere exists. Fock's work will be dealt with in detail in subsequent chapters. Here we wish only to call attention to one of its key features. In his analysis, Fock used a stereographic projection that connects points on a hypersphere in a four-dimensional Euclidean space to points in a three-dimensional Euclidean space. **Figures 1.2** illustrate the analogous projection connecting points on a sphere in three-dimensional space to points in its equatorial plane. They illustrate the mapping of points from the northern and southern hemispheres onto the equitorial plane. The other figures depict the stereographic projection of spherical harmonics with $l = 1$. The projections of the $\mathrm{P_x}$ function and the $\mathrm{P_z}$ function onto the plane yield functions that correspond to degenerate p and s type wave functions of the planar analog of a hydrogen atom.[10] The relationship between the projected functions and the spherical harmonics is directly analogous to the relationship Fock established for real hydrogen atoms.

Note that rotating the $\mathrm{P_x}$ spherical harmonic through 90° around the z axis converts it into a $\mathrm{P_y}$ harmonic, and carries out the same rotation of its projection in the plane. In contrast, the $\mathrm{P_z}$ harmonic, and its projection, are invariant under rotations about the z axis. No rotation in the plane can

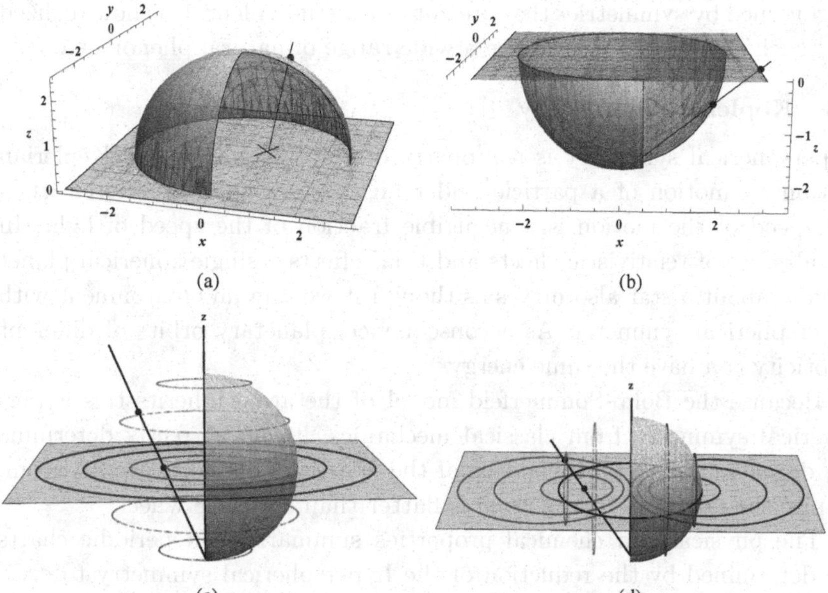

Fig. 1.2 (a) Stereographic projection of point on upper hemisphere. (b) Stereographic projection of point on lower hemisphere. (c) Stereographic projection of circles of constant latitude onto an equatorial plane. (d) Stereographic projection obtained when sphere in Fig. 1.2(c) is rotated 90° about the y axis.

interconvert the *p* and *s* type functions. However, they are indirectly interconverted by rotations in the three-space that carry a P_x or P_y function into a P_z function. Such a rotation carries a *p* type function in the plane into an *sp* type hybrid function, and then into a pure *s* type function. This illustrates a key feature underlying Fock's analysis: stereographic projections can interconvert functions not related by rotations in three-dimensional Euclidean space, to functions related by rotations in a higher dimensional Euclidean space.

Fock's use of a stereographic projection to explain the degeneracy of the hydrogen atom is subject to an evident objection: the projection could be said to produce, rather than uncover, a symmetry. However, Bargmann quickly gave a Lie-algebraic treatment which established the hyperspherical symmetry without the use of any coordinate transformations.[11] The work of Fock and Bargmann provides an impressive demonstration that degeneracies can be due to symmetries that are definitely not symmetries in ordinary space. For natural scientists it was an early hint that natural phenomena can be governed by symmetries that are not at all self-evident. It is now realized that such symmetries can govern a wide range of natural phenomena.

1.3 Keplerian Symmetry

Hyperspherical symmetry is a property of the dynamics of all Keplerian motion — motion of a particle under an inverse square law force when the speed of the motion is a negligible fraction of the speed of light. In the absence of relativistic effects and tidal effects a single spherical planet moving about a star also moves as though it were in an environment with hyperspherical symmetry. As a consequence, planetary orbits of different ellipticity can have the same energy.

Because the Bohr–Sommerfeld model of the atom inherits this hyperspherical symmetry from classical mechanics, it can correctly determine the degeneracy and n-dependence of the spectrum of the hydrogen atom, despite the fact that Bohr's atom is flatter than a bicycle wheel.

The physical and chemical properties summarized in periodic charts are determined by the reduction of the hyperspherical symmetry to ordinary symmetry to ordinary spherical symmetry not by ordinary spherical symmetry itself. The hyperspherical symmetry has a persistent effect in organizing the energy levels of many electron atoms. It also affects the organization of energy levels in molecules. In independent-electron models of polyatomics it produces level crossings that would not otherwise be present. In real molecules, these level crossings causes inflections in potential energy curves, and as noted below, are instrumental in determining the shapes of molecules. In short, the hyperspherical symmetry, by influencing the organization of energy levels in atoms and molecules, affects much of physics, chemistry, geology and biology.

The hyperspherical symmetry governing the motion of a single planet has similar consequences; it does much to govern the response of the planet to the perturbing influences of other planets and satellites.

How do these symmetries of objects in higher dimensional spaces get into our three-dimensional world? How is that they govern motions in ordinary three-dimensional Euclidean space? We begin now to address these questions.

1.4 Dynamical Symmetry

The Hamiltonian expression whose value is the energy, E, of a particle subject to gravitational or electrostatic force is

$$\mathrm{H} = \mathrm{p}^2/2\mathrm{m} - \mathrm{K}/|\mathbf{r}|. \tag{1.4.1}$$

The potential energy term, $K/|\mathbf{r}|$, with K a constant, has ordinary rotational symmetry; and the kinetic energy term, $p^2/2m$, with the mass, m, constant, has ordinary rotational symmetry. This statement remains true in Schrödinger quantum mechanics, where p is converted to $-i\hbar\nabla$. The hyperspherical symmetry is not a symmetry of the potential energy term in the Hamiltonian nor is it a symmetry of its kinetic energy term, it is a symmetry of the entire Hamiltonian. Operations of a group of transformations interconvert the kinetic and potential energy terms of the Hamiltonian H while leaving E, the value of H, unchanged. These operations transform the position vector $\mathbf{r}$ to $\mathbf{r}' = f(\mathbf{r}, \mathbf{p})$, and the momentum vector $\mathbf{p}$ to $\mathbf{p}' = g(\mathbf{r}, \mathbf{p})$, so as to leave only the entire expression $p^2/2m - K/|\mathbf{r}|$, and hence E, invariant. For a two-dimensional atom, the operation that interconverts the kinetic energy and potential energy can be put in one-to-one correspondence with the $P_x \leftrightarrow P_z$ rotations on the sphere of **Fig. 1.2**, i.e. the rotations that interconvert the *s* type and *p* type projections in the plane.

The hyperspherical symmetry of Kepler Hamiltonians is a truly *dynamical* symmetry — a symmetry present when there is motion. It is a symmetry that exists only when motion is allowed. It is extensively investigated in Chapter 8.

An analogous situation occurs with the harmonic oscillator Hamiltonian

$$H = p^2/2m + kr^2/2. \tag{1.4.2}$$

Harmonic oscillators have energy degeneracies larger than expected. Their energy is also unchanged by transformations of the form $\mathbf{r} \rightarrow \mathbf{r}' = f(\mathbf{r}, \mathbf{p})$, and $\mathbf{p} \rightarrow \mathbf{p}' = g(\mathbf{r}, \mathbf{p})$. The shell model of nuclear structure based on the Hamiltonian (1.4.2) develops extensive physical consequences of these degeneracies and their removal by internuclear forces.[12]

These dynamical symmetries are symmetries of systems allowing motions. The dynamical symmetry of a system may, or may not, be greater than its evident geometrical symmetry. The phrase *hidden symmetry* is sometimes used to signify the presence of dynamical symmetry greater than ordinary geometrical symmetry. In general, *no analysis of forces or potentials alone will find dynamical symmetries.* Methods for uncovering them are discussed in Chapters 7 and 9.

In recent years it has been realized that in classical mechanics dynamical symmetries are, in the simplest cases, geometrical symmetries in the combined space of positions, q, and momenta, p, i.e. *symmetries in the phase space* of the physicist. The observation, that the classical analog of the hyperspherical symmetry of the hydrogen atom is a hyperspherical geometric symmetry in phase space, was not at all evident in Fock's paper, nor in that of Bargmann.[11] Such a connection was difficult to interpret at the time, because the uncertainty principle states that one cannot know both the position and the momentum of a particle with arbitrary accuracy. In Chapter 9 a Lie-algebraic extension of Dirac's correspondence principal is stated and used to establish the relation between the dynamical symmetry of quantum mechanical systems and that of the corresponding classical systems, which are geometrical symmetries in phase space. It enables one to correctly exploit classical analogies when interpreting the dynamical symmetries in quantum mechanics.

1.5 Dynamical Symmetries Responsible for Degeneracies and Their Physical Consequences

If no analysis of forces or potentials alone can expose the symmetry and consequent degeneracy of the hydrogen atom, one would expect surprises to continually emerge in attempts to uncover other organizing features in the natural sciences, unless, of course, only atoms, nuclei, elementary particles, or other comparatively *simple* systems are directly affected by dynamical symmetries. Because most physical systems are much more complex, and furthermore, exact or approximate spatial symmetries are rather rare in nature, it has long been supposed that *hidden symmetry* is rarer still. However, direct physical evidence establishes that large systems can have large degeneracy groups. In statistical mechanics, the entropy of a macroscopic system is a measure of its degeneracy, for it is equal to Boltzmann's constant times the logarithm of the number of equally likely microscopic states of the same energy. A great host of operations can redistribute the energy among the constituents of the system, and the corresponding symmetry groups are enormous. Dynamical symmetry is ubiquitous, but often well hidden.

Whenever a quantum mechanical system is degenerate, an infinitesimal perturbation can make a finite alteration in its wave function. *This has the physical consequence that minute forces and minute displacements that involve a minute, even infinitesimal, amount of work can make finite, even large, changes in the physical properties of initially degenerate systems.* The connection between symmetries of a Hamiltonian and energy degeneracies imparts additional importance to symmetries in the quantum mechanics of bound states, for which the energy levels that differ in E are separated by energy gaps. As a consequence, among the bound states only the degenerate ones are exquisitely sensitive to perturbations. Degeneracies not directly due to hyperspherical symmetry, and the subtleties and consequences of their removal, also have many consequences in chemistry and physics:

Forces between nucleons produce an average binding potential which may be approximated by that in radially symmetric harmonic oscillators. Removal of the accidental oscillator degeneracies organizes much nuclear physics.[13]

In producing the resonance stabilization so ubiquitous in chemical compounds, at least two degenerate, or nearly degenerate, electronic wave functions represented by different valence-bond pictures interact to produce wave functions, molecular orbitals, of lower energy than either.[14]

In the superconductive materials dealt with in the theory of Bardeen, Cooper, and Schrieffer, electrons with initially degenerate one-electron translational and spin wave functions have this degeneracy removed, and a new degeneracy is produced by minute velocity-dependent forces stemming from interactions between the electrons and vibrating atomic nuclei.[15]

Minute forces which act on nearly degenerate translational wave functions can also develop other coordinated, gross, electronic motions in solids, motions such as Davydov excitons[16] and charge-density waves.[17]

Sudden polarization[18] occurs in molecules when nuclear motions cause electronic energy levels to cross, producing degeneracies that allow small molecular asymmetries to produce very large polarizations.

Iterative algorithms that approximate wave functions of molecules and solids may fail to converge for some assumed nuclear configurations. (This quite often occurs in Hartree–Fock calculations.) In attempting to choose an improved approximate wave function, the computer cycles between several different electronic wave functions whose energy cannot be distinguished. Calculations on cuprate lattices exhibit this effect, and even $Cu(OH_2)^{3+}$ has states whose degeneracy is not due to spatial symmetry.[19] Failure of Hartree–Fock calculations to converge can provide a very useful signal that unexpected physical or chemical phenomena are waiting to be discovered. *Hartree–Fock symmetry breaking* occurs when such degeneracy leads to one or more nuclear configurations with less energy and less symmetry in three-space than the initial nuclear configuration.[20]

Approximate dynamical symmetries of rotating and vibrating molecules have been found to organize their rotational and vibrational energy states.[21] This has made it possible to express rovibrational energies as functions of far fewer parameters than spectroscopists had previously thought possible, cf. Chapter 13.

1.6 Dynamical Symmetries When Energies Can Vary

In the late 1960's it was found that one-electron atoms and their gravitational analogs possess more symmetry than that of the hypersphere of (1.2.1). When energy, E, is considered variable, Keplerian systems behave as though they have the symmetry of the surface defined by the equation[22]

$$y_1^2 + y_2^2 + y_3^2 + y_4^2 - y_5^2 - y_6^2 = 0. \tag{1.6.1}$$

Here, as before, the y's are Cartesian cordinates of a point.

What is the physical interpretation of this symmetry? In classical Hamiltonian mechanics, the operations of a group that alters the energy of a system express relationships between the position, momenta, energy, and time variables of the system. The dynamical symmetries defined by the invariance group of (1.6.1) are symmetries in the *extended phase space* of positions, momenta, energy, and time, an eight-dimensional *PQET* space. In quantum mechanics the symmetry of Eq. (1.6.1) becomes a symmetry of the *time-dependent* Schrödinger equation of hydrogen-like atoms, and governs their dynamics.[23] Operations of this dynamical group can be used to change energy states. The group can be a *spectrum-generating group* for the atom. Chapters 9–13 deal with such dynamical symmetries of quantum mechanical systems.

Dynamical symmetries of time-dependent Schrödinger equations, $H\psi = i\partial\psi/\partial t$, may have surprising consequences even for simple systems. Because of a dynamical symmetry of their time-dependent Schrödinger equation, hydrogenic atoms obey surprising selection rules that preserve l as a good quantum number.[24] This, together with the related hyperspherical symmetry of the atom, has the consequence that when a nucleus of charge Z is (conceptually) pulled apart to form the protomolecule termed a *nearly-united atom*, there is no mixing of wave functions of different orbital angular momentum quantum number l. If one does not recognize the presence of a hidden symmetry, the united atom–nearly united atom energy level correlation diagrams may appear to violate the noncrossing rule of Wigner and von Neumann.[25] As shown in Chapter 12, this *hidden symmetry* ensures that nearly-united atoms obey the same rules that govern the geometrical symmetry of actual triatomic molecules.[26] These rules are due to Walsh, who first pointed out that the ordinary spatial symmetry of triatomic and tetraatomic molecules can be anticipated by simply counting the number of valence electrons in the molecules.[27]

As the distance between atoms varies during the course of chemical reactions, degeneracies, or near degeneracies, are often produced. Though often considered accidental, many can be predicted from energy-level correlation diagrams. Profound and rapid changes may take place at such crossings. This happens, for example, when a neutral alkali atom approaches a neutral halogen atom from afar. An electron jumps from the metal to the halogen during a change of distance on the order of 10^{-2} angstroms.[28]

Finally, we would note that even when the entropy and other equilibrium properties of highly degenerate, weakly interacting, systems are unaffected by weak interactions, these interactions can govern energy flow between subsystems. The resulting dynamical symmetry breaking then becomes relevant in nonequilibrium statistical mechanics.

1.7 The Need for Critical Reexamination of Concepts of Physical Symmetry. Lie's Discoveries

The observations made in the preceding paragraphs illustrate that simple relations in complicated systems may be overlooked by familiar modes of thought. Surprises develop for several reasons:

First of all, even in simple classical systems, dynamical symmetries are symmetries in a phase space of many dimensions.

Secondly, *phase space is not a Euclidean space.* For this reason, the concepts of physical symmetry appropriate to classical and quantum mechanics, the concepts of dynamical symmetry, are profoundly different than the usual concepts appropriate to Euclidean space. Chapter 2 begins our discussion of the difference.

Thirdly, symmetries in phase space have no immediate obvious interpretation as symmetries in quantum mechanics. Nevertheless, as previously noted, dynamical symmetries in Schrödinger mechanics and Hamiltonian mechanics are closely related, and the relationship is established in Chapter 9.

> Dynamical symmetries are, in general, so different from ordinary symmetries that understanding them requires a critical reexamination of the concept of symmetry itself.

The beginnings of such a critical analysis originated in the mind of the mathematician Sophus Lie, apparently during the winter of 1871–2.[29] As noted in the Appendix at the end of this chapter, Lie was attempting to develop a unified theory of solution for ordinary differential equations. He discovered that the solutions of differential equations can be both defined by, and interconverted by, operations of *continuous* groups of transformations. Lie's groups are termed continuous because their operations are labeled by continuously variable parameters, for example the angles of rotation. As will be seen in Chapters 4–6 of this book, *differential equations have group structure simply because*

$$\int_a^b df + \int_b^c df = \int_a^c df. \tag{1.7.1}$$

This observation, though at first sight trivial, profoundly connects the study of differential equations to studies of groups of transformations. In fact, Lie used first-order ordinary differential equations, and sets of first-order partial differential equations, to *define* continuous groups. He devoted much of his life to developing the extensive mathematical implications of the connection between differential equations and continuous groups. In doing so, he gave us much of what is now called the theory of Lie groups.

The operations that convert solutions of a differential equation into themselves and into other solutions, define the *invariance group* of the differential equation. These groups are primarily Lie groups, and they can be of great utility. In fact, most of the differential equations of the sciences have Lie group structure which neatly organizes and simplifies their solution. Unfortunately, for much of the past century the utility of the connection between Lie groups and differential equations went unrecognized, and was seldom exploited by mathematicians interested in solving differential equations. Several monographs that are outstanding exceptions to this statement have been listed at the end of the Preface. In this book, the group structure of the equations of dynamics is used to relate dynamics to geometry by relating dynamical symmetries to geometrical symmetries.

The relevance of Lie's work to the natural sciences is direct and fundamental. Whenever a continuous succession of *causes and effects* governs natural phenomena, it becomes possible to formulate differential equations which govern the phenomena. Approximate or exact integration of the equations then produces solutions that can correlate and predict the results of observations. And whenever differential equations describe nature, their group structure describes relationships amongst natural phenomena. As earlier sections of this chapter have pointed out, when the symmetries are unrecognized, the relationships they imply may lie unrecognized — even when the solutions are static in time, and some of the natural phenomenon are known and much studied. An example is provided by the history of superconductivity. Onnes discovered the phenomenon in 1911.[30a] Despite intensive investigations carried out by brilliant investigators, it was not until 1957, 31 years after the advent of quantum mechanics, that Bardeen, Cooper, and Schreifer found the cause of superconductivity.[30b] As previously mentioned they thereby uncover a symmetry breaking that creates a new symmetry.

In evolving systems, the consequences of group structure can also be profound and the causes well hidden. Quantum mechanics provides many such examples. The wave functions of quantum mechanical systems, evolving in accord with equations such as the time-dependent Schrödinger equation, may smoothly lead to degenerate states utterly unstable to perturbations. At such points, symmetries may suddenly be enlarged, decreased, or simply changed. This behavior illustrates a general property of nature: *stable states may smoothly evolve into states so unstable that minute perturbations can produce gross changes in the state of the system. A discontinuously produced state may then result, and become the initial state of a new continuous causal chain.* A few examples follow:

> The temperature and pressure of a gas reach critical values and crystals form: a system with continuous translational and rotational symmetry suddenly produces systems with discrete translational symmetry. If the crystals are metastable, subsequent phase changes may then occur when they smoothly evolve into a state whose energy coincides with that of a state with different crystallographic symmetry. What will be the symmetry of the initial crystal lattice? Of subsequent ones? Current science makes astonishingly few predictions.
>
> Diffusion of the components of oscillatory chemical reactions can develop chemical concentration waves in initially homogeneous media.[31,32] In a spherical container, standing concentration waves with $s, p, d, \ldots$ type spatial symmetries can develop. Such reactions make it possible for a fertilized egg to produce a set of identical

daughter cells which in a coordinated manner then produce further daughter cells of different types. Subsequent intertwined chains of continuous causality, punctuated by symmetry breaking events, make it possible for "a sphere to become a horse".[32] Even at the few-cell stage where reaction–diffusion phenomena are likely to dominate cell differentiation, the biochemistry involved remains wondrously obscure as this is written.

The differential equations governing fluid flow describe the causal chains that produce smooth streamline flow, and allow for the pulsating, swirling, chaotic, eddies and vortex rings that may break the translational symmetry of a flow. Investigations of the invariance groups of the differential equations, and the symmetries they imply, have produced powerful tools of analysis utilized by aerodynamic engineers.[33] But some phenomenon defy detailed prediction; it is still a good idea to determine the stall speed of a new wing using a wind tunnel.

A homogenous fluid contained in a vertical circular cylinder and heated from below by a cylindrical heat source may develop instabilities which lead to stable Benard convection cells. Their vertical boundaries appear at unpredictable locations, and break the initial cylindrical symmetry of the system.[34] The differential equations which govern motions of the earth's atmosphere develop instabilities that limit the reliability of weather predictions. The very complex equations involved are elegantly modeled in the much simpler equations of Lorentz. They predict circumstances that produce unpredicability.[35]

Geologists believe that oceanic rifts and plate tectonic motions are produced by convection cells in the magma below the earth's crust. Even minute perturbations in a sphere filled with fluid and centrally heated with perfect spherical symmetry can develop symmetry breaking convection cells.[36] Though plate tectonics apparently rests on the ultimately unpredictable, its sudden general acceptance in 1967–8 has enabled today's geologists to accept, and greatly extend, earlier correlations of global geology.[37] This most recent of scientific revolutions has established an ever-growing body of causal connections between geological events both past and present. And this is being accomplished in a branch of natural

> science in which it can be very difficult to distinguish between the causal and the adventitious.[38]

These are but a few examples of how the continuous causal chains that organize so much of science typically develop instabilities, that in turn produce versions of the *punctuated evolution* of biologist Steven Jay Gould.[39] Previously expressed relationships are broken, and new relationships develop.

In many of the examples just given, symmetries in ordinary Euclidean space have changed. However, as we have emphasized, by no means all physically relevant symmetry is symmetry in ordinary Euclidean space. As will be seen in latter chapters, it is the differential equations of the physical sciences that determine the physically relevant spaces.

Early in the twentieth century, Lie's mathematical insights began to stimulate new physical insights. We have already mentioned Noether's work of 1918. Yet even earlier, in 1909, Cunningham and Bateman[40] showed that the transformation group which interconverts solutions of Maxwell's equations directly requires that the spacetime that propagates light is non-Euclidean and has the properties discovered by Einstein. In Chapter 14 it will be pointed out that the group has other, surprising, properties as well. Though, as we have seen, relativity theory requires a rethinking of the meaning of physical symmetry, historically it is the 1935 work of Fock and Bargmann which first established that physically consequential symmetries may be symmetries in spaces other than spacetime.

The dynamical symmetries that are the subject of this book arise because the differential equations governing classical and quantum dynamics have solutions that can be defined, time-evolved, and interconverted by, operations of groups. The operations of a dynamical group have direct physical interpretation, and geometric interpretation as operations of a symmetry group. The resulting connection between geometry and dynamics is the organizing principle of the book. However, the connection is more general than can be dealt with herein. The general situation is as follows:

The equations describing a great variety of physical systems are differential equations. All have invariance groups and an associated geometry in which transformations that interconvert solutions of the equations can be interpreted as geometric operations. The space in which these geometric operations act is usually a space in which geometric symmetries are not just those possible in ordinary space or spacetime. And, when differential equations describe the natural world their geometric symmetries may, or may not, be physically stable, they may persist, or be altered in ways made more understandable by Lie's work.

The next two chapters informally explore fundamental conceptual relationships between the invariance groups of equations and their geometric and physical interpretations. To follow the resulting connection between geometry and physics one must consistently conceive of symmetries in spaces of higher dimension than three, and one must consistently conceive of symmetries in spaces that are not necessarily Euclidean. These introductory chapters are intended to develop the required conceptions in a manner that will enable readers to expand, and continue to exploit, their intuitive geometric insights. Subsequent chapters more formally develop and apply concepts introduced in Chapters 1–3.

Appendix A. Historical Note

Sophus Lie was born in the Norwegian village of Nordfordeid, and spent his childhood there as a son of the local Lutheran pastor. He was thirty years old when he began to develop his theory of differential equations, a theory he saw as analogous to Galois' theory[41] of solution of algebraic equations. He quickly established that a first-order ordinary differential equation can be integrated directly, or with the aid of an integrating factor, if and only if there is a continuous group of transformations that leaves the differential equation invariant.[42] Knowledge of the group determines whether an integrating factor is necessary, and if it is, the group determines the integrating factor (and *vice versa*). Subsequently Lie expanded his theory of first-order equations into a unified theory of solution of sets of ordinary differential equations of arbitrary order. Several texts on the subject appeared at the beginning of the twentieth century.[43] Even before his death from pernicious anemia in 1899, Lie's ideas began to have an extraordinary influence in pure mathematics. However, two world wars, practical difficulties, and several historical accidents delayed the use of Lie's work by applied mathematicians for some fifty years. There also appeared to be theoretical impediments limiting extensions of Lie's ideas to partial differential equations.[44] The theoretical impediments were found to be largely illusory,[45] and it is now known that partial differential equations possess much more group structure than was previously realized.

During the 1940's Bargmann[46] and Wigner[47] developed the representation theory of the Poincare group that underpins relativistic quantum kinematics. In the mid 1950's attempts to generalize Gell-Mann's eight-fold way[48] led physicists to a renewed interest in the symmetry of the hydrogen atom and to an interest in Lie groups in general. The 1960's and 70's saw the recognition of the dynamical symmetries of a number of systems governed by Schrödinger equations. These symmetries are discussed in several chapters of this book. The final chapter of the book is devoted to symmetries of Maxwell's equations. Special attention is given to the consequences of inversion symmetry that enlarges the Poincare group.[49]

By the turn of the twentieth century, the development of symbolic computation algorithms, and elegant applied mathematics due to Reid,[50] removed much of the practical difficulty in applying Lie's ideas. Computer programs now make it possible to systematically uncover the symmetries of increasingly complex systems governed by differential equations. And studies of the symmetries of partial differential equations have led to new methods for obtaining their solutions.[42] Kumei[51] initiated the use of Lie methods to obtain solutions of nonlinear partial differential equations, a field in which they are proving particularly useful.

The connection between dynamics and geometry that is emphasized in this book has impressive antecedents. For example, Synge in his article on dynamics in *Handbuch der Physik* long ago pointed out that Hamiltonian dynamics allows a complete geometrization of classical mechanics.[52] This conception of a geometrodynamics, applied via tensor analysis rather than Lie theory, is extensively developed in Meisner, Thorn, and Wheeler's monograph on general relativity.[53]

References

[1] J. Kepler, *Tertius Interveniens, Gesammelte Werke*, edited by W. van Dycke, M. Caspar; translated by A. Koestler, *The Watershed* (Doubleday, Garden City, N.Y. 1960), pp. 65, 226.

[2] J. Kepler, *Mysterium Cosmographicum* (Tuebingen, 1596).

[3] E. Noether, *Nachr. Koenig. Gesell. Wissen. Goettigen, Math.-Phys. Kl* **235** (1918).

[4] J. C. Maxwell, Phil. Mag., I, **48** (1860); cf. *The Scientific Papers of James Clerk Maxwell*, ed. W. D. Niven, Vol. 1, p. 379 (Librairie Scientific J. Hermann, Paris, 1890).

[5] a) J. C. *Maxwell, Cambridge Phil. Soc. Trans.*, **X**, pt.1, (1855–6) *Scientific Papers*, 155.
b) J. C. Maxwell, *Phil. Trans. Roy. Soc.* (London) **CLV** (1864) *Scientific Papers*, 526.

[6] a) A. Einstein, *Ann. der Physik* 17, 891 (1905); translated by R. W. Lawson, *Relativity, The Special and General Theory*, (Henry Holt, New York, 1921).

b) A. S. Eddington, *Space Time and Gravitation* (Cambridge University Press, 1920).

[7] G. E. Euhlenbeck, S. Goudsmit, *Naturwiss.* **13** (1925) **13**.

[8] H. Weyl, *Gruppenntheorie und Quantenmechanik*, 1930; cf. Hermann Weyl, *Theory of Groups and Quantum Mechanics*, translated by (H. P. Robertson, Dover, N.Y., 1950).

[9] V. Fock, Z. Physik, **98** (1936) 145; translated by V. Griffing, *J. Phys. Chem.* **61**, (1957) 11.

[10] T. Shibuya, C. E. Wulfman, *Am. J. Phys.* **33**, (1965) 376.

[11] V. Bargmann, *Z. Physik*, **99**, (1936) 576.

[12] M. Goeppert Mayer, J. H. D. Jensen, *Elementary Theory of Nuclear Shell Structure* (Wiley, N.Y., 1955).

[13] a) G. A. Baker, Jr, *Phys. Rev.* **103**, (1956) 1119.
b) J. P. Elliott, in *Topics in Atomic and Nuclear Theory* (University of Canterbury, Christchurch, New Zealand, 1970); deals with the breaking of the degeneracy group of isotropic oscillators.

[14] C. A. Coulson, *Valence* (Clarendon Press, Oxford, 1952), Ch. VI.

[15] J. R. Schrieffer, *Theory of Superconductivity*, revised edn. (Benjamin, N.Y., 1983).

[16] A. S. Davydov, *Theory of Molecular Excitons* (Plenum, N.Y., 1977); cf. also *Physica Scripta* **20** (1979) 387.

[17] R. M. White and T. H. Geballe, *Long Range Order in Solids* (Academic Press, N.Y., 1979); analyzes the effects of many symmetry breaking interactions.

[18] a) C. E. Wulfman, S. Kumei, *Science* **172** (1971) 1061;
b) L. Salem, *Isr. J. Chem.* **12** (1979) 8;
c) C. E. Wulfman and G. C. Hyatt *Proc. Indian Acad. Sci.* (*Chem. Sci.*) **107**, No. 6, (1995).

[19] C. E. Wulfman, Larry Curtis, *Int. J. Mod. Phys. B.* **3** (1989) 1287.

[20] a) J. Paldus, A. Veillard, *Chem. Phys. Lett.* **50** (1977) 6.
b) M. Ozaki, *Prog. Theor. Phys.* **67** (1982) 83.
c) J. L. Stuber, J. Paldus, in *Fundamental World of Quantum Chemistry* (Kluver, Dordrecht, 2003), pp. 67–139.

[21] F. Iachello, R. D. Levine, *Algebraic Theory of Molecules* (Oxford University Press, Oxford, 1995).

[22] a) E. C. G. Sudarshan, N. Mukunda, L. O'Raifertaigh, *Phys. Rev. Lett.* **15** (1965) 1041; *ibid.* **19** (1965) 322;
b) A. O. Barut, P. Budini, C. Fronsdal, *Proc. Roy. Soc.* (London) A291, 106 (1966);
c) H. Bacry, *Nuovo Cimento* **41** (1966) 223;
d) A. Bohm, *Nuovo Cimento* **43** (1966) 665;
e) A. O. Barut, H. Kleinert, *Phys. Rev.* **156** (1967) 1541;
f) M. Bednar, *Ann. Phys.* (NY) **75** (1973) 305.

[23] S. Kumei, M.Sc. Thesis, University of the Pacific, Stockton, CA, 1972; R. L. Anderson, S. Kumei, C. E. Wulfman, *Phys. Rev. Lett.* **28** (1972) 988.

[24] a) S. Pasternack, R. M. Sternheimer, *J. Math. Phys.* **3** (1962) 1280;

b) N. V. V. J. Swami, R. G. Kulkarni, L. C. Biedenharn, *J. Math. Phys.* **11** (1970) 1165.
[25] J. von Neumann, E. Wigner, *Physik Z.* **30** (1929) 467.
[26] C. Wulfman, *J. Chem. Phys.* **31** (1959) 381.
[27] A. D. Walsh, *Disc. Faraday Soc.*, Series 2, **18** (1947); *J. Chem. Soc.* (1953) 2260, 2266, 2278, 2296, 2300, 2306, 2318, 2325, 2330.
[28] R. S. Berry, *J. Chem. Phys.* **27** (1957) 1288.
[29] E. Strom, Marius Sophus Lie, in *Proceedings of the 1992 Sophus Lie Memorial Conference*, eds. O. A. Laudal, B. Jahren, (Scandinavian University Press, Oslo, 1994).
[30] a) H. K. Onnes, *Comm. Phys. Lab. Univ. Leiden*, 119, 120, 122 (1911);
b) J. Bardeen, L. N. Cooper, J. R. Schrieffer, *Phys. Rev.* **108** (1957) 1175.
[31] a) A. M. Weinberg, *Growth*, **2** (1938) 81;
b) A. M. Weinberg, *Bull. Math. Biophysics* **1** (1939) 19.
c) A. M. Zabotinsky, *Biophysics*, **9** (1965) 329; (translated from *Of Biofizika* **2** (1964) 306).
[32] A. M. Turing, *Phil. Trans. Roy. Soc.* London, **B237** (1952) 37.
[33] B. J. Cantwell, *Introduction to Symmetry Analysis* (Cambridge, 2002).
[34] H. L. Swinney, J. P. Golub, *Hydrodynamic Instabilities and the Transition to Turbulence*, Springer-Verlag, N.Y. (1981).
[35] a) E. Lorenz, *J. Atmospheric Sci.* **20** (1963) 130–141;
b) V. I. Arnol'd, *Catastrophe Theory*; translated by G. S. Wassermann, R. K. Thomas, (Springer-Verlag, Berlin, 1984, p. 23);
c) D. Ruelle, Strange Attractors, *Mathematical Intelligencer*, **2**, no. 3 (1980) 126.
[36] F. H. Busse, *Ann. Rev. Fluid. Mech.* **32** (2000) 383.
[37] a) A. Wegener, *Die Entstelbung der Kontinent und Ozeane* (1915); cf. *The Origin of the Continents and Oceans*, a translation of the 4-th edition by J. Biram (Methuen, London);
b) A. Du Toit, *Our Wandering Continents* (Oliver and Boyd, London, 1937);
c) S. Warren Carey *et al.*, *Continental Drift, A Symposium*, (University of Tasmania, Hobart, 1956).
[38] D. Yuen, *Chaotic Processes in the Geological Sciences*, (Springer-Verlag, 1992).
[39] S. Jay Gould, *The Structure of Evolutionary Theory* (Harvard U. Press, Cambridge, Massachusetts, 2002a). (Punctuated evolution is seen as a consequence of punctuated biological equilibria.)
[40] a) E. Cunningham, *Proc. Math. Soc. (London) Ser. 2* **8** (1910) 77;
b) H. Bateman, *ibid.* pp. 228, 469.
[41] F. Cajori, *Theory of Equations* (Macmillan, N.Y., 1928), Chapters XIII–XVI.
[42] a) G. W. Bluman, S. Kumei, *Symmetries and Differential Equations* (Springer, 1989);
b) Brian J. Cantwell, *Introduction to Symmetry Analysis* (Cambridge U. Press, 2002).

[43] a) J. M. Page, *Ordinary Differential Equations with an Introduction to Lie's Theory of the Group of One Parameter* (Macmillan, London, 1897);
b) A. Cohen, *An Introduction To The Lie Theory of One-parameter Groups* (1911); reprinted by G. E. Stechert, New York, 1931.

[44] E. A. Muller, K. Matschat, *Miszellanen der Angewandten Mechanik* (Berlin, 1962), p. 190.

[45] a) S. Kumei, M.Sc. Thesis, University of the Pacific (Stockton, CA, 1972);
b) N. H. Ibragimov, R. L. Anderson, *Dokl. Akad. Nauk SSSR* **227** (1976) 539.

[46] V. Bargmann, *Ann. of Math.* **48** (1947) 568.

[47] E. Wigner, *Ann. of Math.* **40** (1939) 149.

[48] M. Gell Mann, *Phys. Rev.* **125** (1962) 1067.

[49] E. Cunningham, *Proc. Math. Soc. (London) Ser. 2.* **8** (1909) 77.

[50] G. J. Reid, *J. Phys. A*, **23** (1990) 2853; *Eur. J. Appl. Math.* **2** (1991) 293, 319.

[51] S. Kumei, *J. Math. Phys.* **16** (1975) 2461; *ibid.* **18** (1977) 256.

[52] J. L. Synge, *Classical Dynamics*, in Encyclopedia of Physics, Vol. III/1, S. Flugge (Springer, Berlin, 1960).

[53] C. W. Meisner, K. S. Thorn, J. A. Wheeler, *Gravitation* (W.H. Freeman, San Francisco, 1971).

CHAPTER 2

Physical Symmetry and Geometrical Symmetry

In Newtonian physics it is supposed that $\mathbf{r}$, $d\mathbf{r}$ and $d(d\mathbf{r})$ are approximable by measurements of $\mathbf{r}$, $\Delta\mathbf{r}$ and $\Delta(\Delta\mathbf{r})$ respectively in a local Cartesian reference frame in Euclidean space. Because forces are, in principle, defined via measurements of displacements, Newtonian mechanics inherits consequences of the presuppositions (axioms) of Euclidean geometry. Newton also presupposed that the time, t, and its derivatives dt, d(dt), can be approximated by measurements of Δt and $\Delta(\Delta t)$ using local clocks. Quantum mechanics inherits these presuppositions about space and time measurements in a restricted way, under the supposition that the position vector $\mathbf{r}$ of a particle of fixed mass and its velocity vector $d\mathbf{r}/dt$ cannot both be approximated by increasingly refined measurements. However, the correspondence principle enables one to carry over into quantum mechanics some of the geometric intuition one develops from studying Euclidean geometry.

As suggested in the previous chapter, understanding by concepts of symmetry the manner in which natural phenomena are interrelated might require a reconception of relationships between geometry and measurements of $\mathbf{r}$ and $\Delta\mathbf{r}$, and also t and Δt. The rethinking required goes beyond that suggested by the remarks in Chapter 1. With this in mind we shall proceed with more care than what might at times seem necessary.

2.1 Geometrical Interpretation of the Invariance Group of an Equation; Symmetry Groups

The equation $G(z) = c$, where c is a constant, is by definition invariant under a transformation $z \to z' = f(z)$ if and only if $G(z') = c$ for all values of z that satisfy the original equation. If substituting of z' for z carries all solutions of an equation into solutions of that equation, it defines an invariance transformation of the equation. Conversely, a transformation

$z \to z' = f(z)$ that leaves the equation invariant is a transformation that converts solutions of the equation to solutions of the equation.

Let us investigate some common assumptions on geometric interpretation of invariance transformations by considering the invariance transformations of an algebraic equation. Consider, for example, the equation

$$w^4 = 1. \tag{2.1.1}$$

It is invariant under the transformation $w \to w' = f(w)$ if and only if

$$w'^4 = 1, \quad \text{i.e.} \quad (f(w))^4 = 1, \tag{2.1.2}$$

for all values of w satisfying (2.1.1). The substitution $w \to -w$ leaves w^4 unchanged and transforms a solution of Eq. (2.1.1) into a solution. The substitution which carries $w \to iw$ has the same property. Compounding these two substitutions carries $w \to -w \to -iw$. As $w = 1$ is a solution of (2.1.1), it follows that -1, i and $-i$, are also solutions. **Figure 2.1**, in which $w = x + iy$, represents the four solutions by points with coordinates (x, y) in a plane. For clarity, the points are connected by straight lines. Though four points are sufficient to represent the solutions, more than four geometric operations carry solutions into solutions — all such operations,

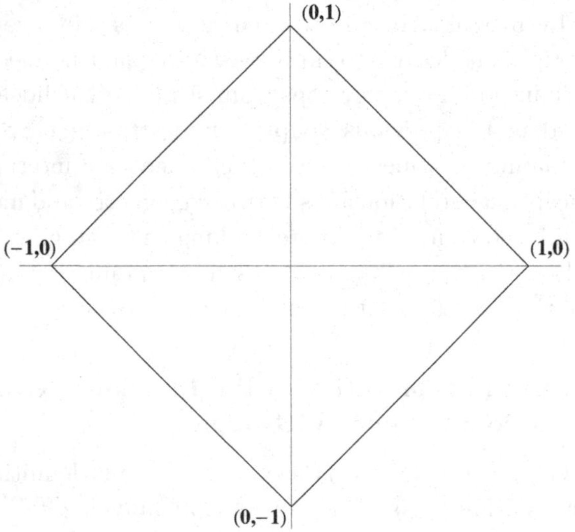

Fig. 2.1 A plot of the roots of Eq. (2.1).

Table 2.1. Operations that carry the square into itself and interconvert solutions of $w^4 = 1$, $w = x + iy = 1, i, -1, -i$.

A. Rotations:

$R(0,1) = (1,0), \quad R(1,0) = (0,-1),$
$R(0,-1) = (-1,0), \quad R(-1,0) = (0,1).$

B. Reflections:

$S_1(x,y) = (x,-y), \quad S_2(x,y) = (-x,y).$

C. Inversions:

$J_1(x,y) = (y,x), \quad J_2(x,y) = (-y,-x).$

and their action on a coordinate vector $\mathbf{r} = (x, y)$, are listed in Table 2.1. Each operation corresponds to a substitution that leaves (2.1.1) invariant.

We wish to determine whether these operations give rise to a group of operations. A set of operations comprise a group of operations iff the following hold:

(a) The application of any operation T_a, followed by the application of any operation T_b, is an operation in the set, say T_c. Symbolically,

$$T_b T_a = T_c. \tag{2.1.3a}$$

(b) The set contains the identity operation, I, which satisfies

$$T_a I = I T_a = T_a. \tag{2.1.3b}$$

(c) Every operation T_a has an inverse $T_{a'}$ in the set:

$$T_a T_{a'} = I = T_{a'} T_a. \tag{2.1.3c}$$

(d) The operations obey the associative law

$$T_a(T_b T_c) = (T_a T_b) T_c. \tag{2.1.3d}$$

Here, the inverse of a compound operation $(T_a T_b)$ is $T_{b'} T_{a'}$.

The operations of Table 2.1 give rise to the operations A' in the first column of Table 2.2. Their action on the operators A of the top row of the table is indicated at the intersection of the corresponding rows and columns. With the aid of the table the reader may verify that, taken together, the operations are those of a group. Note that $S_1 S_1 = I$, $S_2 S_2 = I$, $J_1 J_1 = I$ and $J_2 J_2 = I$. It follows that each of these operations is its own inverse, e.g. $S_1^{\text{inv}} = S_1$. Consequently the operations S_1 and I are themselves those of a group. As this group is contained in the larger group, it is termed a

Table 2.2. Composition of the Operations A′ and A to give A′A.

A′A\ A	I	R	R^2	R^3	S_1	S_2	J_1	J_2
I	I	R	R^2	R^3	S_1	S_2	J_1	J_2
R	R	R^2	R^3	I	J_1	J_2	S_2	S_1
R^2	R^2	R^3	I	R	S_2	S_1	J_2	J_1
R^3	R^3	I	R	R^2	J_2	J_1	S_1	S_2
S_1	S_1	J_2	S_2	J_1	I	R^2	R	R^3
S_2	S_2	J_1	S_1	J_1	I	R^2	R	R^3
J_1	J_1	S_1	J_2	S_2	R^3	R	I	R^2
J_2	J_2	S_2	J_1	S_1	R	R^3	R^2	I

subgroup of the later. Each of the operations of reflection and inversion are members of a subgroup of the larger group. Because $R^4 = I$, R^3 so $R = I$, R^3 is the inverse of R. Similarly, R^2 is its own inverse. These rotations also comprise a subgroup of the full group.

Because the operations carry solutions of Eq. (2.1.1) into solutions of the same equation, it is left unchanged, i.e. it is invariant under the operation of the group. Given our geometric assumptions, the transformations in Table 2.2 are also those of the symmetry group of the square of **Fig. 2.1**.

There are geometrical operations that do not leave Eq. (2.1.1) invariant, but do leave the symmetry of the square unchanged. These include arbitrary rotations of the square about any point, translations of the square in any direction, and the uniform *dilatations*, which enlarge or shrink the square.

Rotations about the origin carry a point with coordinates (x, y) to the point with coordinates (x', y'), where

$$x' = x\cos(\theta) - y\sin(\theta), \quad y' = x\sin(\theta) + y\cos(\theta), \tag{2.1.4}$$

and θ is the angle of rotation. Operations of the group will carry a circle with center at the origin into itself. The group is termed a *continuous group* because the operations of the group depend upon a parameter, θ, that can vary continuously. The rotations that carry the square of **Fig. 2.1** into itself comprise a discrete subgroup of this continuous group of rotations.

Translations carry a point with coordinate (x, y) to the points at (x', y'), where

$$x' = x + \alpha, \quad y' = y + \beta. \tag{2.1.5}$$

These transformations also comprise a continuous group.

Composing the operations of (2.1.4) and (2.1.5) one obtains E(2), the Euclidean group of the plane. If we let $T(\alpha)x$ represent the operation that converts x to $x + \alpha$, the application of $T(\alpha_1)$ followed by $T(\alpha_2)$ is a transformation

$$T(\alpha_2)T(\alpha_1)x = T(\alpha_2)(x + \alpha_1) = x + \alpha_1 + \alpha_2 = T(\alpha_3), \qquad (2.1.6)$$

with

$$\alpha_3 = \alpha_1 + \alpha_2.$$

The group E(2) is termed a three-parameter group because its operations depend on the parameter θ, the angle of rotation, the parameter α, the distance a point is translated parallel to the x axis, and the parameter β, the distance it is translated parallel to the y axis.

Note that the operations of E(2) do not alter the distances D_{ij} separating points i and j.

Uniform dilatations in the plane carry out the transformations

$$x \to e^{\mu}x, \quad y \to e^{\mu}y, \quad -\infty < \mu < \infty. \qquad (2.1.7)$$

Together with the operations of E(2), they comprise the operations of a continuous group termed the *similitude group* of the plane. The operations of this four-parameter group may be used to superpose objects of the same symmetry — bring them into coincidence; they are unable to bring objects of different symmetry into coincidence.

Operations of the similitude group that do not carry the square into itself, do however, leave its symmetry unchanged. They might be said to alter the perspective of a viewer of the square, changing the relation of the square to this viewer, by actively rotating, translating, or *dilatating* it — altering its apparent size. Operations of the similitude group of the plane do not alter the relative distances D_{ij} separating points i and j, and $D_{i'j'}$ separating points i′ and j′, because both are multiplied by e^{β}.

Now let us compare the effects of the discrete operations of inversion and reflection applied to the vertices in **Fig. 2.1**.

The *inversions* through the origin transform $\mathbf{r} \to -\mathbf{r}$. The same effect can be obtained by rotating the vector $\mathbf{r}$, with components x, y, through 180°. Inversions are operations contained in the similitude group.

However, the *reflection* $x \to -x$, $y \to y$, and the reflection $y \to -y$, $x \to x$, may alter the symmetry of a "left-handed" or "right-handed" figure in the plane, interconverting the one into the other. None of these reflections is an inversion. Also, reflections are not special cases of rotations.

Reflections, though they leave invariant the separation between points, cannot be carried out by any operation of the continuous similitude group of the plane because there is no continuous motion of the similitude group that can move a point $P_{x,y}$ through all intermediate points P between $P_{x,y}$ and $P_{-x,y}$, or between $P_{x,y}$ and $P_{x,-y}$. In short, it is impossible for the similitude group acting in the plane to *continuously* deform a right-handed object in the plane into a left-handed object.

A situation that is analogous to the Euclidean plane holds in three-dimensional Euclidean space: operations of the corresponding continuous similitude group may be used to rotate, translate, and enlarge or shrink objects and thereby superpose geometrical objects of the same symmetry. Extending the group by including the separate reflections, $x \to -x$, $y \to -y$, $z \to -z$, one obtains the group whose subgroups define symmetries in the space. Again, the effect of the inversions $\mathbf{r} \to -\mathbf{r}$ can be obtained by operations of rotation and reflection. Right-handed and left-handed objects have different symmetry, and reflections may interconvert them, i.e., alter their symmetry. However, there is no continuous operation acting in the three-dimensional space that will deform a right-handed object so as to convert it to a left-handed object, while at the same time leaving the separation between points in three-dimensional Euclidean space unchanged, uniformly enlarged, or uniformly shrunk.

These observations on the symmetry of objects call attention to general principles that are true in both two- and three-dimensional Euclidean geometry:

1. *The group defining the symmetry of any figure or lattice is a subgroup of the group of all similitude and inversion transformations.*
2. *The symmetry of any figure or lattice is unaltered by any operation of the continuous group of similitude operations acting on the entire figure or lattice.*

The human mind is able to recognize that two objects in space are essentially identical if an operation of the similitude group of three-space can convert the one into the other. Apparently our minds habitually and rapidly rotate, translate, dilatate, and then compare memories of visual images without us being conscious of this. When our minds impart such structures in group theory, we may not notice that we have seen similar objects from different perspectives.

2.2 On Geometric Interpretations of Equations

A number of arbitrary decisions were made in using **Fig. 2.1** to give a geometric meaning to the algebraic equation $w^4 = 1$. These affect the geometric interpretation of the operations that leave $w^4 = 1$ invariant. In particular:

1. The real part of w was plotted on one axis, and the coefficient of its imaginary part on the other. The coordinates of a point, (x, y), were thus chosen to be real numbers.
2. The x and y axes were, by choice, made perpendicular to each other so as to provide an orthogonal coordinate system. It is conceivable that there might be relations between x and y which invalidate this supposition. Gibbs phase diagrams provide an example of this.
3. Though four points do not necessarily lie in a Euclidean plane, the four roots were plotted in a plane. One could have plotted them in any two-dimensional space, such as that provided by the surface of a sphere or a hyperboloid in three-dimensional Euclidean space. And on each surface one could have chosen both x and y to be measured along the axes of a variety of coordinate systems. Some examples are depicted in **Figs. 2.2**. In each case one can imagine the points labeled a, b, c, d to

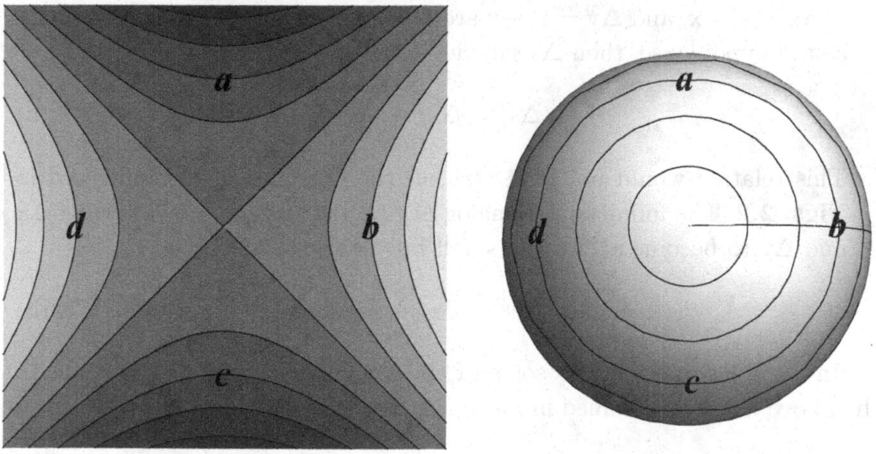

(a) (b)

Fig. 2.2 Plots of roots of Eq. (2.1) on a hyperboloid (a) and on a sphere (b).

correspond to the solutions $i, 1, -i, -1$. Depending upon the physical (or other) interpretations given to the roots, a plot on one of these surfaces might be more appropriate than a plot in the plane.

4. One could have supposed the four points to be points in some space of dimension greater than three.
5. We supposed a fixed coordinate system and moved points about it. The relation between a point and a coordinate system is, however, a relative one. We could just as well have considered the points to remain fixed and carried out the inverse motion on the coordinate system. For example, we rotated points by 90° about the point $(0, 0)$ in the plane, but the resulting relation of point to coordinate system is the same as that obtained by rotating the coordinate system through $-90°$. We imagined uniformly enlarging the square, but the resulting relation to the coordinate system could have been obtained by uniformly shrinking units of measurement on the axes. (Considering an operation to act on a geometrical figure is the *active* interpretation of the operation; considering it to act on a coordinate system is its *passive* interpretation.)
6. By plotting the roots of (2.1.1) in a Euclidean plane equipped with an orthogonal coordinate system, we accepted a concept of measurement in which the distance, Δs, between a point with Cartesian coordinates (x, y) and a point with Cartesian coordinates (x', y') is predetermined; if $\Delta x = x' - x$, and $\Delta y = y' - y$ are two sides of a right triangle with Δs as the hypotenuse, then Δs satisfies

$$\Delta s^2 = \Delta x^2 + \Delta y^2. \tag{2.2.1a}$$

This relation would not hold true on the curved surfaces indicated in **Figs. 2.2**. The infinitesimal analog of (2.2.1a), obtained by allowing Δx and Δy to become arbitrarily small but nonzero, is:

$$ds^2 = dx^2 + dy^2. \tag{2.2.1b}$$

In the following sections some of the assumptions and observations in the above list are examined in more detail.

2.3 Geometric Interpretations of Some Transformations in the Euclidean Plane

In this section we begin by considering the relation between the active and passive interpretations of the transformations of E(2).

Let a transformation T be defined by

$$x \to x' = F(x, y), \quad y \to y' = G(x, y), \tag{2.3.1a}$$

and let it have inverse, T^{inv}, defined by

$$x' \to x = F^{inv}(x', y'), \quad y' \to y = G^{inv}(x', y'). \tag{2.3.1b}$$

The transformations we will consider in this section have the property that the transformed variables may be inhomogeneous linear functions of the original variables. That is, in (2.3.1a,b)

$$\begin{aligned} x' &= a_0 + a_1 x + a_2 y, \quad y' = b_0 + b_1 x + b_2 y, \\ x &= a'_0 + a'_1 x' + a'_2 y', \quad y = b'_0 + b'_1 x' + b'_2 y'. \end{aligned} \tag{2.3.1c,d}$$

Inhomogeneous linear transformations, together with homogeneous linear transformations, are termed *affine* transformations. Not all affine transformations in the plane are operations of the similitude group of the plane.

Let us suppose that x,y and x', y' are variables in a Euclidean plane with Cartesian coordinates defined by orthogonal unit vectors $\mathbf{x}^c = 1\mathbf{i}$, $\mathbf{y}^c = 1\mathbf{j}$ satisfying the usual relations $\mathbf{i}'\mathbf{i} = 1$, $\mathbf{j}'\mathbf{j} = 1$, $\mathbf{i}'\mathbf{j} = 0$. If the transformation is interpreted in the active sense, T may be considered to move points about in a plane equipped with a single system of Cartesian coordinates. A point is moved from $(x, y) = x\mathbf{i} + y\mathbf{j}$ to $(x', y') = x'\mathbf{i} + y'\mathbf{j}$. T^{inv} is considered to move the point back again. In **Fig. 2.3.1a** the transformation T is a rotation, interpreted as an active transformation in which

$$x' = x\cos(\theta) - y\sin(\theta), \quad y' = x\sin(\theta) + y\cos(\theta). \tag{2.3.2a}$$

The point follows the indicated path as θ varies continuously from an inital value of zero. For each value of θ the inverse transformation is obtained by changing θ to $-\theta$ and applying the transformation to the final point. That is,

$$x = x'\cos(-\theta) - y'\sin(-\theta), \quad y = x'\sin(-\theta) + y'\cos(-\theta). \tag{2.3.2b}$$

In **Fig. 2.3.1b**, the rotation has been followed by a translation. **Figure 2.3.1c** depicts a uniform dilation of **Fig. 2.3.1b** by a factor of $e^{1/2}$.

In the passive interpretation of a rotation, the relation between the point and the coordinate system in **Fig. 2.3.1a** is obtained by rotating the coordinate system through an angle $-\theta$. This is illustrated in **Fig. 2.3.2a**. In it, the notation XX and YY denote the transformed axes X and Y. If we

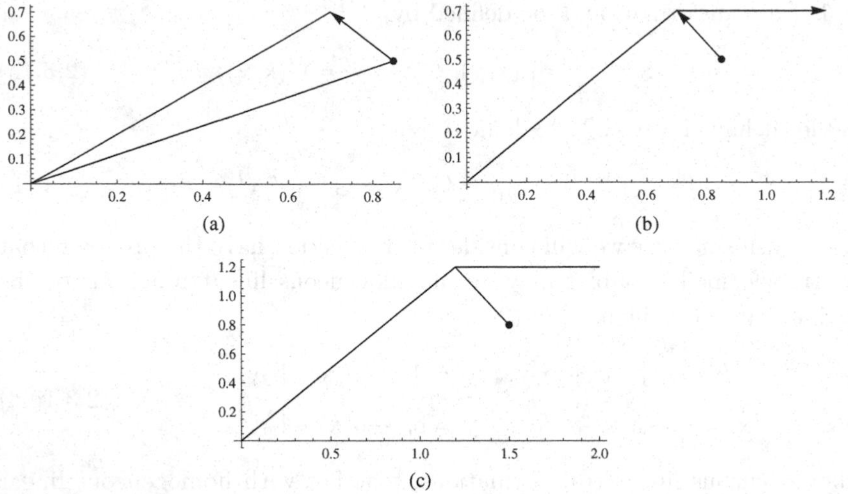

Fig. 2.3.1 Rotation followed by translation followed by dilatation.

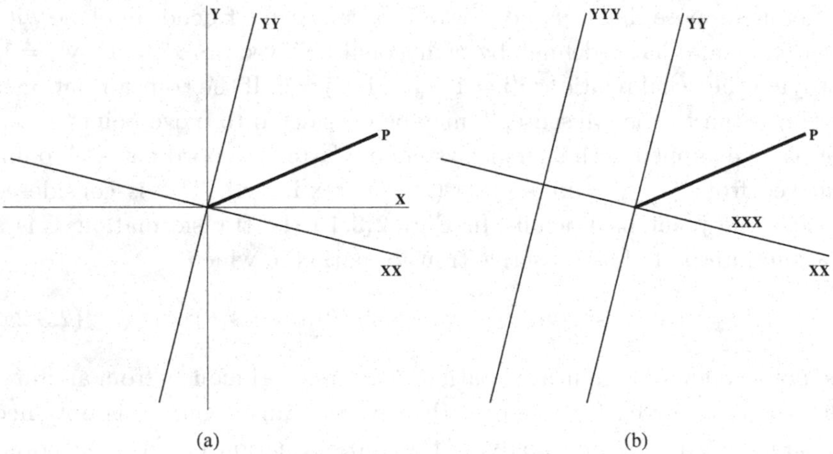

Fig. 2.3.2 A passive rotation followed by a passive translation.

use the superscript c to denote points on the coordinate axes, the passive interpretation of an active rotation through an angle θ is defined by

$$\mathrm{xx}^{\mathrm{c}} = \mathrm{x}^{\mathrm{c}}\cos(-\theta) - \mathrm{y}^{\mathrm{c}}\sin(-\theta), \quad \mathrm{yy}^{\mathrm{c}} = \mathrm{x}^{\mathrm{c}}\sin(-\theta) + \mathrm{y}^{\mathrm{c}}\cos(-\theta). \tag{2.3.3a}$$

If x^{c}, y^{c}, are of unit length, then in vector notation, (2.3.3a) becomes

$$\mathrm{xx}^{\mathrm{c}} = \mathbf{i}\cos(-\theta) - \mathbf{j}\sin(-\theta), \quad \mathrm{yy}^{\mathrm{c}} = \mathbf{i}\sin(-\theta) + \mathbf{j}\cos(-\theta). \tag{2.3.3b}$$

It follows that $\mathrm{xx^c xx^c} = 1$, $\mathrm{yy^c yy^c} = 1$, $\mathrm{xx^c yy^c} = 0$, so $\mathrm{xx^c}$ and $\mathrm{yy^c}$ are orthogonal unit vectors, and we write

$$\mathbf{ii} = \mathbf{i}\cos(-\theta) - \mathbf{j}\sin(-\theta), \quad \mathbf{jj} = \mathbf{i}\sin(-\theta) + \mathbf{j}\cos(-\theta). \tag{2.3.3c}$$

Figure 2.3.2b illustrates the passive interpretation of the combined tranformations of **Fig. 2.3.1b**. Note the passive version of the transformations of **Fig. 2.3.1b** is a transformation in which the final transformation, the translation, is inverted first; then the inverse rotation is applied to the result.

Figures 2.3.3 depicts the passive and the active interpretations of a dilation. As the passive interpretation represents a change of units of length, it would be confusing and inconsistent to superimpose **Figs. 2.3.3a,b**. On the other hand, no confusion results from superimposing **Fig. 2.3.3a** and the circle of **Fig. 2.3.3c** obtained by is active dilatation.

Finally, we consider an affine transformation which is not an operation of the similitude group of the plane. It is a *pseudo-rotation* or *hyperbolic rotation*, a transformation in which

$$\mathrm{x}' = \mathrm{x}\cosh(\gamma) + \mathrm{y}\sinh(\gamma), \quad \mathrm{y}' = \mathrm{x}\sinh(\gamma) + \mathrm{y}\cosh(\gamma). \tag{2.3.4a}$$

Its interpretation as an active transformation acting on a point is illusrated in **Fig. 2.3.4a**. The path illustrated is obtained by having γ run continuously from 0 to 2/3. **Figure 2.3.4b** depicts the action of a hyperbolic rotation, with $\gamma = 1/4$, on a unit circle. The inverse of a transformation with any value of the parameter γ is obtained by changing that value of γ to $-\gamma$ in (2.3.4a), and applying the transformation to x', y', rather than x, y:

$$\mathrm{x} = \mathrm{x}'\cosh(-\gamma) + \mathrm{y}'\sinh(-\gamma), \quad \mathrm{y} = \mathrm{x}'\sinh(-\gamma) + \mathrm{y}'\cosh(-\gamma). \tag{2.3.4b}$$

To obtain the passive interpretation of hyperbolic rotations we apply the inverse transformation to the coordinate vectors $\mathrm{x^c}$ and $\mathrm{y^c}$ and obtain

$$\begin{aligned}\mathrm{xx^c} &= \mathrm{x^c}\cosh(-1/4) + \mathrm{y^c}\sinh(-1/4),\\ \mathrm{yy^c} &= \mathrm{x^c}\sinh(-1/4) + \mathrm{y^c}\cosh(-1/4).\end{aligned} \tag{2.3.5a}$$

Figure 2.3.4c depicts the effect of this passive transformation on the unit vectors i, j, and the corresponding alteration of a Cartesian coordinate lattice, when $\gamma = 1/4$. Care must be exercised when interpolating the coordinates of points in this system of coordinates because separations between points with coordinates $(\mathrm{x}'_1, \mathrm{y}'_1)$ and $(\mathrm{x}'_2, \mathrm{y}'_2)$ do not obey the Pythagorean

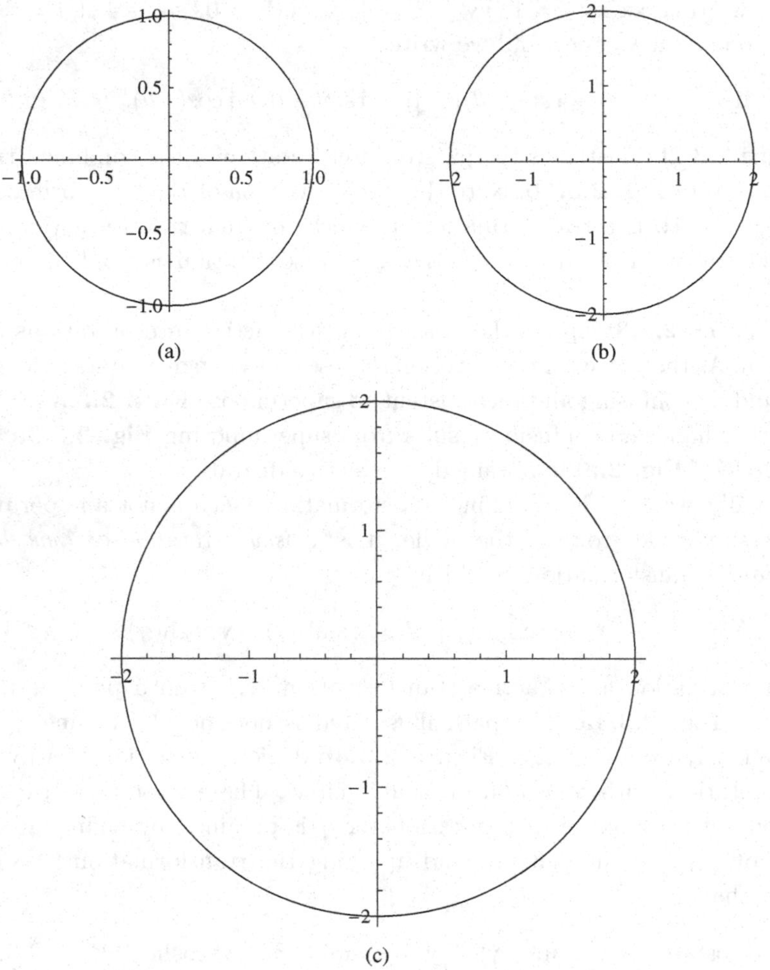

Fig. 2.3.3 Passive and active interpretations of a uniform diltation.

theorem. Though, $\Delta r^2 = \Delta x^2 + \Delta y^2$, $\Delta r'^2 \neq \Delta x'^2 + \Delta y'^2$. In fact, using the relation $\cosh(\gamma)^2 - \sinh(\gamma)^2 = 1$ one finds it

$$\Delta s^2 \equiv \Delta x^2 - \Delta y^2 = \Delta x'^2 - \Delta y'^2, \tag{2.3.6}$$

unchanged by pseudo-rotations.

When studying the illustrations of the previous figures one naturally supposes the paper to be held fixed, and may unconsciously use this perception in interpreting the transformations being illustrated. A geometric

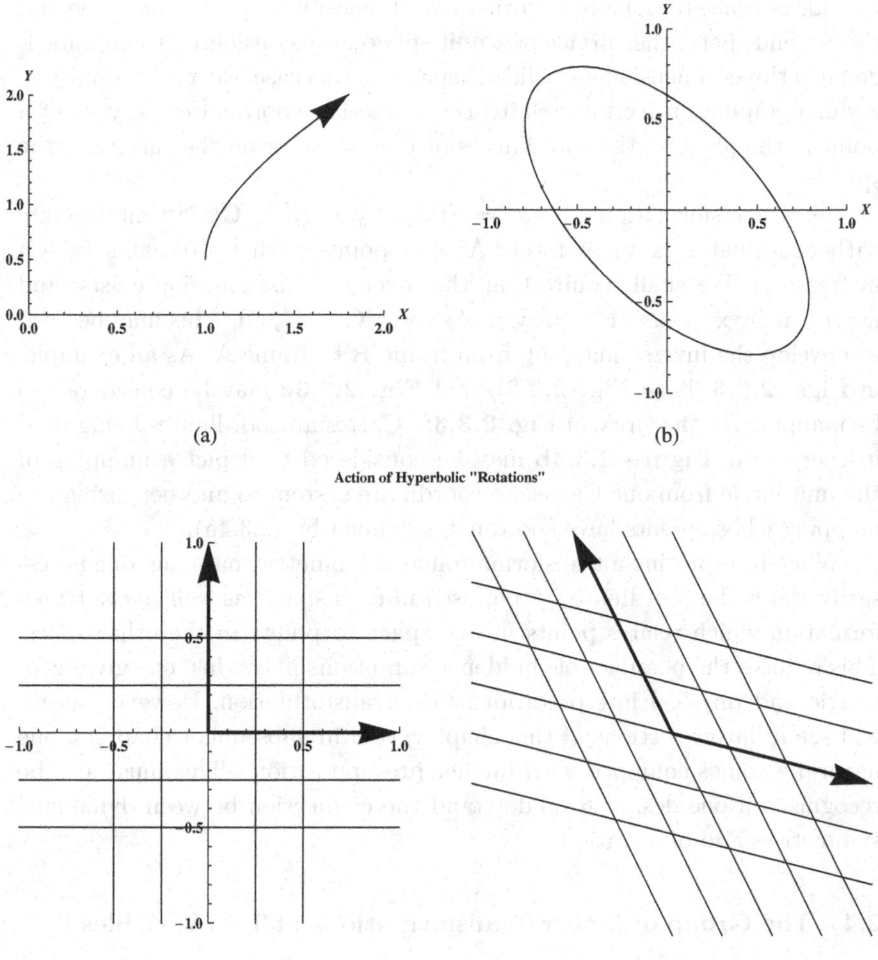

Fig. 2.3.4 Transformation of Cartesian coordinates by a hyperbolic rotation.

interpretation of transformations which helps to avoid confusion may be obtained by using a rule of transformation to relate points in one coordinate system of choice to points in another freely chosen coordinate system. If one wishes, one may consider the mapping to be from one space to another, and one may consider the two spaces as surfaces in a higher dimensional space. Points in one system are *mapped* into points in the other. **Figure 1.2** depicts a mapping that connects points in a two-dimensional

Euclidean space to points in another two-dimensional space. The plane and this second space, the surface of a unit sphere, are considered to be imbedded in a three-dimensional Euclidean space. In this case, the transformation, a stereographic projection, relates the Cartesian coordinates $(\mathrm{x}, \mathrm{y}, \mathrm{z})$ of a point in the plane to the coordinates (θ, ϕ) of a point on the surface of the sphere.

Let a transformation $\mathrm{x} \to \mathrm{x}' = \mathrm{F}(\mathrm{x}, \mathrm{y})$, $\mathrm{y} \to \mathrm{y}' = \mathrm{G}(\mathrm{x}, \mathrm{y})$ map points with coordinates (x, y) in frame A to a points with coordinates $(\mathrm{x}', \mathrm{y}')$ in frame B. We shall require that the inverse transformation exists, and express it by $\mathrm{x}' \to \mathrm{x} = \mathrm{F}^{\mathrm{inv}}(\mathrm{x}', \mathrm{y}')$, $\mathrm{y}' \to \mathrm{y} = \mathrm{G}^{\mathrm{inv}}(\mathrm{x}', \mathrm{y}')$. This may be used to develop the inverse mapping from frame B to frame A. As an example, in **Figs. 2.3.3**, both **Fig. 2.3.3b** and **Fig. 2.3.3c** may be considered to be mappings of the circle of **Fig. 2.3.3a**, Cartesian coordinates being used in every case. **Figure 2.3.4b** may be considered to depict a mapping of the unit circle from one Cartesian coordinate system to another, using the mapping whose points have coordinates defined by (2.3.4a).

When interpreting a transformation as a geometric mapping, one necessarily states the coordinate system used in each space, as well as the transformation which relates points in one space to points in the other space. This reduces the possible role hidden assumptions play when one gives geometric and physical interpretations to a transformation. However, as we will see in latter sections of this chapter, and in subsequent chapters, one naturally comes equipped with further presuppositions. This must also be recognized if one desires to understand the connection between dynamical symmetries and observations.

2.4 The Group of Linear Transformations of Two Variables

Rotation and pseudo-rotation transformations are homogeneous linear transformations, as are the dilation transformations $\mathbf{x} \to \mathbf{x}' = \exp(\beta)\ \mathbf{x}$, $\mathbf{y} \to \mathbf{y}' = \exp(\gamma)\ \mathbf{y}$. Homogeneous linear transformations of two variables x_1 and x_2 can be expressed as $\mathbf{X}' = \mathbf{MX}$, where X represents column vector that is the transpose of the row vector $(\mathbf{x}_1, \mathbf{x}_2)$, $\mathbf{X}'$ is the transpose of $(\mathrm{x}_1', \mathrm{x}_2')$ and $\mathbf{M}$ is a 2×2 matrix with elements m_{11}, m_{12}, m_{21}, m_{22} that may depend on parameters. The inhomogeneous transformations arising from translations cannot be expressed in this same manner. Taken together, translations and homogeneous linear transformations are those of the affine group. The affine transformation

$$(\mathrm{x}, \mathrm{y}) \to (\mathrm{x}', \mathrm{y}') = (\mathrm{x} + \mathrm{a} + \mathrm{bx} + \mathrm{cy},\ \mathrm{y} + \mathrm{d} + \mathrm{ex} + \mathrm{fy}) \tag{2.4.1}$$

converts the difference vectors

$$(\Delta x, \Delta y) = (x_2, y_2) - (x_1, y_1) = (x_2 - x_1, y_2 - y_1)$$

to

$$(\Delta x', \Delta y') = (\Delta x + b\ \Delta x + c\ \Delta y, \Delta y + e\ \Delta x + f\ \Delta y). \qquad (2.4.2)$$

These are homogeneous linear transformations, independent of the parameters a and d. In matrix and vector notation (2.4.2) may be written $\Delta\mathbf{X}' = \mathbf{M}\Delta\mathbf{X}$. For the same reason one may write $d\mathbf{X}' = \mathbf{M}d\mathbf{X}$.

Altogether, the rotation, pseudo-rotation, dilation, reflection, and inversion transformations, form a group of transformations, the *general linear group* in two real variables, denoted Gl(2, R). Any homogeneous linear transformation of two real variables is a transformation of this group. It contains a four-parameter subgroup, the continuous group Sl(2, R). If the rotations and pseudo-rotations depend on the group parameters θ, and γ, and the dilatations of x, y depend on α, β, then in the general transformation of the group, $\mathbf{X}' = \mathbf{MX}$, or equivalently

$$\begin{aligned} x_1' &= m_{11}(\alpha,\beta,\gamma,\theta)\ x_1 + m_{12}(\alpha,\beta,\gamma,\theta)x_2, \\ x_2' &= m_{21}(\alpha,\beta,\gamma,\theta)\ x_1 + m_{22}(\alpha,\beta,\gamma,\theta)x_2. \end{aligned} \qquad (2.4.3)$$

The functional dependence of the matrix elements m_{ij} upon the group parameters $\alpha, \beta, \gamma, \theta$ depends upon the order in which the transformations are carried out. Sl(2,R) differs from the similitude group because it includes pseudo-rotations; the similitude group is a subgroup of Sl(2, R).

Generalizing the observations of this section, one finds that the group of affine transformations of N real variables x_i yields the group Gl(N, R) of linear transformations of the corresponding Δx_i and dx_i.

2.5 Physical Interpretation of Rotations

Suppose that one has

$$\mathbf{X}' = \mathbf{HX}, \quad \mathbf{H} = \begin{pmatrix} h_{11} & h_{12} \\ h_{21} & h_{22} \end{pmatrix}, \quad \mathbf{X} = (x_1, x_2)^T \qquad (2.5.1a)$$

Now let

$$\mathbf{R} = \mathbf{R}(\theta) = \begin{pmatrix} \cos(\theta) & -\sin(\theta) \\ \sin(\theta) & \cos(\theta) \end{pmatrix} \qquad (2.5.1b)$$

actively rotate $\mathbf{X}'$. To determine the relation between the operation of $\mathbf{R}$ on $\mathbf{X}'$ and the operation of $\mathbf{R}$ on $\mathbf{X}$, we use the equations

$$\mathbf{RX}' = \mathbf{R}(\mathbf{HX}) = \mathbf{RHR}^{-1}\mathbf{RX} = \mathbf{H}'\mathbf{RX} \qquad (2.5.2a)$$

with

$$\mathbf{H}' = \mathbf{RHR}^{-1}. \tag{2.5.2b}$$

Here $\mathbf{R}^{-1}$, the inverse of $\mathbf{R}(\theta)$, is $\mathbf{R}(-\theta)$. Carrying out the matrix multiplications in (2.5.2b) and using trigonometric identities to simplify the result, one finds that the elements of $\mathbf{H}'$ are

$$\begin{aligned}
h'_{11}(\theta) &= 1/2\ h_{11}(1+\cos(2\theta)) + 1/2\ h_{22}(1-\cos(2\theta)) \\
&\quad - 1/2\ (h_{12}+h_{21})\sin(2\theta), \\
h'_{12}(\theta) &= 1/2\ h_{12}(1+\cos(2\theta)) - 1/2\ h_{21}(1-\cos(2\theta)) \\
&\quad + 1/2\ (h_{11}-h_{22})\sin(2\theta), \\
h'_{21}(\theta) &= 1/2\ h_{21}(1+\cos(2\theta)) - 1/2\ h_{12}(1-\cos(2\theta)) \\
&\quad + 1/2\ (h_{11}-h_{22})\sin(2\theta), \\
h'_{22}(\theta) &= 1/2\ h_{22}(1+\cos(2\theta)) + 1/2\ h_{11}(1-\cos(2\theta)) \\
&\quad + 1/2\ (h_{12}+h_{21})\sin(2\theta).
\end{aligned} \tag{2.5.3}$$

The components of the vector $\mathbf{RX}$ have the dependence on θ given by $\mathbf{R}(\theta)(x_1, x_2)^T = (x_1\cos(\theta) - x_2\sin(\theta),\ x_1\sin(\theta) + x_2\cos(\theta))^T$. *In contrast, as θ varies continuously from 0 to π, the matrix elements $h'_{ij}(\theta)$ return to their initial values h_{ij}, while the vector $\mathbf{RX}'$ reverses direction*!

A further 180° rotation cycles the matrix elements $h'_{ij}(\theta)$ through their full range a second time and brings the vector back to its original orientation. Alternatively, if $\phi = 2\theta$ and we let $\mathbf{H}' = \mathbf{H}'(\phi)$, then $\mathbf{H}'(2\pi) = \mathbf{H}'(0)$, but $\mathbf{RX}(2\pi) = -\mathbf{X}$. If for physical or mathematical reasons θ has a range of 2π, then $\mathbf{X}$ can be a vector in a Euclidean plane. A passive, as well as an active, rotation through 360° will restore the coordinates of this vector to its original value. If it is ϕ that has a range of 2π, then active and passive rotations of $\mathbf{X}$ establish different relations between it and a coordinate system, so $\mathbf{X}$ cannot be a vector in a Euclidean plane.

A geometric illustration, of vectors that are carried into their negative by a 360° rotation, is provided by vectors normal to the girdle of a Moebius strip. **Figure 2.5.1a** depicts such a band. **Figure 2.5.1b** depicts a portion of the surface swept out by a vector normal to the Moebius band when its "root" moves around the circle formed by the intersection of the band with the x–y plane. An active 360° rotation reverses the direction of the normal. On the other hand, the passive rotation obtained by walking once around the z axis of the strip while carrying a coordinate system, brings the normal back to its original relation to the observer and coordinate system.

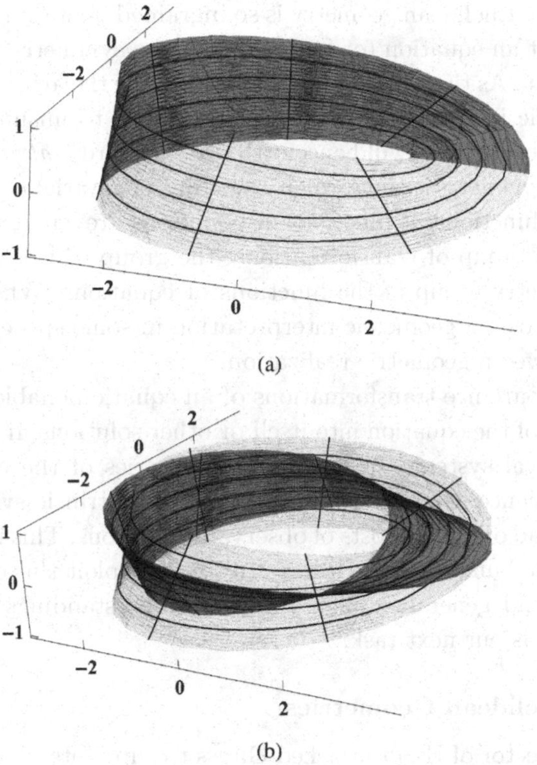

Fig. 2.5.1 (a) Moebius band. (b) Surfaces swept through by the normal to the band.

As both bands in **Fig. 2.5.1b** are Moebius strips, vectors in both behave similarly.

2.6 Intrinsic Symmetry of an Equation

In Sec. 2.1 the roots of the equation $w^4 = 1$ were plotted as the four vertices of figures on three different surfaces. The vertices can, in each case, be converted into themselves by operations of a group determined by the group composition relations summarized in Table 2.2. Though the composition relations are the same in all three cases, in only one case is the figure a square. The group composition laws were determined by the equation $w^4 = 1$, not by its geometric realizations.

The invariance group of an equation is an intrinsic property of the equation; its geometric interpretation involves arbitrary choices. Nevertheless, perhaps because Euclidean geometry is so ingrained in our minds, it is commonly said that an equation (or function) has the symmetry defined by its invariance group. As this usage of words can obscure the arbitrariness inherent in geometric interpretations, it can easily lead to misunderstandings. To avoid this problem, we will henceforth use the words *intrinsic symmetry* and *intrinsic symmetry groups* when speaking of invariance properties of equations and functions. If functions or equations are left invariant by the operations of a group of transformations, the group will be said to be an intrinsic symmetry group of the functions or equations. When the invariance group is given a geometric interpretation in some space, the intrinsic symmetry is given a geometric realization.

Knowing invariance transformations of an equation enables one to convert a solution of the equation into itself or other solutions. If the equations govern a physical system, the intrinsic symmetries of the equation have physical significance. Geometric interpretation of intrinsic symmetries can then suggest, and organize, hosts of observable relations. This is a marvelous capability of the human mind. However, to fully exploit this capability one must expand and generalize one's intuitive understandings of Euclidean geometry. This is our next task.

2.7 Non-Euclidean Geometries

In 1820 The Elector of Hanover asked Gauss to superintend a cartographic survey of his kingdom. Gauss, being by nature as much a mathematician as a civil servant, the Elector's request stimulated a general investigation of the metrical properties of curved surfaces.[1]

In Euclidean three-space, an infinitesimal displacement **dr** with Cartesian components dx, dy, and dz gives rise to points separated by a distance ds, with

$$\mathrm{ds}^2 = \mathrm{dr}^2 = \mathrm{dx}^2 + \mathrm{dy}^2 + \mathrm{dz}^2. \tag{2.7.1}$$

Except in the Harz mountain region, Hanover has a fairly *flat* terrain and for many purposes it is sufficient to suppose the earth is a sphere of fixed radius r. One can then use the colatitude θ and longitude ϕ as coordinates of points on the sphere. The infinitesimal displacements $\mathrm{d}\theta$, $\mathrm{d}\phi$ lie in the surface r = constant, and one has

$$\begin{aligned}
\mathrm{dx} &= \mathrm{d}(\mathrm{r}\sin\theta\cos\phi) = \mathrm{r}\{\cos\theta\cos\phi\,\mathrm{d}\theta - \sin\theta\sin\phi\,\mathrm{d}\phi\},\\
\mathrm{dy} &= \mathrm{d}(\mathrm{r}\sin\theta\sin\phi) = \mathrm{r}\{\cos\theta\sin\phi\,\mathrm{d}\theta + \sin\theta\cos\phi\,\mathrm{d}\phi\},\\
\mathrm{dz} &= \mathrm{d}(\mathrm{r}\cos\theta) = -\mathrm{r}\sin\theta\,\mathrm{d}\theta.
\end{aligned} \tag{2.7.2}$$

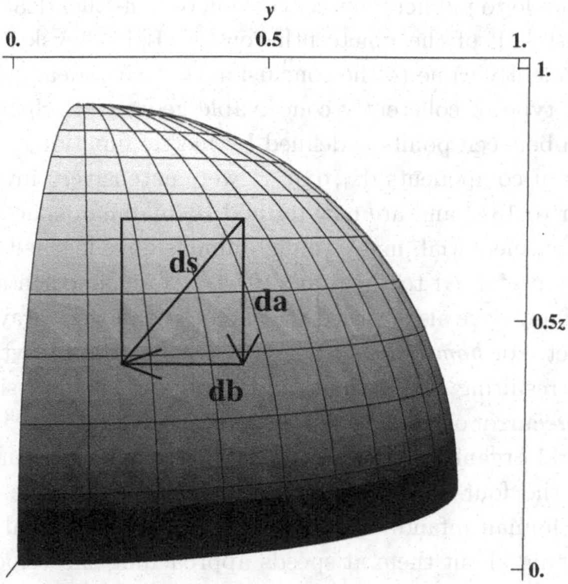

Fig. 2.7.1 Displacements that determines the metric on a sphere.

Substituting the right-hand side of relations (2.7.2) into (2.7.1) yields

$$ds^2 = g_{11}d\theta^2 + g_{12}d\theta d\phi + g_{22}d\phi^2, \quad (2.7.3a)$$

with

$$g_{11} = r^2, \quad g_{12} = 0, \quad g_{22} = r^2 \sin^2\theta. \quad (2.7.3b–d)$$

The geometric meaning of (2.7.3) is illustrated in **Fig. 2.7.1**.

In the figure, $|da| = rd\theta$, $|db| = r\sin\theta\, d\phi$, and $ds^2 = da^2 + db^2$. Because da and db are orthogonal, g_{12} is zero.

Equations such as (2.7.3) express metrical properties of the coordinate system θ, ϕ on the two-dimensional surfaces r = constant. Gauss generalized the equations to other surfaces and became the first person to take the further, highly imaginative step of considering that *surfaces with defined metrical properties are spaces in their own right. They may, but need not be, thought of as surfaces embedded in spaces of higher dimension.* As the axioms and theorems of plane Euclidean geometry do not all hold true on non-planar surfaces, the imaginative jump of Gauss has the consequence of introducing spaces with non-Euclidean geometries. The right-hand side of an equation such as (2.7.3a) is then considered to *define* the metric in the geometry, the measure of the separation of points in the space.

Gauss did little to publicize his conception of non-Euclidean geometries, and in the first half of the nineteenth century Lobachevsky, Bolyai, and others independently came to the conclusion that Euclidean geometry was only a special type of coherently conceivable geometry.[1] Spaces in which the separation between points is defined by metric functions $g_{ij}(x_1, x_2, \ldots)$ and quadratic in components $dx_i, dx_j, \ldots$ were extensively investigated by Riemann prior to 1854 and are now termed Riemannian spaces.[2]

From time immemorial, many young animals have learned to recognize that an object presented to them at different angular orientations is the *same* object. They have also learned that an object moved away and back is the *same* object. For *homo sapiens* who are beneficiaries of Pythagoras and Descartes the resulting unconscious, and enormous, mental simplification becomes the precursor of the abstract relation $\Delta r^2 = \Delta x^2 + \Delta y^2 + \Delta z^2$. However, the mental organization of sense data has not lead humans to intuit and visualize the four-dimensional non-Euclidean spacetime of Einstein. The minds of human infants do not have to cope with visual impressions of parents zipping about them at speeds approaching the velocity of light.

In the 1860's Helmholtz recognized that our brains make use of geometric presuppositions of which we are not aware.[3] In profound studies of human visual perceptions he independently obtained some of the key results of Riemann. However, prior to 1905, mankind's Euclidean geometric presuppositions led physicists to interpret the results of experiments such as those of Michelson and Morley[4] by supposing that objects moving at constant velocity with respect to an observer shrunk in the direction of motion, and that the mechanism of any type of clock attached to the moving object slowed down. It required Einstein's inspired logical analysis of physical observations and presuppositions to recognize that this strange behavior is not a property of the objects themselves, but a property of the relationship between the observer and the observed.[5] Einstein then demonstrated that if one adopted an appropriate geometry of spacetime, the observed precession of the orbit of Mercury followed.

Suppose one observes a pair of events taking place at times t_1 and t_2 at points with Cartesian coordinates (x_1, y_1, z_1), and (x_2, y_2, z_2). Then one will be dealing with events separated by a time interval $\Delta t = \int dt$ measured by one's clock, and at points separated from each other by a displacement $\Delta \mathbf{r} = \int d\mathbf{r} = (\int dx, \int dy, \int dz)$.[6] With this understanding the metric of special relativity can be written as

$$ds^2 = ds^2_{sp} = (dx^2 + dy^2 + dz^2) - c^2 dt^2 \qquad (2.7.4a)$$

or as

$$ds^2 = ds_t^2 = c^2dt^2 - (dx^2 + dy^2 + dz^2), \tag{2.7.4b}$$

where c is the velocity of light in vacuo. The metric in (2.7.4a) is said to be *spacelike*, and the metric in (2.7.4b) is said to be *timelike*. All observers find for c a value of approximately 2.99776×10^8 meters/sec. For a point at (x, y, z) moving to $(x + dx, y + dy, z + dz)$ with the velocity of light, $dx^2 + dy^2 + dz^2 = c^2\, dt^2$, and so in both cases $ds = 0$. We use ds_{sp} to denote the real-valued ds that is obtained when the spatial separation $dr = (dx^2 + dy^2 + dz^2)^{1/2}$ is greater than c dt, the time it takes for light to travel between them; and ds_t to denote the real-valued ds that arises when $dr < c\,dt$. Some authors use only one definition of the metric, allowing ds to take on imaginary as well as real values. By preference we require ds be real-valued, and shift between the spacelike ds_{sp} and the timelike ds_t to keep ds real.

The Lorentz transformations of special relativity leave invariant the ds^2 of Eqs. (2.7.4) and relate observations of the same pair of events perceived by two observers moving at constant velocity relative to one another, but using their own Cartesian spacetime coordinate systems, coordinate systems in which they are at rest. These observers will not agree on the spatial distance dr separating events nor upon the time interval dt between them, but they will agree upon the spacetime interval ds separating them.

In the simplest case, an observer A and another observer B at rest with respect to one another arrange their clocks and meter sticks to agree as closely as possible. They align their x, y, and z axes as closely as possible with those of the other, and then set themselves, together with their measuring systems, in relative motion with constant velocity vectors parallel to their x axes. If v_x is the velocity A measures for B's system, then B finds that A's system has velocity $-v_x$. A and B observe the same set of events and each measures the spatial separation and time intervals between the events, then transmits the results to the other observer. If one finds that two events are separated by Δx and $\Delta\tau = c\Delta t$, the other finds them separated by $\Delta x'$ and $\Delta\tau' = c\Delta t'$, with

$$\Delta x' = \Delta x \cosh(\alpha) - \Delta\tau \sinh(\alpha), \quad \Delta\tau' = \Delta\tau \cosh(\alpha) - \Delta x \sinh(\alpha),$$
$$\tanh(\alpha) = v/c, \quad \alpha = \ln((1 + v/c)/(1 - v/c))^{1/2}, \quad v = \pm v_x. \tag{2.7.5a,b,c}$$

Each observer could conclude that the other's clock has slowed, and meter stick contracted, in the x direction. If, however, they plot their observed

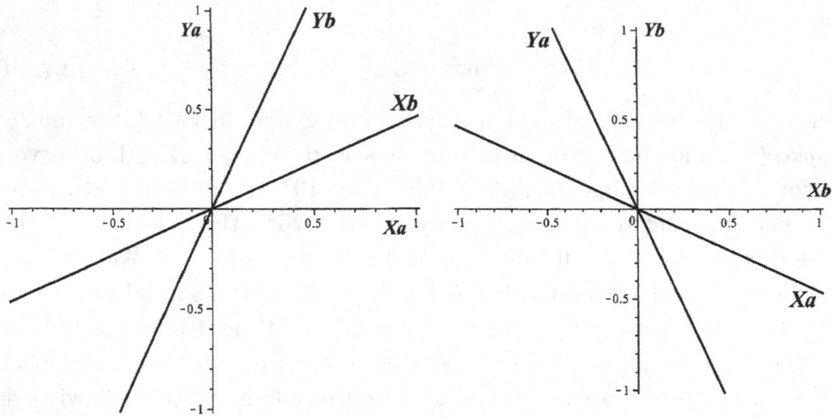

Fig. 2.7.2 (a) Projection of B's coordinate system onto the cartersian system of A; (b) Projection of A's coordinate system onto the cartersian system of B.

positions and times of occurrence of the events using Cartesian coordinates in a Euclidean plane, they may come to a less desirable situation. **Figures 2.7.2a,b** illustrate the relationship that develops between coordinate systems when observes attached to them have a relative speed of 0.46c. On plotting other's observations reported, each may think the other is using non-orthogonal coordinates. The transformation $(\Delta x', \Delta\tau') \rightarrow (\Delta x, \Delta\tau)$ is the inverse of the transformation $(\Delta x, \Delta\tau) \rightarrow (\Delta x', \Delta\tau')$, because changing v_x to $-v_x$ in (2.7.5c) changes α to $-\alpha$. The figures may hence be thought of as depicting projections onto each observer's Euclidean plane — projections of the spacetime coordinate system of the other observer. One may consistently consider that the two observers view the same events from different *perspectives.* The perspectives are such that the observers agree on the spacetime intervals, Δs, because equations a, b, of the Lorentz transformations (2.7.5) imply that

$$\Delta x'^2 - c^2\Delta t'^2 = \Delta x^2 - c^2\Delta t^2. \tag{2.7.6}$$

Figures. 2.7.3 and **2.7.4** display lines of constant Δs in a Euclidean plane with orthogonal Δx and $c\Delta t$ axes. Observers in uniform relative motion who observe the same events will assign them points with different Δx and $\Delta(ct)$ coordinates in the figures, but the points will lie on the same curve $\Delta s = \text{const}$. Each curve is carried into itself by the Lorentz pseudo-rotation transformations (2.7.5). The point $(\Delta x, \Delta\tau) = (0, 0)$ is invariant under the transformation. If both observers measure $(\Delta x, \Delta\tau)$ from their

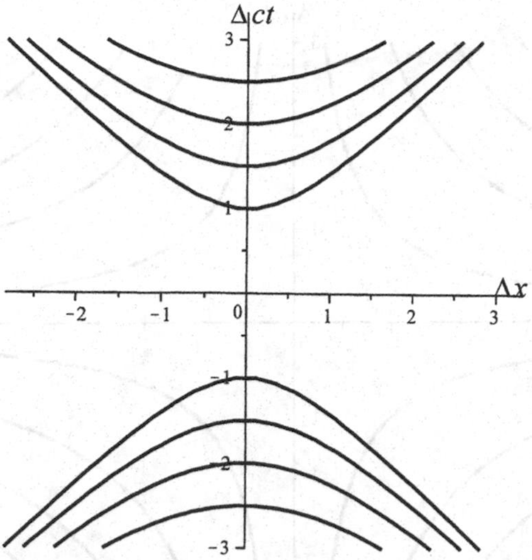

Fig. 2.7.3 Curves of constant $\Delta s, \Delta x < \Delta ct$.

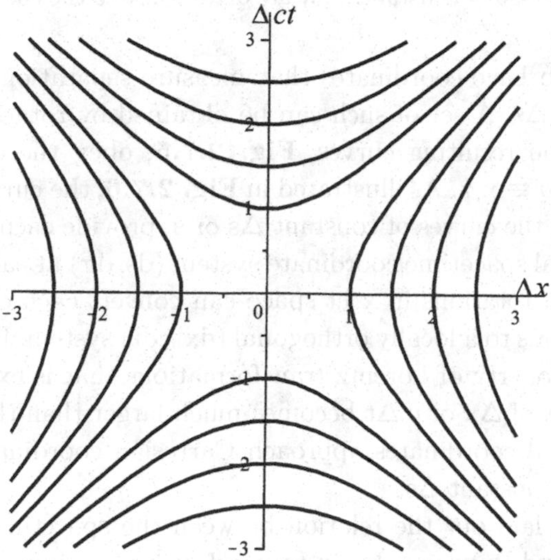

Fig. 2.7.4 Curves of constant Δs.

own $(0, 0)$ *point*, then both may replace $(\Delta x, c\,\Delta t)$ by (x, t). Points on the straight lines in the figures represent intervals measured for motions taking place at the speed of light in vacuo, c. These lines separate regions where $\Delta s = \Delta s_t$, from the top and bottom regions where $\Delta s = \Delta s_{sp}$.

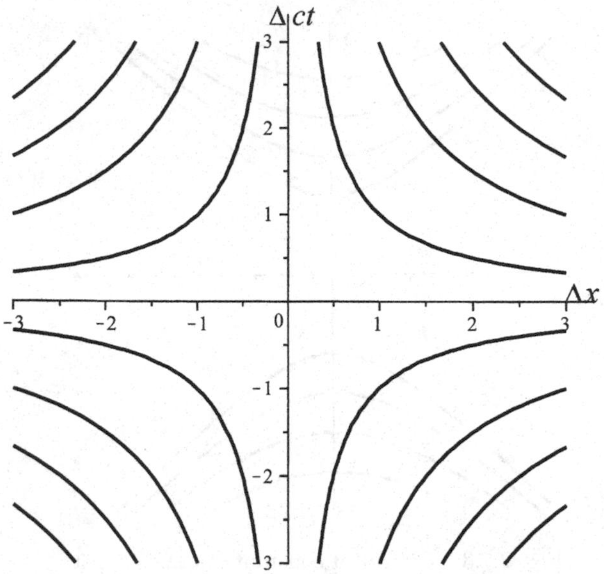

Fig. 2.7.5 Curves of σ = constant; they are orthogonal to the curves of Fig. 2.7.4.

It is useful to have coordinates that measure separations along curves of constant s or Δs. A set of such can be obtained by rotating **Fig. 2.7.4** through 90°. The resulting curves, **Fig. 2.7.5**, obey the equation. σ = constant, where $\sigma \equiv \mathrm{x} \cdot \tau$. As illustrated in **Fig. 2.7.6**, the curves of constant σ, together with the curves of constant Δs or s, provide each observer with a local orthogonal spacetime coordinate system (ds, dσ) at each point (s, σ). Translations and rotations in x, ct space can convert each of these locally orthogonal systems to a locally orthogonal (dx, cdt) system. Finally we note an important property of Lorentz transformations that is expressed in the figures. As either of Δx or c Δt becomes much larger than the other, these locally orthogonal coordinates approach Cartesian coordinates on almost parallel lines of constant Δs.

Now, having laid out the relation between the spacetime observations of observers A and B, we can begin to ponder the meaning of symmetry in their spacetime. We start with two-dimensional spacetime.

Fig. 2.7.7a plots the results of three (x, ct) observations made at x = 0 by an observer, A, who has clocked flashes of light from two sources, both observed at t = 0, one from x = 0, the other from x = 2. To this observer the flashes have Δx values of 0, +2, respectively, and have the spacelike Δs values of 0 and 2. The third observation is of a flash from x = 1, observed

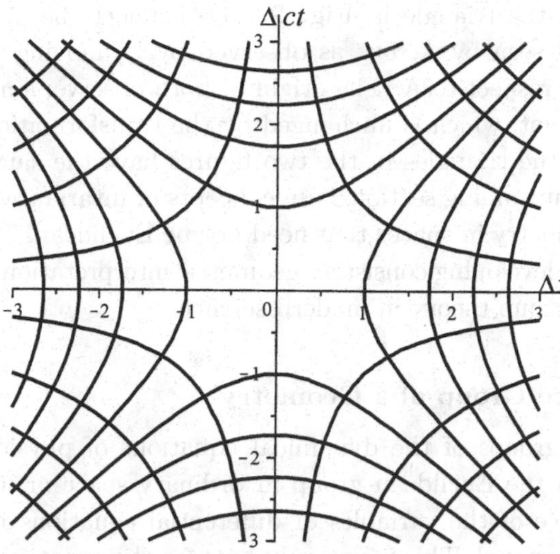

Fig. 2.7.6 A locally orthogonal coordinate system: s = constant, σ = constant.

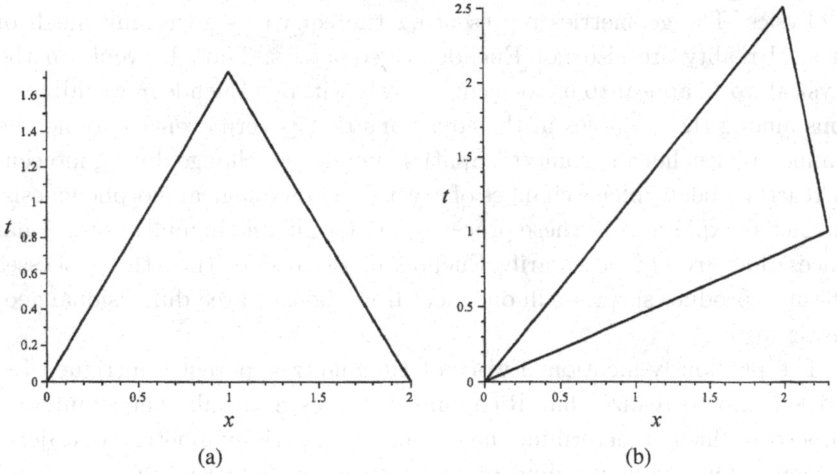

Fig. 2.7.7 (a) Triangle connecting three events, through observer A's frame of references; (b) triangle connecting same events, through observer B's frame of reference.

at time $t = \sqrt{3}$ in observer A's system. In this case $\Delta x = 1$, $\Delta t = \sqrt{3}$, so Δs is timelike, and Δs_t has the value $((\sqrt{3})^2 - 1^2)^{1/2} = \sqrt{2}$. Connecting the observations by straight lines yields the equilateral triangle shown in the figure.

The sides of the triangle in **Fig. 2.7.7b** connect the observations of the same events seen by A, but as observed by B, moving with velocity $v/c = 0.46$ with respect to A. The origin of both observer's measurements is their (0, 0) point, which is unchanged by the transformation (2.7.5). To decide whether the triangles in the two figures have the same symmetry, we shall, in the upcoming sections, use concepts of invariance to define the meaning of symmetry in spaces that need not be Euclidean. This turns out to be the key in developing consistent geometric interpretations of extensive applications of group theory in modern science.

2.8 Invariance Group of a Geometry

The invariance groups of the dynamical equations of physics and chemistry are seldom the Euclidean group in ordinary space or its subgroups. In fact, the space of the variables of differential equations is often more than three-dimensional. This is not only true for the equations of relativistic mechanics, it is true for the differential equations of classical mechanics as well as quantum mechanics, electrodynamics, and many other branches of physics. The geometries representing the equations governing much of physical reality are also not Euclidean geometries. Though events in the physical world appear to us to occur entirely within a Euclidean world, relations among the variables in the equations that govern events may not be confined by Euclidean geometry. Spatial symmetries change during molecular rearrangments, phase changes of crystals, embryological morphogenesis, and stellar explosions — these processes are formulated in multidimensional spaces that are not necessarily Euclidean. Because of this, the processes oft-times produce surprises in our mental Euclidean three-dimensional geometric world.

The previously mentioned work of Helmholtz apparently intrigued Lie and led him to realize that Riemannian spaces have inherent symmetric properties which predetermine the possible geometric symmetries of objects in them.[7] Our understanding of the connection between symmetry and geometry was further advanced by Lie's friend and mentor, Klein.[8]

The metric of an n-dimensional Euclidean geometry is

$$ds^2 = dx_1^2 + \cdots + dx_i^2 + \cdots + dx_n^2. \qquad (2.8.1)$$

The transformation groups that characterize Euclidean symmetry in an n-dimensional space are transformations of real variables that leave invariant these metrics. These groups have continuous subgroups, Euclidean groups termed E(n). These contain a subgroup of translations $T(\alpha_1, \ldots, \alpha_i, \ldots, \alpha_n)$ that carries points with cartesian coordinates x_i to points with cartesian coordinates $x_i + \alpha_i$; and they contain a subgroup group of rotations $R(\theta_{1,2}, \ldots, \theta_{i,j}, \ldots, \theta_{n-1,n})$ that rotate points through angles $\theta_{i,j}$ about the origin in the two-planes which contain the x_i and x_j axes. The invariance group of the Euclidean metric (2.8.1) also contains a discrete subgroup of n reflections $x_j \rightarrow -x_j$. A reflection of all of the x's together produces an inversion through the center of the coordinate system.

The geometry of the spacetime of special relativity is defined by the group of transformations of real variables that leave invariant the metrics (2.7.4). The continuous subgroup of this invariance group is called the *Poincare group.* The Poincare group contains the operations of translation and rotation of the group E(3) acting in the subspace of x, y, z, together with the operations of time translation, $t \rightarrow t + \tau$, and the subgroup of hyperbolic pseudo-rotations

$$\begin{aligned} x' &= x\cosh(\alpha) - ct\sinh(\alpha), & ct' &= ct\cosh(\alpha) - x\sinh(\alpha), \\ y' &= y\cosh(\beta) - ct\sinh(\beta), & ct' &= ct\cosh(\beta) - y\sinh(\beta), \\ z' &= z\cosh(\gamma) - ct\sinh(\gamma), & ct' &= ct\cosh(\gamma) - z\sinh(\gamma). \end{aligned} \tag{2.8.2}$$

The group parameters are functions of the component of the vector $V = (v_x, v_y, v_z)$ which expresses the relative velocity of the (x', y', z', t') frame of reference to that of the unprimed frame. In the resulting Lorentz transformations, one has, for example

$$\gamma = \ln\left(\frac{(1 + |v_z|/c)}{(1 - |v_z|/c)}\right)^{1/2}. \tag{2.8.3}$$

The Poincare group of spacetime translations, rotations and Lorentz pseudo-rotations is a Lie group whose operations depend upon ten parameters. Four parameters determine translations, three determine rotations, and three determine hyperbolic rotations; none can have infinite values. The spacetime metric of special relativity is also left invariant by the inversions $x_j \rightarrow -x_j$ and $t \rightarrow -t$. Maxwell's equations are invariant under the actions of the Poincare group and these space and time inversions. Maxwell's equations are also left invariant by the larger group which includes uniform dilatations of spacetime. And they are left invariant by a still larger group, a *conformal group*, which will be discussed in Chapter 14.

2.9 Symmetry in Euclidean Spaces

We have previously noted that the symmetry of a particular figure or lattice in a two- or three-dimensional Euclidean space is defined by a subgroup of the group of all translations, rotations, reflections and inversions — the subgroup that converts the object into itself. And we have noted that the symmetry of the object is unchanged by all operations of the continuous group of arbitrary translations and rotations when these are applied to the entire figure or lattice. All these operations leave invariant

$$ds^2 = dx_1^2 + dx_2^2, \tag{2.9.1a}$$

or

$$ds^2 = dx_1^2 + dx_2^2 + dx_3^2. \tag{2.9.1b}$$

We have also noted that the symmetry of an object in these Euclidean spaces is left invariant by any operation of the continuous similitude groups which leave (2.9.1a,b) invariant, or convert them to $ds'^2 = e^{2\alpha}ds^2$. As the ds^2 of (2.1.9a,b) are special cases of the corresponding ds'^2, in which $\alpha = 0$, the continuous goups which leave invariant the metrics

$$ds'^2 = e^{2\alpha}(dx_1^2 + dx_2^2), \quad \text{or} \quad ds'^2 = e^{2\alpha}(dx_1^2 + dx_2^2 + dx_3^2), \tag{2.9.1c,d}$$

leave invariant the symmetries of objects in these spaces.

If in an n-dimensional space, displacements along n mutually orthogonal axes are denoted dx_j, the space is Euclidean if its metric is

$$ds^2 = dx_1^2 + \cdots + dx_i^2 + \cdots + dx_n^2. \tag{2.9.2}$$

In this space, the operations of the continuous group E(n) carry out n linearly independent translations, one parallel to each of its n Cartesian axes, and rotations in each of $n(n-1)/2$ planes that contain two of these axes. The operations of E(n) consequently depend on $n(n+1)/2$ independent parameters. The metric is also left invariant by the discrete group of inversions and reflections.

The symmetry of any particular object in the space is defined by the subgroup of the group that leaves invariant the metric. The symmetry of the object is invariant under all the operations of the continuous similitude group of the space. These leave invariant

$$ds'^2 = e^{2\alpha}(dx_1^2 + \cdots + dx_i^2 + \cdots + dx_n^2), -\infty < \alpha < \infty. \tag{2.9.3}$$

The operations of the continuous similitude group of n-dimensional Euclidean space consequently depends on one more independent parameter.

2.10 Symmetry in the Spacetime of Special Relativity

We next consider what is to be meant by symmetry when the metric is the

$$ds^2 = \pm(dx^2 + dy^2 + dz^2 - c^2\, dt^2) \qquad (2.10.1a)$$

that is left invariant by all transformations of the Poincare group. We shall call the continuous group that leaves (2.10.1a) invariant or converts it to

$$ds'^2 = \pm e^{2\alpha}(dx^2 + dy^2 + dz^2 - c^2\, dt^2), \quad -\infty < \alpha < \infty, \qquad (2.10.1b)$$

the *Poincare similitude group.*

The metrics are, first of all, left invariant by all transformations in ordinary space which separately leave invariant the Euclidean metric $dr^2 = dx^2 + dy^2 + dz^2$, and the metric dt^2. The latter is unchanged by time translations and the inversion $t \to -t$. Observers not moving relative to an object in space can agree on both dt^2 and dr^2, and hence agree on the spatial symmetry of an object, and the time span over which it is observed. However, as was illustrated in **Fig. 1.1**, observers moving relative to one another may not agree on the Euclidean symmetry of an object in spacetime. *Observers moving at different, but constant, velocity with respect to an object will agree on the symmetry of the object if and only if they change their concept of symmetry.*

The alternative to making such a change is to believe, in the sense understood prior to 1905, that Fitzgerald contractions and time dilatations are physical effects, i.e., that objects moving at constant velocity compress in the direction of their motion, and that clocks moving at constant velocity tick at a different rate. If, on the other hand, one adopts Einstein's viewpoint, *the unit cells of* **Figs. 1.1** *and the two spacetime triangles exhibited in* **Figs. 2.7.7** *must be considered as the same physical objects viewed from different spacetime perspectives.*

Objects in spacetime are to be considered identical if their only apparent difference is due to a change in spacetime perspective that can be produced by an operation of the Poincare group. This produces a further general conclusion: objects in spacetime that can be interconverted by operations of the Poincare similitude group have the same spacetime symmetry. We therefore make the definitions which follow.

Definition 2.1. Let a set of *events* determine a figure or lattice in spacetime with metric

$$ds^2 = \pm(dx^2 + dy^2 + dz^2 - c^2\, dt^2). \qquad (2.10.2)$$

The symmetry of the figure or lattice is invariant under the action of the continuous group of transformations of the Poincare similitude group, which leaves invariant the real-valued spacetime intervals defined by (2.10.2), or multiplies these by a real number greater than zero.

Definition 2.2. The symmetry of any figure or lattice in spacetime is defined by the subgroup of the group of all transformations which leave spacetime intervals invariant — the subgroup of the Poincare group that carries the spacetime figure or lattice into itself.

In the next few paragraphs we briefly consider implications of these definitions.

The Euclidean similitude group of space is a subgroup of the Poincare similitude group of spacetime. Its operations can therefore be used to define symmetries in spacetime, but its operations only alter spatial perspectives. Because Lorentz transformations act in spacetime they can define symmetries in spacetime, and alter spatial perspectives as well as spacetime perspectives.

Figures 2.10.1–2.10.3 display examples of these observations. **Figure 2.10.1** plots the coordinates of four light flashes recorded by observer A, with $(x, y) = (0, 0)$, as having occured at $ct = -1/2$. The points all lie on the on the circle $x^2 + y^2 = r^2 = 1/2$ and the hyperboloid

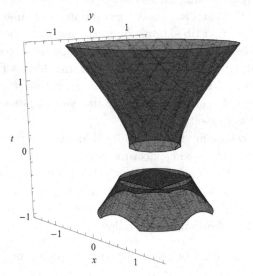

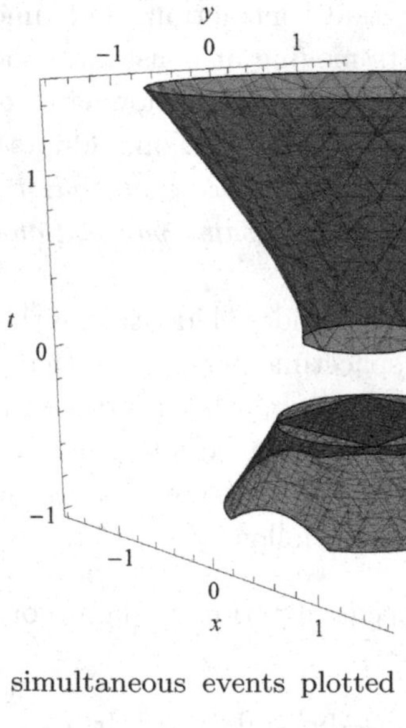

Fig. 2.10.1 Four simultaneous events plotted on a hyperholoid in spacetime x, y, ct.

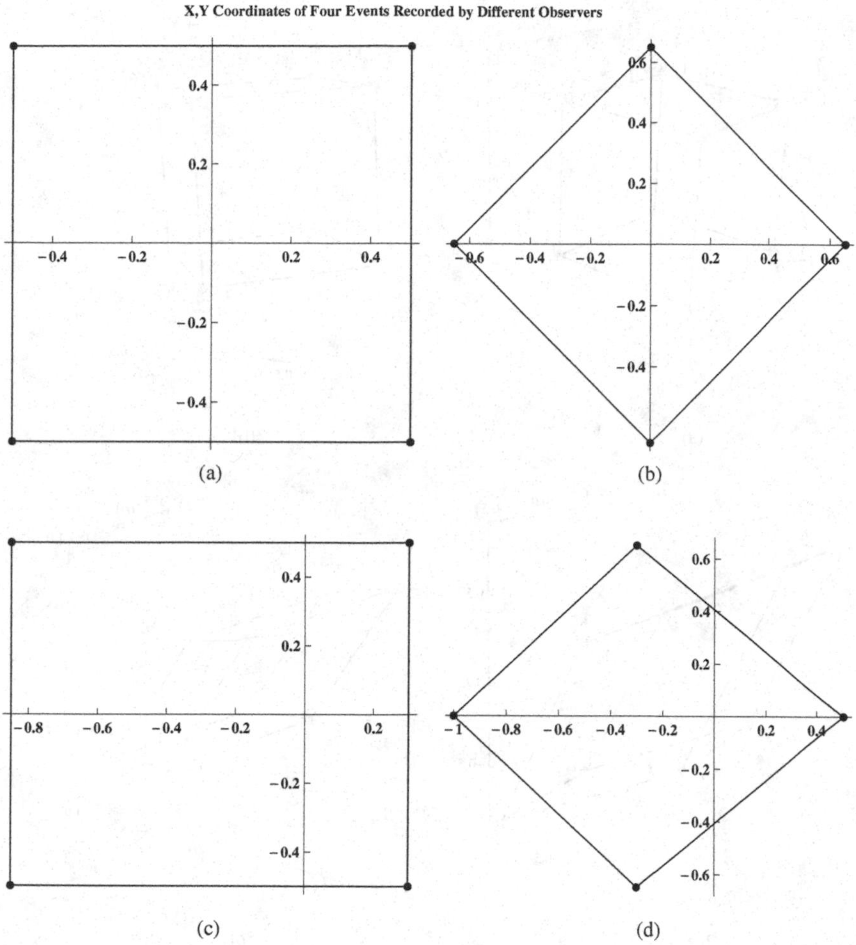

Fig. 2.10.2 Events from the perspectives of A, B, C and D in (a), (b), (c) and (d) respectively.

$x^2 + y^2 - c^2t^2 = s^2 = 1/4$. If A resets his clock to convert $ct = 1/2$ to $ct = 0$, then **Figure 2.10.1** projects onto the $(x, y, 0)$ plane to yield **Fig. 2.10.2a**, for which $r^2 = s^2 = 1/2$. **Figures 2.10.2b,c,d** plot observations of the same flashes by three observers B,C,D who have clocks initially syncronized with A's (say, in the manner described in Sec. 2.7). In **Figs. 2.10.3**, θ is the angle of rotation and α, the "angle" of pseudo-rotation, which relate the observer's coordinate system to the coordinate system of A. The value $\alpha = 1/2$ arises when two observers are moving

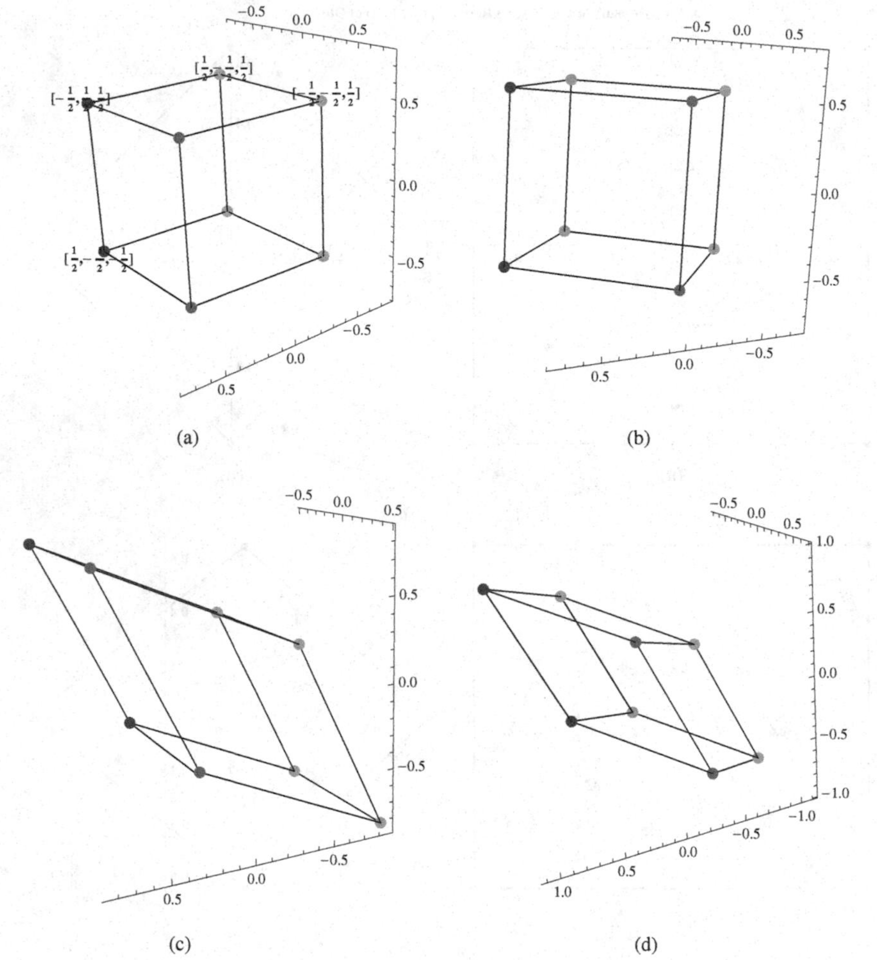

Fig. 2.10.3 Relation between eight events in spacetime from perspectives of four observers in x, y, ct coordinates. (a) Events in A's frame; (b) events in B's frame, $\alpha = 0, \theta = \frac{\pi}{2}$; (c) events in C's frame, $\alpha = \frac{1}{2}, \theta = 0$; (d) events in D's frame, $\theta = \frac{1}{2}, \alpha = \frac{1}{2}$.

with relative velocity 0.46c. All observers find $s^2 = 1/2$. Observers A, with $(\theta = 0,\ \alpha = 0)$, and B, with $(\theta = \pi/4,\ \alpha = 0)$, find the flashes occur at the corners of a square. Observer C moving relative to A with $(\theta = 0,\ \alpha = 1/2)$ observes the events at corners of a rectangle; and observer D, with $(\theta = \pi/4,$ $\alpha = 1/2)$, finds them to be at the vertices of a rectangular parallelepiped. An analogous situation occurs in Euclidean geometry. If a square in an x–y plane through z = 1/2 is rotated about an axis in the plane, its projections on the x–y plane through z = 0 can be similar to the projections in **Figs. 2.10.2**.

As the physical relationship defined by the four events depicted in **Fig. 2.10.2a** has not itself changed, observers A,B,C,D are all recording events defining a quadrilateral with all the spatial symmetry of the square determined in the analysis of **Fig. 2.1** and tabulated in Table 2.1. Stated in another way, figures with the apparently different spatial symmetries pictured in **Figs. 2.10.2** have physical symmetries that are indistinguishable in a spacetime with metric (2.10.2).

Figures 2.10.3 illustrate the spacetime perspectives of observers A, B, C, D on two sets of four events, each defining quadrilaterals like those in **Figs. 2.10.2**. For simplicity it is supposed that the clocks are so imaginatively set that for A the events occur at ct = −1/2 and ct = 1/2. All the observers find $s^2 = 1/4$. Again, all the figures are interconverted by actions of the Poincare group, so their symmetries in spacetime are identical.

Figures 2.10.1 and **2.10.2** display a symmetry defined by an E(2) spatial subgroup of the Poincare group: the E(2) operations act in x–y planes with fixed t. **Figures 2.10.3** display this same symmetry, and also relations defined in planes parallel to a t axis. To deal with these, we consider briefly symmetries defined by operations of the Poincare group that are not contained in its spatial subgroups.

Hyperbolic symmetry is defined by the invariance of curves of constant Δs or s under the one-parameter subgroup of hyperbolic pseudo-rotations. Hyperbolic symmetry can be thought of as a spacetime analog of rotational symmetry: the Minkowski transformations $z \to ict$, $r^2 \to \pm s^2$ convert circles and spheres defined by

$$y^2 + z^2 = r^2, \quad x^2 + y^2 + z^2 = r^2, \qquad (2.10.3a,b)$$

to hyperbolae and hyperboloids defined by

$$y^2 - c^2\,t^2 = \pm s^2, \quad x^2 + y^2 - c^2\,t^2 = \pm s^2. \qquad (2.10.3c,d)$$

In **Figs. 2.10.3**, the Lorentz transformations move points along hyperbolae on the hyperboloid $x^2 + y^2 - c^2t^2 = 1/4$. All faces of the rhomboids of **Figs. 2.10.3c,d** have symmetries defined by discrete subgroups of the Poincare group. The symmetries of the faces lying in planes in which t varies are defined by discrete subgroups of a continuous group of Lorentz transformations. The operations of these subgroups move points along hyperbolae rather than circles. The same is true of the discrete operations which act on the *top* and *bottom* faces of **Figs. 2.10.3a,b** to yield **Figs. 2.10.3c,d**.

For a conclusion, we note that arguments of the form we have used to define the meaning of geometric symmetry both in Euclidean spaces and in spacetime may be used to define geometric symmetry in any space with a metric. We shall be using them later. In each case geometrical interpretations will depend upon physical interpretations.

2.11 Geometrical Interpretations of Nonlinear Transformations: Stereographic Projections

The geometrical and physical interpretation of nonlinear transformations raises issues that we have not yet considered. In Sec. 2.3 we investigated active and passive interpretations of linear transformations. Linear transformations were also considered as transformations from one space to another. In this section we will quickly find that the interpretation of nonlinear transformations introduces geometrical and physical considerations that make the active and passive interpretations *inequivalent* and make the passive interpretation extremely misleading.

Let (x, y) be the Cartesian coordinates of a point in a Euclidean plane. A transformation of these variables defined by

$$\mathrm{x} \rightarrow \mathrm{x}' = \mathrm{f}(\mathrm{x}, \mathrm{y}), \quad \mathrm{y} \rightarrow \mathrm{y}' = \mathrm{g}(\mathrm{x}, \mathrm{y}) \tag{2.11.1}$$

is nonlinear if at least one of the functions f or g has a nonlinear dependence on at least one of the variables x, y. It follows that in the equations

$$\mathrm{dx}' = (\partial \mathrm{f}/\partial \mathrm{x})\,\mathrm{dx} + (\partial \mathrm{f}/\partial \mathrm{y})\,\mathrm{dy}, \quad \text{and} \quad \mathrm{dy}' = (\partial \mathrm{g}/\partial \mathrm{x})\,\mathrm{dx} + (\partial \mathrm{g}/\partial \mathrm{y})\,\mathrm{dy}, \tag{2.11.2}$$

at least one of the derivatives is not a constant. However, if x', y' are to be Cartesian coordinates in a single Euclidean plane it is necessary that

$$\mathrm{dx}'^2 + \mathrm{dy}'^2 = \mathrm{k}^2(\mathrm{dx}^2 + \mathrm{dy}^2), \quad \mathrm{k} \text{ real}. \tag{2.11.3}$$

In the case of uniform dilations, k is constant. Transformations of conformal groups, similar to the group discussed in Chapter 14, can produce k's that are functions of x.

Given an n-dimensional *Euclidean* space of variables z, and a nonlinear transformation of these variables, one may consider the transformation to be an active one that moves points in the space, but it is seldom easy to believe with consistency the transformation is a passive transformation of coordinates in the same Euclidean space.

However, the coordinates of points moving in an n-dimensional non-Euclidean space may be related by nonlinear one-to-one transformations to coordinates of points moving in a Euclidean space of higher dimension. This happens when the non-Euclidean space can be considered to be a surface in the higher dimensional space. Such transformations, and their inverse transformations, are said to be *projections*. In this section several transformations of physical interest will be used to illustrate a number of properties of projections.

The most common projections are those that map points on the surface of the earth onto flat paper. As previously mentioned, one of these, the stereographic projection, is also used in crystallography to map points on the surfaces of a three-dimensional crystal onto a Euclidean plane. In his discussion of the degeneracy of the hydrogen atom, Fock used a four-dimensional generalization of this transformation to map points on the surface of a hypersphere in a four-dimensional Euclidean space to points in a three-dimensional Euclidean space.

The following relations define a stereographic projection connecting points with Cartesian coordinates (x, y) in a Euclidean plane to points with Cartesian coordinates (z_1, z_2, z_3) in Euclidean three-space:

$$\begin{aligned} z_1 &= 2ax/(a^2 + r^2) = 2(x/a)/(1 + (r/a)^2), \\ z_2 &= 2ay/(a^2 + r^2) = 2(y/a)/(1 + (r/a)^2), \\ z_3 &= (a^2 - r^2)/(a^2 + r^2) = (1 - (r/a)^2)/(1 + (r/a)^2), \quad (r^2 = x^2 + y^2). \end{aligned} \tag{2.11.4}$$

It follows from these equations that

$$z_1^2 + z_2^2 + z_3^2 = 1. \tag{2.11.5}$$

Thus the projection connects points at $\mathbf{r}$ in the Euclidean plane to points at $\mathbf{z}$ on the surface of a unit sphere in the Euclidean space of the variables z_j with center at the origins $\mathbf{z} = 0$, $\mathbf{r} = 0$.

The transformation of coordinates inverse to (2.11.4) is

$$x/a = z_1/(1 + z_3), \quad y/a = z_2/(1 + z_3). \tag{2.11.6}$$

The variable a establishes the scale of the mapping. In particular, if $z_3 = 0$, then $r = (x^2 + y^2)^{1/2} = a$, so in the x, y coordinate system of the plane, a is the radius of the circle where the sphere intersects the plane. On this circle $z_1 = x/a$, $z_2 = y/a$. Note that both r and the scale parameter a must have identical units, as the variables z_j satisfy (2.11.5). Equations (2.11.4) define the stereographic projection illustrated in **Figs. 1.2**, the second set of figures in Chapter 1.

Figures 1.2a,b depict the manner in which points on the Northern and Southern hemispheres are mapped into points in the plane. Points in the Northern hemisphere map into points that lie inside the sphere, and points in the Southern hemisphere map to points outside the sphere. This property is independent of the value of the scale factor a. In the illustrated case, the scale factor is unity, so the intersection of the sphere with the x–y plane is a unit circle in both the x and y coordinates, and the z_j coordinate systems. When a is not unity, the circle of intersection is defined by $z_1^2 + z_2^2 = 1$ in the one coordinate system, and by $x^2 + y^2 = a^2$ in the other.

Using spherical polar coordinates (θ, ϕ), with $0 \leq \theta \leq \pi$ and $0 \leq \phi \leq 2\pi$, one has on a unit sphere

$$z_1 = \sin(\theta)\cos(\phi), \quad z_2 = \sin(\theta)\sin(\phi), \quad z_3 = \cos(\theta). \tag{2.11.7}$$

Using (2.11.4), (2.11.7) one finds that stereographic projection connects the x, y coordinates to angular coordinates on the sphere via the relations:

$$\begin{aligned} x &= \frac{a\sin(\theta)\cos(\phi)}{(1+\cos(\theta))} = a\cos(\phi)\tan(\theta/2) \equiv X(\theta,\phi), \\ y &= \frac{a\sin(\theta)\sin(\phi)}{(1+\cos(\theta))} = a\sin(\phi)\tan(\theta/2) \equiv Y(\theta,\phi). \end{aligned} \tag{2.11.8}$$

It follows that

$$r/a = \tan(\theta/2). \tag{2.11.9}$$

In the x–y plane, $r = a$, $\theta = \pi/2$, and (2.11.8) requires

$$x = r\cos(\phi), \quad y = r\sin(\phi). \tag{2.11.10}$$

Stereographic projections relate measurements of coordinates x, y and displacements dx, dy in a plane to measurements of coordinates θ, ϕ and displacements $d\theta$, $d\phi$ in the curved two-dimensional space, termed S(2), that may be considered to be the surface of the sphere in three-space. In the Euclidean space of (z_1, z_2, z_3), the metric is

$$ds^2 = dz_1^2 + dz_2^2 + dz_3^2. \tag{2.11.11}$$

On the unit sphere this becomes

$$ds^2 = d\theta^2 + \sin^2(\theta)d\phi^2. \tag{2.11.12}$$

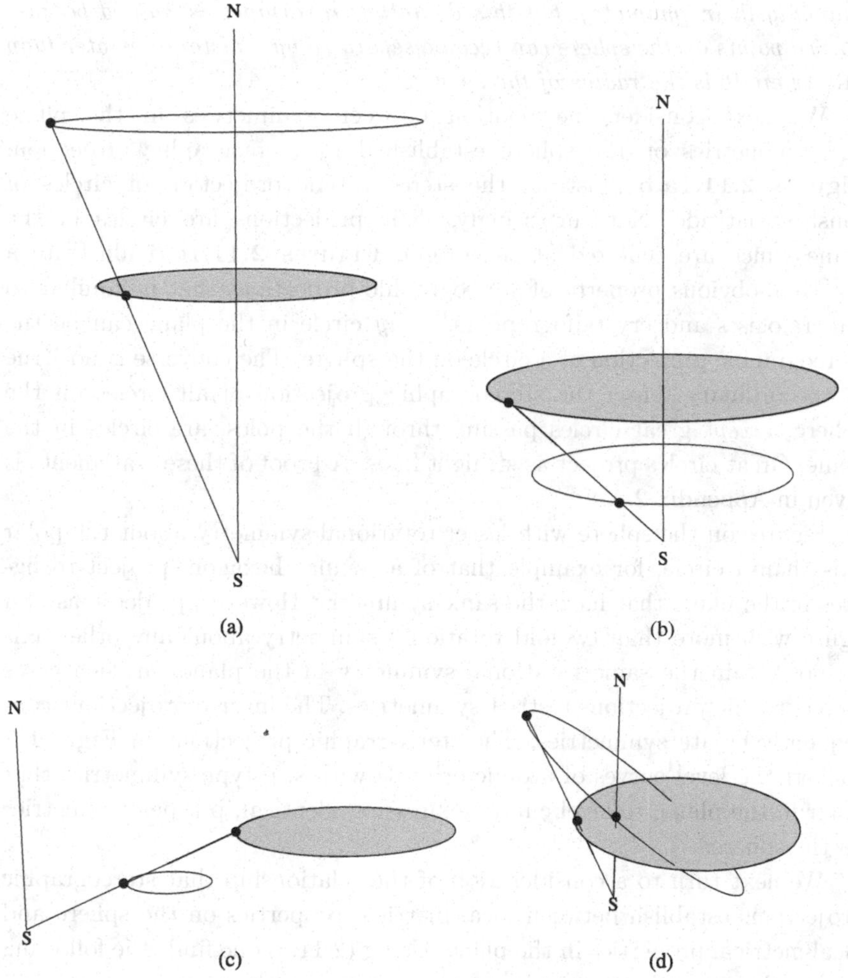

Fig. 2.11.1 The projection shaded in each figure is onto the equatorial plane.

This metric is invariant under the group of rotations about the center of the sphere, inversions $z \to -z$, and reflections $z_j \to -z_j$. The group, denominated O(3), is termed an orthogonal group because it transforms orthogonal vectors into orthogonal vectors. Both the metric and the sphere are left invariant by the transformations of O(3).

Subgroups of the invariance group of the sphere will consequently define the symmetry of a figure on the sphere. The symmetry is not altered by rotations of the entire figure. Figures may be shrunk or enlarged without changing their symmetry, but this dilatation invariance is limited because no two points on the sphere can become separated by a distance greater than πR, *where* R *is the radius of the sphere.*

We next consider the relation between symmetries in the plane and symmetries on the sphere established by stereographic projection. **Figures 2.11.1a,b** illustrate the stereographic projection of circles of constant latitide. Not surprisingly, their projections are circles in the plane which are centered at the origin. **Figures 2.11.1c,d** illustrate a less-than-obvious property of stereographic projections that is familiar to mineralogists and crystallographers.[9] *Any* circle in the plane can be the stereographic projection of a circle on the sphere. The converse is not true in the ordinary sense: the stereographic projection of all circles on the sphere, *except* great circles passing through the poles, are circles in the plane. Great circles project as straight lines. A proof of these statements is given in Appendix 2.A.

Figures on the sphere with lesser rotational symmetry about the polar axis than a circle, for example that of a regular hexagon, project to figures in the plane that have the same symmetry. However, projections of a figure with more than twofold rotational symmetry about any other axis do not retain the same rotational symmetry in the plane. In such cases stereographic projections restrict symmetries. The inverse projections consequently create symmetries. The stereographic projections in **Figs. 1.2** convert the level curves of atomic orbitals with s, p type symmetries that differ in the plane, to level curves expressing identical, p type, symmetries on the sphere.

We next turn to a consideration of the relationship that stereographic projections establish between local metrical properties on the sphere and local metrical properties in the plane. Using (2.11.8) one finds the following relation between the metric ds^2 on the sphere and the metric

$$\mathrm{dr}^2 = \mathrm{dx}^2 + \mathrm{dy}^2 \tag{2.11.13}$$

in the plane:

$$\mathrm{ds}^2 = \mathrm{dr}^2(1 + \cos(\theta))^2. \tag{2.11.14}$$

As $\cos(\theta)$ ranges in value from 1 at the North pole of the sphere, to -1 at its South pole, $(1 + \cos(\theta))^2$ ranges from 4 to 0. As one expects from the figures, separations in the Northern hemisphere project into separations in the plane that are smaller than those obtained for points in the plane that lie on the equator of the sphere, where $\cos(\theta) = 0$. Then, by making

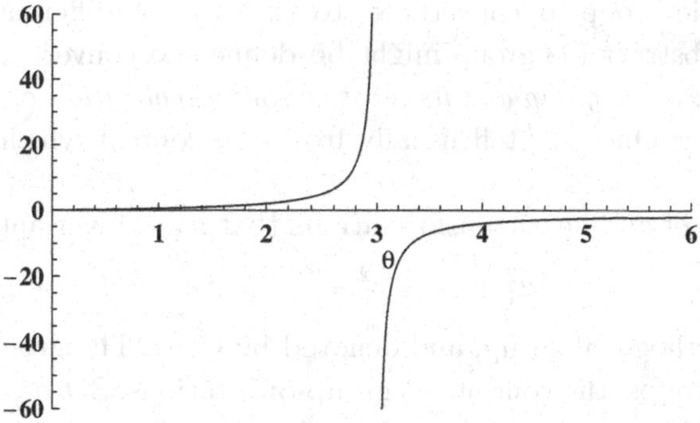

Fig. 2.11.2 The function $\tan(\frac{1}{2}\theta)$.

increasingly large local-scale-transformations as r increases, the projection fits all the remaining points in the plane onto the Southern hemisphere of the sphere. In short, as θ approaches π, r approaches infinity, and ds approaches zero.

Before concluding this discussion of stereographic projections we must deal a little more extensively with the behaviour of $x = a\tan(\theta/2)\cos(\phi)$, and $y = a\tan(\theta/2)\sin(\phi)$ as $\theta \to \pi$. **Figure 2.11.2** illustrates a well known behavior of the tangent function. Both x and y shift from $+\infty$ to $-\infty$, as θ passes π, and the reverse shift in θ produces the reverse shift in x, y. This happens with a minute shift in θ, and at a point where $ds^2 = 0\,dr^2$. The stereographic mapping has the effect of treating $-\infty$ as though it differed minutely from $+\infty$! In subsequent chapters we will deal with other mappings which have this property, and with physical consequences that result from it.

2.12 Continuous Groups that Leave Euclidean and Pseudo-Euclidean Metrics Invariant

In this section we consider some generalizations of the metrics of ordinary Euclidean space and spacetime considered in earlier sections.

All of the continuous groups that leave these metrics invariant are special cases of Lie groups. Each is obtained by defining the manner in which a Lie group transforms the variables of interest. A Lie group acting on a particular set of variables is said to be a *realization* of an *abstract* Lie group. Thus, for example, E(1) the transformation group which converts a Cartesian coordinate x in Euclidean space to x + a, is a realization of an abstract Lie group. It converts x^2 to $(x + a)^2$. A different realization of the same abstract Lie group might be defined to convert x^2 to $x^2 + a$. *Unfortunately, a Lie group and its most common realization are often given the same name.* One may tell usually from the context which meaning is intended.

The group of all linear transformations that leave invariant

$$z_1^2 + \cdots + z_i^2 + \cdots + z_n^2 \tag{2.12.1}$$

is called an orthogonal group, and denoted by O(n). The group O(n) contains, as subgroups, the continuous group of rotations, SO(n), and the discrete group of inversions and reflections that arise from letting

$$z_i \to -z_i. \tag{2.12.2}$$

The continuous group of linear transformations that leaves invariant

$$dz_1^2 + \cdots + dz_i^2 + \cdots + dz_n^2 \tag{2.12.3}$$

is denoted by E(n) or ISO(n), which stands for *Inhomogeneous Special Orthogonal group.* Its operations carry out translations $z_j \to z_j + a_j$, as well as rotations. When the latter Lie group is extended to include the continuous group of uniform *dilatations*, $z_j \to e^{\alpha} z_j$, $j = 1, \ldots, n$, the group becomes the corresponding similitude group. When ISO(n) is extended to include the discrete inversions $z_j \to -z_j$, the group becomes the inhomogenous orthogonal group denoted IO(n).

The discussion of these groups is greatly simplified if one makes use of the concept of vector introduced by Gibbs.[10]

For an n-dimensional space there is a direct generalization of Cartesian vectors in ordinary two- and three-dimensional space. Let $\mathbf{z} = (z_1, \ldots, z_n)$ represent such a vector.

For $\mathbf{z}$ and $\mathbf{z}'$ to be vectors it is required that:

(i) There is a law of vector addition which states that $\mathbf{z}+\mathbf{z}' = \mathbf{z}''$, meaning

$$\begin{aligned}(z_1, z_2, \ldots, z_N) &+ (z_1', z_2', \ldots, z_N') \\ &= (z_1 + z_1', z_2 + z_2', \ldots, z_N + z_N') = \mathbf{z}''.\end{aligned} \tag{2.12.4}$$

where $\mathbf{z}''$ is also a vector.

(ii) There is a law of multiplication of z by a real number, c, (a *scalar*) defined by

$$\mathbf{z} \to c\mathbf{z} = (cz_1, cz_2, \ldots, cz_N) = \mathbf{z}', \tag{2.12.5}$$

where z' is also a vector.

(iii) Vectors satisfy the distributive law

$$c(\mathbf{z}+\mathbf{z}') = c\mathbf{z} + c\mathbf{z}' = \mathbf{z}''. \tag{2.12.6}$$

(iv) Finally, vectors obey an associative law

$$(\mathbf{z}+\mathbf{z}') + \mathbf{z}'' = \mathbf{z} + (\mathbf{z}' + \mathbf{z}'') = \mathbf{z}'''. \tag{2.12.7}$$

Vectors will henceforth be denoted by bold-face roman characters. A list of variables contained within parentheses will be considered a vector if the parentheses in boldface.

Vectors may be subject to further operations. Taking the scalar product of two vectors $\mathbf{z}$ and $\mathbf{z}'$ is one such operation. In a Euclidean space it is defined as

$$\mathbf{z}\cdot\mathbf{z}' = z_1 z_1' + \cdots + z_n z_n'. \tag{2.12.8a}$$

The *magnitude* of the vector $\mathbf{z}$ is then defined to be $|\mathbf{z}| = (\mathbf{z}\cdot\mathbf{z})^{1/2}$. The cosine of the angle α, between two vectors $\mathbf{z}$ and $\mathbf{z}'$, is

$$\cos(\alpha) = \mathbf{z}\cdot\mathbf{z}'/|\mathbf{z}||\mathbf{z}'|. \tag{2.12.8b}$$

All operations of the orthogonal group O(n) leave this quantity invariant. Orthogonal groups leave orthogonal vectors orthogonal, hence their name. Unit vectors $\mathbf{v}_j = (0,\ldots,0,v_i,0,\ldots,0)$, with $v_j = 1, j = 1,\ldots,n$, may be attached to a set of n orthogonal axes. The component c_j of the vector $\mathbf{r} = \Sigma c_j \mathbf{v}_i$ on the axis j is then defined by $c_j = \mathbf{r}\cdot\mathbf{v}_i/|\mathbf{v}_i|$. While unit vectors remain so when dilatations are considered to be active transformations, they change length when dilatation transformations are interpreted passively.

In the most common realization of the group SO(n), the $n(n-1)/2$ group parameters are the angles that define rotation operations. These act in $n(n-1)/2$ mutually orthogonal two-planes in an n-dimensional Euclidean space. Rotation of a vector z through an angle θ_{jm} about the origin in the z_j, z_m two-plane only affects the coordinates (z_j, z_m). It is defined by

$$\begin{aligned} z_j \to z_j' &= z_j\cos(\theta_{jm}) - z_m\sin(\theta_{jm}),\\ z_m \to z_m' &= z_j\sin(\theta_{jm}) + z_m\cos(\theta_{jm}). \end{aligned} \tag{2.12.9}$$

All the transformations preserving Euclidean symmetries preserve angles between lines. One can erect a fixed set of n mutually orthogonal axes in n-dimensional Euclidean spaces, and these orthogonal axes remain orthogonal when transformations of the groups O(n) and SO(n) are interpreted passively or actively.

In spacetime the scalar product may be defined to be plus or minus

$$z_1z_1' + z_2z_2' + z_3z_3' - z_4z_4', \quad z_4 = ct. \tag{2.12.10}$$

It is invariant under operations of the continuous group SO(3,1). The operations of its SO(3) subgroup leave spatial angles invariant, while Lorentz transformations do not. However, as noted in Sec. 2.7, there exist well defined locally orthogonal spacetime coordinate systems.

The group of all linear transformations that leave invariant

$$(z_1^2 + \cdots + z_m^2) - (z_{m+1}^2 + \cdots + z_n^2) \tag{2.12.11}$$

is denoted $O(m, n-m)$. The Lie group $SO(m, n-m)$ is a subgroup of $O(m, n-m)$. (It is commonplace to suppose $m \geq n-m$.) $SO(m, n-m)$ has $SO(m)$ and $SO(n-m)$ subgroups which operate in the subspace of $z_1 \cdots z_m$ and $z_{m+1} \cdots z_n$, respectively. In $O(m, n-m)$ the scalar product of two vectors is defined by

$$(z_1z_1' + \cdots + z_mz_m') - (z_{m+1}z_{m+1}' + \cdots + z_nz_n'). \tag{2.12.12}$$

It is invariant under all the operations that leave (2.12.11) invariant. Orthogonality, Cartesian coordinates, and unit vectors in these spaces are defined using this scalar product.

Interpretating hyperbolic pseudo-rotations as mappings, one is allowed to use Cartesian coordinates to interpret their action. This was done in the figures of Sec. 2.3, and throughout Sec. 2.7. In the general case, the operations of $SO(m, n-m)$ can be then considered to carry out $m(n-m)$ active hyperbolic pseudo-rotations about mutually perpendicular axes. The group has two subgroups $SO(n)$ and $SO(n-m)$, whose operations carry out active rotations as described above. Passive interpretation of hyperbolic pseudo-rotations destroys Euclidean orthornormality. One may interpret all the operations of these groups as mappings that relate the coordinates of points in one orthogonal coordinate system to those of points in another, *equivalent* orthogonal coordinate system.

The transformations of the continuous group $ISO(m, n-m)$ supplement those of $SO(m, n-m)$ with m+n translation operations $z_j \to z_j + \alpha_j$ and leave invariant

$$ds^2 = dz_1^2 + \cdots + dz_m^2 - dz_{m+1}^2 - \cdots - dz_n^2. \tag{2.12.13}$$

On including the discrete inversions $z_j \to -z_j$, the group becomes $IO(m, n-m)$. Supplementing these operations of $ISO(m, n-m)$ with uniform continuous dilatations yields the corresponding similitude group that multiplies the ds^2 of (2.12.13) by a positive constant. Minkowsk's *trick* $\mathbf{z} \to \mathbf{iz}$ may be used to convert these non-Euclidean spaces to Euclidean spaces.

The transformation groups discussed in this section carry out linear transformations on a set of variables z_i, and leave metrics invariant or multiply them by a constant. The *sphere* of action of the groups may be greatly extended by considering nonlinear transformations of the Cartesian variables z, transformations which do not preserve the ISO(n) or ISO(m, n) metrics or multiply them by constants. As an example, one may observe that the stereographic projection converts rotations on a sphere in Euclidean three-space into operations in the plane, some of which are not rotations, and do not leave Euclidean distances in the plane invariant.

2.13 Geometry, Symmetry, and Invariance

In the previous sections we have used mathematical properties of transformations of position and time variables, and also invariance properties of Euclidean space and physical spacetime, to demonstrate that for one to understand the role of symmetry in the physical sciences, it is necessary to first investigate the geometric interpretation imposed on observations. This lead us into the realm of non-Euclidean geometries and symmetries. These symmetries can be related to ordinary symmetries in a manner that enhances, rather than undermines, one's Euclidean insights. The greatest surprises occur when nonlinear mappings relate one space to another.

We have seen that geometry in a Euclidean space may be characterized by the group of invariance transformations of its metric, ds^2. The symmetry of a geometric object in the space is characterized by the group that transforms the object into itself. This is a subgroup of the invariance group of the metric. Geometric objects are considered equivalent, i.e. congruent, if operations of the group can superpose them. Objects also have the same symmetry if they may be superposed by operations of an enlarged group that contains dilations which can uniformly stretch or compress all distances by multiplying ds^2 with a positive constant.

Geometry and congruence in any Riemannian space may be similarly characterized. However, as we have seen in the case of the spherical surface S(2), the role of dilatations must be restricted if Δs has a restricted range of values.

Let us re-emphasize three other key points made in this chapter:

(1) Equations possess intrinsic symmetries, defined by the transformation groups that leave them invariant. Even in simple cases, a variety of

assumptions are made when one interprets these intrinsic symmetries as geometric symmetries.

(2) When equations describe physical relationships, intrinsic symmetries of the equations are given physical interpretation. Geometric interpretation of these physical relationships involves assumptions about observations and measurements as well as their geometric interpretation. Geometric interpretations of relations between observations of natural phenomena are not dictated solely by the relations themselves.

(3) Because spacetime is non-Euclidean, Euclidean conceptions of geometric symmetry require revision when considering physical symmetries in spacetime.

These observations are of central importance in much of what follows. In Chapter 3 we introduce the conception of symmetry — symmetry in phase-space — that is appropriate to systems governed by the ordinary differential equations of Hamiltonian mechanics. In Chapter 9 we will describe the manner in which this conception applies to systems governed by the partial differential equations of Schrödinger mechanics.

Differential equations govern a great variety of physical and chemical systems, and differential equations are left invariant by groups of transformations of their variables. For each system, the variables can be considered to reside in a space which has an associated metric. The metric is itself invariant under a group of transformations. This allows one to define the space of the physical variables and the meaning of symmetry in the space. Once this space has been defined, the transformations that leave the differential equation invariant can be given geometrical interpretations. Because the operations of the group carry solutions of the equation into solutions, the operations interrelate observations and have physical interpretation. This provides the connection between geometry and the natural world when natural phenomena are related by continuous causal chains.

Appendix A: Stereographic Projection of Circles

We begin by setting up a description of circles on a unit sphere. When a plane, $\mathrm{P_s}$, parallel to the $\mathrm{z_1}$–$\mathrm{z_3}$ plane, intersects the unit sphere of Sec. 2.11, the intersection is a circle of radius s whose points are defined by the vector

$$\mathbf{Z} = (\mathrm{z_1}, \mathrm{z_2}, \mathrm{z_3}) = (\mathrm{s}\cos(\tau), (1-\mathrm{s}^2)^{1/2}, \mathrm{s}\sin(\tau)), \tag{2.A.1}$$

with s lying between 0 and 1. The polar axis of the sphere is $\mathrm{z_3}$. The $\mathrm{z_2}$ coordinate in $\mathbf{Z}$, $(1-\mathrm{s}^2)^{1/2}$, is the distance to the $\mathrm{P_s}$ plane, measured along

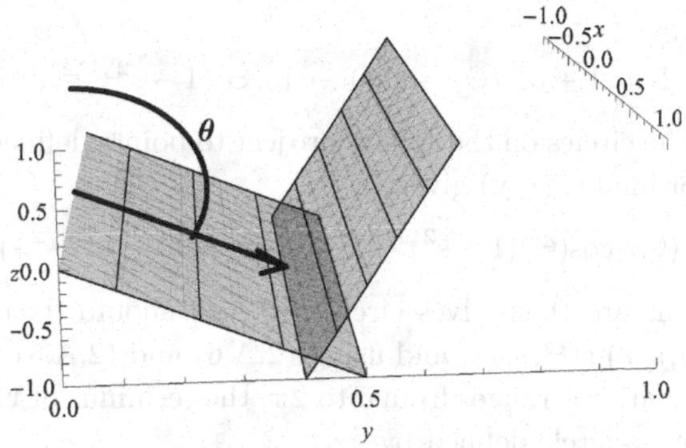

Fig. 2.A.1 The vector **C** and the plane normal to it.

a vector **C**, a normal to the plane from the origin. **Figure 2.A.1** depicts the relationship between **C**, and the plane normal to it when **C** has been rotated about the z_1 axis through an angle Θ, measured from the positive z_3 axis. Equation (2.A.1) applies when $\Theta = 90°$. In this case, $\mathbf{C} = (0, (1-s^2)^{1/2}, 0)$. As τ ranges from 0 to 2π the terminus of the vector **Z** sweeps along the circle at the intersection of the plane with the unit sphere. For all values of s, the terminus of **C** is the center of the circles swept out by **Z**.

When the plane P_s is tilted by altering Θ, the normal from the origin becomes

$$\mathbf{C}(\Theta) = (0, \cos(\Theta)(1-s^2)^{1/2}, -\sin(\Theta)(1-s^2)^{1/2}). \tag{2.A.2}$$

The vector **Z** of (2.A.1) then becomes a vector $\mathbf{Z}(\Theta, s, \tau)$ with coordinates

$$\begin{aligned} z_1 = s\cos(\tau), \quad z_2 = \cos(\Theta)(1-s^2)^{1/2} + \sin(\Theta)s\sin(\tau), \\ z_3 = \cos(\Theta)s\sin(\tau) - \sin(\Theta)(1-s^2)^{1/2}. \end{aligned} \tag{2.A.3}$$

As τ ranges from 0 to 2π, the terminus of this vector describes circles at the intersection of the tilted plane with the sphere, and center at $\mathbf{C}(\Theta)$.

Consider, then, the projection of these circles onto the x–y plane. The three-vector $\mathbf{Z}(\Theta, s, \tau)$ is projected onto the x–y plane, i.e. z_1–z_2 plane, using the relations

$$x = az_1/(1+z_3), \quad y = az_2/(1+z_3). \tag{2.A.4}$$

This projection yields the two-vector $(x, y) = \mathbf{r}(\Theta, s, \tau)$ whose components are

$$\begin{aligned} x &= a\,s\cos(\tau)/D, \\ y &= a\{\cos(\Theta)(1-s^2)^{1/2} + \sin(\Theta)s\sin(\tau)\}/D, \end{aligned} \tag{2.A.5}$$

where

$$D = 1 + \cos(\Theta) s \sin(\tau) - \sin(\Theta)(1 - s^2)^{1/2}.$$

The centers of the circles on the sphere project to points defined by (2.A.4), which have coordinates (x, y) given by

$$(x_0, y_0) = (0, a\cos(\Theta)(1 - s^2)^{1/2}/\{1 - \sin(\Theta)(1 - s^2)^{1/2}\}). \qquad (2.A.6)$$

If the projections are themselves circles, (2.A.6) should fix their centers. Expanding $\mathbf{r}(\Theta, s, \tau)\cdot\mathbf{r}(\Theta, s, \tau)$, and using (2.A.6) and (2.A.5) to relate x, y to τ, one finds that as τ ranges from 0 to 2π, the terminus of $\mathbf{r}(\Theta, s, \tau)$ does indeed sweep out a circle defined by

$$x^2 + (y - y_0)^2 = R^2,$$

with radius

$$R = a s/\{1 - \sin(\Theta)(1 - s^2)^{1/2}\}, \qquad (2.A.7)$$

and

$$y_0 = a\cos(\Theta)(1 - s^2)^{1/2}/\{1 - \sin(\Theta)(1 - s^2)^{1/2}\}.$$

Both y_0 and R become infinite when $\{1 - \sin(\Theta)(1 - s^2)^{1/2}\}$ becomes zero, which happens for a great circle passing through the poles. This projection of the circle is one that is indistinguishable from a straight line.

The circles defined by (2.A.3) lie in planes whose normal through the origin lies in the y–z plane. Rotating this normal, **C**, about the z_3 axis, and varying Θ, allows one to obtain planes with arbitrary orientation. By varying s one can then obtain all possible circles produced by the intersection of planes with the sphere. The rotation of **C** about the z_3 axis through an angle α produces a corresponding rotation of the vector (x_0, y_0) and the circles of (2.A.7). As all circles inscribed on the sphere can be produced by varying s, Θ, and α, it follows from (2.A.7) that the stereographic projection of any circle on the sphere is also a circle.

References

[1] B. A. Rosenfeld, *A History of Non-Euclidean Geometry*, translated by A. Shenitzer, (Springer, New York, 1988).
[2] B. Riemann, *Gesammelte mathematische Werke und wissenschaftlicher Nachlass*, Herausg. von H. Weber unter Mitwerkung von R. Dedekind (Dover., N.Y., 1953).
[3] H. Helmholtz, *Wissenschaftlich Abhandlungen*, Bd. 2 (Barth, Leipzig, 1882).
[4] A. A. Michelson, E. W. Morley, *Silliman's. J.*, **34** (1887) 333.

[5] A. Einstein, *Annalen der Physik*, **17** (1905) 891, cf. "On The Electrodynamics of Moving Bodies", in *The Principle of Relativity*, W. Perrett, translated by G. B. Jeffery, (Dover, 1923).

[6] This involves a presupposition about the continuity of space and time which is open to challenge. It also involves a presupposition that the infinitesimal quantities can be integrated to give unique finite quantities. Cf. A. S. Eddington, *The Mathematical Theory of Relativity* (Cambridge University Press, 1924), pp. 198–9.

[7] S. Lie, *Gesammelte Abhandlungen*, Vol. 2, pp. 374–479 (Teubner, Leipzig, 1935) (Johnson Reprint, New York, 1973).

[8] F. Klein, *Elementary Mathematics From An Advanced Standpoint*, Vol. II, Geometry (Dover, New York, 1939).

[9] E. S. Dana, *A Textbook of Mineralogy*, 4th edn. revised by W. E. Ford (Wiley, New York, 1932) pp. 49–56.

[10] E. B. Wilson, *Vector Analysis Founded upon the Lectures of J. Willard Gibbs* (1901).

CHAPTER 3

On Symmetries Associated With Hamiltonian Dynamics

3.1 Invariance of a Differential Equation

The reader will find that the connections made in this chapter are guided by the conceptions due to Sophus Lie. Properties of motions in one spatial dimension will be used to clarify the manner in which invariance properties of Hamilton's differential equations bring concepts of geometry and symmetry into dynamics. Systems of higher dimension are considered in Chapters 7 and 8. As Hamiltonian dynamics is the classical dynamics most directly relevant to quantum mechanics, concepts developed in this chapter will also underpin the discussions from the beginning of Chapter 9 to the end of Chapter 14.

The differential equations of different Hamiltonian systems may have different intrinsic symmetries. It will be shown that all have a common intrinsic symmetry that determines an associated invariant metric. Investigating the geometric and dynamical expressions of this metric will enable us to define the meaning of *dynamical symmetry* in classical Hamiltonian mechanics.

Any transformation that leaves a differential equation invariant must also convert a solution of the differential equation into a solution that can be the same solution or a different one. In this respect the intrinsic symmetries of differential equations have consequences similar to intrinsic symmetries of algebraic polynomial equations. However such algebraic equations possess a discrete set of solutions, while differential equations typically possess families of solutions. Members of these families may be distinguished by continuously variable parameters, or in the case of partial differential equations, continuously variable functions. For ordinary differential equations, ODEs, changing the values of the parameters corresponds to changing initial values of the variables, which for Hamilton's equations are positions and momenta. Because the solutions of differential equations can be labeled by continuous variables, most operations that interconvert the solutions are operations of continuous groups, rather than that of a discrete group such as that admitted by the equation $w^4 = 1$ of the previous chapter. The differential equation

$$dx/dt = 1 \tag{3.1.1a}$$

has, for example, the particular solution $x = t$. The continuous group of translations in time,

$$t \to t' = t + \alpha, \tag{3.1.1b}$$

converts the particular solution into the one-parameter family of solutions

$$x = t + \alpha, \tag{3.1.1c}$$

which is the general solution of the equation. The differential equation

$$d^2x/dt^2 = -x, \tag{3.1.2a}$$

has the particular solution $x = \cos(t)$. The continuous group of time-translations converts this to the one-parameter family of solutions

$$x = \cos(t + \alpha). \tag{3.1.2b}$$

The group of dilatations

$$x \to x' = e^{\beta}x, \tag{3.1.2c}$$

converts (3.1.2b) to the general solution

$$x = e^{\beta}\cos(t + \alpha). \tag{3.1.2d}$$

In (3.1.1), α is the group parameter, while in (3.1.2) the group parameters are α and β.

As noted in Chapter 1, it was Lie's investigations of the (intrinsic) symmetries of differential equations that led him to develop what is now known as the theory of Lie transformation groups. Lie's theory of the intrinsic symmetries of differential equations involves several conceptual subtleties. In the usual terminology, one says that the differential equations (3.1.1a) and (3.1.2a) involve one independent variable, the *argument* t, and one dependent variable, f(t). This statement can be misleading because, when f(t) denotes a solution of a differential equation it is a function whose value only becomes fixed when one fixes the value of further variables, *parameters of integration*. Thus (3.1.1a) itself only relates dx to dt; it does not define a definite functional relation between x and t. The solution is a function of two variables, the t and α of (3.1.1c). The solution of the second-order differential equation (3.1.2a) establishes a functional relation between the four variables, the x, t, α and β of (3.1.2c). However, in each case, only t appears in the differential equation as an argument.

It is important to keep in mind the distinction between relations defined by definite functions, say $y = f(t)$, and relations defined by differential equations, such as $dy = g(t)\,dt$. In the latter, y and t remain independently variable. This distinction is fundamental to Lie's work, to d'Alembert's principle, to Hamiltonian mechanics, to all variational treatments of mechanics, and to all of science that is governed by differential equations.

The relation between transformations of functions, arguments, derivatives, and solutions of differential equations, will be more thoroughly dealt with in Chapters 5–7. The connection is often subtle enough to make it difficult to anticipate the intrinsic symmetry of differential equations. Very fortunately, the intrinsic symmetries of Hamilton's equations are easy to characterize. These intrinsic symmetries become geometric symmetries in the *phase–space* of physics — a non-Euclidean metric space. In the following pages the reader will find that the non-Euclidean geometry of this phase-space allows free reign to one's Euclidean geometric imagination.

3.2 Hamilton's Equations

Hamilton's mechanics treats position and momentum as initially independent variables on an equal footing. It deals with variables that may, for example, be linear combinations of ordinary position and momentum variables. For this reason it is convenient to use dimensionless variables. In the following discussion ordinary type will be used to denote dimensionless variables in equations. The italic letters l, m, t will denote the units in which length, mass and time, respectively, are measured. Thus:

$$\mathrm{q} = \mathrm{length}/l, \quad \mathrm{t} = \mathrm{time}/t, \quad \mathrm{m} = \mathrm{mass}/m, \quad \mathrm{p} = \mathrm{momentum}/lmt^{-1},$$

$$\mathrm{E} = \mathrm{energy}/ml^2t^{-2}, \quad \mathrm{L} = \mathrm{angular\ momentum}/l^2mt^{-1}.$$

We begin with the relation

$$\mathrm{E} = \mathrm{H}(\mathrm{q},\mathrm{p}), \tag{3.2.1}$$

where H(q,p) is the Hamiltonian function of a particle with energy E, q is the Cartesian position coordinate of the particle of mass m, and p = mdq/dt represents the particle's momentum.[a] Hamilton's equations of motion are

$$dq_j/dt = \partial H/\partial p_j, \quad dp_j/dt = -\partial H/\partial q_j, \tag{3.2.2}$$

with dt representing a time interval.[b] A familiar example is provided by a mass of magnitude m oscillating under the influence of a potential $(k/2)\, q^2$. (We suppose that k, like m, is dimensionless.) For this harmonic oscillator,

$$H = p^2/2m + (k/2)\, q^2. \tag{3.2.3}$$

Equations (3.2.2) become

$$dq/dt = p/m, \quad dp/dt = -kq. \tag{3.2.4a,b}$$

The well known family of solutions of these equations can be written

$$Q(t) = Q(0)\cos(\omega t) + (P(0)/m\omega)\sin(\omega t),$$
$$P(t) = P(0)\cos(\omega t) - m\omega\, Q(0)\sin(\omega t), \quad \omega = (k/m)^{1/2}. \tag{3.2.4c–e}$$

To gain further insight into the implications of these equations, it is helpful to simplify their appearance by setting $\omega t = \tau$, and then defining the new variables $x = q\,(km)^{1/4}$, and $y = p/(km)^{1/4}$. This changes (3.2.3) to

$$H = (\omega/2)(y^2 + x^2), \tag{3.2.5}$$

and Eqs. (3.2.4a,b) become

$$dx/d\tau = \omega y, \quad dy/d\tau = -\omega x. \tag{3.2.6a,b}$$

Solving these last equations converts x to $X(\tau)$, y to $Y(\tau)$, and determines the two-parameter family of solutions

$$X(\tau) = X(0)\cos(\omega\tau) + Y(0)\sin(\omega\tau), \tag{3.2.6c}$$

$$Y(\tau) = Y(0)\cos(\omega\tau) - X(0)\sin(\omega\tau). \tag{3.2.6d}$$

Several of these solutions, circles of radius $(\omega/2)^{1/2}$, are plotted in **Figs. 3.2.1**. In the figures, the distinguished points may be considered to be points with coordinates (X(0), Y(0)). If one views X(0) and Y(0) as arbitrary values of x and y, then on referring back to the discussion of **Fig. 2.2.1** one is led to consider Eqs. (3.2.6) as equations that define a transformation group

[a]This momentum variable is said to be *conjugate* to the q. The general definition of the *conjugate momentum*, p, that must be used in Hamilton's equations, and the equations themselves, are derived from Lagrange's equations; cf. Chapter 7.

[b]Though dt is a time interval, identifying t with physical time can be misleading, even in nonrelativistic mechanics. Here, for example, t is a cyclic variable (cf. Sec. 5.7).

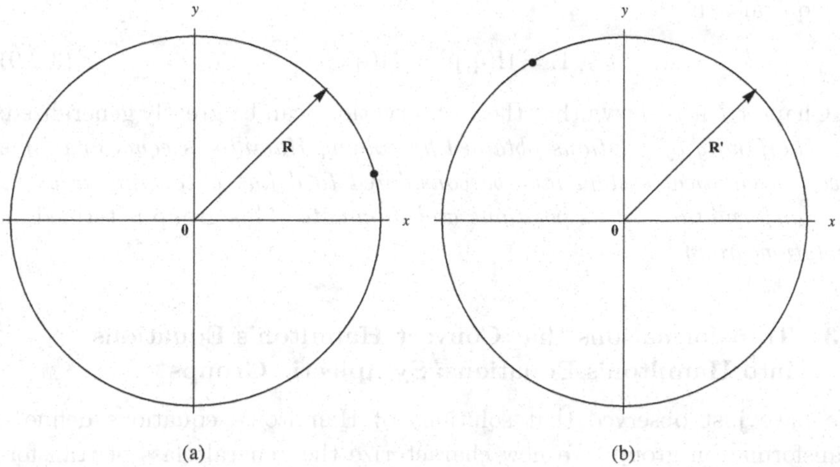

Fig. 3.2.1 (a) A solution of Hamilton's equations (3.2.6). (b) A solution of Eqs. (3.2.6) with different initial conditions.

$$\begin{aligned} \mathrm{x} \to \mathrm{x}' &= \mathrm{X}(\tau) = \mathrm{x}\cos(\tau) + \mathrm{y}\sin(\tau), \\ \mathrm{y} \to \mathrm{y}' &= \mathrm{Y}(\tau) = \mathrm{y}\cos(\tau) - \mathrm{x}\sin(\tau), \end{aligned} \tag{3.2.7a,b}$$

or, equivalently,

$$\mathrm{q} \to \mathrm{q}' = \mathrm{Q}(\tau), \quad \mathrm{p} \to \mathrm{p}' = \mathrm{P}(\tau).$$

The point moves through a space of dimensionless position and momentum variables, developing a trajectory in the phase–space as the physical system evolves. The inverse motion is obtained by changing t to $-$t. At every point along a trajectory, the motion of q and p is restricted by the equation H(q,p) = E. The fully evolved trajectory is sometimes termed a *phase portrait* of the motion. In this case, the phase portrait can be sketched by rotating the representing point at (q,p) about the origin (0,0), with the angle of rotation, α, equal to ωt.

The transformations (3.2.7) convert Eqs. (3.2.6a,b) to

$$\mathrm{dx}'/\mathrm{d}\tau = \mathrm{y}', \quad \mathrm{dy}'/\mathrm{d}\tau = -\mathrm{x}', \tag{3.2.8a,b}$$

that is, they leave Hamilton's equations invariant. They also leave invariant the energy of the system:

$$\mathrm{E} = (\omega/2)(\mathrm{y}^2 + \mathrm{x}^2) = (\omega/2)(\mathrm{y}'^2 + \mathrm{x}'^2)$$

or, equivalently,

$$E = H(q, p) = H(q', p'). \tag{3.2.9}$$

In Chapter 7 it is shown that these observations can be greatly generalized:

The family of solutions obtained by solving Hamilton's equations for a given mechanical system may be considered to define a continuous group of transformations of its positions and momenta. This group is termed an *evolution group.*

3.3 Transformations that Convert Hamilton's Equations into Hamilton's Equations; Symplectic Groups

We have just observed that solutions of Hamilton's equations define a transformation group. We now characterize the general class of transformations that convert Hamilton's equations into Hamilton's equations when the equations govern the evolution of a particle with only one position coordinate and one momentum coordinate.

Transformations that leave invariant the equation H(q,p) = E, even for a fixed value of E, do not necessarily convert Hamilton's equations of motion into Hamilton's equations of motion. The transformation utilized to convert (3.2.4a,b) to (3.2.5a,b) does both. On the other hand, the transformation $q \to q' = p$, $p \to p' = q$, leaves (3.2.3) invariant, but it does not leave (3.2.4a,b) invariant. If a transformation does not leave Hamilton's equations invariant, the interpretation of E as an energy need no longer be valid.

When investigating transformations that convert Hamilton's equations into Hamilton's equations one is dealing with functions that must possess derivatives. We shall use the now standard term *C^k function* to denote a function whose derivatives of k'th order are all themselves continuous functions for all real values of their arguments. For $k > 0$, a C^k function is necessarily C^{k-1}; e.g., if a function is C^5, then it is C^4, ..., C^0. If k is infinite, the function is said to be *smooth.* Let

$$q \to q' = Q(q, p), \quad p \to p' = P(q, p), \tag{3.3.2a}$$

and

$$q' \to q = Q^{inv}(q', p'), \quad p' \to p = P^{inv}(q', p'), \tag{3.3.2b}$$

define a one-to-one transformation and its inverse. If Q and P are C^k functions of q and p, then (3.3.2a) is said to define a C^k transformation; and if Q^{inv} and P^{inv} are $C^{k'}$ functions of q', p', then (3.3.2b) defines a $C^{k'}$ transformation. Together, the four equations define a *C^k diffeomorphism,* with k being the smaller of k and k'.[c] Diffeomorphisms transform (q + dq, p + dp) to (q'+ dq', p'+ dp') with

$$\begin{aligned} dq' &= (\partial Q(q,p)/\partial q)\, dq + (\partial Q(q,p)/\partial p)\, dp, \\ dp' &= (\partial P(q,p)/\partial q)\, dq + (\partial P(q,p)/\partial p)\, dp. \end{aligned} \tag{3.3.2c}$$

The diffeomorphisms also transform (q'+ dq', p'+ dp') to (q + dq, p + dp) with

$$\begin{aligned} dq &= (\partial Q^{inv}(q',p')/\partial q')\, dq' + (\partial Q^{inv}(q',p')/\partial p')\, dp', \\ dp &= (\partial P^{inv}(q',p')/\partial q')\, dq' + (\partial P^{inv}(q',p')/\partial p')\, dp'. \end{aligned} \tag{3.3.2d}$$

For reasons to be explained in Chapter 7, we require transformations of position and momentum variables be C^2 diffeomorphisms. These convert the Hamiltonian equations of motion

$$dq_j/dt = \partial H/\partial p_j, \quad dp_j/dt = -\partial H/\partial q_j, \tag{3.3.3a}$$

into the Hamiltonian equations of motion

$$dq'/dt = \partial H'(q',p')/\partial p', \quad dp'/dt = -\partial H'(q',p')/\partial q', \tag{3.3.3b}$$

iff

$$\partial q'/\partial q\ \partial p'/\partial p - \partial p'/\partial q\ \partial q'/\partial p = 1. \tag{3.3.4}$$

Mathematicians term a transformation of q and p that satisfies (3.3.4) a *symplectic* transformation.

To ensure that Hamilton's equations of motion produce evolution transformations that are C^2, their Hamiltonian must be a C^3 function of q and p. *Unless otherwise noted, we henceforth require that H(q,p) is* C^3.

Equations involving expressions similar to that in (3.3.4) appear so often in Hamiltonian mechanics that it is convenient to introduce the *Poisson bracket* of any two functions A(**q,p**), B(**q,p**), written {A,B}, and defined by

$$\{A,B\} = \Sigma_j(\partial A/\partial q_j\ \partial B/\partial p_j - \partial B/\partial q_j\ \partial A/\partial p_j). \tag{3.3.5}$$

In the exercises at the end of the chapter, the reader is asked to establish that when q → q' = A, p → p' = B, then if {A,B} = 1, the variables q' and

[c] C^0 transformations do not define diffeomorphisms, they can *define homeomorphisms.* A homeomorphism is a transformation which carries adjacent points with coordinates (q_1,p_1), (q_2,p_2) to adjacent points with coordinates (q_1',p_1'), (q_2',p_2'), and has an inverse which carries out the reverse transformation. In a C^0 homeomorphism, the transformation and its inverse are continuous functions. In the Euclidean plane, a homeomorphism can interconvert a square and a circle. A diffeomorphism cannot do so, because the derivatives required in (3.3.2e,f) do not exist at the corners of the square.

p' are also conjugate variables. When this happens, one may treat (q',p') in the same way one treats (q,p) in Hamilton's equations.

The following one-parameter groups of transformations are all groups of smooth (i.e., C^∞) symplectic diffeomorphisms, and hence canonical diffeomorphisms:

$$q \to q' = q + a, \quad p \to p' = p + b, \tag{3.3.6a, b}$$

$$q \to q' = e^{\beta} q, \quad p \to p' = e^{-\beta} p, \tag{3.3.6c}$$

$$q \to q' = q\cos(\alpha) - p\sin(\alpha), \quad p \to p' = p\cos(\alpha) + q\sin(\alpha), \tag{3.3.6d}$$

$$q \to q' = q\cosh(\gamma) + p\sinh(\gamma), \quad p \to p' = p\cosh(\gamma) + q\sinh(\gamma). \tag{3.3.6e}$$

Figures 2.3.1a,c, 2.3.3c and 2.3.4a in Chapter 2 depict smooth symplectic transformations if one considers one of the variables x,y to be a 'q', and the other to be its conjugate, 'p'.

The symbolism Sp(2n,R) is used to denote a group of homogeneous linear symplectic transformations of 2n real variables. Taken together, transformations (3.3.6c,d,e) comprise the 3-parameter group Sp(2,R). Any real homogeneous linear transformation of (q,p) that is symplectic may be expressed as a succession of the three transformations (3.3.6c–e). If the translations (3.3.6a,b) are included, the group becomes an inhomogeneous linear symplectic group, ISp(2,R). Many useful symplectic transformations are linear, and many are nonlinear.

The symplectic transformations of the mathematician are a more restricted class of transformations than those often called canonical transformations by physicists. This is because physicists, in interconverting units of measurement, often use scale transformations $q \to q' = aq$, $p \to p' = bp$ with a,b not equal to 1. Then (3.3.4) becomes ab rather than 1.

3.4 Invariance Transformations of Hamilton's Equations of Motion

Symplectic transformations

$$q \to q' = Q'(q, p), \quad p \to p' = P'(q, p), \tag{3.4.1a,b}$$

that convert H(q,p) to H(q',p') and convert Hamilton's equations into Hamilton's equations, do not in general leave the Hamiltonian function, H(q,p), unchanged. To determine whether H(q,p) is unchanged, one first substitutes $Q'(q,p)$ for q', and $P'(q,p)$ for p' in $H(q',p')$. Then

$$H(q,p) \to H(q',p') = H(Q'(q,p), \quad P'(q,p)) = H'(q,p). \tag{3.4.1c}$$

A symplectic transformation is an *invariance* transformation of Hamilton's equations with Hamiltonian $H(q,p)$ *iff* $H'(q,p) = H(q,p) = E$. Then

$$dq/dt = \partial H'(q,p)/\partial p \to dq/dt = \partial H(q,p)/\partial p,$$

and

$$dp/dt = -\partial H'(q,p)/\partial q \to dp/dt = -\partial H(q,p)/\partial q. \tag{3.4.2}$$

As a consequence, Hamilton's differential equations possess the wonderfully simple property that they are left invariant by a transformation of q and p *iff* the transformation is a symplectic one which leaves the Hamiltonian function itself invariant:

If t *is not transformed, the study of the invariance transformations of Hamilton's equations of motion reduces to a study of invariance transformations of the equation* $E = H(q,p)$ *under symplectic transformations of the variables* q *and* p.

The invariance transformations may be different for different values of E. The characterization of invariance when t is transformed will be considered in Sec. 3.7, below, and in Chapter 7.

The evolution transformations produced by Hamilton's equations are themselves invariance transformations of the equations of motion. They carry every point on a solution curve into a point on the same curve; they carry solutions into themselves. Other invariance transformations of the equations of motion may interconvert solution curves.

Before proceeding further let us consider a further example that illustrates some of the foregoing discussion. The Hamiltonian of a free particle of mass m is $H = p^2/2m$. Hamilton's equations of motion become

$$dq/dt = p/m, \quad dp/dt = 0. \tag{3.4.3a,b}$$

For each value of $E > 0$ they determine the two separate straight-line trajectories in q,p space defined by $q = q_0 + (p/m)t$, with $p = (2mE)^{1/2}$, and $p = -(2mE)^{1/2}$. The two trajectories, illustrated in **Fig. 3.4.1**, are interconverted by the symplectic transformation $p \to -p$, $q \to -q$, which corresponds to a 180° rotation in q,p space.

Letting $q \to q' = q + a$, also does not alter dq or $p\partial/\partial q$, so the free-particle Hamiltonian, H(p), the energy, and the equations of motion are all invariant under this transformation. The symplectic transformation

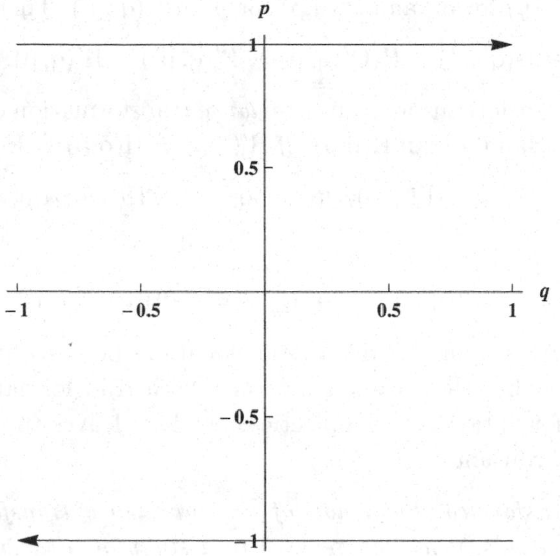

Fig. 3.4.1 Trajectories of a free particle in phase space.

$p \to p' = e^{\beta}p$, $q \to q' = e^{-\beta}q$, converts the Hamiltonian equations of motion into Hamiltonian equations of motion, but it converts $H(p)$ into $H(p') = p'^2/2m = e^{2\beta}\, p^2/2m$. As the transformation does not leave H invariant it is not an invariance transformation of the equations of motion for a free particle of mass m. It is, however, a class of scaling transformations that it is useful in investigations of isotope effects and *effective* masses in quantum mechanics.

3.5 Geometrization of Hamiltonian Mechanics

In Newtonian mechanics one thinks of motions as taking place in a Euclidean space of at most three dimensions. Position vectors, $\mathbf{r}$, are defined in the space of Newton, as are infinitesimal displacement vectors $\mathbf{dr}$, velocity vectors $\mathbf{v} = \mathbf{dr}/dt$, and momentum vectors $\mathbf{p} = m\mathbf{v}$. Symmetries are often defined in this Euclidean space, or in its relativistic generalization, space–time. However, the intrinsic symmetries of Hamilton's equations of motion are all symmetries defined by symplectic transformations. Assigning geometric expression to these intrinsic symmetries requires the development of the geometry of spaces whose points have coordinates (q_j, p_j) that may be altered by symplectic transformations.

If one considers symplectic transformations to define active transformations in a phase–space, one can erect a fixed cartesian coordinate system in it. Let us denote vectors in p,q space by **s**. For the moment, consider the case n = 3, and let $\mathbf{s} = (q_1, q_2, q_3, p_1, p_2, p_3)$ be a vector with cartesian components q_j and p_j in the space. Defining the unit vectors $\mathbf{i_q}$, $\mathbf{j_q}$, $\mathbf{k_q}$, one has $\mathbf{q} = (q_1, q_2, q_3, 0, 0, 0) = q_1\mathbf{i_q} + q_2\,\mathbf{j_q} + q_3\mathbf{k_q}$. Similarly, a momentum vector $\mathbf{p} = (0, 0, 0, p_1, p_2, p_3) = p_1\,\mathbf{i_p} + p_2\,\mathbf{j_p} + p_3\,\mathbf{k_p}$. The components of any vector in this space can be obtained by determining its dot product with six orthogonal unit vectors which are $\mathbf{i_q}$, $\mathbf{j_q}$, $\mathbf{k_q}$, $\mathbf{i_p}$, $\mathbf{j_p}$ and $\mathbf{k_p}$. From a mathematical standpoint, a momentum vector in ordinary space may be obtained from this momentum vector by a mapping, such as a rotation through 90°, that projects $\mathbf{i_p}$ to $\mathbf{i_q}$, *etc.*

For ordinary three-dimensional vectors one defines the cross product of two vectors, of which $\mathbf{L} = \mathbf{r} \times \mathbf{p}$ is a prime example. An inversion of Cartesian coordinate axes through their origin has the effect of negating all the Cartesian components of an ordinary vector **V**, which in the inverted system becomes expressed as $-\mathbf{V}$. Both **r** and **p** behave in such a manner, and as a consequence **L** does not do so. For this reason **L** is sometimes termed a *pseudo-vector*. As the cross product is only defined in three-dimensional space, in Hamiltonian mechanics we shall replace $L_x, L_y, L_z = L_1, L_2, L_3$ by

$$L_k = \varepsilon_{ijk}\, q_i p_j, \quad i,\, j,\, k = 1, \ldots, 3. \tag{3.5.1}$$

Summation over repeated indices is assumed, and $\varepsilon_{ijk} = +1$ for clockwise cyclic permutations of i, j, k = 1, . . . , 3, and equals −1 for counterclockwise cyclic permutations. For noncyclic permutations of the indices, $\varepsilon_{ijk} = 0$. Thus, e.g., $\mathbf{L}_3 = r_1\, p_2 - r_2\, p_1$. The $\mathbf{L}_k$ should not be considered components of a vector in the six-dimensional phase–space of positions and momenta. (They are components of a tensor.)

We next turn to an investigation of the elements of a plane geometry in which only symplectic transformations of $\mathbf{s} = [\mathbf{q},\mathbf{p}]$ are allowed. We will then settle on the concept of symmetry that this geometry entails. Many other key points will emerge from a study of the properties of this two-dimensional real space, and it will be the subject of the remainder of this chapter. In a Cartesian system, $\mathbf{s} = \mathbf{q}\mathbf{i_q} + \mathbf{p}\mathbf{i_p}$, with $\mathbf{i_q} \cdot \mathbf{i_p} = 0$, and both q and p real numbers. A moving mass point will be represented by the motion of a point at the terminus of **s**. Hamiltonian equations of motion govern canonical evolutionary motions of points in this space. **Figs. 3.2.1** and **3.4.1** have illustrated this.

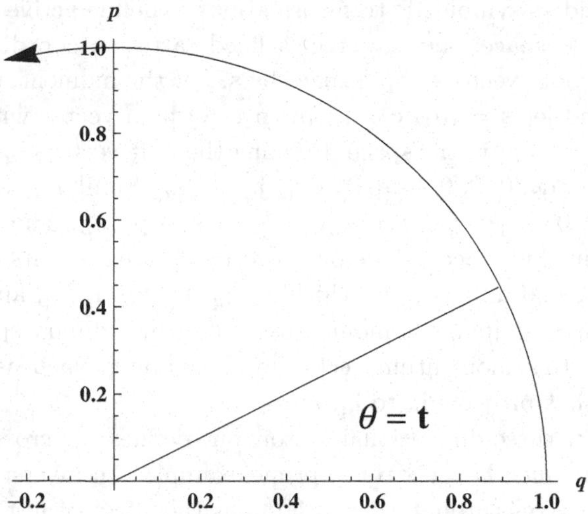

Fig. 3.5.1 Trajectory of an oscillator in phase space if k/m = 1.

Because the transformations we are considering are characterized only by being C^2 symplectic diffeomorphisms, if our space of q,p contains a given point, it must contain all others that can be obtained from the given point by any C^2symplectic diffeomorphism. Mathematicians call a space whose points are all connected by symplectic diffeomorphisms, a *symplectic space.* In two-dimensional symplectic space, the relation

$$\{Q(q,p), \quad P(q,p)\} = 1 \tag{3.5.2a}$$

implies

$$\{Q^{inv}(q,p), \quad P^{inv}(q,p)\} = 1. \tag{3.5.2b}$$

Symplectic space differs from the phase–space of most theoretical mechanics texts because, as previously noted, when physicists compare observations they often utilize scale transformations that are not symplectic.

Figure 3.5.2a depicts unit vectors i, j with q lying along i and p lying along j. In **Fig. 3.5.2b**, q and p have been transformed to q′ and p′ by a symplectic transformation. **Figure 3.5.2c** illustrates the effect an inversion through the origin has on vectors **A**, **B**, **C** that define a triangle traversed in a clockwise direction. The inversion is a symplectic transformation. In **Fig. 3.5.2d**, the vectors **A**, **B**, **C** are reflected through the p axis along j. This transformation is not a symplectic one. Note that in **Fig. 3.5.2c** the

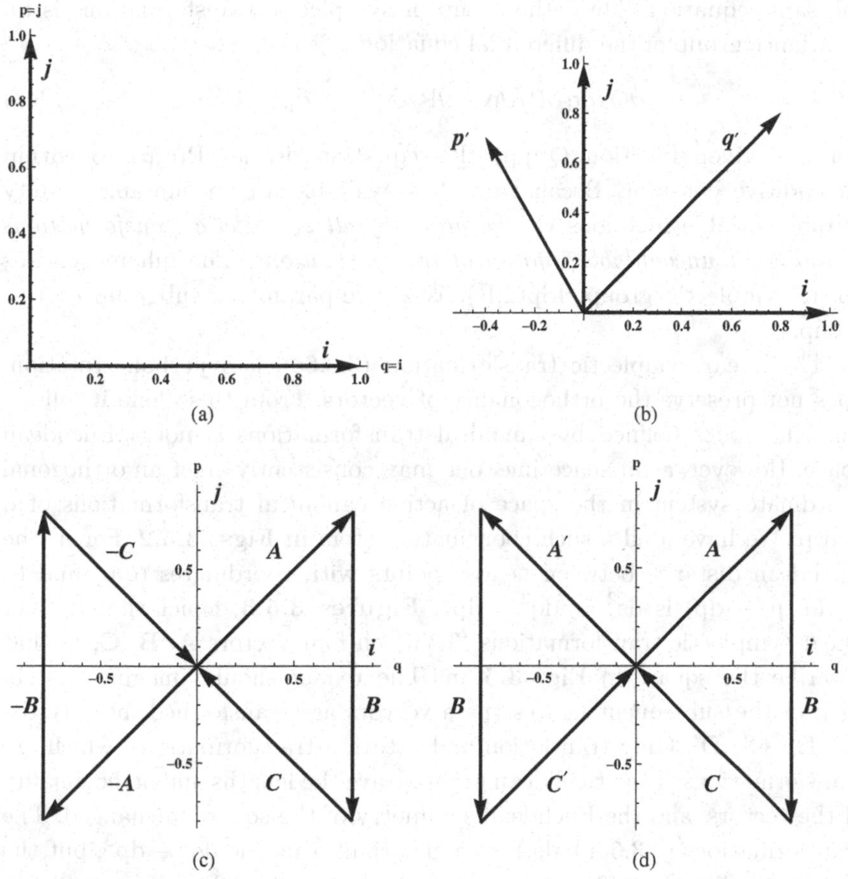

Fig. 3.5.2 (a), (b) A sympletic transformation of p and q to p′ and q′. (c) Inversion of A, B, C. (d) A reflection of A, B, C.

inverted vectors **A**, **B**, **C** define a triangle that is traversed in the same, clockwise, sense as the triangle defined by **A**, **B**, **C**. On the other hand, the reflected vectors in **Fig. 3.5.2d** define a triangle that is traversed in a counterclockwise direction.

The successive application of any two symplectic transformations, carries a point in symplectic space into a point in the space, and is necessarily a C^1 canonical transformation with a C^1 inverse. As the identity transformation is a symplectic transformation, taken together, symplectic diffeomorphims comprise a group of transformations. All the transformations convert solutions of the differential equation $\{Q(q,p), P(q,p)\} = 1$ into solutions of the same equation. Thus, the group of symplectic transformations is an invariance group of the differential equation

$$\partial Q/\partial q\, \partial P/\partial p - \partial P/\partial q\, \partial Q/\partial p = 1. \qquad (3.5.3)$$

For any given function Q(q,p) this equation defines P(q,p) to within an additive constant. Because one has available an uncountable infinity of differentiable functions Q, *the group of all symplectic transformations contains an uncountable infinity of transformations.* The inhomogeneous linear symplectic group, ISp(2,R), is a five-parameter subgroup of this group.

The linear symplectic transformation (3.3.6e), a hyperbolic rotation, does not preserve the orthogonality of vectors. From this alone it follows that the space defined by canonical transformations is not a Euclidean space. However, as in spacetime, one may consistently erect an orthogonal coordinate system in the space of active canonical transformations of q and p. We have used a such coordinate system in **Figs. 3.5.2**. For it, the Euclidean distance between nearby points with coordinates (q,p) and (q + dq, p + dp) is $ds^2 = dq^2 + dp^2$. **Figures 3.5.3** depict the effect of linear symplectic transformations (3.3.6) on four vectors **A**, **B**, **C**, **D** that describe the square in **Fig. 3.5.3a**. The reader should imagine that in each of the subsequent figures these vectors are transformed into vectors **A′**, **B′**, **C′**, **D′**. Only translation and rotation transformations, which are transformations of the Euclidean group, leave the lengths and orthogonality of the vectors, and the Euclidean symmetry of the square unchanged. The transformations (3.3.6 a,b,d,e) leave invariant a metric $dq^2 - dp^2$, but the symplectic dilatation (3.3.6c) does not. And, though (3.3.6c,d) leaves dq dp invariant, (3.3.6e) does not.

In short, symplectic transformations do not in general preserve a Euclidean metric, nor a metric analogous to that of special relativity, nor dq dp. However, there is a quantity that is left invariant by transformations of ISp(2,r). We shall shortly use it to define a metric in symplectic p,q space.

The cross product of Euclidean three-vectors may be used to define the oriented area of each of **Figs. 3.5.3** as $(\mathbf{A} \times \mathbf{B})/k = \mathbf{A}_q\mathbf{B}_p - \mathbf{B}_q\mathbf{A}_p$ where $\mathbf{k} = \mathbf{i} \times \mathbf{j}$. The oriented area of the square is $(\mathbf{A}\times\mathbf{B})/\mathbf{k} = \mathbf{A}_q\mathbf{B}_p = 1$. In the general case, if $\theta_{a\to b}$ is the angle between **A** and **B** measured clockwise from **A** to **B**, then $(\mathbf{A}\times\mathbf{B})/k = |\mathbf{A}||\mathbf{B}|\sin(\theta_{a\to b})$. The dimensionless area defined here can be obtained by dividing the corresponding physical quantity, an *action*, by the units of action, ml^2t^{-1}. These are units of energy, ml^2t^{-2}, multiplied by the units, t, of time.

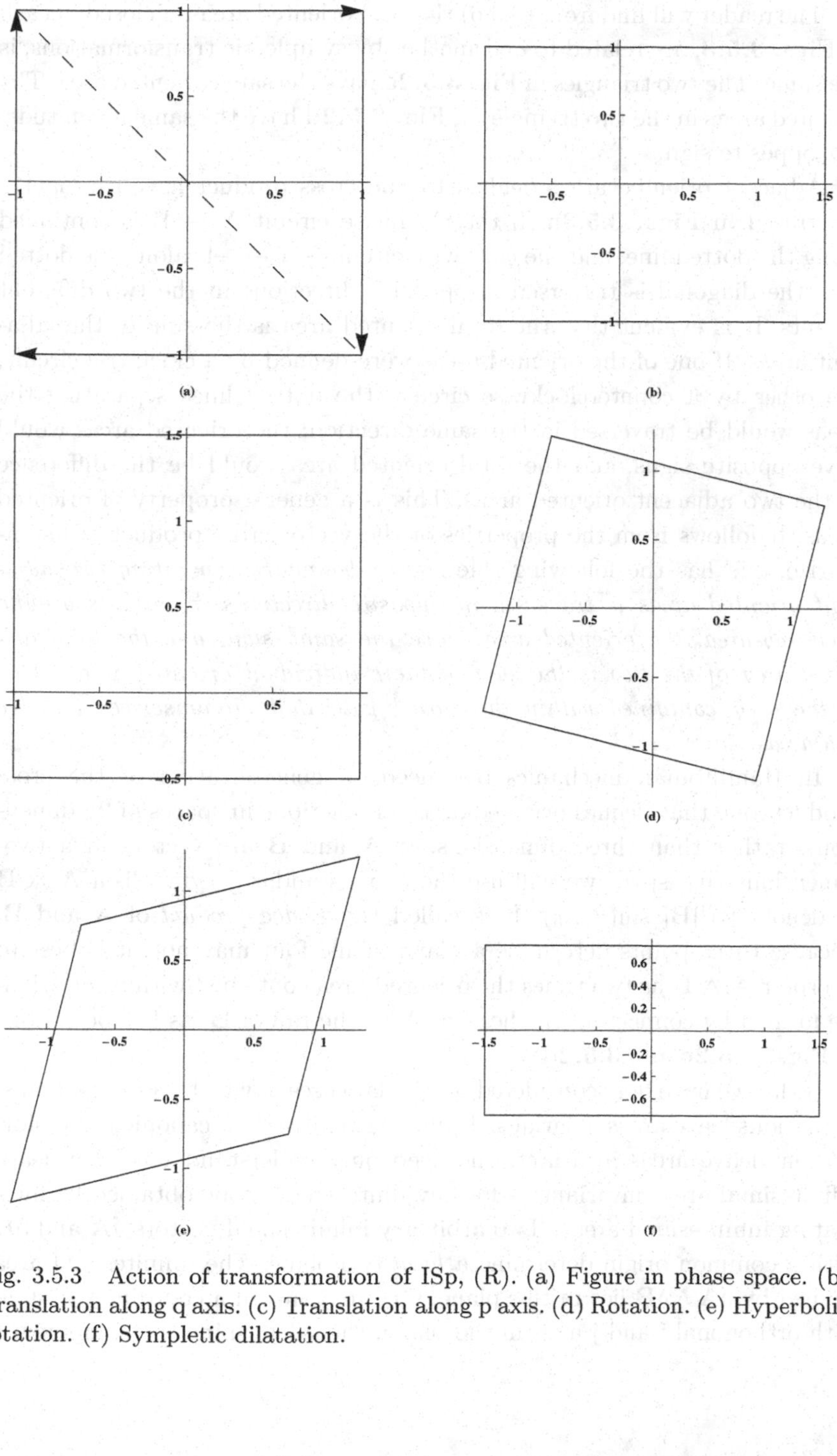

Fig. 3.5.3 Action of transformation of ISp, (R). (a) Figure in phase space. (b) Translation along q axis. (c) Translation along p axis. (d) Rotation. (e) Hyperbolic rotation. (f) Sympletic dilatation.

The reader will find from (3.3.6) that the oriented areas enclosed in each of **Figs. 3.5.3**, are related to one another by symplectic transformations, is the same. The two triangles in **Fig. 3.5.2c** have the same oriented area. The oriented areas in the two triangles in **Fig. 3.5.2d** have the same magnitude, but opposite sign.

Adjacent oriented areas defined by the cross product have a property illustrated in **Figs. 3.5.3a**. If the clockwise circuit A $\rightarrow$ B is continued along the dotted line, and the clockwise circuit $-$A $\rightarrow$ $-$B along the dotted line, the diagonal is traversed in opposite directions in the two different circuits. It is evident that the total oriented area is the sum of the adjacent areas. If one of the oriented areas were defined by a clockwise circuit, the other by a counterclockwise circuit, the dotted lines separating the areas would be traversed in the same direction; the oriented areas would have opposite signs, and the total oriented area would be the difference of the two adjacent oriented areas. This is a general property of oriented areas. It follows from the properties of the vector cross product. To summarize, one has the following rule: *If the boundary separating two adjacent oriented areas is traversed in opposite directions in circuits around each sub-area, the oriented areas have the same sign, and the total oriented area of the two is the sum of their individual oriented areas. This is the area contained within the closed path that circumscribes the two sub-areas.*

In Hamiltonian mechanics one needs a generalization of the cross product, one that defines oriented areas and actions in spaces of 2n dimensions, rather than three dimensions. If **A** and **B** are vectors in a two-dimensional p,q space we will use the (now standard) symbolism $\mathbf{A} \wedge \mathbf{B}$ to denote $|\mathbf{A}||\mathbf{B}| \sin(\theta_{a\rightarrow b})$. It is called the *wedge product* of **A** and **B**. Because three points determine a plane, while four may not, it is best to interpret $\mathbf{A} \wedge \mathbf{B}$ as two times the oriented area contained within the triangle formed by connecting the head of **A** to the tail of **B**, as has been done in **Figs. 3.5.2c** and **3.5.2d**.

So far we have only considered particular cases in which canonical transformations leave areas invariant. To understand why *all* canonical transformations leave areas invariant, one need only understand how they leave infinitesimal areas invariant — for any finite area can be obtained by integrating infinitesimal areas. Two arbitrary infinitesimal vectors $\delta\mathbf{A}$ and $\delta\mathbf{B}$ with a common origin determine a *local* two-plane. The infinitesimal area defined by $\delta\mathbf{A} \wedge \delta\mathbf{B}$ lies in this plane. One may install a coordinate system with orthogonal i and j axes in the plane, then determine the components of $\delta\mathbf{A}$ and $\delta\mathbf{B}$ on these axes. Let us apply these observations to equations (3.3.2). They state that if a diffeomorphism converts

$$\mathrm{q} \to \mathrm{Q}(\mathrm{q},\mathrm{p}), \quad \mathrm{p} \to \mathrm{P}(\mathrm{q},\mathrm{p}), \tag{3.5.4a,b}$$

then

$$\mathrm{dq} \to \mathrm{a}(\mathrm{q},\mathrm{p})\ \mathrm{dq} + \mathrm{b}(\mathrm{q},\mathrm{p})\ \mathrm{dp}, \tag{3.5.4c}$$

$$\mathrm{dp} \to \mathrm{c}(\mathrm{q},\mathrm{p})\ \mathrm{dq} + \mathrm{d}(\mathrm{q},\mathrm{p})\ \mathrm{dp}, \tag{3.5.4d}$$

with

$$\mathrm{a} = \partial\mathrm{Q}/\partial\mathrm{q}, \quad \mathrm{b} = \partial\mathrm{Q}/\partial\mathrm{p}, \quad \mathrm{c} = \partial\mathrm{P}/\partial\mathrm{q}, \quad \mathrm{d} = \partial\mathrm{P}/\partial\mathrm{p}. \tag{3.5.4e}$$

One may choose axes such that the untransformed dq, which we shall term $\mathrm{dq_o}$, lies along $\mathbf{i}$, and $\mathrm{dp_o}$, the untransformed dp, lies along $\mathbf{j}$. The right-hand side of (3.5.4c) can be considered to define a vector, $\mathbf{dA}$, and the right-hand side of (3.5.4d) can be considered to be $\mathbf{dB}$. And neither of these need lie along $\mathbf{i}$ or $\mathbf{j}$. Equations (3.5.4c,d) can then be written

$$\mathbf{dq_o} \to \mathbf{dq} = \mathrm{a}\,\mathbf{dq_o} + \mathrm{b}\,\mathbf{dp_o}, \quad \mathbf{dp_o} \to \mathbf{dp} = \mathrm{c}\,\mathbf{dq_o} + \mathrm{d}\,\mathbf{dp_o}, \tag{3.5.5a}$$

with

$$\mathbf{dq_o} = \mathrm{dq_o}\mathbf{i}, \quad \mathbf{dp_o} = \mathrm{dp_o}\mathbf{j}. \tag{3.5.5b}$$

These equations imply that $\mathbf{dq}$ and $\mathbf{dp}$ are not, in general, orthogonal. **Figures 3.5.4** depict the active interpretation of such a transformation, one in which, at the point of interest, the derivatives in (3.5.4c,d) have numerical values such that

$$\begin{aligned}\mathbf{dq} &= (3/2)^{1/2}\mathbf{dq_o} + (1/2)^{1/2}\mathbf{dp_o},\\ \mathbf{dp} &= (1/2)^{1/2}\mathbf{dq_o} + (3/2)^{1/2}\mathbf{dp_o}.\end{aligned} \tag{3.5.6}$$

The oriented area defined by $\mathbf{dq_o} \wedge \mathbf{dp_o}$ in **Fig. 3.5.4a** is a special case; the area defined by $\mathbf{dq} \wedge \mathbf{dp}$ in **Fig. 3.5.4b** represents the more general case, for which $\mathbf{dq}$ and $\mathbf{dp}$ are not orthogonal. As the point at $(\mathbf{q},\mathbf{p})$ is continuously shifted, this $\mathbf{dq} \wedge \mathbf{dp}$ will change continuously to $\mathbf{dq'} \wedge \mathbf{dp'}$, the transformation of $\mathbf{dq}$ and $\mathbf{dp}$ being defined by (3.5.4). The oriented areas $\mathbf{dq'} \wedge \mathbf{dp'}$ and $\mathbf{dq} \wedge \mathbf{dp}$ are related via (3.5.5) and (3.5.6), with the

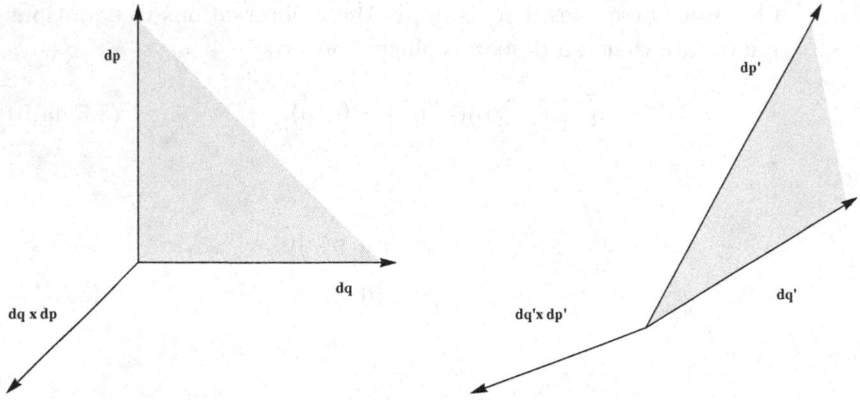

Fig. 3.5.4 Sympletic transformation of $\mathbf{dq} \times \mathbf{dp}$ to $\mathbf{dp'} \times \mathbf{dp'}$.

general relation

$$\begin{aligned} \mathbf{dq'}\wedge\mathbf{dp'} &= (\mathrm{a}\,\mathbf{dq} + \mathrm{b}\,\mathbf{dp}) \wedge (\mathrm{c}\,\mathbf{dq} + \mathrm{d}\,\mathbf{dp}) \\ &= (\mathrm{ad} - \mathrm{bc})\mathbf{dq}\wedge\mathbf{dp} \qquad (3.5.7a) \\ &= (\partial\mathrm{Q}/\partial\mathrm{q}\,\partial\mathrm{P}/\partial\mathrm{p} - \partial\mathrm{P}/\partial\mathrm{q}\,\partial\mathrm{Q}/\partial\mathrm{p})\mathbf{dq}\wedge\mathbf{dp}. \end{aligned}$$

If the transformation is symplectic, $\partial\mathrm{Q}/\partial\mathrm{q}\,\partial\mathrm{P}/\partial\mathrm{p} - \partial\mathrm{P}/\partial\mathrm{q}\,\partial\mathrm{Q}/\partial\mathrm{p} = 1$, so

$$\mathbf{dq'}\wedge\mathbf{dp'} = \mathbf{dq}\wedge\mathbf{dp}. \qquad (3.5.7b)$$

Hence, symplectic diffeomorphisms do not alter oriented infinitesimal areas.

A symplectic diffeomorphism that interconverts $\mathbf{dq} \wedge \mathbf{dp}$ and $\mathbf{dq'} \wedge \mathbf{dp'}$ will interconvert a region R and a region R′, and each area element $\mathbf{dq} \wedge \mathbf{dp}$ within R will be converted to a corresponding area element $\mathbf{dq'} \wedge \mathbf{dp'}$ within R′, and vice versa. From (3.5.7) and the definition of the sum of oriented areas, it follows that

$$\iint_{\mathrm{R'}} \mathbf{dq'}\wedge\mathbf{dp'} = \iint_{\mathrm{R}} \mathbf{dq}\wedge\mathbf{dp}. \qquad (3.5.8)$$

Thus, symplectic diffeomorphisms do not alter oriented finite areas.

The transformation from (q,p) to (r, θ) defined by q = r cos(θ), p = r sin(θ), is not symplectic, since {r,θ} = 1/r, and the area element is r dr dθ. On the other hand, the transformation defined by q = $\sqrt{r}$ cos(2θ), $p = \sqrt{r}$

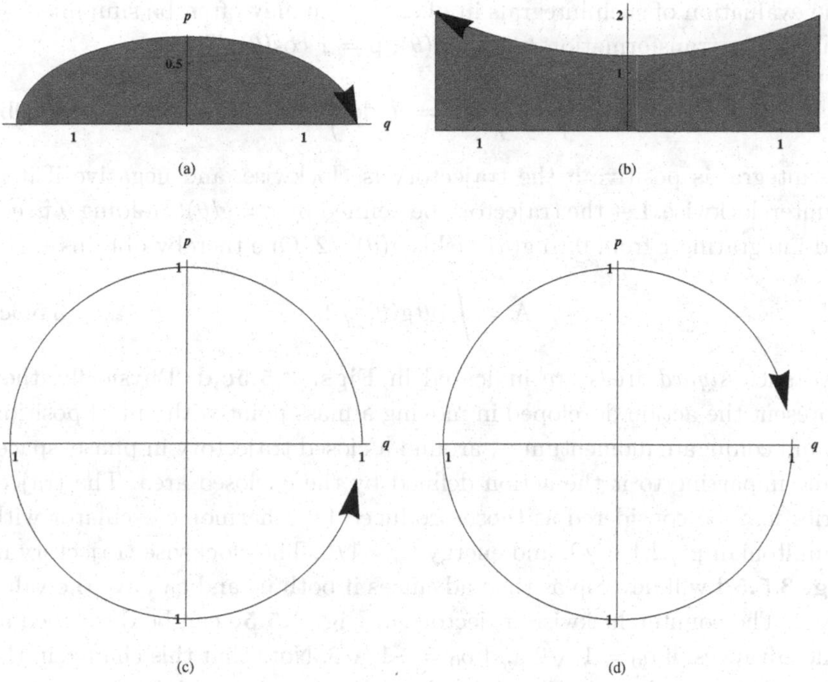

Fig. 3.5.5 (c) Positive area. (d) Negative area.

$\sin(2\theta)$, has $\{r,\theta\} = 1$, so it is a symplectic, and therefore area-preserving, transformation.

As an application of (3.5.8), consider the action developed when the points with coordinates (q,p) lie along a smooth curve. **Figures 3.5.5a,b** display oriented areas below two phase trajectories. The gray areas may be considered as defined by integrations. In **Fig. 3.5.5b** the area under the trajectory must be assigned a negative value. The trajectory in **Fig. 3.5.5a** may be extended to produce a clockwise elliptical trajectory containing positive area. The total oriented area within the ellipse will be determined by the addition rule given above. The oriented area and action can be obtained directly by evaluating the double integral over the enclosed region:

$$A = \iint_R \mathbf{dq} \wedge \mathbf{dp}. \tag{3.5.9a}$$

The evaluation of such integrals in phase–space may often be simplified by making the transformation $q = r \sin(\theta)$, $p = r \cos(\theta)$. Then

$$A = \oint d\theta \int r dr = \oint d\theta \int d(r^2/2), \tag{3.5.9b}$$

the integral is positive if the trajectory is clockwise, and negative if it is counterclockwise. Let the trajectory be defined by $r = g(\theta)$. Holding θ fixed, and integrating r from 0 to $g(\theta)$ yields $g(\theta)^2/2$. One thereby obtains

$$A = \oint d\theta \, g(\theta)^2/2. \tag{3.5.9c}$$

Two such *signed* areas are indicated in **Figs. 3.5.5c,d**. Physically, they represent the action developed in moving a mass point with initial position q_a and conjugate momentum p_a around a closed trajectory in phase–space, while imparting to it the action defined by the enclosed area. The trajectories may be considered as those produced by a harmonic oscillator with Hamiltonian $p^2/2 + q^2/2$, and energy $E = 1/2$. The clockwise trajectory in **Fig. 3.5.5d** will develop as time advances if both p_0 and q_0 have the value $1/\sqrt{2}$. The counterclockwise trajectory in **Fig. 3.5.5c** can be developed as time advances, if $q_0 = 1/\sqrt{2}$ and $p_0 = -1/\sqrt{2}$. Note that this change in the sense of rotation has been brought about by changing an initial value, and not by changing t to $-t$.

Now let us return to Eqs. (3.5.7), and consider a more general geometric significance of the fact that symplectic transformations do not alter the value of $\mathbf{dq} \wedge \mathbf{dp}$. The relationship between symplectic transformations and the invariant $\mathbf{dq} \wedge \mathbf{dp}$ is analogous to the relationship between Euclidean transformations in an x–y plane and the Euclidean metric $\mathbf{ds} \cdot \mathbf{ds} = dx^2 + dy^2$.

In fact, $\mathbf{dq} \wedge \mathbf{dp}$ *is the metric that characterizes geometry in the* q,p *space termed a symplectic space. The geometry of this space is the geometry of a space with an invariant metric that is a measure of the oriented area that represents action.*

This result establishes a direct connection between *symplectic geometry*, and the classical mechanics of Hamilton. The concept of action took on increased physical importance in 1913 when Paul Ehrenfest[1] argued that the magnitude of the action developed by any closed trajectory in phase–space is quantized. It appeared that it must be an integer multiple of Plank's constant divided by 2π:

$$\int\int_R |\mathbf{dq} \wedge \mathbf{dp}| = n\,h/2\pi. \tag{3.5.10}$$

This straightforward connection between physics and symplectic geometry persisted until 1927, when Heisenberg demonstrated the impossibility of measuring both positions and momenta with arbitrary accuracy.[2] The conceptual foundations of the old quantum theory then had to be abandoned, and were replaced by a quantum mechanics in which conjugate position and momentum variables satisfying $\{q,p\} = 1$ are replaced by operators q_{op} and p_{op} satisfying the operator relation

$$q_{op}p_{op} - p_{op}q_{op} = i\,h/2\pi. \tag{3.5.11}$$

The *correspondence principle* establishes connections between quantum dynamics and classical dynamics that exist when the classical variables q and p are replaced by operators that satisfy this relation.[3] As stated by Dirac, the principle was only applicable to polynomial functions of p′s and q′s. And as mentioned previously, in Chapter 9 we show that Dirac's correspondence principle can be generalized in a manner which enables one to directly establish the relation between invariance groups of Schrödinger equations and invariance groups of Hamilton's equations.

3.6 Symmetry in Two-Dimensional Symplectic Space

Knowing that the metric of two-dimensional symplectic space is $\mathbf{dq} \wedge \mathbf{dp}$ enables one to deal with the questions: "how are symmetries in this space defined? When do two geometric objects in the space have the same symmetry?" Let us first deal with the first question.

We have seen that the symmetry of an object in Euclidean space is defined by the subgroup of the Euclidean group, E(2), that carries it into itself. And we have seen that analogous statements hold for three-dimensional space, and for spacetime. Applying the same program to Hamiltonian mechanics, we shall define the symmetry of an object in symplectic space by that subgroup of all symplectic transformations which carries the object into itself.

Now let us turn to the second question. Two objects in Euclidean space have identical symmetry if operations of the Euclidean group, and/or uniform dilatations can superimpose the objects. These operations do not alter the symmetry of an object, they simply change the relationship of the object to the coordinate system of an observer. An analogous statement was found to hold for spacetime. However, in contrast to the metric in a Euclidean space, the spacetime metric could be both positive and negative — a property of real physical significance.

We shall suppose that two objects must certainly have identical symplectic symmetry if they can be superimposed by symplectic transformations. As translations and rotations of a rigid figure do not change the area enclosed by it, they may be used to test for superimposibility of two objects in symplectic space. However, this by no means exhausts the catalog of operations that can be used to superpose figures in symplectic space. All symplectic diffeomorphisms can be considered to change the relationship of an object to the coordinate system of an observer, without altering the symplectic symmetry of the object. Before going on to say anything about the geometric roles of uniform dilatations of $\mathbf{dq} \wedge \mathbf{dp}$, and of transformations that change its sign, we will, in the next section, consider the physical significance of these operations.

3.7 Symmetry in Two-Dimensional Hamiltonian Phase Space

The phase space of mechanics differs from symplectic space in several respects. As we have already noted, changing initial values p_0 and q_0 can convert a closed clockwise trajectory into a counterclockwise one, and do so without changing the magnitude of the action. Now p_0 and q_0 are special values of p and q, similiar to p and q, they may take on a continuous range of real values. We conclude that, for this reason alone, transformations that change the sign of $\mathbf{dq} \wedge \mathbf{dp}$ should be allowed in the p,q space of Hamiltonian mechanics. There is another reason for admitting the transformation $\mathbf{dq} \wedge \mathbf{dp} \to -\ \mathbf{dq} \wedge \mathbf{dp}$ as an operation that moves points in Hamiltonian phase–space. On inspection, one will find that reversing the direction of a motion described by $\mathbf{A} \to \mathbf{B}$ requires that one reverse the direction of the vectors, and also traverse them in the reverse order. This converts $\mathbf{A} \times \mathbf{B}$ to $-\mathbf{B} \times -\mathbf{A}$, that is to $-\mathbf{A} \times \mathbf{B}$. For the same reason, reversing the direction of motion converts $\mathbf{dq} \wedge \mathbf{dp}$ to $-\mathbf{dq} \wedge \mathbf{dp}$. As reversing the direction of a motion may be brought about by the transformation $t \to -t$, this time reversal also changes the sign of $\mathbf{dq} \wedge \mathbf{dp}$. Time is a parameter in evolution transformations in (p,q) space, just as is an angle is a parameter in rotation transformations. Time reversal is an operation that leaves kinetic energies, potential energies, and H(p,q) invariant. Time reversal defines a symmetry of every time-independent Hamiltonian that is a quadratic function of velocities.

There is also another reason, mentioned previously, for distinguishing physical phase–space from symplectic space. Observers necessarily choose units of measurement when measuring positions and momenta. When one converts measurements from one system to another, the transformation can be considered to be a dilatation, but it is seldom a symplectic one. For example, in the conversion between MKS and CGS systems, the numerical relations are $q^{CGS} = 10^2 q^{MKS}$, $m^{CGS} = 10^3 m^{MKS}$, $t^{CGS} = t^{MKS}$, $p^{CGS} = 10^5 p^{MKS}$, and the relation between the corresponding units of action is $q^{CGS}\, p^{CGS} = 10^7 q^{MKS}\, p^{MKS}$.

Suppose, then, that a change of units multiplies q by a and p by b, both a and b being positive real numbers. Such a transformation may always be considered as the composition of a symplectic dilation and a uniform dilation of q and p. To see this, let a symplectic dilatation convert q to $q' = e^{\alpha}\, q$, and convert p to $p' = e^{-\alpha} p$. Let $a = e^{\nu} e^{\alpha}$, and $b = e^{\nu} e^{-\alpha}$. Then $e^{2\nu} = ab$, and $a/b = e^{2\alpha}$. Consequently, the change of units is the composition of a uniform dilatation with parameter $\nu = \ln(ab)/2$, and a symplectic dilation with parameter $\alpha = \ln(a/b)/2$. For example, in changing MKS units to CGS units, $ab = 10^7$, $a/b = 10^{-2}$, so $\nu = (7/2)\ln(10)$, and $\alpha = -\ln(10)$.

Though scale changes may not convert H(p,q) to H(p′q′), they are allowed operations in Hamiltonian mechanics, and we conclude they must be allowed operations in the phase–spaces of Hamiltonian mechanics. We therefore allow uniform dilatations as well as symplectic dilatations to carry points in the phase–space of physics into points in the phase–space.

With these observations in hand we are prepared to define our two-dimensional Hamiltonian phase–space, and the meaning of symmetry in it.

Definition of Two-dimensional Hamiltonian Phase–Space:

The *Hamiltonian phase-space* of the canonically conjugate variables q, and p, is the space containing all points that may be carried into one another by uniform dilatations, and C^2 diffeomorphisms which leave invariant the metric

$$ds^2 = |\mathbf{dq} \wedge \mathbf{dp}|. \tag{3.7.1}$$

Henceforth, we will use *canonical transformation* to denote a symplectic transformation which is C^2. We will use *canonical diffeomorphism* to denote a symplectic transformation with an inverse, *both of which are* C^2.

Definition of Symmetry in Two-Dimensional Hamiltonian Phase–Space:

The symmetry of an object in two-dimensional Hamiltonian phase–space is defined by the group of canonical diffeomorphisms which leaves ds invariant, and carries the object into itself.

Two objects have the same symmetry in two dimensional Hamiltonian phase–space if they can be superposed by canonical diffeomorphisms that leave the magnitude of $\mathbf{dq} \wedge \mathbf{dp}$ *invariant, or multiply it by a positive real number.*

The transformations $\mathrm{dq} \rightarrow -\mathrm{dq}$ and $\mathrm{dp} \rightarrow -\mathrm{dp}$ each preserve the value of $\mathrm{dp}^2 + \mathrm{dq}^2$ and $|\mathbf{dq} \wedge \mathbf{dp}|$, but not $\mathbf{dq} \wedge \mathbf{dp}$ itself. These *local reflections*, and their counterparts $\mathrm{q} \rightarrow -\mathrm{q}$, $\mathrm{p} \rightarrow -\mathrm{p}$ which define ordinary reflections in Euclidean space, interconvert clockwise and counterclockwise motions in Hamiltonian phase–space. The symplectic symmetries which they interconvert are identical symmetries in the Hamiltonian phase space.

Circles of different area have different symmetries in symplectic space, and cannot be interconverted by symplectic transformations. However, they have the same symmetry in the Hamiltonian phase–space because they can be interconverted by uniform dilatations of p,q.

Rotations, the inversion defined by $\mathrm{q} \rightarrow -\mathrm{q}$, $\mathrm{p} \rightarrow -\mathrm{p}$ taken together, the transformation $\mathrm{q} \rightarrow \mathrm{p}$, $\mathrm{p} \rightarrow -\mathrm{q}$, and the transformation $\mathrm{q} \rightarrow -\mathrm{p}$, $\mathrm{p} \rightarrow \mathrm{q}$ — they all preserve the value of both $\mathrm{dp}^2 + \mathrm{dq}^2$ and $\mathbf{dq} \wedge \mathbf{dp}$. These transformations, which define symmetries in the Euclidean plane, can also define symmetries in two-dimensional symplectic space. The two Euclidean reflections $\mathrm{q} \rightarrow -\mathrm{q}$, and $\mathrm{p} \rightarrow -\mathrm{p}$, leave $\mathrm{dp}^2 + \mathrm{dq}^2$ and $|\mathbf{dq} \wedge \mathbf{dp}|$ invariant. Consequently, *symmetries in a Euclidean space of* (p,q) *are also symmetries in the Hamiltonian phase–space of* (p,q).

As hyperbolic pseudo-rotations in p,q space are also symplectic transformations, they too can be used to define symmetries in Hamiltonian phase–space, in the same way as they define symmetries in two-dimensional spacetime.

Treating transformations in Hamiltonian phase–space as passive transformations, and figures in the space as fixed objects, provides a direct way to investigate the manner in which the transformations alter the relation of an object to the (q,p) coordinate system of an observer. The uniform dilatations, along with the reflections, inversions, and continuous transformations of E(2) leave the object undisturbed, and simply change the perspective from which it is observed. Clockwise and counterclockwise motions then become identical motions observed from *right-handed* and *left-handed* coordinate systems. Symplectic transformations in phase space can interconvert Euclidean symmetries, just as Lorentz transformations in spacetime can interconvert Euclidean symmetries. One must adopt a non-Euclidean perspective if one wishes to consider the symmetry of an object to be unchanged by transformations that are linear and non-Euclidean, or nonlinear and non-Euclidean. In the first case, it is simplest to adopt an active viewpoint. In general, the choice between active and passive viewpoints must be determined by the physical context, and when this is not decisive, the transformation is best considered to be a mapping. As neither the concept of symmetry in phase space, nor its physical implications, are intuitively obvious, the remainder of this chapter is devoted to developing examples that establish connections between the geometry and the physics.

3.8 Symmetries Defined by Linear Symplectic Transformations

We begin by considering three systems whose Hamiltonians *cannot* be interconverted by symplectic transformations. *Each has a different symmetry in symplectic space and Hamiltonian phase–space.* Define free-particle, oscillator, and repulsive oscillator Hamiltonians, by respectively:

$$H_{fp} = p^2/2, \quad H_{OSC} = p^2/2 + q^2/2, \quad H_{rep} = p^2/2 - q^2/2. \qquad (3.8.1)$$

H_{fp} is invariant under the translations $q \to q + a$. H_{OSC} is invariant under rotations about the origin (q,p) = (0,0), and H_{rep} is invariant under hyperbolic pseudo-rotations about this origin.

The three Hamiltonians are all invariant under symplectic inversions and the non-symplectic reflections allowed in Hamiltonian phase–space. **Figure 3.8.1a** depicts the direction of motion along trajectories determined by H_{fp} when E = 1/2. It is designed to be used by the reader to determine the consequences of reflections in the q axis and the p axis. **Figures 3.8.1b,c** similarly depict trajectories determined by H_{rep} for E = 1/2 and E= −1/2, respectively. **Figures 3.5.5c,d** may be used to determine the manner in which reflections affect trajectories of H_{OSC}. Note that in every case, reflection interconverts clockwise and counterclockwise motions, as one expects, since each reflection changes the sign of $\mathbf{dq \wedge dp}$.

For each Hamiltonian in (3.8.1), the equations H(q,p) = E directly determine non-directional phase portraits. The phase portraits for H_{fp} have Euclidean translational symmetry in q, these for H_{OSC} have Euclidean circular symmetry, and those for H_{rep} have non-Euclidean hyperbolic symmetry. The directional portraits are determined by the equations of motion.

The successive composition of any of the transformations of the linear symplectic group ISp(2,R) will transform Hamilton's equations into Hamilton's equations. Any object obtained by action of a symplectic

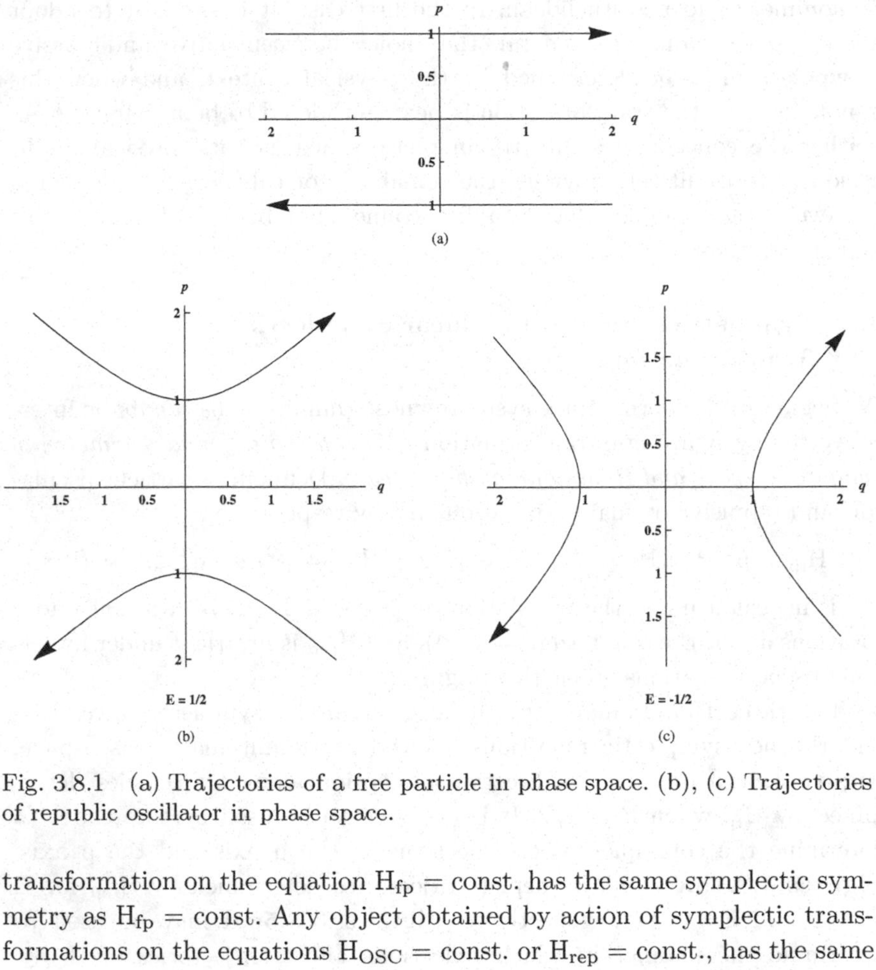

Fig. 3.8.1 (a) Trajectories of a free particle in phase space. (b), (c) Trajectories of republic oscillator in phase space.

transformation on the equation H_{fp} = const. has the same symplectic symmetry as H_{fp} = const. Any object obtained by action of symplectic transformations on the equations H_{OSC} = const. or H_{rep} = const., has the same symmetry as H_{OSC} = const. or H_{rep} = const., respectively.

The effect of several linear symplectic transformations that alter the phase portrait of a free particle with $H_{fp} = E = 1/2$ is displayed in **Figs. 3.8.2**. All have the same symplectic symmetry as **Fig. 3.8.2a**.

Figures 3.8.3b,c illustrate the effect of two linear symplectic transformations that alter the phase portrait of an oscillator with $H_{OSC} = E = 1/2$. All these portraits in **Figs. 3.8.3** are considered as having the circular symplectic symmetry of **Fig. 3.8.3a**.

Linear symplectic transformations that alter the phase portrait of a repulsive *oscillator* with $H_{rep} = E = 1/2$ are depicted in **Figs. 3.8.4b,c**. Again, the portraits have the same symplectic symmetry as the first portrait, **Fig. 3.8.4a**.

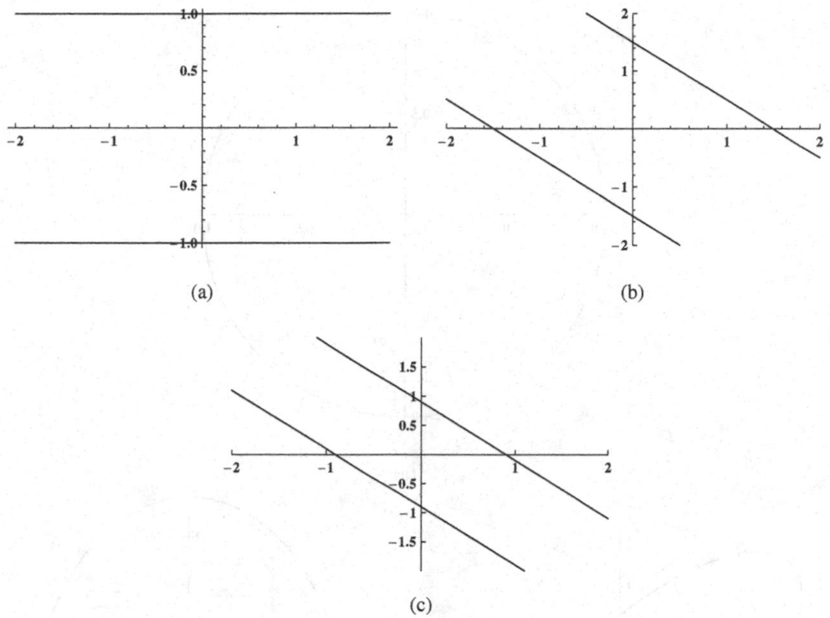

Fig. 3.8.2 Sp(2) transformation of phase portaits of systems with E = 1/2. (a) Free particle: Grid [{{a,b}}, Spacings → {6,0}]. (b) Rotation of 2a. (c) Hyperbolic rotation of 2a: Grid [{{c,d}}, Spacings → {6,0}].

In this section we have considered all transformations as active. The reader interested in geometrodynamics may wish to also consider their passive interpretation in some detail.

3.9 Nonlinear Transformations in Two-Dimensional Phase Space

Because Hamilton's equations are differential equations, they directly relate the properties of adjacent points in phase–space. This is reflected in the Poisson criterion for a canonical transformation; it is a local criterion. As a consequence, Hamiltonian mechanics allows local symplectic changes in scale, changes that vary continuously throughout phase–space. These canonical transformations involve nonlinear transformations of q and p. The action of any symplectic diffeomorphism, linear or nonlinear, will transform Hamilton's equations into Hamilton's equations, and will not alter the symplectic symmetry of the Hamiltonian or its phase portraits.

To obtain some of the consequences of these observations, consider, for example, the transformations

$$\mathrm{q} \rightarrow \mathrm{q}' = \mathrm{q}\exp(\alpha\mathrm{qp}), \mathrm{p} \rightarrow \mathrm{p}' = \mathrm{p}\exp(-\alpha\mathrm{qp}). \tag{3.9.1a}$$

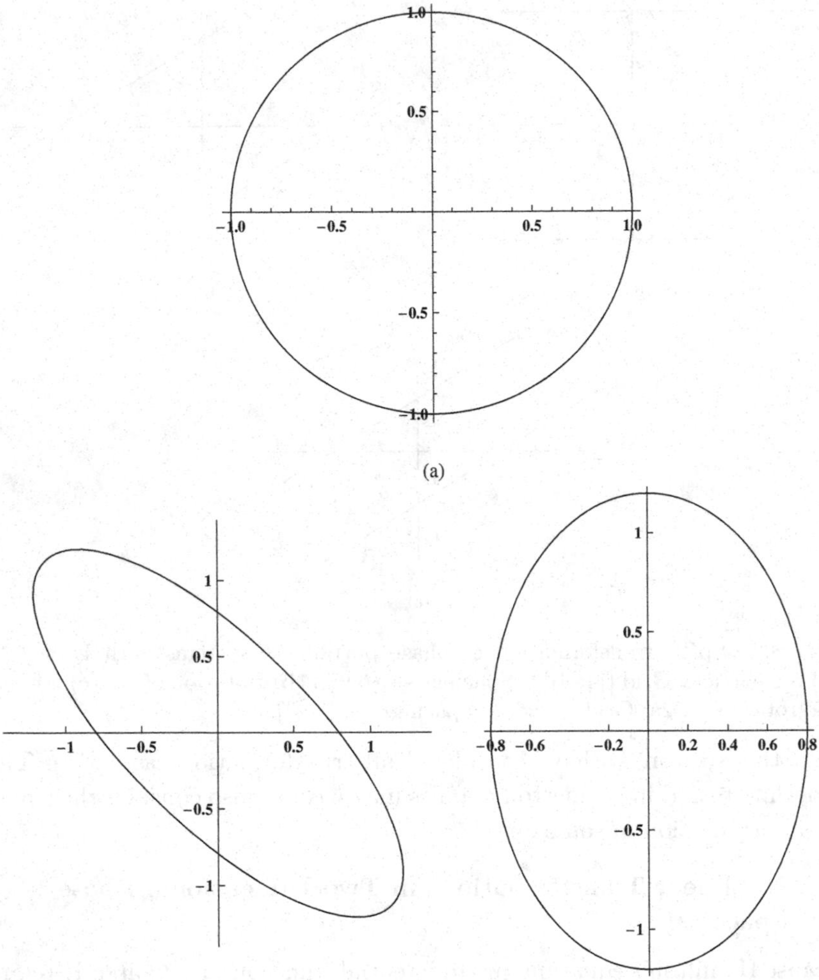

Fig. 3.8.3 (a) Harmonic oscillator. (b) Hyperbolic rotation of 3a. (c) Sympletic dilatation of 3a.

For them,

$$\{\mathrm{p}', \mathrm{q}'\} = 1, \tag{3.9.1b}$$

and

$$\mathrm{q} = \mathrm{q}'\exp(-\alpha \mathrm{q}'\mathrm{p}'), \quad \mathrm{p} = \mathrm{p}'\exp(\alpha \mathrm{q}'\mathrm{p}'). \tag{3.9.1c}$$

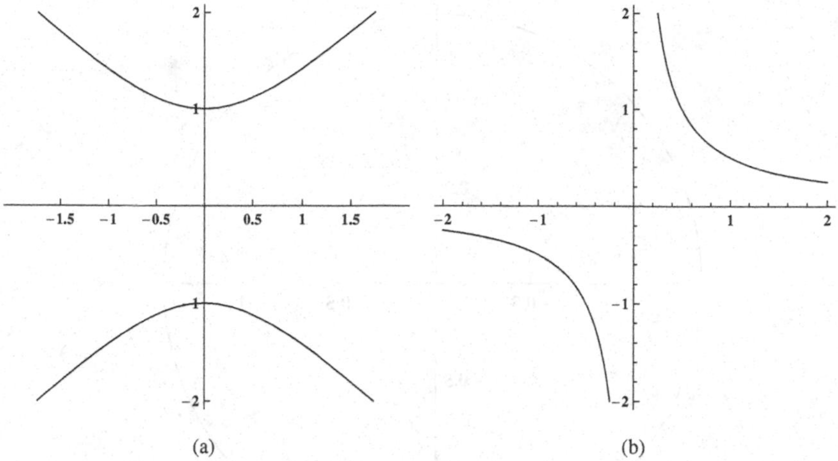

Fig. 3.8.4 (a) Repulisive "oscillator". (b) Rotation of 4a.

Note that

$$p'q' = p\,q. \tag{3.9.1d}$$

The transformations are smooth canonical diffeomorphisms for all real values of α. They convert the oscillator Hamiltonian of (3.8.1) to

$$H'(q,p) = (p^2/2 + q^2/2)\cosh(2\alpha\ q \cdot p) - (p^2/2 + q^2/2)\sinh(2\alpha\ q \cdot p). \tag{3.9.2}$$

Figures 3.9.1–3.9.4. illustrate the phase portraits that arise when $E = 1$, and α takes on the values 0, 1/2, 3/2, and 2.

The symmetry of all these figures in phase–space is the same: it is that of a circle. The action developed in each trajectory is the same. In the old quantum theory, the energy of each system would be the same.

If one chose to interpret the transformation (3.9.1a) in the passive sense, one would plot the orbits in a coordinate system Q^s, P^s related to a Q,P Cartesian system by the inverse of (3.9.1a), that is by (3.9.1c). An observer using these coordinate systems would, for each value of a, find all the orbits to be circles. From the standpoint of the Q,P system, the coordinates would have unit vectors whose orientations and lengths vary from point to point. From this standpoint, no symplectic diffeomorphism can alter the Euclidean symmetry of the circles of **Figs. 3.5.5** and **3.9.1**. From this standpoint, all canonical transformations will simply provide the coordinate systems defining different perspective snapshots of the circles.

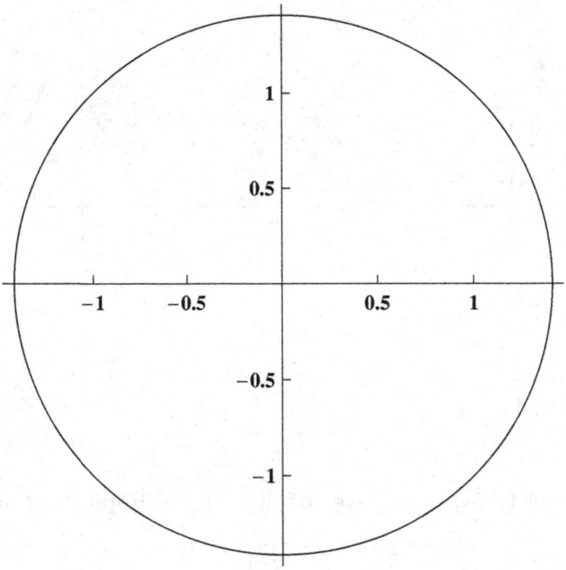

Fig. 3.9.1 $q' = qe^{aqp}$, $p' = pe^{-aqp}$, $a = 0$.

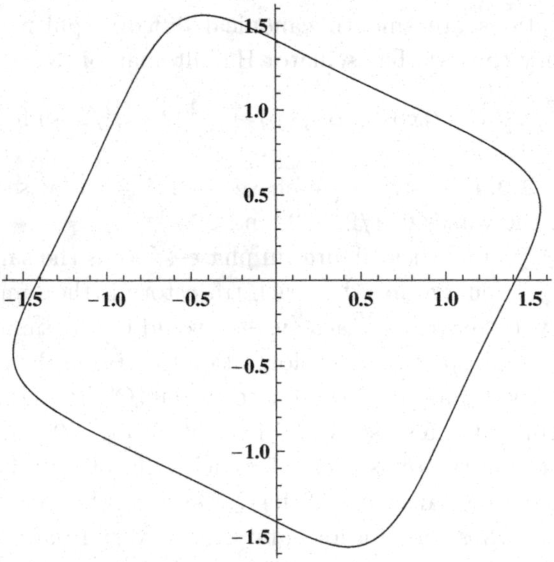

Fig. 3.9.2 $q' = qe^{aqp}$, $p' = pe^{-aqp}$, $a = \frac{1}{4}$.

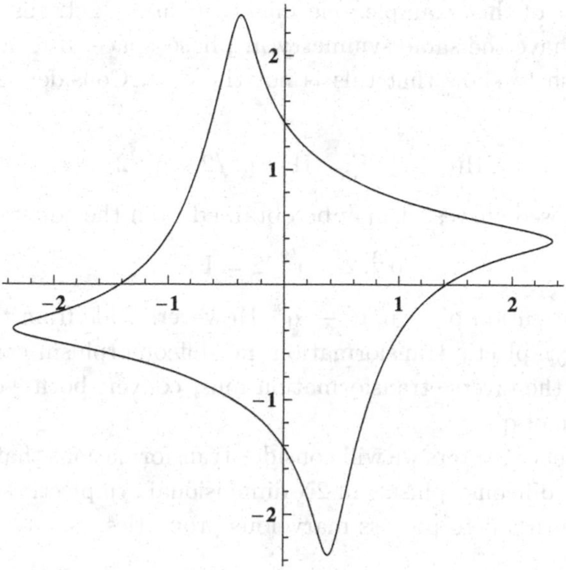

Fig. 3.9.3 $q' = qe^{aqp}$, $p' = pe^{-aqp}$, $a = \frac{3}{4}$.

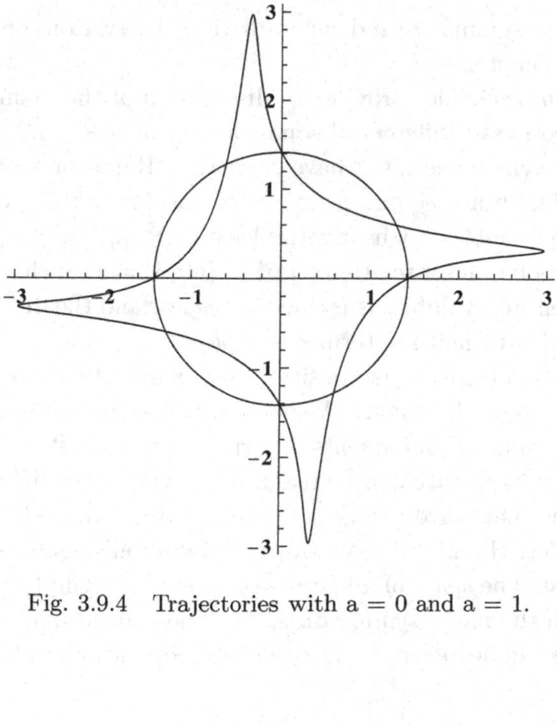

Fig. 3.9.4 Trajectories with a = 0 and a = 1.

In the light of this example, one might, at first sight, suspect that all closed curves have the same symmetry in phase–space. In concluding this section we wish to show that this is not the case. Consider, for example, the equation

$$H(q, p) = E, \quad H = p^2/2 + q^4/2, \tag{3.9.3}$$

that defines closed curves. It may be obtained from the equation

$$p'^2/2 + q'^2/2 = E, \tag{3.9.4}$$

by the transformation $p' \to p$, $q' \to q^2$. However, aside from the fact that this is not a symplectic transformation, no diffeomorphism can convert q' to q^2 because the inverse transformation must convert both $+q$ and $-q$ to the same value of q'.

In subsequent chapters we will consider transformations that are nonlinear symplectic diffeomorphisms in 2N dimensional symplectic spaces. Some of these will turn out to possess marvelous properties.

3.10 Dynamical Symmetries as Intrinsic Symmetries of Differential Equations and as Geometric Symmetries

In this section we summarize and generalize the observations of the previous sections of the chapter.

Hamiltonian mechanics provide an illustration of the manner in which intrinsic symmetries of differential equations are defined, and then realized as geometrical symmetries. We have seen that Hamilton's equations are converted to Hamilton's equations by diffeomorphisms that convert a pair of variables p,q to another pair of variables $p' = P(q,p)$, $q' = Q(q,p)$, which satisfy the Poisson bracket relation $\{q,p\} = \{q',p'\} = 1$. Such a pair of position and momentum variables is termed *conjugate* and the diffeomorphisms that leave $\{q,p\}$ invariant are termed *symplectic.*

Geometrization begins when positions and momenta are represented by coordinates of a point in a space. Vectors **q, p** and **dq**, **dp** are introduced, and considered to have components on orthogonal axes. It is then required that all points in the space can be reached by symplectic diffeomorphisms, transformations that leave invariant a metric, $\mathbf{dq} \wedge \mathbf{dp}$. The transformations that convert Hamilton's equations to Hamilton's equations leave this metric invariant. The space of all points connected by symplectic diffeomorphisms is given the name *symplectic space.* The transformations of Hamilton's equations can be interpreted as geometric operations in this space. For a given Hamiltonian, a symplectic transformation of Hamilton's equations either leaves them invariant, or transforms them into those corresponding to a different Hamiltonian. Symplectic transformations need not preserve the orthogonality of vectors, so symplectic space is non-Euclidean. However the set of all symplectic diffeomorphisms contains a Euclidean subgroup.

In order to compare physically significant intrinsic symmetries of equations arising from different Hamiltonians, it is necessary to allow dilatations and reflections, as well as symplectic diffeomorphisms, to alter positions and momenta. In the resulting space, which we have termed *Hamiltonian phase space*, intrinsic symmetries of Hamilton's equations become geometric symmetries.

In Chapter 7 this geometrization of Hamiltonian mechanics will be extended to equations involving n different q's and their conjugate p's, a 2n-dimensional *PQ space*. The transformation from Hamiltonian dynamics to geometry will thereafter be completed by treating time itself as a position-like variable. Energy (or rather, its negative) then becomes a momentum-like variable. The enlarged space is termed *PQET space*. Evolving trajectories in PQ space become stationary curves in PQET space. In this phase space all dynamics becomes geometry, and every intrinsic dynamical symmetry can be given geometrical interpretation.

Both the meaning and role of symmetry in Hamiltonian mechanics is evidently larger than physical experience might seem to suggest. This richness of meaning, and of implication, has its origin in a property of nature, unknown until the time of Newton and Liebniz: the laws governing the spatial and temporal development of natural processes can often be expressed by differential equations. It was over two centuries after the time of Newton and Leibniz when Lie first realized that differential equations possess, and define, symmetries. The path taken through the subject matter of this chapter is a path guided by Lie's conceptions.

The intrinsic symmetries of most differential equations governing physical systems are larger, and different, than would be expected from a knowledge of symmetries in ordinary two- and three-dimensional space, or four-dimensional spacetime. There are fundamental reasons for this:

(1) Differential equations may have more independent variables than arguments because, without initial conditions or boundary conditions, differential equations do not establish fixed functional relations. This has the consequence that the spaces in which dynamical symmetries find

geometrical expression can be spaces of dimension greater than that of ordinary space or spacetime.

(2) The intrinsic symmetries of differential equations are in part determined by purely local relations, relations between dx's, dt's, etc., that do not of themselves establish, or depend upon, fixed functional relations.

(3) As these local intrinsic symmetries often define metrics that are not Euclidean metrics, the symmetries often find geometric realization in non-Euclidean spaces.

The next three chapters lay the mathematical groundwork needed to understand and exploit Lie's discoveries in studies of the intrinsic symmetries of dynamical systems governed by differential equations.

Exercises:

1. Does the transformation $q \to p$, $p \to -q$, determine a symplectic diffeomorphism?
2. Find the phase–space trajectories determined by the Hamiltonian, $H = q\,p$, and plot representative examples of them.
3. After inspecting phase–space trajectories, show that no diffeomorphism in p q space can interconvert the Hamiltonian of the harmonic oscillator and: (a) that of the previous exercise; (b) the free-particle.
4. Let $r = (q^2 + p^2)^{1/2}$ be a position variable. Show that if f(q,p) is the corresponding momentum variable, it satisfies the differential equation
$$(1/r)(q\,\partial/\partial p - p\,\partial/\partial q)f = 1,$$
and find a function f(q,p) that satisfies this equation. What is its general solution?
5. Let (3.3.2c) and (3.3.2d) arise from a canonical diffeomorphism relating (q,p) and (q′,p′). Show that
$$\partial Q/\partial q\,\partial P/\partial p - \partial P/\partial q\,\partial Q/\partial p = 1$$
implies
$$\partial Q'/\partial q'\;\partial P'/\partial p' - \partial P'/\partial q'\;\partial Q'/\partial p' = 1.$$

References

[1] P. Ehrenfest, *Verh. Deutsch Physikal. Ges.* **15** (1913) 451; *Physik. Zeitschr.* **15** (1914) 657; *Phil. Mag.* **33** (1917) 500.
[2] W. Heisenberg, *Z. Physik* **43** (1927) 172; cf., W. Heisenberg, *Physical Principles of the Quantum Theory* (University Chicago Press, 1930).
[3] P. A. M. Dirac, *Proc. Roy. Soc. A* **109** (1926) 642.

CHAPTER 4

One-Parameter Transformation Groups

4.1 Introduction

Previous chapters have considered examples of continuous groups of transformations — groups such as those of translation, rotation, hyperbolic rotation, and dilatation, which act on a set of real variables x with transformations that depend upon a set of continuous real variables, termed group parameters. The parameters are not members of the set of variables x. This chapter deals with fundamental properties possessed by groups of transformations whose operations are labeled by one continuously variable real parameter. In this and the next six sections we will sometimes distinguish between initial values, final values, and *running values* lying between the initial and final values of variables. Primes will be used to denote running values and asterisks to denote final values. Variables with neither primes nor asterisks will denote initial values. We begin with examples that illustrate key points.

Figure 4.1.1 depicts an active interpretation of the dilatation transformation

$$\mathrm{x} \to \mathrm{x}' = \mathrm{T}(\alpha')\mathrm{x}, \quad \text{with} \quad \mathrm{T}(\alpha')\mathrm{x} = \mathrm{x}\exp(\alpha'). \tag{4.1.1a}$$

As the *running parameter* α' increases from 0, a point P, with initial coordinate x, takes on coordinates x$'$, and moves along curves to a final point at which $\mathrm{x}' \equiv \mathrm{x}^*$, $\alpha' \equiv \alpha$:

$$\mathrm{x}^* = \mathrm{T}(\alpha)\mathrm{x} = \mathrm{x}\exp(\alpha). \tag{4.1.1b}$$

The single-valuedness of the function $\mathrm{x}\exp(\alpha')$ has an important geometric consequence: it ensures that only one curve passes through each point.

In Chapter 2 we assumed that the transformations (4.1.1) are those of a continuous group. Let us prove that this is so. For $\mathrm{T}(\alpha)$ to define a

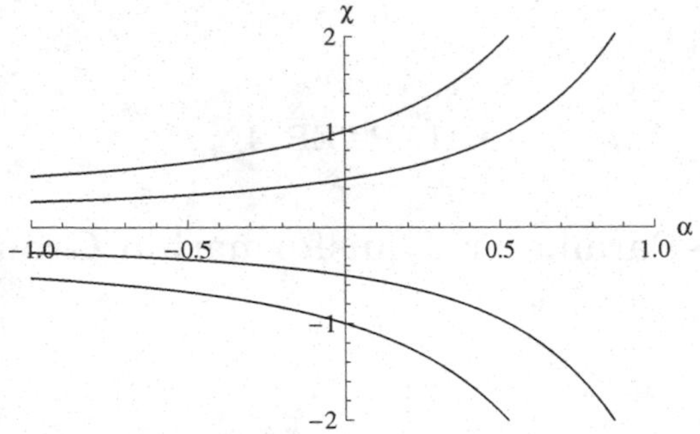

Fig. 4.1.1 Action of Dilatation Transformation $x' = xe^{\alpha'}$.

continuous group of transformations of x that depends on a parameter, α, it is first of all necessary that there be an identity transformation, I, for some value of the parameter, α_0, such that

$$T(\alpha_0)x = I\,x = x. \tag{4.1.2a}$$

In this case, $\alpha_0 = 0$. Next, let the transformation $T(\alpha)x$ yield x_α, and let it be followed by a transformation of x_α given by $T(\beta)$ i.e.

$$T(\beta)x_\alpha = T(\beta)T(\alpha)x. \tag{4.1.2b}$$

For the transformations to comprise a group it is necessary that

$$T(\beta)T(\alpha)x = T(\gamma)x, \quad \gamma = \phi(\beta, \alpha). \tag{4.1.2c}$$

Since (4.1.2a) and (4.1.2b) hold for a range of x, they state a more abstract property of the group which may be expressed as

$$T(\beta)T(\alpha) = T(\gamma), \quad \gamma = \phi(\beta, \alpha). \tag{4.1.2d}$$

The expression $\gamma = \phi(\beta, \alpha)$ is said to define the composition law of the group.

The requirement that every transformation of a group have an inverse is the requirement that for every value of α there must be a value, β such that

$$T(\alpha)T(\beta) = I; \quad \phi(\alpha, \beta) = \alpha_0. \tag{4.1.2e}$$

The associative law

$$T(\gamma)(T(\beta)T(\alpha)) = (T(\gamma)T(\beta))T(\alpha), \tag{4.1.2f}$$

requires that

$$\phi(\gamma, \phi(\beta, \alpha)) = \phi(\phi(\gamma, \beta), \alpha). \tag{4.1.2g}$$

For the transformation (4.1.1), with

$$\mathrm{x} \to \mathrm{T}(\alpha)\mathrm{x} = \mathrm{x}\exp(\alpha) = \mathrm{x}^*, \tag{4.1.3}$$

the identity transformation is obtained by setting $\alpha = 0$. For the composition of two transformations one has

$$\mathrm{T}(\gamma)\mathrm{x} = \mathrm{T}(\beta)(\mathrm{T}(\alpha)\mathrm{x}) = \mathrm{T}(\beta)\mathrm{x}\exp(\alpha) = \mathrm{x}\exp(\beta)\exp(\alpha) = \mathrm{x}\exp(\alpha+\beta). \tag{4.1.4}$$

The composition law is $\gamma = \alpha + \beta$. The identity transformation is produced if $\beta = -\alpha$; consequently the inverse of $\mathrm{T}(\alpha)$ is $\mathrm{T}(-\alpha)$. Note that this inverse transformation, from x^* to x in (4.1.3) may also be obtained by solving equation (4.1.3) for x.

The reader can directly verify that all of the relations (4.1.2) hold true for a continuous range of the parameters α, β, γ. Once the reader also verifies that the associative law, Eq. (4.1.2f) holds, it will become evident that Eq. (4.1.3) defines a continuous transformation group.

So far, in Eq. (4.1.2), we have not considered the range allowed for the parameters α, β, and γ. A continuous transformation group acting on real variables x is termed *global* if the group parameter may run through the full range of the reals, $-\infty < \alpha' < \infty$, without carrying real variables x outside of the range of the reals. In global groups, α' and α' mod some number, e.g. 2π, may give rise to the same transformation. The reader may verify that the transformation group defined in this example is a *global* one-parameter continuous group.

Figure 4.1.2 illustrates curves determined by the transformations $\mathrm{x}' = \mathrm{T}(\alpha')\mathrm{x}$ with

$$\mathrm{x}' = \mathrm{x}/(1 - \alpha'\mathrm{x}). \tag{4.1.5}$$

The identity transformation is obtained when $\alpha' \to 0$. A transformation with parameter α, followed by one with parameter β, gives

$$\mathrm{x}'' = \mathrm{T}(\beta)\mathrm{T}(\alpha')\mathrm{x} = \mathrm{T}(\beta)\mathrm{x}', \tag{4.1.6a}$$

so

$$\mathrm{x}'' = \mathrm{x}'/(1 - \beta\mathrm{x}'), \quad \text{with} \quad \mathrm{x}' = \mathrm{x}/(1 - \alpha\,\mathrm{x}). \tag{4.1.6b}$$

Thus

$$\mathrm{x}'' = \frac{\{\mathrm{x}/(1 - \alpha\mathrm{x})\}}{(1 - \beta\{\mathrm{x}/(1 - \alpha\mathrm{x})\})}. \tag{4.1.6c}$$

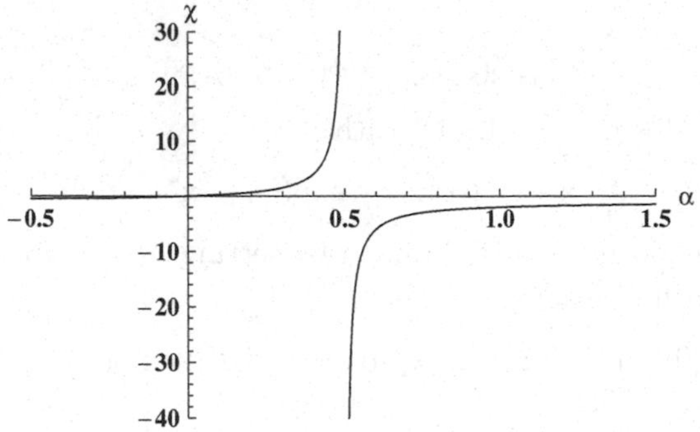

Fig. 4.1.2 Action of the projective transformation $x' = \frac{1}{1-\alpha' x}$.

On rationalizing the denominators one finds

$$x'' = x/(1-\gamma x) \quad \text{with} \quad \gamma = \alpha + \beta. \tag{4.1.6d}$$

It follows that the identity transformation is obtained if $\beta' = -\alpha'$. The reader may verify that all the group requirements are met if x and the group parameter, say α', do not take on values such that $1 - \alpha' x = 0$.

A *local* continuous transformation group of real variables is a transformation group whose group parameter may have to be restricted in range to ensure that it transforms real variables into real variables. The transformation $x \to x' = x/(1-\alpha' x)$ defines a local transformation group, but not a global one.

These concepts of real *local* and *global* groups may be extended by allowing the variables x and/or the parameters α to become imaginary or complex.

4.2 Finite Transformations of a Continuous Group Define Infinitesimal Transformations and Vector Fields

Let the Greek letter α denote a real group parameter that yields the identity transformation when it equals 0. It may take on running values α' between 0 and α, in a range lying between $-\infty$ and $+\infty$. Let x stand for a set of real variables x_j that may also take on values x'_j between $-\infty$ and $+\infty$. Consider all variables to be running ones, and write

$$x''_j = F_j(x', \alpha'). \tag{4.2.1a}$$

The requirement that the $F_j(x', \alpha')$ are continuous functions of their arguments implies that as $\alpha' \to 0$,

$$x_j'' = F_j(x', \alpha') \to x_j' = F_j(x', 0). \tag{4.2.1b}$$

Let us also require that the derivatives

$$dx_j''/d\alpha' = dF_j(x', \alpha')/d\alpha', \quad j = 1, 2, \ldots, n, \tag{4.2.2a}$$

are continuous real functions of x', α'. Define

$$f_j(x') = dF_j(x', \alpha')/d\alpha')|_{\alpha'=0}, \tag{4.2.2b}$$

then, as $\alpha' \to 0$, Eqs. (4.2.2a) become

$$dx_j'/d\alpha' = f_j(x'), \quad j = 1, 2, \ldots, n. \tag{4.2.2c}$$

Given these requirements, the transformation group determines a set of first-order ordinary differential equations in which $-\infty < f_j(x') < +\infty$.

One may suppose that (4.2.2c) relates *infinitesimal* changes, $\delta\alpha'$ in the parameter α' and *infinitesimal* changes $\delta x'$ in the x_j', via the n equations

$$\delta x_j' = f_j(x')\delta\alpha', \quad j = 1, 2, \ldots, n. \tag{4.2.3a}$$

From a physical standpoint one may consider that, in some system of measurement, these equations extrapolate an observed relationship

$$\Delta x_j' = f_j(x')\Delta\alpha' \tag{4.2.3b}$$

that continues to hold within experimental error as the observer makes increasingly refined observations. As an example of this, consider the group of mathematically defined functional relations

$$x^* = x\cos(\alpha) - y\sin(\alpha), \quad y^* = y\cos(\alpha) + x\sin(\alpha). \tag{4.2.4}$$

For running values of the variables one has

$$(dx''/d\alpha')|_{\alpha'=0} = (-x'\sin(\alpha') - y'\cos(\alpha'))|_{\alpha'=0} = -y', \tag{4.2.5}$$

and

$$(dy''/d\alpha')|_{\alpha'=0} = (-y'\sin(\alpha') + x'\cos(\alpha'))|_{\alpha'=0} = x'.$$

Also, when $\alpha' \to 0$, the variables $x'' \to x'$, and $y'' \to y'$, so one may write

$$dx'/d\alpha' = -y', \quad dy'/d\alpha' = x'. \tag{4.2.6}$$

An experimentalist studying rotations in the plane might make measurements of increasingly minute values of $\Delta x'$, $\Delta y'$, and a change in angle, $\Delta\alpha'$.

After taking account of experimental errors in the observations, the experimentalist might extrapolate the results to

$$\delta x'/\delta\alpha' = -y',\ \delta y'/\delta\alpha' = x', \quad \text{or} \quad \delta x' = -y\,\delta\alpha',\ \delta y' = x'\delta\alpha', \qquad (4.2.7)$$

and imagine $\delta x'$, $\delta y'$, $\delta\alpha'$ to represent arbitrarily small, but nonzero, *infinitesimal* quantities. In obtaining these relations, the experimentalist, in contrast to the mathematician, has made no assumptions about functional relations and mathematical limits.

However, after obtaining relations (4.2.7) for many repeated experiments, the experimentalist would conclude that functional relations exist between x', y', α'. For a point $P_{x',y'}$ that is rotated about the origin, either set of equations, (4.2.6) or (4.2.7) can define *infinitesimal transformations.* Keisler has established the conditions under which infinitesimal quantities such as δx are well defined, without necessarily being obtained by the limiting process mathematicians use to define differentiation of a function.[1] Physicists often switch between the two views. We shall do so as well.

The geometric consequences of (4.2.6) and (4.2.7) are suggested by **Figs. 4.2.1a,b**. In them, the infinitesimal increments $\delta x'$, $\delta y'$, $\delta\alpha'$ are replaced by small finite increments $\Delta x'$, $\Delta y'$, $\Delta\alpha'$. The *vector fields* illustrated in **Figs. 4.2.1a,b**, may be considered to be composed of vectors tangent to circles and spirals such as those depicted in **Figs. 4.2.1c,d**. The figures could just as well represent electric or magnetic force fields determined by solving Maxwell's equations, or determined by measurements such as those Maxwell describes in his monograph *Electricity and Magnetism.*[2]

Taking a mathematical approach, one finds by differentiating (4.1.3) that the infinitesimal transformation associated with the dilatation transformation can be expressed either as

$$dx'/d\alpha' = x', \quad \text{or as } \delta x' = x'\delta\alpha'. \qquad (4.2.8)$$

In the same manner, one finds that the infinitesimal transformation associated with the finite transformation (4.1.5) is

$$\delta x' = (x')^2\,\delta\alpha'. \qquad (4.2.9)$$

A further geometric example is provided by the transformations depicted by the paths and vector fields of **Figs. 4.2.2**. The infinitesimal transformations that define the vector fields are:

$$\delta x' = \delta\alpha', \quad \delta y' = (y'^2 - x')\delta\alpha'. \qquad (4.2.10)$$

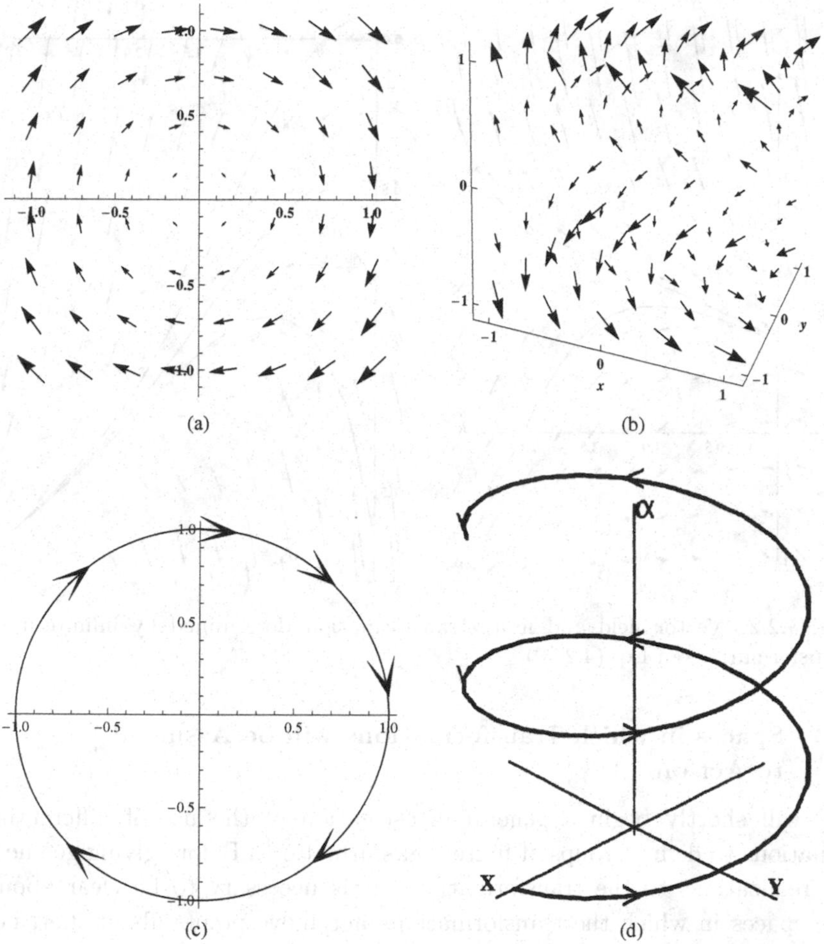

Fig. 4.2.1 (a) Vector field $\Delta \mathrm{r} = [\Delta \mathrm{x},\ \Delta \mathrm{y}]$ defined by infinitesimal rotations; (b) vector field $[\Delta \mathrm{x},\ \Delta \mathrm{y},\ \Delta \alpha]$ defined by infinitesimal rotations; (c) path curves of successive finite rotations in the x–y plane; (d) path curve in the space of $(\mathrm{x},\ \mathrm{y},\ \alpha)$.

The finite transformations that move points along the indicated lines can be approximated by *connecting the arrows.* For the figure, this has been done by numerically integrating Eqs. (4.2.10), beginning at an initial point on each line. In this last example we have not produced a general equation that defines the finite transformations. The reason for this will become apparent in Sec. 4.8 below.

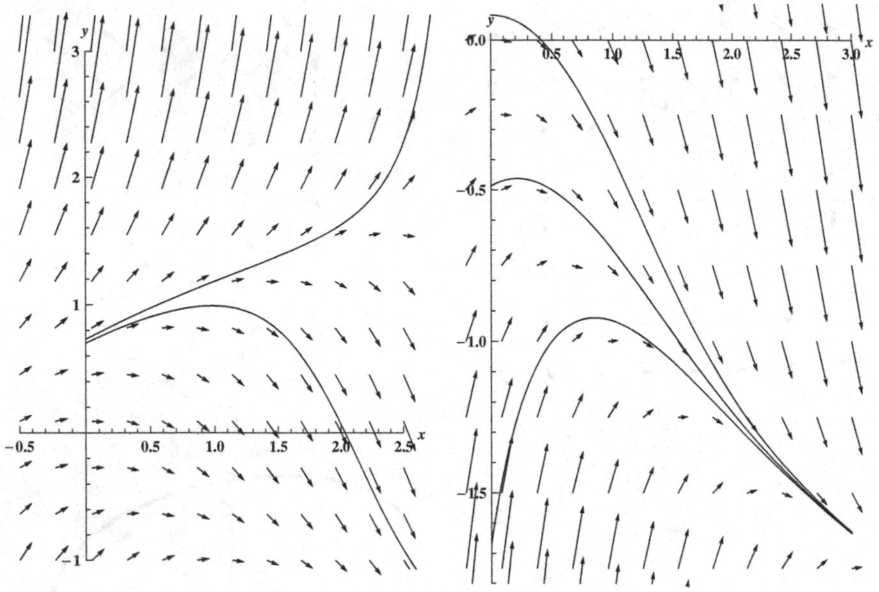

Fig. 4.2.2 Vector field and finite transformation determined by infinitesimal transformations of Eq. (4.2.10).

4.3 Spaces in which Transformations will be Assumed to Act on

We will shortly begin a general investigation of the use of differential equations to define groups of finite transformations. Before giving geometric realizations to the transformations, it is necessary to be clear about the spaces in which the transformations act/move points about. Let, i.e. $\mathrm{x} = (\mathrm{x}_1, \mathrm{x}_2, \ldots, \mathrm{x}_\mathrm{n})$ represent a collection of n real variables x_j, and let a be a real parameter. Consider transformations that convert $\mathrm{x} \to \mathrm{x}^* = (\mathrm{x}_1^*, \mathrm{x}_2^*, \ldots, \mathrm{x}_\mathrm{n}^*) = \mathrm{X}(\mathrm{x}, \alpha)$, i.e., for $\mathrm{j} = 1, 2, \ldots, \mathrm{n}$, one has $\mathrm{x}_\mathrm{j}^* = \mathrm{X}_\mathrm{j}(\mathrm{x}, \alpha)$.

As before, we will write $\mathrm{x}^* = \mathrm{T}(\alpha)\mathrm{x}$ when we are concerned only with the relation between initial and final values of x and the parameter, and use the notation $\mathrm{x}' = \mathrm{T}(\alpha')\mathrm{x}$ when we wish to emphasize that x' and α' can vary continuously. The parameters α, α' will always be allowed to vary over all the reals.

We will deal with transformations that may be nonlinear. As the transformation (4.1.5) demonstrates, such transformations have properties which make it necessary to proceed with mathematical caution. Unless otherwise noted, throughout this book we will be dealing with the set $\mathbb{R}^\mathrm{N}$ of N continuous real variables z_i that may be independently varied over the full open range of the reals, i.e. $-\infty < z_i < \infty$. Variations of z_i that are arbitrarily small but nonzero, that is, *infinitesimal*, will hereafter be denoted by dz_i *or* δz_i. *Except when otherwise noted, the variables* z_i *will represent Cartesian coordinates of points* P_Z *in a Euclidean space with metric*

$$\delta s^2 = \delta z_1^2 + \delta z_2^2 + \cdots + \delta z_n^2. \tag{4.3.1}$$

This space is termed E^N. Real variables that do not necessarily have all the properties of the Cartesian variables z_i, will be generically denoted as x_i's. Often the x's will be geometrically interpreted as defining points on a manifold in a Euclidean space. Thus the angular coordinates (θ, ϕ), of a point on the unit sphere defined by $z_1^2 + z_2^2 + z_3^2 = 1$, are x's, as are the z's, but θ, ϕ, are not z's. The time coordinate in spacetime will be considered an 'x', except when otherwise stated.

We follow current mathematical usage and require that, by definition, a *function* is single-valued. Thus, for example, the relation $z_3 = \pm(z_1^2 - z_2^2)^{1/2}$ does not define a function; it defines *two* functions which have misleadingly been denoted by the same symbol, z_3.

As previously noted we will be dealing with transformations which, like stereographic projections and canonical transformations, are more general than the linear transformations of vector algebra. Letting x represent a set of variables, $x_1, \ldots, x_n$, their transformation from x to x′ will be defined by sets of functions:

$$x_1' = F_1(x), \quad x_2' = F_2(x), \ \ldots\ , x_n' = F_n(x). \tag{4.3.2a}$$

The inverse transformation will be denoted by

$$x_1 = F_1^{inv}(x'), \quad x_2 = F_2^{inv}(x'), \ \ldots, x_n = F_n^{inv}(x'). \tag{4.3.2b}$$

The transformation $x \to x' = x/(1 - \alpha x)$ has as inverse the transformation $x' \to x = x'/(1 + \alpha' x')$. These equations define one diffeomorphism when $1 - \alpha x < 0$, or $1 - \alpha' x' < 0$, and another when $1 - \alpha x > 0$, or $1 - \alpha' x' > 0$. They do not define a diffeomorphism when $1 \pm \alpha x$ or $1 \pm \alpha' x = 0$.

Lines, surfaces, and hypersurfaces in E^N may be defined by sets of diffeomorphisms stating functional relations amongst Cartesian coordinates z_i. The number of independent coordinates required to uniquely fix a point on a manifold is the dimension of the manifold. A manifold is a *smooth manifold* if the functions defining it possess partial derivatives of all orders at it. The two-dimensional surface of a sphere of fixed finite radius is, for example, related to the Cartesian coordinates in Euclidean 3-space by fixing the value of r in the equations

$$z_1 = r\sin(\theta)\cos(\phi), \quad z_2 = r\sin(\theta)\sin(\phi), \quad z_3 = r\cos(\theta). \tag{4.3.3}$$

Derivatives of the z's with respect to θ and ϕ are well defined functions however high their order. The surface of the sphere is thus a smooth manifold. Theorems due to Whitney ensure that any smooth n-dimensional manifold may be embedded in a Euclidean space of no more than 2n dimensions.[3] In this case, a two-dimensional manifold has been embedded in three-dimensional space. Whitney's embedding theorems ensure that one's natural tendency to use Cartesian coordinates in Euclidean spaces need not necessarily lead one astray.

The equations defining a transformation may, or may not, establish a one-to-one correspondence between points P_z and $P_{z'}$ in E^N, or between points in E^N and points on a manifold of the same or lesser dimension. For example, the stereographic projections illustrated in Chapters 2 and 3, though differentiable, establish a one-to-one transformation between all points in the *infinite* plane and points on the surface of a sphere, *except* the point at the south pole. Objects that can be interconverted by a diffeomorphism are said to be diffeomorphic. In the Euclidean plane, a circle and an ellipse are diffeomorphic; a circle and a figure eight are not. Some further concepts related to the concept of diffeomorphism are discussed in the Appendix.

4.4 The Defining Equations of One-Parameter Groups of Infinitesimal Transformations. Group Generators

Now, let us return to the problem of determining the relation between infinitesimal and finite transformations. We first note that the set of equations

$$\delta x'_j = \delta\alpha' \xi_j(x'), \quad j = 1, \ldots, n, \tag{4.4.1a}$$

or, equivalently, the set of equations

$$dx'_j/d\alpha' = \xi_j(x'), \quad j = 1, \ldots, n, \tag{4.4.1b}$$

define a one-parameter *group of infinitesimal transformations* $T(\delta\alpha)$. This follows from the following observations. One has:

1. $T(\delta\alpha)T(\delta\beta) = T(\delta\gamma),\ \delta\gamma = \delta\alpha + \delta\beta$; (a law of composition)
2. $T(0) = I$; (an identity operation)

3. $T(-\delta\alpha)T(\delta\alpha) = I;$ (an inverse for every transformation)
4. $T(\delta\alpha)(T(\delta\beta)T(\delta\gamma)) = (T(\delta\alpha)T(\delta\beta))T(\delta\gamma).$ (associativity)

(4.4.2a–d)

Associated with the infinitesimal transformation defined by Eqs. (4.4.1) is the *group generator*, the operator

$$U = \Sigma_1^n \xi_j(x')\partial/\partial x'_j. \tag{4.4.3a}$$

The generator, U, of rotations, determined from (4.2.6), is $x'\partial/\partial y' - y'\partial/\partial x'$. As each x' signifies a variable which may take on all values from x through x^*. For U to act on x means that $Ux \equiv Ux'|_{x'=x}$. This is equivalent to letting

$$U \to \Sigma_i^n \xi_j(x)\partial/\partial x_j. \tag{4.4.3b}$$

One has

$$(1 + \delta\alpha\ U)x_i = x_i + \delta\alpha\ \xi_i(x) = x'_i; \quad x'_i = x_i + \delta x_i. \tag{4.4.4}$$

The operator $(1 + \delta\alpha\ U)$ carries out infinitesimal transformations: it is an expression for $T(\delta\alpha)$.

Taken in the active sense, $(1 + \delta\alpha U)$ moves a point with coordinates x to the point with coordinates $x' = x + \delta x = x + \delta\alpha Ux$. A point so moved is termed an ordinary point. However, if all the functions $\xi_i(x)$ vanish at a point, P_x, the point is not moved by the transformation, and is termed an *invariant point* or *critical point.* The point at the origin $x = 0$ is an invariant point of the transformation determining **Fig. 4.4.1**, and of two of the transformations used as examples in Sec. 4.3.

If one supposes the variables x_i to be Cartesian variables z_i in a Euclidean space, then

$$Uz_i = \xi_i(\mathbf{z}), \tag{4.4.5a}$$

and one may write

$$U = \boldsymbol{\xi} \cdot \boldsymbol{\nabla}, \quad \boldsymbol{\nabla} = (\partial/\partial z_1, \partial/\partial z_2, \ldots, \partial/\partial z_n). \tag{4.4.5b}$$

At each point P_z the vector $\boldsymbol{\xi}$ points in the direction that the infinitesimal transformation moves P_z. The vector fields of $\boldsymbol{\Delta}$r's in previous figures are depictions of vectors, $\boldsymbol{\xi}$. The path curves in **Fig. 4.4.1** represent transformations in Euclidean space generated when $\boldsymbol{\xi} = (y'z', -x'z', x'y')$. The tangent vectors to the path curves at the points $P_{x',y',z'}$ are $\Delta\alpha$ times the $\boldsymbol{\xi}$ at the respective points.

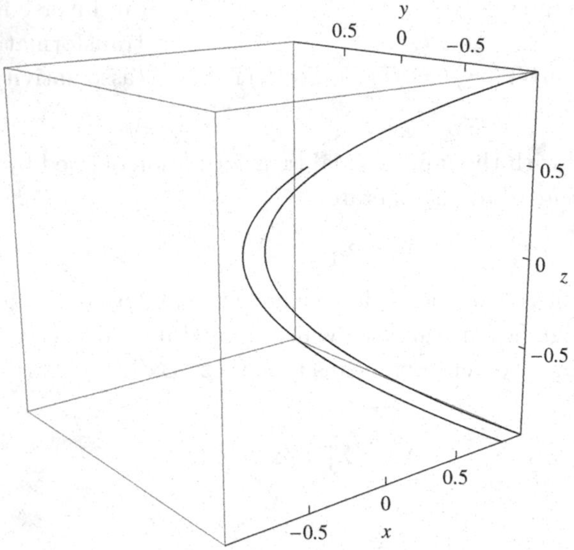

Fig. 4.4.1 Two finite transformations of a one-parameter group with generators $V = \mathbf{X}\cdot\mathbf{\Delta}$, $X = (zy,\ -zx,\ xy)$.

Because the generators U are first-order differential operators they possess the property of derivatives

$$\mathrm{Uf(x)g(x) = g(x)Uf(x) + f(x)Ug(x).} \tag{4.4.6}$$

Making use of (4.4.4), the composition of two successive transformations of x with parameters $\delta\alpha_1$ and $\delta\alpha_2$ may be expressed as

$$(1 + \delta\alpha_2 \mathrm{U})(1 + \delta\alpha_1 \mathrm{U})\mathrm{x} = (1 + (\delta\alpha_2 + \delta\alpha_1)\mathrm{U})\mathrm{x} + \delta\alpha_2\delta\alpha_1 \mathrm{UUx}. \tag{4.4.7a}$$

Thus, if UUx is finite,

$$(1 + \delta\alpha_2 \mathrm{U})(1 + \delta\alpha_1 \mathrm{U})\mathrm{x} = (1 + (\delta\alpha_2 + \delta\alpha_1)\mathrm{U})\mathrm{x} + \mathrm{O}(\delta^2). \tag{4.4.7b}$$

This equation suggests a number of relations that will be investigated in the remainder of the chapter.

4.5 The Differential Equations that Define Infinitesimal Transformations Define Finite Transformation Groups

In this section we establish the conditions required for the set of equations

$$\delta\alpha' = \delta \mathrm{y}'_1/\xi_1(\mathrm{y}') = \delta \mathrm{y}'_2/\xi_2(\mathrm{y}') \cdots = \delta \mathrm{y}'_\mathrm{n}/\xi_\mathrm{n}(\mathrm{y}'), \tag{4.5.1}$$

which define infinitesimal transformations of the y'_j, to also define a group of finite transformations. We begin by observing that for each value of j from 1 to n, Eqs. (4.5.1) imply that

$$\int_a^b \delta\alpha' = \Delta_{ba} = b - a = \int_{y_a}^{y_b} \delta y'_j/\xi_j(y'). \tag{4.5.2a}$$

Similarly,

$$\int_b^c \delta\alpha' = \Delta_{cb} = c - b = \int_{y_b}^{y_c} \delta y'_j/\xi_j(y'). \tag{4.5.2b}$$

Path curves developed by successive integrations in a three-dimensional space are illustrated in **Figs. 4.5.1a,b**.

Because the process of integration possesses the composition property

$$\int_b^c dg + \int_a^b dg = \int_a^c dg, \tag{4.5.3}$$

it follows that

$$\int_a^b \delta\alpha' + \int_b^c \delta\alpha' = \int_a^c \delta\alpha', \tag{4.5.4a}$$

just as

$$(b - a) + (c - b) = c - a. \tag{4.5.4b}$$

Using this observation one easily establishes that integration can be considered an operation that defines a group in which the law of composition is addition.

Because each of the terms $\delta y'_j/\xi_j(y')$ in (4.5.1) satisfy the relation $\delta y'_j/\xi_j(y') = \delta\alpha'$, one also has

$$\int_{y_a}^{y_c} \delta y'_j/\xi_j(y') = \int_{y_a}^{y_b} \delta y'_j/\xi_j(y') + \int_{y_b}^{y_c} \delta y'_j/\xi_j(y'). \tag{4.5.4c}$$

From the geometric standpoint illustrated in **Fig. 4.5.1c**, the transformation in which α' ranges from a to b carries the coordinates of the point P_y from $y = y_a$ to $y = y_b$. This transformation may be expressed as $y_b = T(\Delta_{ba})y_a$. The transformation in which α' ranges from b to c carries y from y_b to y_c and may be expressed as $y_c = T(\Delta_{cb})y_b$. The transformation in which α' ranges from a to c carries y from y_a to y_c and may be

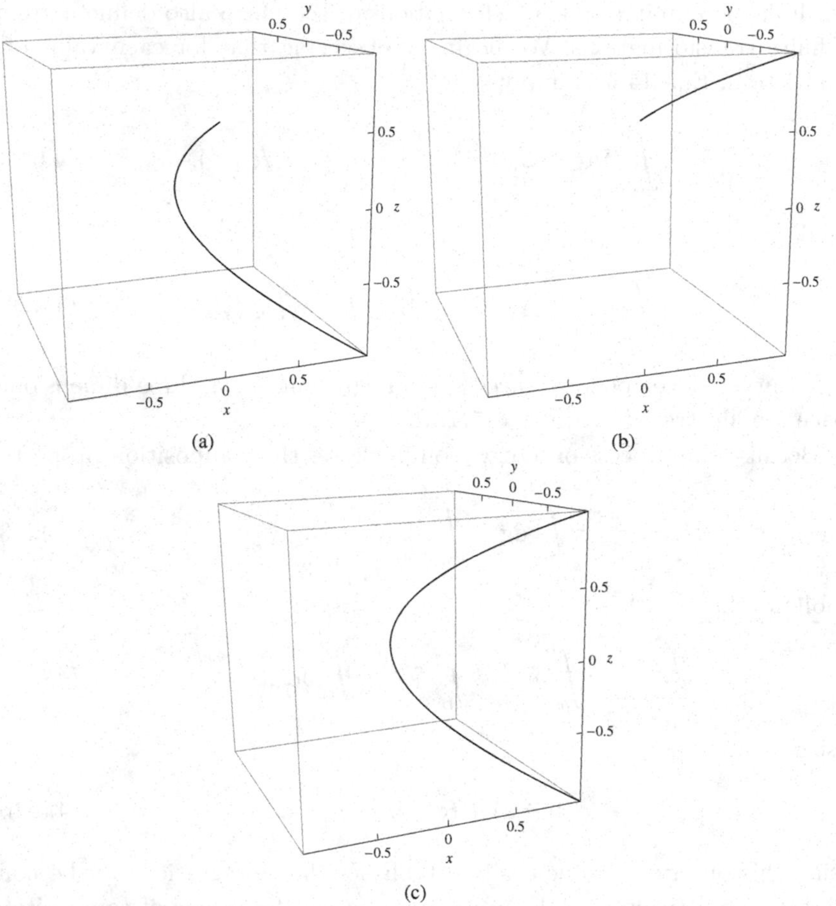

Fig. 4.5.1 Addition of the finite transformation of a one-parameter group.

expressed as $y_c = T(\Delta_{ca})y_a$. In terms of the notation of Sec. 4.1, we have, from (4.5.4):

$$T(\Delta_{cb})T(\Delta_{ba})y = T(\Delta_{ca})y, \quad \Delta_{ca} = \Delta_{cb} + \Delta_{ba}. \tag{4.5.5}$$

From (4.5.4) that it also follows that

$$T(0) = I, \tag{4.5.6}$$

the identity transformation. One easily also finds from (4.5.4) that

$$T(\Delta)T(-\Delta) = I, \tag{4.5.7}$$

and that

$$T(\Delta_3)(T(\Delta_2)T(\Delta_1)) = (T(\Delta_3)T(\Delta_2))T(\Delta_1). \tag{4.5.8}$$

Consequently, if the differential equations (4.5.1) that define a group of infinitesimal transformations are, in principle, integrable, they also define a group of finite transformations.

The transformations above may be put in standard form by replacing the parameter changes Δ by a parameter, signified by a lower case Greek letter, that yields the identity transformation when it is set equal to zero. The composition law (4.5.5) then becomes, e.g.,

$$\mathrm{T}(\alpha)\mathrm{T}(\beta)\mathrm{y} = \mathrm{T}(\gamma)\mathrm{y}, \quad \gamma = \beta + \alpha. \tag{4.5.9}$$

An example of this replacement is provided by the transformation $\mathrm{x}^* = \mathrm{ax}$. Replacing a by $\exp(\alpha)$, one obtains the transformation $\mathrm{x}^* = \exp(\alpha)\mathrm{x}$, which becomes the identity transformation when α approaches 0.

These results lead to the following observations:

1. *The group property of the operation of integration imposes finite group structure on sets of first-order ordinary differential equations if the equations possess solutions that may be expressed as functions of the "independent variable," which becomes the group parameter.*

The solution curves of the set of differential equations (4.5.1) are path curves of a transformation group $\mathrm{T}(\alpha)$. The operations of the group, integrations, *evolve* solutions from initial data. Acting at any point on a given solution curve, they carry the point along the solution curve. They may be said to convert the solution curve into itself. Consequently:

2. *The differential equations define a group structure, and an intrinsic symmetry, which they themselves possess.*

This intrinsic symmetry becomes a geometric symmetry when, as in the previous figures, the variables are given a geometric interpretation. The intrinsic symmetry, and the geometric symmetry, becomes a dynamical symmetry when the variables represent dynamical variables.

4.6 The Operator of Finite Transformations

The function e^{w} has the well known power series expansion

$$\mathrm{e}^{\mathrm{w}} = \Sigma_0^{\infty} \mathrm{w}^{\mathrm{n}}/\mathrm{n}!. \tag{4.6.1}$$

If one formally replaces w by $\alpha\mathrm{U}$ in it one obtains

$$\exp(\alpha\mathrm{U}) = \Sigma_0^{\infty} (\alpha\mathrm{U})^{\mathrm{n}}/\mathrm{n}!. \tag{4.6.2}$$

It is also true that

$$\exp(\mathrm{w}) = \lim_{\mathrm{n}\to\infty} (1 + \mathrm{w}/\mathrm{n})^{\mathrm{n}}, \tag{4.6.3a}$$

and one expects that

$$\exp(\alpha \mathrm{U}) = \lim_{\mathrm{n}\to\infty} (1 + \alpha\mathrm{U}/\mathrm{n})^{\mathrm{n}}. \tag{4.6.3b}$$

If $\mathrm{U} = \Sigma_{\mathrm{j}}^{\mathrm{n}}\ \xi_{\mathrm{j}}(\mathrm{x}')\partial/\partial \mathrm{x}'_{\mathrm{j}}$, then when $\exp(\alpha\mathrm{U})$ acts on functions of the x'_{j}, (4.6.3b) represents its action as the limit of a *continuous succession of infinitesimal transformations* each with operator $(1 + \delta\alpha\mathrm{U})$, in which $\delta\alpha = \alpha/\mathrm{n}$. With this observation as a motivation, let us explore the idea that the finite transformation

$$\mathrm{x} \to \mathrm{x}^* = \mathrm{T}(\alpha)\mathrm{x}, \quad \mathrm{x} = (\mathrm{x}_1, \ldots, \mathrm{x}_{\mathrm{n}}), \tag{4.6.4}$$

defined by

$$\mathrm{x}_{\mathrm{j}} \to \mathrm{x}_{\mathrm{j}}^* = \mathrm{X}_{\mathrm{j}}(\mathrm{x}, \alpha), \quad \mathrm{j} = 1, \ldots, \mathrm{n},$$

can be expressed as

$$\mathrm{x}_{\mathrm{j}} \to \mathrm{x}_{\mathrm{j}}^* = \exp(\alpha\mathrm{U})\mathrm{x}_{\mathrm{j}}, \tag{4.6.5a}$$

with

$$\mathrm{X}_{\mathrm{j}}(\mathrm{x}, \alpha) = \exp(\alpha\mathrm{U})\mathrm{x}_{\mathrm{j}}. \tag{4.6.5b}$$

Because the transformation depends continuously upon x and α, it is necessary that Eq. (4.6.5) hold true for all x' ranging from x to x^*. Because of the differentiabilty requirement on (4.6.4), one must have

$$\mathrm{dx}'_{\mathrm{j}}/\mathrm{d}\alpha = \mathrm{dX}_{\mathrm{j}}(\mathrm{x}', \alpha)/\mathrm{d}\alpha \tag{4.6.6a}$$

so one must also have

$$\begin{aligned} \mathrm{dx}'_{\mathrm{j}}/\mathrm{d}\alpha &= \mathrm{d}(\exp(\alpha\mathrm{U})\mathrm{x}'_{\mathrm{j}})/\mathrm{d}\alpha, \\ &= \exp(\alpha\mathrm{U})\mathrm{U}\mathrm{x}'_{\mathrm{j}}. \end{aligned} \tag{4.6.6b}$$

The group property requires that these equations must hold in the limit $\alpha \to 0$. Now

$$(\mathrm{dX}_{\mathrm{j}}(\mathrm{x}', \alpha)/\mathrm{d}\alpha)_{|\alpha=0} = \xi_{\mathrm{j}}(\mathrm{x}'), \tag{4.6.7a}$$

and

$$(\exp(\alpha\mathrm{U})\mathrm{U}\mathrm{x}'_{\mathrm{j}})_{|\alpha=0} = \mathrm{U}\mathrm{x}'_{\mathrm{j}}. \tag{4.6.7b}$$

Thus if

$$\mathrm{U} = \Sigma_1^{\mathrm{n}}\ \xi_{\mathrm{j}}(\mathrm{x}')\partial/\partial\mathrm{x}'_{\mathrm{j}}, \tag{4.6.7c}$$

then $\exp(\alpha U)x_j$ satisfies the differential equations satisfied by the $X_j(x, \alpha)$. Since both $\exp(\alpha U)x_j$ and $X_j(x, \alpha)$ yield the identity transformation when $\alpha \to 0$, it follows that, as conjectured,

$$X_j(x, \alpha) = \exp(\alpha U)x_j. \tag{4.6.8}$$

Thus, a transformation with generator U can be expressed as

$$T(\alpha)x = \exp(\alpha U)x. \tag{4.6.9}$$

Putting these last two equations together, and recalling that each $X_j(x, \alpha)$ is defined by integrating Eqs. (4.4.1), one concludes that in (4.6.9) *the operation of integration has been converted into one of exponentiation.*

It is important to note, however, that for the series and product expansions of $\exp(\alpha U)x$ to be valid, the functions $\xi_j(x')$ in (4.6.7c) must be infinitely differentiable, and the expansion must converge.

When the group parameter α is so related to physical time, t, that $d\alpha = dt$, the group becomes the evolution group of the dynamical system governed by the differential equations that define the group. As an example of this, consider Hamiltonian's differential equations (3.3.3), *viz.*,

$$dq_j/dt = \partial H(q, p)/\partial p_j, \quad dp_j/dt = -\partial H(q, p)/\partial q_j. \tag{4.6.10a}$$

They can be rewritten in the form

$$dt = dq_j/(\partial H(q, p)/\partial p_j) = dp_j/(-\partial H(q, p)/\partial q_j), \tag{4.6.10b}$$

and define the group generator

$$U = -(\partial H/\partial q_i\ \partial/\partial p_i - \partial H/\partial p_i\ \partial/\partial q_i). \tag{4.6.11}$$

This generator is often written as $\{-H\cdot\}$, where, for any function $A(q, p)$, the operator

$$\{A\cdot\} = \Sigma_i(\partial A/\partial q_i\ \partial/\partial p_i - \partial A/\partial p_i\ \partial/\partial q_i), \tag{4.6.12}$$

is a "*Poisson Bracket waiting to happen*". The corresponding evolution operator $\exp(t\{-H\cdot\})$ acts on q, p to give $q^* = \mathbf{Q}(q, p, t), p^* = \mathbf{P}(q, p, t)$, the values of the position and momentum vectors at t, given their initial values q, p at $t = 0$. If $F(q, p)$ is an analytic function of q, p, then its value evolves to $F(\mathbf{Q}(q, p, t), \mathbf{P}(q, p, t)) = \exp(t\{-H\cdot\})F(q, p)$.

As the motions of a system of particles express the action of a one-parameter group upon the positions and momenta of the particles, they express a dynamical symmetry of the system. In subsequent chapters, operators $\exp(\alpha\{A\cdot\})$ that convert solutions of the equations of motion into different solutions will be shown to express dynamical symmetries of the system in addition to the symmetry expressed by $\exp(t\{-H\cdot\})$.

When the functions $\xi_j(x')$ are sufficiently differentiable, approximating $\exp(\alpha U)$ by a finite product $(1 + \alpha U/n)^n$ can provide a convenient way of programming computers to approximately integrate sets of autonomous first-order ordinary differential equations. The expansion $\exp(\alpha U) = \Sigma_0(\alpha U)^n/n!$, which is also used to develop approximate series expansions of the solutions of differential equations, is sometimes termed a *Lie series* expansion. Consider for example the equations

$$dx'/dt' = y', \quad dy'/dt' = x'. \tag{4.6.13a}$$

Here

$$U = y'\partial/\partial x' + x'/\partial y'. \tag{4.6.13b}$$

One has

$$Ux' = y', Uy' = x'; \; U^2x' = x', \; U^2y' = y', \ldots. \tag{4.6.14}$$

If x, y are the values of x', y' at $t' = 0$, the Lie series expression for the final value of x' at $t' = t$, is

$$\begin{aligned} x^* = \exp(t'U)x' &= x'(1 + t'^2/2! + t'^4/4! + \cdots) + y'(t' + t'^3/3! + \cdots) \\ &\to x\cosh(t') + y\sinh(t'). \end{aligned} \tag{4.6.15a}$$

The corresponding Lie series expansion of $\exp(t'U)y'$ produces

$$y^* = y\cosh(t') + x\sinh(t'). \tag{4.6.15b}$$

In most cases the series expansion one obtains is not that of easily recognized functions, and one simply retains a finite number of terms in the expansion and so obtains approximate solutions of differential equations.

4.7 Changing Variables in Group Generators

It is often useful to change variables in the generator U. Let the diffeomorphism

$$x'_j = f_j(x), \quad x_j = f_j^{in}(x'), \tag{4.7.1}$$

determine a change of coordinates $x \leftrightarrow x'$. Using the chain rule one sees that U remains a first-order differential operator, and one may write

$$U = \Sigma_j \; \xi_j(x)\partial/\partial x_j \to \Sigma'_k \; \xi'_k(x')\partial/\partial x'_k. \tag{4.7.2}$$

To determine the form U takes in the x′ coordinate system one need only determine the functions $\xi'_k(x')$. In the x′ system one has

$$\xi'_k(x') = Ux'_k, \tag{4.7.3}$$

with $x'_j = f_j(x)$. Hence

$$Ux'_k = U\, f_k(x) = \Sigma_j\ \xi_j(x)\partial f_k(x)/\partial x_j. \tag{4.7.4}$$

Denoting $\partial f_k(x)/\partial x_j$ by $f_{k,j}$ (x), one uses the inverse transformation $x = f^{inv}(x')$, to convert the right-hand side of (4.7.4) to a function of x′, obtaining

$$\xi'_k(x') = Ux'_k = \Sigma_j\ \xi_j(f^{inv}(x'))f_{k,j}(f^{inv}(x')). \tag{4.7.5}$$

As an example, let us express the generator of rotations in the z_1–z_2 plane, *viz.*,

$$U = z_1\ \partial/\partial z_2 - z_2\ \partial/\partial z_1 \tag{4.7.6}$$

in the polar coordinates defined by

$$x_1 = r = (z_1^2 + z_2^2)^{1/2}, \quad x_2 = \theta = \arctan(z_2/z_1). \tag{4.7.7a}$$

The inverse transformation functions, f^{inv}, are given by

$$z_1 = r\cos(\theta), \quad z_2 = r\sin(\theta). \tag{4.7.7b}$$

In the left-hand side of (4.7.2) one has $\xi_1 = -z_2, \xi_2 = z_1$, so the fundamental Eqs. (4.4.1) take on the form

$$dz'_j/d\alpha' = \xi(z')_j, \tag{4.7.8}$$

and require

$$dz'_1/d\alpha' = -z'_2, \quad dz'_2/d\alpha' = z'_1. \tag{4.7.9}$$

Applying U to r and θ, one finds

$$\xi'_r = Ur = 0, \quad \xi'_\theta = U\theta = 1. \tag{4.7.10}$$

Consequently, in the polar coordinate system, $U = \partial/\partial\theta$, and Eqs. (4.7.9) are replaced by

$$d\theta/d\alpha = 1, \quad dr/d\alpha = 0. \tag{4.7.11}$$

In this example, because the ξ' turned out to be constants rather than functions of the original variables, it is not necessary to use the inverse transformation to express them as functions of the new variables. The exercises contain examples that require the reader to use the inverse transformation to obtain the final results.

In the example just given, the change of variables converted the differential equations (4.7.9) into the much simpler form (4.7.11). This simplification is a special case of a very general result to be described in the next section. (In it we will cease using primes to distinguish running values of variables.)

4.8 The Rectification Theorem

The following theorem has been termed by Arnold *the fundamental theorem of the theory of ordinary differential equations.*[4]

Theorem 4.8.1. *Let x represent a set of real variables. If the functions $\xi_j(x)$ in the equations*

$$dx_j/d\alpha = \xi_j(x), \quad j = 1, \ldots, n < \infty, \tag{4.8.1}$$

are differentiable functions of class $C^k, k > 0$, then in the neighborhood of every point P_x that is not a critical point, there exist open regions in which local diffeomorphisms $x \to y = \phi(x_1, x_2, \ldots)$ convert these equations to

$$dy_1/d\alpha = 1, \quad dy_j/d\alpha = 0, \; j = 2, \ldots, n. \tag{4.8.2}$$

The adjective *'local'* implies that different diffeomorphisms may be required in different regions. The adjective *'open'* means that each x_j and each y_j lies in an open region between two real numbers a_j and b_j, e.g., $a_j < x_j < b_j$.

From the theorem it follows that the generator $U = \Sigma_1^n \xi_j(x)\partial/\partial x_j$ *of the group defined by* (4.8.1) *can be converted by one or more diffeomorphisms to the form* $U = \partial/\partial y_1$, *except in the region of a critical point, where* $U = 0$.

Canonical coordinates are coordinates $y = (y_1, y_2, \ldots)$ in which U takes on the form $\partial/\partial y_1$.

Figures 4.8.1 illustrate the result of converting Eqs. (4.7.9) to the corresponding equations in canonical coordinates, (4.7.11). They are designed to emphasize that canonical coordinates are not necessarily Cartesian coordinates in a Euclidean space, but can, for example, be locally orthogonal coordinates of the sort discussed in Sec. 2.7.

The most direct general method for converting a group generator $U = \Sigma_1^n \xi_j(x)\partial/\partial x_j$ to the rectified form $U = \partial/\partial y_1$ is to integrate Eqs. (4.8.1) in the form

$$dx_1/\xi_1(x) = dx_2/\xi_2(x) \ldots = dx_N/\xi_N(x) = d\alpha. \tag{4.8.3}$$

This method was illustrated in the previous section. It often requires making a number of substitutions and finding a number of integrating factors

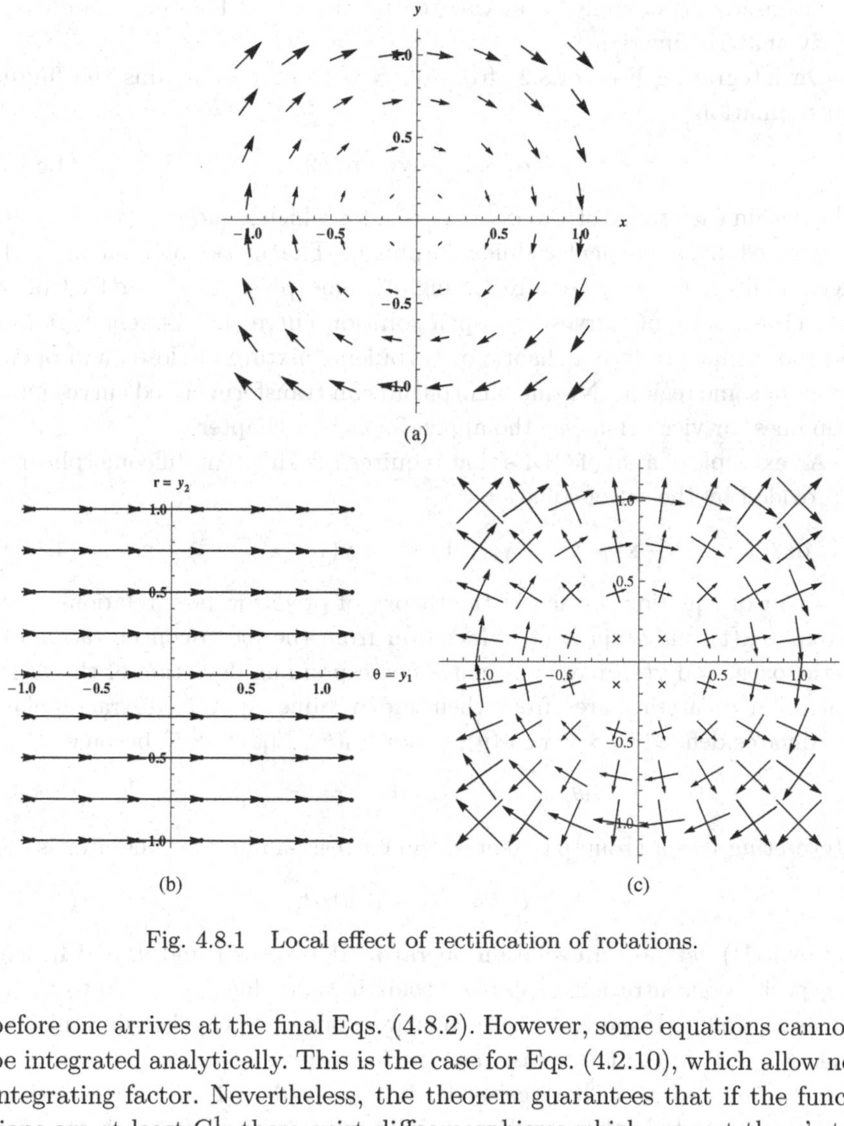

Fig. 4.8.1 Local effect of rectification of rotations.

before one arrives at the final Eqs. (4.8.2). However, some equations cannot be integrated analytically. This is the case for Eqs. (4.2.10), which allow no integrating factor. Nevertheless, the theorem guarantees that if the functions are at least C^1, there exist diffeomorphisms which convert the x's to local or global canonical coordinates y.

In physical applications, the most common functions that are not at least C^1 are *step functions*, which are discontinuous and have discontinuous first derivatives, and *sawtooth functions*, which have discontinuous first derivatives. However in physically relevant equations, such as those of electric circuit theory, these functions approximate to smoother functions. They are themselves commonly approximated by truncated Fourier expansions, i.e. by analytic functions.

On integrating Eqs. (4.8.2) from $\alpha' = 0$ to α, one obtains the finite transformation

$$y_1 \to y_1 + \alpha, \quad y_j \to y_j, \; j = 2, \ldots, N, \tag{4.8.4}$$

valid within each range of $x = (x_1, \ldots, x_N)$, for which x and $y = (y_j, \ldots, y_N)$ are related by a particular diffeomorphism. The most common natural effects limiting the range of a rectifying diffeomorphism are those that produce closed solution curves and open solution curves in different regions, and those that produce a chaotic or turbulent mixture of closed and open curves in some regions. No diffeomorphism can transform closed curves into open ones, or vice versa (see the appendix of this chapter).

An example of a set of ODEs that requires two different diffeomorphisms is provided by the set of equations

$$dx/dt = x(1 - x^2 - y^2) - y, \quad dy/dt = y(1 - x^2 - y^2) + x. \tag{4.8.5}$$

This pair of equations occurs in the theory of predator–prey relations. For example, x(t), may represent a deviation from the mean concentration of arctic foxes, and y(t) may represent a corresponding deviation of the concentration of arctic hares from their mean value. In the abstract polar coordinates defined by $x = r\cos(\theta)$, $y = r\sin(\theta)$, Eqs. (4.8.5) become

$$d\theta/dt = 1, \quad dr/dt = r - r^3. \tag{4.8.6}$$

Interpreting t as a group parameter, the corresponding Lie generator is

$$U = \partial/\partial\theta + (r - r^3)\partial/\partial r, \tag{4.8.7}$$

and exp(tU) acts as an evolution operator. It evolves r and θ, and hence the species concentrations x, y, from their initial values at $t_0 = 0$ to their values at time t. Inspecting these equations one observes that dr vanishes when $r = 0, 1$. In the latter case, as θ varies from $-\pi$ to $+\pi$, r describes a circle of unit radius. In both cases $dr/dt = 0$, $d\theta/dt = 1$, and the equation is rectified. On the other hand, if initially r lies between 0 and 1, then in (4.8.6) dr/dt is positive, and as t increases, r increases — approaching, but never reaching, 1. Conversely, if r is initially greater than 1, then r decreases as t increases, and r approaches the unit circle. If r initially $= 1$, it remains so, and r rotates with constant *velocity* in the *space* of arctic foxes and rabbits. Solution curves of each type are illustrated in **Figs. 4.8.2** in which x and y are treated as Cartesian variables in a Euclidean plane. Note that

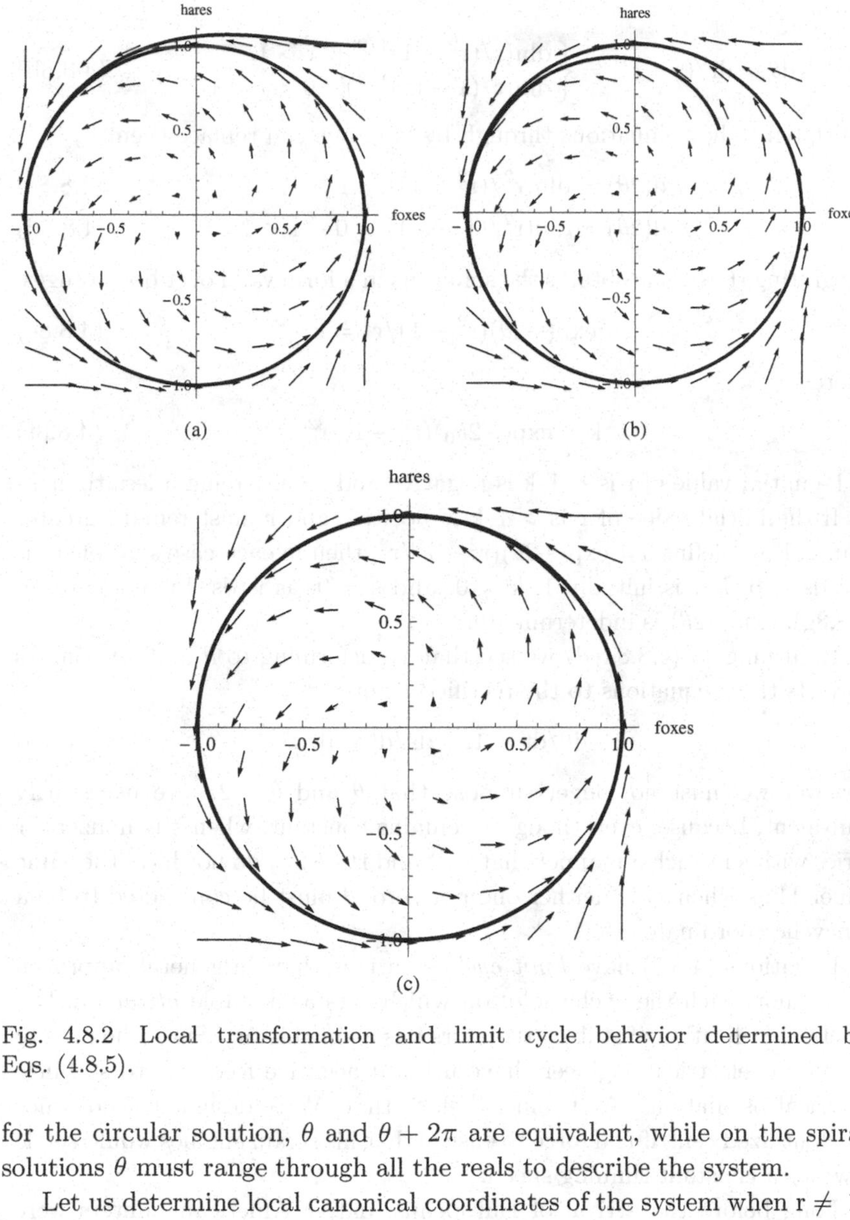

Fig. 4.8.2 Local transformation and limit cycle behavior determined by Eqs. (4.8.5).

for the circular solution, θ and $\theta + 2\pi$ are equivalent, while on the spiral solutions θ must range through all the reals to describe the system.

Let us determine local canonical coordinates of the system when $\mathrm{r} \neq 1$. Different transformations are required when $\mathrm{r} < 1$ and when $\mathrm{r} > 1$. Equations (4.8.6) imply that

$$\mathrm{dr}/\mathrm{d}\theta = \mathrm{r} - \mathrm{r}^3. \tag{4.8.8a}$$

Hence

$$d\theta = dr/(r - r^3) = \begin{cases} d\ln(r/(r^2 - 1)^{1/2}), & r > 1, \\ d\ln(r/(1 - r^2)^{1/2}), & 0 < r < 1. \end{cases} \tag{4.8.8b}$$

Multiplying these equations through by two, one can replace them by

$$d(2\theta) = d\ln(r^2/(r^2 - 1)), \quad r > 1, \tag{4.8.8c}$$

$$d(2\theta) = d\ln(r^2/(1 - r^2)), \quad 0 < r < 1. \tag{4.8.8d}$$

Integrating these, one obtains the solutions in a form valid over both ranges:

$$\exp(-2\theta)(r^2 - 1)/r^2 = k, \tag{4.8.8e}$$

where

$$k = \exp(-2\theta_0)(r_0^2 - 1)/r_0^2. \tag{4.8.8f}$$

If the initial value of r is < 1, k is negative and r must remain less than 1.

If the initial value of r is > 1, k is positive and r must remain greater than 1. If we definc $s = \exp(-2\theta)(r^2 - 1)/r^2$, then in each case s is constant and $ds = 0$. If r is initially 1, $k = 0$, and $s = 0$, as is ds. In this case, in (4.8.8g), $\exp(-2\theta)$ is indeterminate.

Returning to (4.8.8a–d) we see that transforming to θ and s variables converts these equations to the rectified form

$$d\theta/dt = 1, \quad ds/dt = 0. \tag{4.8.9}$$

However we must no longer suppose that θ and $\theta + 2\pi$ are necessarily equivalent, because, even though s remains constant, when s is nonzero, r varies with s in such a manner that $r(\theta)$ and $r(\theta + 2\pi)$ do not have the same value. Thus when r_0 is neither one nor zero, θ must be considered to be a noncyclic coordinate.

Equations (4.8.5) have *limit cycle* solutions that, in general, approach but do not reach the cyclic solution which acts as a stable *attractor*. The practical utilization of stable attractors has a long history. Since the days of De Forest, electrical engineers have utilized negative feedback to stabilize electrical oscillations. And, much before that, Watt designed a governor which utilized centrifugal force to ensure that his steam engines would settle down to a constant running speed.

Long before the arrival of Europeans, native American hunters were aware that populations of predator and prey species oscillate in near synchrony, the peaks of each population being separated from the other by a few years. Eighteenth and nineteenth century fur trade records of the Hudson's Bay Company appear to have led to the realization that these populations obey rate equations with limit cycle solutions.[5]

A variety of chemical kinetic equations that produce concentration oscillations can have the same property — a property which can also be responsible for stable oscillatory behavior in biological systems.

If limit cycle oscillations become very small, the attractor approaches a point, and the corresponding biological system exhibits homeostasis. Oscillatory chemical concentrations can develop when an autocatalytic reaction and an autodepressive one are chemically coupled. They are of central importance in Turing's reaction–diffusion theory in which chemical concentration waves produce cell differentiation and biological morphogenesis.[6]

Equations (4.8.5) were chosen to provide an instructive example of the workings of the rectification theorem when the required change of variables is different for different ranges of the variables x, y, or r, θ.

Differential equations with solution curves of differing topologies are of common occurrence. For example, objects moving under gravitational or electrical attractions typically have both closed orbits and open orbits. In latter chapters it will be shown that this presence of both open and closed orbits in classical Kepler systems affects the dynamical groups of quantum mechanical Kepler systems as well as their classical counterparts.

4.9 Conversion of Non-autonomous ODEs to Autonomous ODEs

The set of non-autonomous equations

$$dx_j/dt = f_j(x,t), \quad j = 1, \ldots, m, \tag{4.9.1}$$

can always be converted to a set of autonomous equations with the same solutions. To accomplish this, one supplements the equations with the further equation $dt/d\tau = 1$, and then replaces them with the autononomous equations

$$dt/d\tau = 1, \quad dx_j/d\tau = f_j(x,t), \; j = 1, \ldots, m. \tag{4.9.2}$$

When using this *autonomous extension* of the non-autonomous Eqs. (4.9.1), one needs to keep in mind that though $dt = d\tau$, the variables t and τ may not only differ by a constant; one may also be confined within a closed interval, while the other is not.

Equations (4.9.2), being autonomous, define a group generator

$$V = \partial/\partial t + \Sigma\ f_j(x,t)\partial/\partial x_j. \tag{4.9.3}$$

$Exp(\tau V)$ acts on initial values of x and t to yield solution curves of Eqs. (4.9.1), defined by the m+1 equations

$$t \to t' = t + \tau, \quad x_j \to x_j' = X_j(x,t;\beta). \tag{4.9.4}$$

If one considers dt and $d\tau$ to represent physical time intervals, it does not follow that t or τ represent physical time; in physics, as currently understood, it is necessary that the variable representing time be noncyclic. If t happens to be a cyclic variable, one may choose τ to be noncyclic. The physical system obeying Eqs. (4.9.2) is then a dynamical system, and its symmetries may be said to be dynamical symmetries. The solutions of (4.9.2) then define an evolution curve in space of variables τ, t, x that includes time. V is the generator of evolution, and $\exp(\tau V)$ is the evolution operator.

Equations (4.9.1) may be expressed in canonical coordinates as

$$dy_1/dt' = 1, \quad dy_j/dt' = 0, \ j = 2, \ldots, m. \tag{4.9.5}$$

The variables y, t' are related to the variables x, t by diffeomorphisms of the form

$$\begin{aligned} t \to t' &= \tau(t,x), \quad x_j \to y_j = \psi_j(x,t); \\ t' \to t &= \tau_{inv}(t',y), \quad y_j \to x_j = \psi_{jinv}(y,t'). \end{aligned} \tag{4.9.6}$$

Supplementing Eqs. (4.9.5) with the equation $dt'/d\beta = 1$, they may be rewritten as

$$dt'/d\beta = 1, \quad dy_1/d\beta = 1, \quad dy_j/d\beta = 0, \quad j = 2, \ldots, m. \tag{4.9.7}$$

The evolution generator is then $V' = \partial/\partial y_1 + \partial/\partial t'$ and the group parameter is β. The solutions of the equations are defined by $t' - \beta = a$, $y_1 - \beta = b$, $y_j = c_j$, $j = 2, \ldots, m$. One may use the first equation to eliminate β from the second, obtaining $y_1 - t' = c_1$, and let the solutions involving y be denoted by the n equations $\phi_j(y,t') = c_j$, $j = 1, \ldots, n$. The discussions of this section lead to the following result:

The solutions of the equations

$$dx_j/dt = f_j(x,t), \quad j = 1, \ldots, m, \tag{4.9.8}$$

with f_j *that are* C^1 *functions may be put in the form*

$$\varphi_j(x,t) = c_j, \quad j = 1, \ldots, m. \tag{4.9.9}$$

This statement follows from the form of the solutions to (4.9.5), which is obtained in the previous paragraph, and the diffeomorphisms (4.9.6); these together yield

$$\varphi_j(x,t) = \phi_j(\psi(x,t), \tau(t,x)). \qquad (4.9.10)$$

4.10 N-th Order ODEs as Sets of First-Order ODEs

A second-order ODE such as

$$d^2y_1/dx^2 = f(y_1, dy_1/dx, x), \qquad (4.10.1)$$

may always be expressed as the pair of first-order ODEs. One simply lets

$$dy_1/dx = y_2, \quad dy_2/dx = f(y_1, y_2, x). \qquad (4.10.2a,b)$$

In a similar manner, any m-th order ODE which can be put in the form

$$d^m y_1/dx^m = f(y_1, \ldots, y_m; x), \qquad (4.10.3)$$

can, on writing $dy_r/dx = y_{r,x}$, be expressed as a set of first-order ODEs:

$$y_{1,x} = y_2, \quad y_{2,x} = y_3, \quad y_{m-1,x} = y_m, \quad y_{m,x} = f(y,x). \qquad (4.10.4)$$

This is a special case of the set of first-order ODEs

$$dy_j/dx - f_j(y,x) = 0, \quad j = 1, \ldots, m. \qquad (4.10.5)$$

4.11 Conclusion

Any finite set of ordinary differential equations defined by sufficiently smooth functions may be converted to a set of first-order autonomous equations

$$\delta y'_j/\delta\alpha' = \xi_j(y'), \quad j = 1, \ldots, n, \qquad (4.11.1a)$$

and expressed as

$$\delta y'_1/\xi_1(y') = \delta y'_2/\xi_2(y') = \cdots = \delta y'_n/\xi_n(y') = \delta\alpha'. \qquad (4.11.1b)$$

The rectification theorem guarantees that if the $\xi_j(y')$ are at least C^1 functions, then these equations at the very least define local one-parameter groups acting within open intervals of the y's, and the equations may define a global transformation group. The generator of a group defined by (4.11.1) is $U = \Sigma\, \xi_j(y)\partial/\partial y_j$.

When Eqs. (4.11.1) are integrated over finite intervals, as the integration proceeds, the variables y'_j change from their initial values $y_j = Y_j(y,0)$ to their final values $y^*_j = Y_j(y,\gamma)$. In each successive interval, $\Delta\alpha$, $\Delta\beta, \ldots$, the integration process defines a finite transformation

$\Delta y = T(\Delta\alpha)y$, $\Delta y = T(\Delta\beta)y, \ldots$, of the variables y_j. The group composition law, $T(\Delta\beta)T(\Delta\alpha)y = T(\Delta\beta + \Delta\alpha)y$, follows from the fact that integration over two successive intervals gives the same result as integration over the total interval.

The finite transformations of the group, $T(\alpha)y_j$, may be expressed as $\exp(\alpha U)y_j$, an expression which formally satisfies the differential equations (4.11.1). The expression can provide a convenient means to evaluate, or approximate, the effect of the transformation.

The operations of the group defined by (4.11.1) develop a solution of the differential equations through any given initial point; they carry each solution of the equation into itself. The equations are thus invariant under the operations of the transformation of the group they define; they possess the intrinsic symmetry so defined. It will be shown in the next chapter that such systems of ordinary differential equations possess further intrinsic symmetries.

Appendix A. Homeomorphisms, Diffeomorphisms, and Topology

As previously noted, a *homeomorphism* is a one-to-one invertible transformation

$$x_j \to x'_j = f_j(x), \quad x'_j \to x_j = f^{in}_j(x)$$

with transformation functions f and f^{in} that depend continuously upon all their variables in the range allowed for these variables. The equation $x \to y = x^2$ has two roots $x = \pm y$, so the inverse transformation $y \to x$ is not one-to-one. In this case, the transformation is not a homeomorphism if y is allowed to have both positive and negative values. Any polynomial equation $y = P(x)$ that defines a transformation over a range of y in which the polynomial has roots with different values, will fail to define a homeomorphism.

Considered in an active sense, because it is defined by continuous functions, a homeomorphism carries contiguous points P_a and P_b into contiguous points P'_a and P'_b. Pairs of geometric objects such as lines, surfaces, and hypersurfaces, that are transformed into each other by a homeomorphism are termed homeomorphic. By definition, homeomorphic geometric objects have the same topology. A square and a circle are homeomorphic, but a circle and a straight line are not.

Diffeomorphisms map noncompact regions into noncompact regions. If a diffeomorphism acts within a compact subregion of the open region over which it is defined, the diffeomorphism will map the compact region into a compact region. A square and a circle in the Euclidean plane are not diffeomorphic because components of the vector $r = (z_1, z_2)$ from the origin to points on the boundary of the square do not change smoothly at the corners of the square.

The following equations define the C^∞ diffeomorphisms indicated:

$$y = \exp(x),\ x = \ln(y),\ -\infty < x < \infty \to 0 < y < \infty;$$
$$y = x/(1-x),\ x = y/(1+y),\ -\infty < x < 1 \to -1 < y < \infty;$$
$$(x, y) \to (x', y') = (xe^a, ye^b), \text{with a, b real.}$$

The following do not define diffeomorphisms for the reasons indicated:

i. $y = \exp(|x|),\ |x| = \ln(y),\ -\infty < x < \infty \to 0 < y < \infty;$
   dy/dx is discontinuous at $x = 0$.
ii. Any transformation from angular coordinates of points on a sphere to angular coordinates on a torus; the topologies of the surfaces are different.

A C^∞ diffeomorphism becomes an analytic transformation iff the functions $f_j(x), f'_j(x)$ can be expressed by convergent power series expansions over the range allowed to x. The expansions need not be about a single point, but may be about a set of points. (If the power series expansion of a function is convergent about all points, the function is said to be an *entire* function.)

Exercises

1. Determine the infinitesimal transformations and generators of the following finite transformations:
   a. $x' = (x^2 + 2\alpha)^{1/2},\ y' = (y^2 - \alpha)^{1/2};$
   b. $x' = x/(1 - \alpha x),\ y' = y/(1 - \alpha x);$
   c. $x' = e^\alpha(x\cos(\alpha) - y\sin(\alpha)),\ y' = e^\alpha(x\sin(\alpha) + y\cos(\alpha)).$
2. Let a be a continuously variable parameter. Can the transformation $x' = 1/(x + a)$, be that of a Lie group with group parameter a, or function of a?
3. Determine the finite transformations produced by the groups with generators
   a. $x(x\partial/\partial y - y\partial/\partial x);$
   b. $x\partial/\partial y - y\partial/\partial x + k\partial/\partial z$, where k is a constant.
4. Does the transformation defined by Eq. (4.2.10) have any critical points?

5. What are the ordinary differential equations that define each of the following group generators?

$$\begin{aligned}
&U_1 = x\partial/\partial y - y\partial/\partial x, U_2 = x\partial/\partial(ct) + (ct)\partial/\partial x, U_3 = xU_1,\\
&U_4 = xU_2, U_5 = ctU_2, U_6 = x\partial/\partial x + y\partial/\partial y + t\partial/\partial t,\\
&U_7 = (x^2 + y^2 - c^2t^2)\partial/\partial x - 2xU_6,\\
&U_8 = (x^2 + y^2 - c^2t^2)\partial/\partial(ct) + 2ctU_6.
\end{aligned}$$

6. Plot the two-dimensional vector fields associated with the Lie generators U_1 through U_5 in problem (5).
7. Plot the three-dimensional vector fields associated with the generators U_6, U_7, U_8, and plot the corresponding fields in the three two-planes in which x, or y, or ct, is zero.
8. Consider the transformation defined by:

$$s_1 = kx/(x^2+y^2-c^2t^2),\ s_2 = ky/(x^2+y^2-c^2t^2),\ s_3 = kct/(x^2+y^2-c^2t^2),$$

where k is a positive constant and c is the velocity of light. Calculate $s_1^2+s_2^2-s_3^2$, and determine the corresponding inverse transformation from $s = (s_1, s_2, s_3)$ to (x, y, ct).
9. Transform the U's of problem 5 to Lie generators of the form $W = \Sigma\xi_j(s)\partial/\partial s_j$.

Note: Problems 5 through 9 all relate to material that will be discussed in subsequent chapters.

References

[1] H. Jerome Keisler, *Foundations of Infinitesimal Calculus* (Prindle, Weber & Schmidt, Boston, 1976).
[2] J. C. Maxwell, *Electricity and Magnetism* (Cambridge, 1873).
[3] H. Whitney, *Collected Papers*, eds. J. Eels, D. Toledo, Vol. II (Birkhauser, Boston, 1992).
[4] V. I. Arnold, *Ordinary Differential Equations*, translated by R. A. Silverman, (MIT Press, Cambridge, 1973).
[5] A. J. Lotka, *Elements of Physical Biology* (Williams and Wilkins, 1925).
[6] A. M. Turing, *Phil. Trans. Roy. Soc. (London)* **237** (1952) 37.

CHAPTER 5

Everywhere-Local Invariance

5.1 Invariance under the Action of One-Parameter Lie Transformation Groups

In this chapter we characterize the invariance of functions, equations, metrics, and ordinary differential equations under the action of Lie transformation groups. All the discussions will deal with invariance that is independent of the value taken on by a group parameter. We will not consider discrete invariance transformations, or invariance that holds only for discrete values of a group parameter, e.g. transformations such as

$$\exp(2\pi\partial/\partial \mathrm{x})\sin(\mathrm{x}) = \sin(\mathrm{x}+2\pi) = \sin(\mathrm{x}). \tag{5.1.1}$$

A function f(x) is invariant under the actions of a Lie transformation group with operator $\exp(\alpha \mathrm{U})$, $\mathrm{U} = \Sigma\ \xi_{\mathrm{j}}(\mathrm{x})\ \partial/\partial \mathrm{x}_{\mathrm{j}}$, iff for all allowed values of x and α,

$$\mathrm{f}(\mathrm{x}) = \mathrm{f}(\mathrm{x}'), \quad \mathrm{x}' = \exp(\alpha \mathrm{U})\mathrm{x}. \tag{5.1.2}$$

If f(x) is differentiable, then invariance requires that $\partial \mathrm{f}(\mathrm{x}')/\partial\alpha = 0$, along each path curve $\mathrm{x}' = \exp(\alpha \mathrm{U})\mathrm{x}$, regardless of its initial point $\mathrm{P_x}$. Now

$$\partial \mathrm{f}(\mathrm{x}')/\partial\alpha = \partial \mathrm{f}(\exp(\alpha \mathrm{U})\mathrm{x})/\partial\alpha = \mathrm{f}(\mathrm{U}(\exp(\alpha \mathrm{U})\mathrm{x})) = \mathrm{Uf}(\mathrm{x}'). \tag{5.1.3}$$

Because of the group property, every point on each path curve may be considered an initial point of the group action as well as a point obtained by an infinitesimal or finite transformation of the group, so one may let $\alpha \to 0$ in (5.1.3) after taking the derivative. Then $\mathrm{x}' \to \mathrm{x}$, and $\mathrm{Ux}' \to \mathrm{Ux}$. The right-hand side of (5.1.3) then becomes Uf(x), and

$$\partial \mathrm{f}(\mathrm{x}')/\partial\alpha|_{\alpha=0} = \mathrm{Uf}(\mathrm{x}). \tag{5.1.4}$$

Thus one obtains:

Theorem 5.1.1. *A differentiable function* f(x) *is invariant under the action of a one-parameter Lie group with generator* U *iff*

$$\mathrm{Uf(x)} = 0 \tag{5.1.5}$$

for all allowed values of x.

This criterion for invariance is purely local. It is equivalent to requiring that for arbitrary $\delta\alpha$,

$$\mathrm{f(x')} = (1 + \delta\alpha\ \mathrm{U})\mathrm{f(x')} \tag{5.1.6}$$

for all allowed values of x′. The group property of the transformation, together with the assumed differentiabilty of f(x) for all values of x, assures that the local invariance holds around each successive point reached by the action of exp(αU) on x. Thus the global invariance of a function under the action of a continuous group is a consequence of local invariance everywhere.

We next characterize the invariance of *equations* under a one-parameter group of transformations. Given an equation of the form g(x) = c, where c is a constant, and g is everywhere differentiable, let f(x) = g(x) − c. The resulting equation f(x) = 0, will be invariant under the transformation $\mathrm{x} \to \mathrm{x'} = \exp(\alpha \mathrm{U})\mathrm{x}$ iff

$$\mathrm{f(x')} = 0 \tag{5.1.7a}$$

when

$$\mathrm{f(x)} = 0, \tag{5.1.7b}$$

for all values of α, and for all values of x such that f(x) = 0. Let it be supposed that f satisfies no other independently valid equations. In particular, suppose that f(x) is not of form $\mathrm{g(x)}^2\ \mathrm{h(x)}$, that contains repeated factors g(x) which themselves vanish. With these restrictions in force, there are only two ways in which exp(αU) can leave the equation invariant. Either:

i) the transformation leaves the function f(x) invariant for all values of x, or,
ii) it leaves the equation f(x) = 0 invariant for values of x satisfying f(x) = 0. (This can only happen if Uf contains f(x) as a factor.)

Both possibilities are subsumed in:

Theorem 5.1.2. *If* f(x) *is a differentiable function containing no repeated factors, satisfying an equation* f(x) = 0 *and no other independent equation, then the equation* f(x) = 0 *is invariant under the action of a one-parameter transformation group with generator* U *iff*

$$0 = \mathrm{Uf(x)}|_{\mathrm{f(x)=0}}. \tag{5.1.8}$$

That is to say Uf(x) = 0 *for all values of* x *such that* f(x) = 0.

Just as in the previous theorem, global invariance under the action of a continuous group is determined by a purely local invariance that is valid for all allowed values of the coordinates x. The group property of the transformation allows x to be considered x′, and vice versa.

There is a hidden aspect to the argument just given: the operator $(1 + \delta\alpha\mathrm{U})$ varies *all* the variables x, which are thus initially considered to vary independently. The functional relation among the variables x, and the functional relation among the running variables x′ is *thereafter* established by a restriction on the form of U. This is imposed by the condition that Uf(x) = 0 for all values of x that satisfy f(x) = 0.

In using Theorem 5.1.2, the reader is reminded to keep in mind the trivial example that develops if U itself contains f as a factor, for then, as mentioned above, $\mathrm{Uf}|_{\mathrm{f}=0}$ will necessarily vanish. Such U's are generally of little interest.

The following application of Theorem 5.1.2 demonstrates that a unit circle is left invariant by a transformation that is not generally a rotation. Let

$$\mathrm{f(x)} = \mathrm{x}_1^2 + \mathrm{x}_2^2 - 1 = 0, \tag{5.1.9a}$$

and let

$$\mathrm{T}(\alpha) = \exp(\alpha\mathrm{U}), \quad \mathrm{U} = \mathrm{x}_1\mathrm{x}_2\partial/\partial\mathrm{x}_1 + (\mathrm{x}_2^2 - 1)\partial/\partial\mathrm{x}_2. \tag{5.1.9b,c}$$

Then, though U itself does not contain $\mathrm{x}_1^2 + \mathrm{x}_2^2 - 1$ as a factor,

$$\mathrm{Uf(x)} = 2\mathrm{x}_1\mathrm{x}_2\mathrm{x}_1 + 2(\mathrm{x}_2^2 - 1)\mathrm{x}_2 = 2\mathrm{x}_2(\mathrm{x}_1^2 + \mathrm{x}_2^2 - 1). \tag{5.1.9d}$$

This expression vanishes on the unit circle defined by $\mathrm{x}_1^2 + \mathrm{x}_2^2 = 1$. Consequently

$$0 = \mathrm{Uf(x)}|_{\mathrm{f(x)=0}}. \tag{5.1.9e}$$

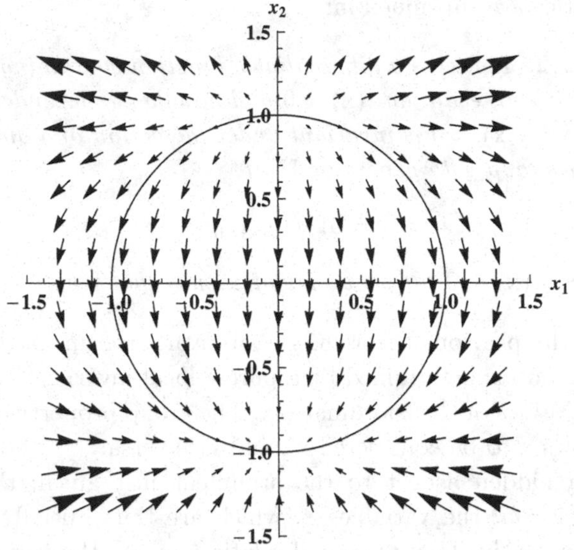

Fig. 5.1.1 Transformation (5.1.9) defines displacement vectors tangent to a unit circle.

Thus, though the transformation $(x_1, x_2) \to (x_1', x_2') = T(\alpha)(x_1, x_2)$ does not leave invariant $x_1^2 + x_2^2$, it carries the unit circle into itself for all values of α. **Figure 5.1.1** illustrates the action of the transformation on points on the unit circle and points off it.

We next turn to a discussion of the invariance of families of curves. The equation

$$f(x, y) = c \tag{5.1.10a}$$

with c a *parameter* rather than a constant, defines a one-parameter family of curves. These families might represent *generalized solutions* of a differential equation

$$M(x, y)dx + N(x, y)dy = 0, \tag{5.1.10b}$$

i.e. of

$$dy/dx = -M(x, y)/N(x, y) \tag{5.1.10c}$$

or

$$dx/dy = -N(x, y)/M(x, y). \tag{5.1.10d}$$

The same family of curves defined by (5.1.10a) may also be defined by

$$F(f(x,y)) = C, \quad C = F(c), \tag{5.1.10e}$$

where F is an arbitrary differentiable function.

Equations (5.1.10a) and (5.1.10e) represent the solutions of the differential Eqs. (5.1.10b–d) in a way which, like Eq. (5.1.10b), does not distinguish independent and dependent variables. (Planetary motion provides an example where this representation is important: the radial position, r, of a planet at time t is not a simple function of t, but t is a simple function of r). **Figure 5.1.2** illustrates the action of a group that interconverts the members of a family of curves that comprise a flow, as defined in Chapter 3. In **Fig. 5.1.2** the transformations of the group are active ones that convert curves of the flow into other curves of the flow. The figure calls attention to the group defining a second flow that intersects the path curves of the original differential equation.

Now let us consider the requirements the generator of an invariance group of Eqs. (5.1.10) must satisfy. A Lie group with generator $U = \xi(x,y)\partial/\partial x + \eta(x,y)\partial/\partial y$ will leave Eqs. (5.1.10) invariant if $0 = Uf|_{f=c}$.

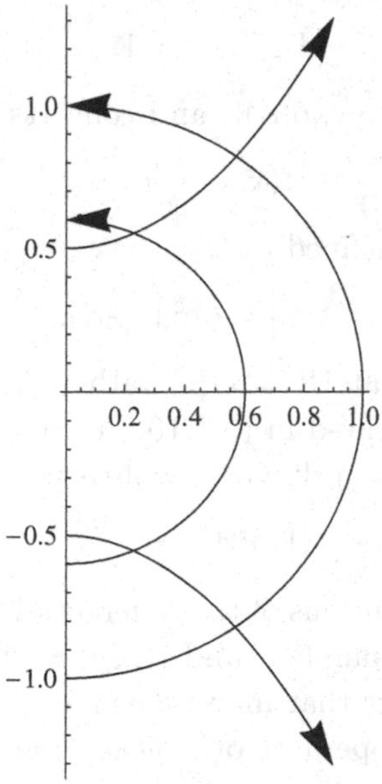

Fig. 5.1.2 Transformation that interconvert curves of a flow defined by $\frac{dx}{da} = x$.

Now if f is analytic, and if

$$\mathrm{Uf} = 0 \tag{5.1.11a}$$

identically, then $\mathrm{f}(\exp(\alpha \mathrm{U})\mathrm{x}, \exp(\alpha \mathrm{U})\mathrm{y}) = \exp(\alpha \mathrm{U})\mathrm{f} = \mathrm{f}$; so if $\mathrm{f}(\mathrm{x}, \mathrm{y}) = \mathrm{c}$, and

$$\mathrm{f}(\exp(\alpha \mathrm{U})\mathrm{x}, \exp(\alpha \mathrm{U})\mathrm{y}) = \mathrm{c}. \tag{5.1.11b}$$

Each curve of the family is thus carried into itself. For a transformation to carry solutions of (5.1.10) into different solutions it is necessary that

$$\mathrm{f}(\exp(\alpha \mathrm{U})\mathrm{x}, \exp(\alpha \mathrm{U})\mathrm{y}) = \mathrm{f}(\mathrm{x}, \mathrm{y}) - \mathrm{c}(\alpha), \tag{5.1.11c}$$

with $\mathrm{c}(\alpha) = \mathrm{c}', \mathrm{c}' \neq \mathrm{c}$. The infinitesimal transformation of f must then be such that $(1 + \delta\alpha \mathrm{U})\mathrm{f}(\mathrm{x}, \mathrm{y})$ can be expressed as $\mathrm{f}(\mathrm{x}, \mathrm{y}) - \mathrm{c}(\delta\alpha)$. This will be true, for example, if

$$\mathrm{f}(\mathrm{x}, \mathrm{y}) + \delta\alpha \mathrm{Uf} = \mathrm{f}(\mathrm{x}, \mathrm{y}) + \delta\alpha \text{ const.}, \tag{5.1.12a}$$

i.e. if

$$\mathrm{Uf} = \text{const.} \tag{5.1.12b}$$

If U satisfies (5.1.12b), then so does U/const. One thus obtains the simple and useful result, that if f is differentiable, and

$$\mathrm{Uf}(\mathrm{x}, \mathrm{y}) = 1, \tag{5.1.13}$$

then $\exp(\alpha\ \mathrm{U})$ converts f to $\exp(\alpha)\mathrm{f}$, and converts a curve defined by

$$\mathrm{f}(\mathrm{x}, \mathrm{y}) = \mathrm{c}, \tag{5.1.14b}$$

into the family of curves defined by

$$\mathrm{f}(\mathrm{x}, \mathrm{y}) = \mathrm{c} \exp(-\alpha). \tag{5.1.14c}$$

If one knows a U such that $\mathrm{Uf} = \mathrm{G}(\mathrm{f})$ rather than a constant, then one may use the freedom indicated in (5.1.10e) to redefine f. The change from f to F(f), defined by $\mathrm{F}(\mathrm{f}) = \int \mathrm{df}/\mathrm{G}(\mathrm{f})$, will then yield a F(f) such that

$$\mathrm{U\ F}(\mathrm{f}) = 1. \tag{5.1.15}$$

The two *families* of solutions of the differential Eqs. (5.1.10b–d) defined by $\mathrm{f}(\mathrm{x}, \mathrm{y}) = \mathrm{c}$, c any real number, and $\mathrm{f}(\mathrm{x}, \mathrm{y}) = \exp(-\alpha)\mathrm{c}$, coincide if the parameter c is a parameter that may take on any real value. When this is possible, $\exp(\alpha \mathrm{U})$ is the operator of a *global* Lie invariance group of the family of solutions of the differential equations. If the range of c or α is restricted, then $\exp(\alpha \mathrm{U})$ is the operator of a *local* Lie invariance group of the family. This must be the case when the solution curves of (5.1.10b–d) do not all have the same topology, for the diffeomorphism carried out by $\exp(\alpha U)$ cannot interconvert curves of different topology. An example of this latter case is provided by the equations of motion of Kepler systems, for they allow both open, hyperbolic and parabolic trajectories, and closed, elliptical and circular trajectories.

5.2 Transformation of Infinitesimal Displacements

The previous paragraph investigated the effect of operators $\exp(\alpha U)$ on the family of solutions of a differential equation. It did not investigate the effect of the $\exp(\alpha U)$ on the differential equation itself. To further characterize Lie transformation groups that act on differential equations and their solutions it is necessary to investigate the effect of the transformation groups on infinitesimal displacements and derivatives. The key to an understanding of transformations of derivatives, metrics, and other expressions involving infinitesimal changes, is an understanding of transformations of infinitesimal displacements.

We begin by considering *finite* transformations of infinitesimal displacements carried out by one-parameter Lie groups. We will then determine the corresponding infinitesimal transformations of these *infinitesimal* displacements.

Let $\mathbf{x} = (x_1, x_2, \ldots, x_n)$ and $\mathbf{x}' = (x'_1, x'_2, \ldots, x'_n)$ be coordinates of two points in E^N, and let $\mathbf{x}' - \mathbf{x}$ become arbitrarily small, that is for all j, let $x'_j - x_j \to dx_j$. We wish to obtain an expression for $\exp(\alpha U)(\mathbf{x}' - \mathbf{x})$ in the limit $\mathbf{x}' - \mathbf{x} \to d\mathbf{x}$. One has

$$T(\alpha)(x'_j - x_j) = X_j(\mathbf{x}', \alpha) - X_j(\mathbf{x}, \alpha). \tag{5.2.1a}$$

As $\mathbf{x}' \to \mathbf{x} + d\mathbf{x}$, this becomes $T(\alpha)(x_j + d\mathbf{x}) - T(\alpha)x_j$, and one has

$$X_j(\mathbf{x} + d\mathbf{x}, \alpha) - X_j(\mathbf{x}, \alpha) = \Sigma_k(\partial X_j(\mathbf{x}, \alpha)/\partial x_k)\, dx_k. \tag{5.2.1b}$$

Thus for a fixed value of α the finite transformation converts

$$d\mathbf{x} = (dx_1, dx_2, \ldots, dx_n)$$

to

$$d\mathbf{x}^* = d(T(\alpha)\mathbf{x}) = (dx_1^*, dx_2^*, \ldots, dx_n^*) \tag{5.2.2}$$

with

$$dx_j^* = d(T(\alpha)x_j) = \Sigma_k(\partial X_j(\mathbf{x}, \alpha)/\partial x_k) dx_k.$$

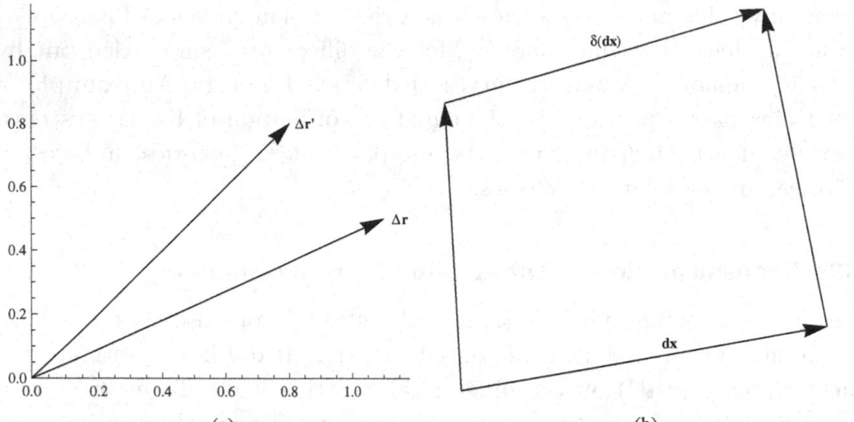

Fig. 5.2.1 (a) Finite transformation of a small displacement. (b) Infinitesimal transformation of an infinitesimal displacement.

This action of a finite transformation on a displacement vector, $\mathbf{dx} = (dx_1, dx_2)$ is illustrated in **Fig. 5.2.1a.**

When the group parameter α becomes arbitrarily small, $T(\alpha) \to T(\delta\alpha)$, and

$$d(T(\alpha)x_j) \to d(T(\delta\alpha)x_j) = d(T(0)x_j + \delta\alpha U x_j). \tag{5.2.3a}$$

On letting $U = \Sigma_j\, \xi_j(\mathbf{x})\partial/\partial x_j$, and noting that $T(0)x_j$ is just x_j, Eq. (5.2.3a) becomes

$$dx_j + \delta\alpha\ d(\xi_j(\mathbf{x})) = dx_j + \delta\alpha\ \Sigma_k(\partial\xi_j(\mathbf{x})/\partial x_k)\ dx_k. \tag{5.2.3b}$$

Consequently, under the infinitesimal transformation with generator U, the dx_j^* of (5.2.2) are given by

$$dx_j^* = dx_j + \delta\alpha\Sigma_k(\partial\xi_j(\mathbf{x})/\partial x_k)\ dx_k. \tag{5.2.4}$$

Let $\delta(dx_j) \equiv dx_j^* - dx_j$ be the infinitesimal change in dx_j due to the infinitesimal transformation. On then obtains the key result:

An infinitesimal transformation with operator $\exp(\delta\alpha\ U)$ *and generator*

$$U = \Sigma_j\ \xi_j(\mathbf{x})\partial/\partial x_j \tag{5.2.5a}$$

alters the infinitesimal displacements dx_j *by the infinitesimal amounts*

$$\delta(dx_j) = \delta\alpha\ \Sigma_k(\partial\ \xi_j(\mathbf{x})/\partial x_k)dx_k. \tag{5.2.5b}$$

Figure 5.2.1b suggests the manner in which a $\mathbf{dx} = (dx_1, dx_2)$, is converted to $\delta(\mathbf{dx}) = (\delta(dx_1), \delta(dx_2))$ by an infinitesimal transformation.

In this case the group generator is $U = x_1(-x_2\partial/\partial x_1 + x_1\partial/\partial x_2)$, so $\xi_1 = -x_1x_2$ and $\xi_2 = x_1^2$, whence $\delta(\mathbf{dx}) = \delta\alpha(-(x_1dx_2 + x_2dx_1),\ 2x_1dx_1)$.

5.3 Transformations and Invariance of Work, Pfaffians, and Metrics

A Pfaffian form is an expression such as

$$\delta v - \Sigma_i\, y_i\, \delta x_i, \quad i = 1, \ldots, n, \tag{5.3.1a}$$

or

$$\Sigma_i\, y_i\delta x_i, i = 1, \ldots, n, \tag{5.3.1b}$$

that is linear in all its variables $\delta v, \delta x_i, y_i$. The variables are *a priori* independent. A Pfaffian equation is an equation in which a Pfaffian form is set equal to zero. An example of a Pfaffian equation is provided by the expression for the element of work, δw, done by a force $\mathbf{F}$ acting on a particle that undergoes a displacement $\delta\mathbf{r}$, *viz.*,

$$\delta w = \mathbf{F} \cdot \delta\mathbf{r}. \tag{5.3.2}$$

When n=1 in (5.3.1a), the Pfaffian equation may be written $dv - ydx = 0$ or $dv/dx = y$, and dv/dx may be interpreted as the derivative of v with respect to x. When n is greater than 1 the situation is not so simple. As is familiar from discussions of work and potential energy, it need not follow from Eq. (5.3.1a) that the y_i are partial derivatives $\partial v/\partial x_i$. However, if for all $i, j = 1, \ldots, n$, one has

$$\partial y_i/\partial x_j = \partial y_j/\partial x_i, \tag{5.3.3}$$

then, and only then, each y_i is the partial derivative $\partial v/\partial x_i$ of a differentiable function $v(x)$, and

$$dv = \Sigma_i(\partial v/\partial x_i)dx_i = d(v(x)). \tag{5.3.4}$$

In this case dv is an *exact differential* and the Pfaffian equation is integrable as it stands.

For a Lie group with generator U to act on a Pfaffian form, it must be considered to act on the variables y, x and v, all initially considered to be independent. Let $T(\alpha) = \exp(\alpha U)$ with

$$U = \Sigma_1^n\, \xi_j(x, y, v)\partial/\partial x_j + \Sigma_1^n\, \pi_j(x, y, v)\partial/\partial y_j + \zeta(x, y, v)\partial/\partial v. \tag{5.3.5}$$

For clarity of notation we will, in the next few lines, use the symbol δ' to signify variations carried out by the infinitesimal transformation with generator U. The infinitesimal transformation of the y_i is then

$$\delta'(y_i) = \delta'\alpha\ \pi_i(x, y, v). \tag{5.3.6}$$

The infinitesimal transformation of the δx_i and δv, in (5.3.1a) is given by (5.2.5) in a form that does not assume integrability of the infinitesimal displacements, i.e. by

$$\delta'(\delta x_j) = \delta\alpha\ \Sigma_k(\partial\xi_j(x)/\partial x_k)\ \delta x_k. \tag{5.3.7}$$

Along with the x_i, the v and the y_i are variables whose values may change infinitesimally. Consequently

$$\begin{aligned}\delta'(\delta x_i) &= \delta'\alpha\{\Sigma_k\ (\partial\xi_i(x, y, v)/\partial x_k)\delta x_k + \Sigma_k\ (\partial\xi_i(x, y, v)/\partial y_k)\delta y_k \\ &\quad + (\partial\xi_i(x, y, v)/\partial v)\delta v\}, \\ \delta'(\delta v) &= \delta'\alpha\{(\partial\zeta(x, y, v)/\partial v)\delta v + \Sigma_k\ (\partial\zeta(x, y, v)/\partial x_k)\delta x_k \\ &\quad + \Sigma_k\ (\partial\zeta(x, y, v)/\partial y_k)\delta y_k\}.\end{aligned} \tag{5.3.8}$$

As an example of the foregoing results, consider the element of work dv, defined by

$$dv - (y_1 dx_1 + y_2 dx_2) = 0, \tag{5.3.9a}$$

where the y_i are understood to be arbitrary functions of the x's. As a test case, suppose that the generator of the transformation $T(\alpha)$ acting on the variables x,y,v is

$$U = x_1\ \partial/\partial x_1 + x_2\ \partial/\partial x_2 + y_1\ \partial/\partial y_1 + y_2\ \partial/\partial y_2 + v\ \partial/\partial v. \tag{5.3.9b}$$

The reader is invited to use (5.3.8) to show that this is the generator of a transformation that leaves (5.3.9a) invariant.

We next turn to an investigation of the action of a continuous transformation group on a metric. Not all metrics define separations between points that are independent of the path taken between the points.[1] Path-dependent metrics are termed *non-integrable*. However the metrics of primary interest here are integrable and the separation between points may be written as ds, with ds defined by

$$ds^2 = \Sigma\Sigma g_{ij}(x)dx_i\ dx_j. \tag{5.3.10}$$

An infinitesimal transformation of infinitesimal displacements dx_j will leave invariant the Euclidean metric

$$ds^2 = \Sigma dx_j^2 \qquad (5.3.11a)$$

iff

$$\Sigma(dx_j')^2 = \Sigma(dx_j + \delta(dx_j))^2 = \Sigma(dx_j)^2, \qquad (5.3.11b)$$

i.e. iff

$$\Sigma 2dx_j\delta(dx_j) + \delta(dx_j)^2 = 0. \qquad (5.3.11c)$$

For $\delta(dx_j)^2$ to vanish one must have

$$\Sigma\ dx_j\ \delta(dx_j) = 0. \qquad (5.3.11d)$$

If the generator of the transformation is

$$U = \Sigma_i\ \xi_i(x)\partial/\partial x_i, \qquad (5.3.12a)$$

then

$$\delta(dx_j) = \delta\alpha\ \Sigma_k(\partial\xi_j(x)/\partial x_k)\ dx_k. \qquad (5.3.12b)$$

Inserting this into (5.3.11d) and requiring that the result vanish for all values of $\delta\alpha$, one determines that the metric is left invariant by the transformation with generator U if

$$\Sigma_j\ \Sigma_k\ dx_j\ dx_k\ \partial\xi_j(x)/\partial x_k = 0. \qquad (5.3.13)$$

The case of two variables, when $ds^2 = dx_1^2 + dx_2^2$, is instructive. One then has

$$\begin{aligned} &\Sigma_j\Sigma_k dx_j dx_k \partial\xi_j(x)/\partial x_k \\ &= dx_1 dx_1 \partial\xi_1(x)/\partial x_1 + dx_1 dx_2\ \partial\xi_1(x)/\partial x_2 \\ &\quad + dx_2 dx_1\ \partial\xi_2(x)/\partial x_1 + dx_2 dx_2\ \partial\xi_2(x)/\partial x_2. \end{aligned} \qquad (5.3.14a)$$

The invariance condition (5.3.13) thus requires that

$$\begin{aligned} &dx_1^2\ \partial\xi_1(x)/\partial x_1 + dx_1 dx_2\ (\partial\xi_1(x)/\partial x_2 + \partial\xi_2(x)/\partial x_1) \\ &\quad + dx_2^2\ \partial\xi_2(x)/\partial x_2 = 0. \end{aligned} \qquad (5.3.14b)$$

Because dx_1 and dx_2 are independently variable, $dx_1^2, dx_1 dx_2$, and dx_2^2 are linearly independent. Consequently, (5.3.14) can vanish for all choices of

dx_1 and dx_2 iff

$$\partial\xi_1(x)/\partial x_1 = 0, \quad \partial\xi_1(x)/\partial x_2 + \partial\xi_2(x)/\partial x_1 = 0, \quad \partial\xi_2(x)/\partial x_2 = 0. \tag{5.3.15}$$

The first of these equations requires that $\xi_1 = \xi_1(x_2)$, and the last requires that $\xi_2 = \xi_2(x_1)$. Consequently, the middle equation can be written

$$\partial\xi_1(x_2)/\partial x_2 + \partial\xi_2(x_1)/\partial x_1 = 0. \tag{5.3.16}$$

Since the first term in this is a function of x_2 only, while the second is a function of x_1 only, and x_1 and x_2 are independently variable, one must have

$$\partial\xi_1(x_2)/\partial x_2 = c = -\partial\xi_2(x_1)/\partial x_1, \tag{5.3.17}$$

where c is an arbitrary constant. Solving these two differential equations one obtains

$$\xi_1 = a + cx_2, \quad \xi_2 = b - cx_1, \tag{5.3.18}$$

where a and b are also arbitrary constants. It follows that the generator of the group that leaves the metric $dx_1^2 + dx_2^2$ invariant, is

$$\begin{aligned} U &= \xi_1\partial/\partial x_1 + \xi_2\partial/\partial x_2 = (a + c\, x_2)\partial/\partial x_1 + (b - c\, x_1)\partial/\partial x_2 \\ &= a\,\partial/\partial x_1 + b\,\partial/\partial x_2 + c(x_2\,\partial/\partial x_1 - x_1\partial/\partial x_2). \end{aligned} \tag{5.3.19}$$

As a, b, and c are arbitrary constants this generator is an arbitrary linear combination of three generators

$$U_a = \partial/\partial x_1, \quad U_b = \partial/\partial x_2, \quad U_c = (x_2\partial/\partial x_1 - x_1\partial/\partial x_2). \tag{5.3.20}$$

The first of these generates translations in the x_1 direction, the second generates translations in the x_2 direction, and the third generates rotations in the x_1–x_2 plane. It will be noted that this analysis implies that the Euclidean metric in the plane of the Cartesian coordinates (z_1, z_2) is invariant under three one-parameter Lie transformation groups acting on (z_1, z_2) — those carrying out translations and rotations in the plane. The resulting three-parameter group here is of course the Euclidean group E(2) investigated in Chapter 2.

A non-Euclidean metric

$$ds^2 = \Sigma\Sigma\; g_{ij}(x)\; dx_i dx_j, \tag{5.3.21}$$

is invariant under the infinitesimal transformation with generator

$$U = \Sigma_1^n \xi_j(x)\partial/\partial x_j, \tag{5.3.22a}$$

such that

$$\delta(dx_j) = \delta\alpha\ \Sigma_k(\partial\xi_j(x)/\partial x_k)dx_k, \tag{5.3.22b}$$

if, to first-order in $\delta\alpha$ one has $ds'^2 = ds^2$, where ds'^2 is given by

$$\Sigma_{ij}((1+\delta\alpha U)g_{ij}(x))(dx_i + \delta\alpha\Sigma_k\ \partial\xi_i/\partial x_k\ dx_k)(dx_j + \delta\alpha\Sigma_m\partial\xi_j/\partial x_m dx_m). \tag{5.3.22c}$$

For this to vanish to first-order in $\delta\alpha$, the following expression must vanish:

$$\Sigma_i\Sigma_j\{(Ug_{ij}(x))dx_i dx_j + g_{ij}(x)(dx_i\ \Sigma_m\ \partial\xi_j/\partial x_m\ dx_m + dx_j\ \Sigma_k\ \partial\xi_i/\partial x_k\ dx_k)\}. \tag{5.3.22d}$$

Examples of the application of this result are developed in the exercises at the end of the chapter.

5.4 Point Transformations of Derivatives

It was pointed out in Chapter 3 that a function f(t) and its derivative df/dt may be varied independently. A variation in f determines a variation in df. Because dt can be varied independently of variations in f and df, the ratio df/dt can be varied independently of f. Thus one can investigate variations in the derivative df/dt by investigating the effect of independent variations of df and dt. We may, in particular, use the results of the previous section to investigate infinitesimal transformations that convert $\delta x_1/\delta x_2$ to $\delta x_1'/\delta x_2'$, ratios which we may also treat as derivatives $dx_1/dx_2 \sim df/dx_2$ and $dx_1'/dx_2' \sim df/dx_2'$. Transformations of a derivative such as df/dx, which depend only upon variables x that define the position of a point are termed *point transformations.* They move the points along curves, without any reference to the slope of the curve other than that resulting from the requirement df = 0.

An infinitesimal point transformation with generator

$$U = \xi_1(x)\partial/\partial x_1 + \xi_2(x)\partial/\partial x_2 \tag{5.4.1}$$

produces variations in dx_1 and dx_2, given by

$$\begin{aligned}\delta(dx_1) &= dx_1^* - dx_1 = \delta\alpha\ ((\partial\xi_1/\partial x_1)\ dx_1 + (\partial\xi_1/\partial x_2)\ dx_2),\\ \delta(dx_2) &= dx_2^* - dx_2 = \delta\alpha\ ((\partial\xi_2/\partial x_1)\ dx_1 + (\partial\xi_2/\partial x_2)\ dx_2).\end{aligned} \tag{5.4.2}$$

This implies

$$\begin{aligned} dx_1^* &= dx_1 + \delta\alpha\ ((\partial\xi_1/\partial x_1)\ dx_1 + (\partial\xi_1/\partial x_2)\ dx_2), \\ dx_2^* &= dx_2 + \delta\alpha\ ((\partial\xi_2/\partial x_1)\ dx_1 + (\partial\xi_2/\partial x_2)\ dx_2). \end{aligned} \quad (5.4.3a)$$

As the group properties of the transformation are governed by terms first-order in the group parameter, it is sufficient to expand this through first-order in $\delta\alpha$. To this order,

$$dx_1^*/dx_2^* = dx_1/dx_2 + \delta\alpha\ \xi_{1,2},$$

with

$$\begin{aligned} \xi_{1,2} = {} & \partial\xi_1/\partial x_1\ dx_1/dx_2 + \partial\xi_1/\partial x_2 \\ & -(dx_1/dx_2)(\partial\xi_2/\partial x_1\ dx_1/dx_2 + \partial\xi_2/\partial x_2). \end{aligned} \quad (5.4.3b)$$

When dx_1/dx_2 is replaced by df/dx_2,

$$\xi_{1,2} = \partial\xi_1/\partial x_1\ df/dx_2 + \partial\xi_1/\partial x_2 - (df/dx_2)(\partial\xi_2/\partial x_1\ df/dx_2 + \partial\xi_2/\partial x_2). \quad (5.4.3c)$$

Note that $\xi_{1,2}$ contains terms that depend upon the zero, first, and second power of df/dx_2.

One may use (5.4.3) to determine the one-parameter Lie groups of point transformations whose action will leave invariant a derivative dx_1/dx_2. It will be invariant iff $\xi_{1,2}$ vanishes for all values of dx_1/dx_2. As $(dx_1/dx_2)^2$, (dx_1/dx_2), and $(dx_1/dx_2)^0$ are linearly independent, this requires

$$\begin{aligned} \partial\xi_1/\partial x_2 &= 0, \\ \partial\xi_1/\partial x_1 - \partial\xi_2/\partial x_2 &= 0, \\ \partial\xi_2/\partial x_1 &= 0. \end{aligned} \quad (5.4.4)$$

The first of these equations implies that ξ_1 must be a function of x_1 only, and the last implies that ξ_2 must be a function of x_2 only. The middle equation therefore requires that

$$\partial\xi_1(x_1)/\partial x_1 = \partial\xi_2(x_2)/\partial x_2. \quad (5.4.5)$$

Because x_1 and x_2 can be varied independently, this is only possible if each of the two derivatives is equal to the same constant, say a. It follows that

$$\xi_1 = a\ x_1 + b, \quad \xi_2 = a\ x_2 + c. \quad (5.4.6)$$

The corresponding generator of the transformations of x_1 and x_2 is

$$U = (a\ x_1 + b)\partial/\partial x_1 + (a\ x_2 + c)\partial/\partial x_2. \quad (5.4.7)$$

Thus U is an arbitrary linear combination of the three generators

$$U_a = x_1\partial/\partial x_1 + x_2\, \partial/\partial x_2, \quad U_b = \partial/\partial x_1, \quad U_c = \partial/\partial x_2. \tag{5.4.8}$$

They generate one-parameter groups of uniform scalings and translations:

$$(x_1, x_2) \to (e^\alpha x_1, e^\alpha x_2), \quad x_1 \to x_1 + \beta, \quad x_2 \to x_2 + \gamma. \tag{5.4.9}$$

As noted above, when x_1 is considered as a function, $f(x_2)$, the transformation yielding (5.4.3) is said to be a *point transformation.* Then

$$\partial\xi_k/\partial x_1\ dx_1/dx_2 + \partial\xi_k/\partial x_2 \to \partial\xi_k/\partial f\, df/dx_2 + \partial\xi_k/\partial x_2 = D\xi_k/Dx_2, \tag{5.4.10}$$

the total derivative of ξ_k with respect to x_2. Consequently,

$$\xi_{1,2} \to \partial\xi_1/\partial f\ df/dx_2 + \partial\xi_1/\partial x_2 - (df/dx_2)(\partial\xi_2/\partial f\, df/dx_2 + \partial\xi_2/\partial x_2), \tag{5.4.11}$$

i.e.

$$\xi_{1,2} = D\xi_1/Dx_2 - (df/dx_2)(D\xi_2/Dx_2).$$

5.5 Contact Transformations

Contact transformations appear in mechanics when one investigates the motion of an object in contact with a surface on which it is able to roll without slipping. In Fig. 5.5.1 the circular wheel of radius r rotates through an angle $d\theta$ about its axis as it rolls along a straight wire. If the center of the wheel advances a distance $\delta x = r\delta\theta$, no slipping has occurred. A contact transformation preserves the form of this condition, converting it to $\delta x' = r\delta\theta'$, and ultimately into $\delta x^* = r\delta\theta^*$. The equivalent equation

$$\delta x - r\delta\theta = 0 \tag{5.5.1a}$$

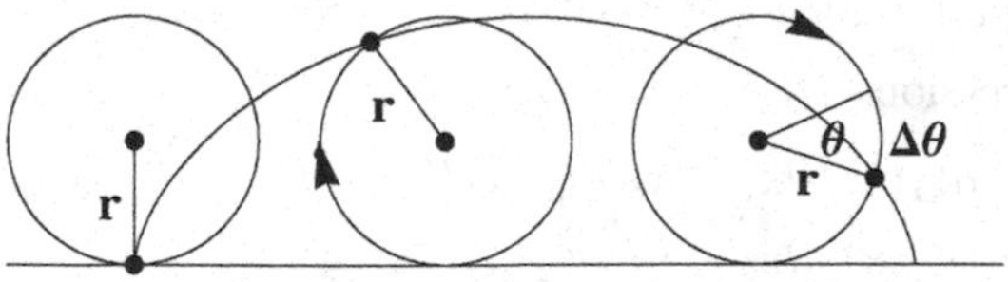

Fig. 5.5.1 $\Delta x = r\Delta\theta$.

is an example of a Pfaffian equation

$$\delta v - \Sigma_i\ y_i\ \delta x_i = 0. \tag{5.5.1b}$$

A transformation that converts the variables in (5.5.1b) to v^*, y_i^*, x_i^*, will leave the equation invariant iff

$$\delta v^* - \Sigma_i y_i^* \delta x_i^* = \rho(v, y_i, x_i)(\delta v - \Sigma_i y_i \delta x_i), \tag{5.5.2}$$

with $\rho \neq 0$ identically. A *contact transformation* is a transformation that transforms v and other variables in the Pfaffian equation (5.5.1b), while leaving the equation invariant.

Equation (5.5.1a) requires that $\delta x/\delta\theta = r$, a constant. It is a special case of the more general equations

$$\begin{aligned} \delta x_1 - x_3\ \delta x_2 &= 0, \\ dx_1 - x_3\ dx_2 &= 0. \end{aligned} \tag{5.5.3a,b}$$

The latter equation is integrable, and establishes that $x_1 = f(x_2)$ plus a constant. In this case

$$\delta x_1/\delta x_2 \rightarrow dx_1/dx_2 \rightarrow df(x_2)/dx_2. \tag{5.5.3c}$$

A contact transformation will convert the derivative $df(x_2)/dx_2$ into a derivative $df^*(x_2^*)/dx_2^*$, and depend upon $df(x_2)/dx_2$.

To determine the form of the generator of a contact transformation of Eqs. (5.5.3), let the generator be

$$U = \xi_1(x_1, x_2, x_3)\partial/\partial x_1 + \xi_2(x_1, x_2, x_3)\partial/\partial x_2 + \xi_3(x_1, x_2, x_3)\partial/\partial x_3 \tag{5.5.4}$$

It generates variations in dx_1, dx_2 and x_3 given by

$$\begin{aligned} \delta(dx_1) &= \delta\alpha\ \Sigma_k\ (\partial\xi_1(x)/\partial x_k)\ dx_k, \\ \delta(dx_2) &= \delta\alpha\ \Sigma_k\ (\partial\xi_2(x)/\partial x_k)\ dx_k, \\ \delta(x_3) &= \delta\alpha\ \xi_3(x). \end{aligned} \tag{5.5.5a–c}$$

To first-order in $\delta\alpha$ this infinitesimal transformation induces the variation $\delta(dx_1 - x_3 dx_2) \equiv \delta\alpha S$, with

$$\begin{aligned} S &= \Sigma_k(\partial\xi_1(x)/\partial x_k)dx_k - x_3\Sigma_k\ (\partial\xi_2(x)/\partial x_k)dx_k - \xi_3(x)\ dx_2 \\ &= A_1 dx_1 + A_2 dx_2 + A_3 dx_3. \end{aligned} \tag{5.5.6a,b}$$

In this last expression

$$\begin{aligned} A_1 &= \partial\xi_1(x)/\partial x_1 - x_3\ \partial\xi_2(x)/\partial x_1, \\ A_2 &= \partial\xi_1(x)/\partial x_2 - x_3\ \partial\xi_2(x)/\partial x_2 - \xi_3(x), \\ A_3 &= \partial\xi_1(x)/\partial x_3 - x_3\ \partial\xi_2(x)/\partial x_3. \end{aligned} \tag{5.5.6c–e}$$

Invariance requires that

$$S = \rho(x)(dx_1 - x_3 dx_2), \tag{5.5.7}$$

because the dx's are independent, invariance will ensue if

$$A_1 \to \rho(x), \quad A_2 \to -x_3\rho(x), \quad A_3 \to 0. \tag{5.5.8a–c}$$

From (5.5.6) one thus obtains

$$\begin{gathered} \partial\xi_1(x)/\partial x_1 - x_3\ \partial\xi_2(x)/\partial x_1 = \rho(x), \\ \xi_3(x) = \partial\xi_1(x)/\partial x_2 - x_3\ \partial\xi_2(x)/\partial x_2 + x_3\rho(x), \\ \partial\xi_1(x)/\partial x_3 - x_3\ \partial\xi_2(x)/\partial x_3 = 0. \end{gathered} \tag{5.5.9a–c}$$

As these equations are compatible with the assumption that $\rho(x)$ is an arbitrary function, they have an uncountable infinity of solutions. Thus an uncountable infinity of Lie transformation groups will leave a derivative dx_1/dx_2 invariant if the transformations are allowed to functionally depend upon the derivative itself. This result contrasts with the result of Sec. 5.4 above, where it was found that a derivative dx_1/dx_2 is left invariant by just three one-parameter groups that carry out point transformations.

Additional properties of contact transformations are uncovered in the exercises at the end of the chapter.

In concluding the discussion in this section we would note that derivative-dependent transformations play a more important, and simpler, role in studies of partial differential equations than they do in studies of ordinary differential equations. This will become evident in subsequent chapters.

5.6 Invariance of an Ordinary Differential Equation of First-Order under Point Transformations; Extended Generators

For notational convenience, let $dy/dx = y_x$ and let the Lie generator that acts on x and y be written

$$U = \xi(x,y)\partial/\partial x + \eta(x,y)\partial/\partial y. \tag{5.6.1}$$

It determines an infinitesimal point transformation

$$x \to x + \delta\alpha\ \xi(x,y), \quad y \to y + \delta\alpha\ \eta(x,y), \tag{5.6.2}$$

and an infinitesimal transformation that acts on y_x. Let the latter be

$$y_x \to y_x + \delta\alpha\ \eta^x. \tag{5.6.3}$$

One can define an *extended generator*

$$U^{(1)} = U + \eta^x \, \partial/\partial y_x. \tag{5.6.4}$$

It is the generator of a Lie group with operator $\exp(\alpha U^{(1)})$ which acts in the space of y_x, x, y. One then has, from (5.4.11),

$$\begin{aligned} \eta^x &= D\eta/Dx - y_x D\xi/Dx. \\ &= \partial\eta/\partial x + y_x(\partial\eta/\partial y - \partial\xi/\partial x) - y_x^2 \, \partial\xi/\partial y. \end{aligned} \tag{5.6.5}$$

Using Theorem 5.1.2 with U interpreted as $U^{(1)}$, one sees that the equation

$$y_x - f(x, y) = 0 \tag{5.6.6}$$

is invariant under the action of a Lie group with extended generator $U^{(1)}$ iff

$$U^{(1)}(y_x - f(x, y)) = 0 \quad \text{when} \quad y_x - f(x, y) = 0. \tag{5.6.7a}$$

This requires that

$$\eta^x - (\xi(x, y)\partial f/\partial x \, + \, \eta(x, y)\partial f/\partial y) = 0 \quad \text{when} \quad y_x - f(x, y) = 0. \tag{5.6.7b}$$

Consequently, for invariance of (5.6.6), one must have

$$\partial\eta/\partial x + f(\partial\eta/\partial y - \partial\xi/\partial x) - f^2\partial\xi/\partial y - Uf = 0. \tag{5.6.8}$$

One may use this analysis to determine the general form of the ODEs $y_x = f(x, y)$ that are left invariant by the transformations of any one-parameter group of physical interest. Consider, for example, the group of rotations about the origin in the x–y plane. Let U be the rotation generator $U = x\partial/\partial y - y\partial/\partial x$. In this case $\xi = -y, \eta = x$. Using (5.6.5) one finds

$$\eta^x = (1 + y_x^2). \tag{5.6.9}$$

If $U^{(1)}$ is to generate an invariance transformation of the equation $y_x = f(x, y)$, then (5.6.7) requires that (5.6.8) becomes

$$(1 + f^2) = Uf. \tag{5.6.10}$$

To solve Eq. (5.6.10) we may change from x, y to $r = \sqrt{(x^2 + y^2)}$, and $\theta = \arctan(y/x)$, the canonical coordinates for U. Then $U = \partial/\partial\theta$. Letting

$$f(x, y) = f(r\cos(\theta), r\sin(\theta)) = F(r, \theta), \tag{5.6.11}$$

Eq. (5.6.10) becomes

$$\partial F/\partial\theta = (1 + F^2). \tag{5.6.12}$$

Its solution is

$$F = -\mathrm{cotan}(\theta + g); \quad g \text{ an arbitrary function of r.} \tag{5.6.13}$$

To express $F(r, \theta)$ as $f(x, y)$ one may use the relations

$$\begin{aligned} \mathrm{cotan}(\theta + g) &= \cos(\theta + g)/\sin(\theta + g), \\ \cos(\theta + g) &= \cos(\theta)\cos(g) - \sin(\theta)\sin(g), \\ \sin(\theta + g) &= \sin(\theta)\cos(g) + \cos(\theta)\sin(g). \end{aligned} \tag{5.6.14}$$

On setting $\cos(\theta) = x/r, \sin(\theta) = y/r$, one finds

$$f = -(ax - by)/(ay + bx), \tag{5.6.15a}$$

with a, b functions of r such that

$$a(r)^2 + b(r)^2 = 1. \tag{5.6.15b}$$

With this restriction, the equation

$$dy/dx = -(ax - by)/(ay + bx) \tag{5.6.16}$$

states the form of the most general ODE that is invariant under the group of rotations carried out by $\exp(\alpha U), U = x\partial/\partial y - y\partial/\partial x$. U itself generates the path curves defined locally by

$$dx/(-y) = dy/x = d\alpha, \tag{5.6.17a}$$

or equivalently, by

$$d\theta = d\alpha, \quad r = \text{const.} \tag{5.6.17b}$$

It follows that along any path curve generated by U,

$$dy/dx = -x/y, \tag{5.6.18a}$$

and

$$xdx + ydy = 0 = d(r^2/2). \tag{5.6.18b}$$

Equation (5.6.18a) coincides with Eq. (5.6.16) if $b = 0$. In all other cases the circular path curves, evolved by the action of $\exp(\alpha U)$ on $\mathbf{r} = (x, y)$, do not coincide with the solution curves of the differential equation (5.6.16). Thus, when $b \neq 0$, in Eq. (5.6.16), rotations interconvert its solution curves.

Lie[1] showed that in such cases, the functions ξ and η in the generator $U = \xi(x, y)\partial/\partial x + \eta(x, y)\partial/\partial y$ determine an integrating factor for the differential equation $y_x = f(x, y)$. This discovery is discussed in Appendix A of this chapter.

5.7 Invariance of Second-Order Ordinary Differential Equations under Point Transformations; Harmonic Oscillators

In the previous section we saw that an infinitesimal point transformation with parameter α and generator

$$U = \xi(x, y)\partial/\partial x + \eta(x, y)\partial/\partial y, \tag{5.7.1}$$

induces a transformation of the derivative y_x to $y_x + \delta\alpha\, \eta^x$. It also induces a transformation of all higher order derivatives. The second derivative y_{xx} is altered to $y_{xx} + \delta\alpha\eta^{xx}$. An argument parallel to that of Sec. 5.4 gives

$$\begin{aligned} \eta^{xx} &= D\eta^x/Dx - y_{xx}\, D\xi/Dx, \\ &= \partial\eta^x/\partial x + y_x\partial\eta^x/\partial y + y_{xx}\partial\eta^x/\partial y_x - y_{xx}(\partial\xi/\partial x + y_x\, \partial\xi/\partial y), \\ &= \eta_{xx} + y_x(2\eta_{xy} - \xi_{xx}) + y_x^2(\eta_{yy} - 2\xi_{xy}) - y_x^3\, \xi_{yy} \\ &\quad + (\eta_y - 2\xi_x)y_{xx} - 3y_x\, y_{xx}\, \xi_y. \end{aligned} \tag{5.7.2a–c}$$

The corresponding *twice extended* generator is

$$U^{(2)} = U + \eta^x\partial/\partial y_x + \eta^{xx}\partial/\partial y_{xx}. \tag{5.7.3}$$

A second-order differential equation

$$g(x, y, y_x, y_{xx}) = 0 \tag{5.7.4}$$

is invariant under the transformation $(x, y) \to (\exp(\alpha U)x, \exp(\alpha U)y)$ iff

$$0 = (U^{(2)}g)|_{g=0}. \tag{5.7.5}$$

As an application of (5.7.5), let us determine the point transformations that leave invariant Newton's equation of motion for a harmonic oscillator. Let this equation of motion be

$$d^2y/dt^2 + y = 0. \tag{5.7.6}$$

Here y represents the position of an oscillating mass and dt corresponds to an infinitesimal time interval. On setting $U = \xi(t, y)\partial/\partial t + \eta(t, y)\partial/\partial y$, one requires

$$U^{(2)}(d^2y/dt^2 + y) \tag{5.7.7}$$

to vanish identically, when $(d^2y/dt^2 + y)$ vanishes. Equations (5.7.2c) and (5.7.6) then yield the determining equation

$$\eta + \eta_{tt} + y_t(2\eta_{ty} - \xi_{tt}) + y_t^2(\eta_{yy} - 2\xi_{ty}) - y_t^3\,\xi_{yy} - (\eta_y - 2\xi_t)y + 3y_t y\,\xi_y = 0. \tag{5.7.8}$$

Here, as required by (5.7.5), we have used (5.7.6) to replace y_{tt} by $-y$. In (5.7.8), ξ and η are functions of y, t, which with y_t are independently variable. Because different powers of y_t are linearly independent, the coefficient of each power of y_t in (5.7.8) must vanish. Consequently,

$$\begin{aligned} &\xi_{yy} = 0, && \eta_{yy} - 2\xi_{ty} = 0, \\ &2\eta_{yt} - \xi_{tt} + 3y\xi_y = 0, && \eta + \eta_{tt} - y(\eta_y - 2\xi_t) = 0. \end{aligned} \tag{5.7.9a–d}$$

To solve these equations, begin with (5.7.9a), from which follows

$$\xi = \xi_0(t) + y\xi_1(t). \tag{5.7.9e}$$

It then follows from (5.7.9b) that

$$\eta_{yy} = 2\partial\xi_1(t)/\partial t. \tag{5.7.9f}$$

Inserting these results in the remaining two equations one finds that the solutions of the equations are linear combinations of eight independent solutions:

$$\eta = \Sigma c_j\,\eta_j, \quad \xi = \Sigma c_j\,\xi_j, \tag{5.7.10}$$

the c_j being arbitrary constants of integration. The solutions may be given the form

$$\begin{aligned} \xi &= c_1 + c_3(-y\cos(t)) + c_4\,y\cos(t) + c_5\,y\sin(t) \\ &\quad + c_6(-y\sin(t)) + c_7\sin(2t) + c_8\cos(2t) \\ \eta &= c_2\,y + c_3\,(1+y^2)\sin(t) + c_4\,(1-y^2)\sin(t) \\ &\quad + c_5\,(1+y^2)\cos(t) + c_6(1-y^2)\cos(t)c_7\,y\cos(2t) + c_8(-y\sin(2t)). \end{aligned} \tag{5.7.11}$$

The resulting generator $U = \xi\partial/\partial t + \eta\partial/\partial y$ is thus of the form $\Sigma c_j U_j$. As there is no choice for the eight constants c_μ such that $\Sigma c_\mu = 0$, there are eight linearly independent generators U_j:

$$\begin{aligned} U_1 &= \partial/\partial t, \\ U_2 &= y\partial/\partial y, \\ U_3 &= (1+y^2)\sin(t)\partial/\partial y - y\cos(t)\,\partial/\partial t, \end{aligned}$$

$$
\begin{aligned}
U_4 &= (1 - y^2)\sin(t)\partial/\partial y + y\cos(t)\ \partial/\partial t,\\
U_5 &= (1 + y^2)\cos(t)\partial/\partial y + y\sin(t)\ \partial/\partial t,\\
U_6 &= (1 - y^2)\cos(t)\partial/\partial y - y\sin(t)\ \partial/\partial t,\\
U_7 &= y\cos(2t)\partial/\partial y + \sin(2t)\ \partial/\partial t,\\
U_8 &= -y\sin(2t)\partial/\partial y + \cos(2t)\ \partial/\partial t.
\end{aligned}
\tag{5.7.12}
$$

In our discussion of point transformations that leave a first-order ODE invariant we found the generators contained arbitrary functions. In contrast, we have just found that our second-order ODE admits point transformations with just eight linearly independent generators, U, none of which contains an arbitrary function. Note however, that we did not obtain the evolution generator of the oscillator given in Chapter 3; it is an explicit function of the momentum, p, which for (5.7.6), is just y_t. If we had allowed our generators in this section to depend upon y_t as well as y and t, that is, to be generators of contact transformations, Eq. (5.7.8) would not have separated into four equations. We would not have obtained a finite number of linearly independent generators each with no free functions. In the general case, a second-order ODE is invariant under contact transformations whose generators may depend upon arbitrary functions.

The harmonic oscillator is one of a small class of second-order ODEs that are invariant under eight different one-parameter groups of point transformations. This is the maximum number allowed any second-order ODE.[3]

5.8 The Commutator of Two Operators

The *commutator*

$$[U, V] \equiv UV - VU \tag{5.8.1}$$

of two operators U and V plays a central role in the study of Lie groups.

If $[U, V] = 0$, then $UV = VU$, and by definition, U and V commute. From the definition, it follows that

$$[U, V] = -[V, U] \tag{5.8.2}$$

and

$$[U_1 + U_2, V] = [U_1, V] + [U_2, V]. \tag{5.8.3}$$

A variable c is a parameter if it commutes with U and V; then

$$[cU, V] = c[U, V] = [U, cV] = [U, V]c. \tag{5.8.4}$$

The operator identity

$$[[U_1, U_2], U_3] + [[U_3, U_1], U_2] + [[U_2, U_3], U_1] = 0 \qquad (5.8.5)$$

may be established by writing the left-hand side out in full and collecting terms. It is known as *Jacobi's relation.*

The following lemma states an important property of the commutator of two first-order differential operators.:

Lemma 5.8.1. *If U and V are first-order differential operators*

$$U = \Sigma\ \xi_i(x)\partial/\partial x_i, V = \Sigma\ \upsilon_j(x)\partial/\partial x_j, \qquad (5.8.6)$$

defined by differentiable $\xi_i(x)$ *and* $\upsilon_j(x)$, *and* $f(x)$ *is a twice differentiable function of its arguments, then*

$$[U, V]f(x) = Wf(x), \qquad (5.8.7a)$$

where

$$W = \Sigma\ \omega_j(x)\partial/\partial x_j \qquad (5.8.7b)$$

is also a first-order differential operator.

The proof is direct. One has

$$\begin{aligned}
&[U, V]f(x)\\
&= (\Sigma_i\ \xi_i(x)\partial/\partial x_i\ \Sigma_j\ \upsilon_j(x)\partial/\partial x_j - \Sigma_j\ \upsilon_j(x)\partial/\partial x_j\ \Sigma_i\ \xi_i(x)\partial/\partial x_i)f(x)\\
&= \Sigma_i\ \Sigma_j(\xi_i(x)\partial/\partial x_i\ \upsilon_j(x)\partial/\partial x_j - \upsilon_j(x)\partial/\partial x_j\ \xi_i(x)\partial/\partial x_i)f(x)\\
&= \Sigma_i\Sigma_j(\xi_i(x)\ \partial\upsilon_j(x)/\partial x_i\ \partial/\partial x_j - \upsilon_j(x)\ \partial\xi_i(x)/\partial x_j\ \partial/\partial x_i)f(x)\\
&\quad + \Sigma_i\Sigma_j(\xi_i(x)\ \upsilon_j(x)\partial/\partial x_i\ \partial/\partial x_j - \upsilon_j(x)\ \xi_i(x)\partial/\partial x_j\ \partial/\partial x_i)f(x).
\end{aligned}$$

Because f(x) is twice differentiable the last line vanishes. On interchanging summation indices in the second term of the previous equality one obtains

$$[U, V]f = Wf, \qquad (5.8.8)$$

with

$$\omega_j(x) = \Sigma_i(\xi_i(x)\ \partial\upsilon_j(x)/\partial x_i - \upsilon_i(x)\ \partial\xi_j(x)/\partial x_i).$$

This result is often written

$$[U, V] = W, \qquad (5.8.9)$$

presupposing that the differential operators U, V, W are to be applied to functions that are at least twice continuously differentiable.

In the Appendix B to this chapter it will be shown that a diffeomorphism that transforms the U, V and W, of (5.8.9) to U', V', W' also transforms $[U', V']$ to W'. In short, *diffeomorphisms do not alter the commutation relations of first-order differential operators.*

5.9 Invariance of Sets of ODEs. Constants of Motion

In this section we use the commutator of differential operators to develop an alternative criterion for determining the generators of groups that leave sets of first-order ODEs invariant. The commutator criterion will be used to characterize the constants of motion of dynamical systems whose evolution is governed by ordinary differential equations.

We begin with the set of autonomous ODEs

$$dx_j/dt = f_j(x), j = 1, \ldots, n, \tag{5.9.1a}$$

or, the equivalent set

$$dt = dx_1/f_1(x) = dx_2/f_2(x) \cdots = dx_n/f_n(x). \tag{5.9.1b}$$

Then, if

$$V = \Sigma_j\ f_j(x)\partial/\partial x_j, \tag{5.9.2}$$

the evolution operator exp(tV) acts on the x_j to convert them from their values at $t = 0$, to their values at $t = t'$.

If all the f's are at least C^1 functions, then, except in regions where (5.9.1) has singular points, local diffeomorphisms

$$\begin{aligned} &x \to y = Y(x), \quad \text{with} \quad y_i = Y_i(x), \\ &y \to x = X(y), \quad x_i = X_i(y) = Y_i^{inv}(x), \end{aligned} \tag{5.9.3}$$

will convert V to $V' = \partial/\partial y_1$, and (5.9.1b) to the canonical form

$$dy_1/dt = 1, \quad dy_j/dt = 0, \quad j = 2, \ldots, m, \tag{5.9.4}$$

with solutions

$$y_1 - t = c_1, \quad y_j = c_j, \quad j = 2, \ldots, m. \tag{5.9.5}$$

As $y_j = Y_j(x)$, the first relation in (5.9.5) establishes that $Y_1(x) = t + c_1$. For $j > 1$, (5.9.5) establishes the relations $Y_j(x) = c_j$. If Eqs. (5.9.1) govern the time evolution of a system of bodies in motion, these $Y_j(x)$ are *constants of the motion* or *integrals of the motion*.[a]

[a] $Y_1(x) - t$ is sometimes called a time-dependent constant of motion. Constants of motion $Y_j(x)$ are divided into two classes: isolating integrals and non-isolating integrals. The distinction is addressed in Appendix C.

To prepare for the statement of the theorem which follows, we append to Eqs. (5.9.2) the equation $\mathrm{dt}/\mathrm{d}\tau = 1$, and use the relation $\mathrm{dt} = \mathrm{d}\tau$ to replace (5.9.1b) by

$$\mathrm{d}\tau = \mathrm{dt}/1 = \mathrm{dx_1}/\mathrm{f_1(x)} = \mathrm{dx_2}/\mathrm{f_2(x)} \cdots = \mathrm{dx_n}/\mathrm{f_n(x)}. \tag{5.9.6}$$

The evolution operator of these equations will be taken to be $\exp(\tau \mathrm{V_t})$,

$$\mathrm{V} = \partial/\partial \mathrm{t} + \mathrm{V} = \partial/\partial \mathrm{t} + \Sigma_\mathrm{j} \mathrm{f_j(x)} \partial/\partial \mathrm{x_j}. \tag{5.9.7}$$

Because the diffeomorphism that converts (5.9.1) to (5.9.4) is independent of t, it converts (5.9.6) to

$$\mathrm{dt}/\mathrm{d}\tau = 1, \quad \mathrm{dy_1}/\mathrm{d}\tau = 1, \quad \mathrm{dy_j}/\mathrm{d}\tau = 0, \quad \mathrm{j} = 2, \ldots, \mathrm{m}. \tag{5.9.8}$$

Thus $\mathrm{V_t}$ is converted to

$$\mathrm{V'_t} = \partial/\partial \mathrm{t} + \partial/\partial \mathrm{y_1}. \tag{5.9.9}$$

$\mathrm{V'_t}$ acts on the solution $\mathrm{y_1 - t = c_1}$ to give zero, and it acts on all the other $\mathrm{y_j}$ to give zero. It follows that if $\phi(\mathrm{y,t}) = \mathrm{c}$ represents *any solution of* (5.9.4), *with or without explicit time dependence*, then

$$\mathrm{V'_t}\, \phi(\mathrm{y,t}) = 0, \tag{5.9.10a}$$

and

$$\exp(\tau \mathrm{V'_t})\, \phi(\mathrm{y,t}) = \phi(\mathrm{y,t}). \tag{5.9.10b}$$

On transforming back to the x system, $\phi(\mathrm{y,t}) \to \Phi(\mathrm{x,t})$, and one obtains the corresponding relations

$$\mathrm{V_t}\Phi(\mathrm{x,t}) = 0, \tag{5.9.11a}$$

and

$$\exp(\tau \mathrm{V_t})\Phi(\mathrm{x,t}) = \Phi(\mathrm{x,t}), \tag{5.9.11b}$$

where $\Phi_1(\mathrm{x,t}) = \mathrm{c}$ defines the solution of (5.9.1) corresponding to $\mathrm{y_1 - t} = \mathrm{c_1}$, and the other Φ_j correspond to $\mathrm{y_j = c_j}$. Let us proceed:

Theorem 5.9.1. *Let the solutions* $\Phi(\mathrm{x,t}) = \mathrm{c}$ *and* $\Phi(\mathrm{x}) = \mathrm{c}$, *of the equations*

$$\mathrm{dx_j}/\mathrm{dt} = \mathrm{f_j(x)}, \mathrm{j} = 1, \ldots, \mathrm{n}, \quad \mathrm{x} = (\mathrm{x_1}, \ldots, \mathrm{x_n}), \tag{5.9.12}$$

be at least twice differentiable. Let

$$V_t = \partial/\partial t + \Sigma_j f_j(x)\partial/\partial x_j, \quad j = 1, \ldots, n. \tag{5.9.13}$$

Require that

$$U = \xi_0(x,t)\partial/\partial t + \Sigma_k \xi_k(x,t)\partial/\partial x_k, \quad k = 1, \ldots, n \tag{5.9.14}$$

satisfy the operator equation

$$[V_t, U] = w(x,t)V_t, \tag{5.9.15}$$

where $w(x,t)$ *is an arbitrary function. Then* U *is the generator of an invariance group of Eqs.* (5.9.12).

Proof. Applying both sides of (5.9.15) to a function $\Phi(x,t)$ or $\Phi(x)$ which defines a solution $\Phi = c$, and using the definition of the commutator, Eq. (5.9.16) becomes

$$(V_t U - U_t V)\Phi = w(x,t)V_t\Phi. \tag{5.9.16}$$

The fact that for any solution, $\Phi = c, V_t\Phi = 0$, requires that the right-hand side of the equation, as well as $UV_t\Phi$, vanish, whence

$$V_t(U\Phi) = 0. \tag{5.9.17}$$

Thus $U\Phi$ is a solution of Eqs. (5.9.12). Consequently, the one-parameter family of functions $\exp(\alpha U)\Phi$ satisfies Eqs. (5.9.12). Q.E.D.

To exemplify the previous discussion, suppose $n = 3$ and that one wishes to have transformations that will convert a given solution of (5.9.4), and hence the differential Eqs. (5.9.1) from which it originated, into any other solution. The solutions of (5.9.4) are $y_1 - t = c_1,\ y_2 = c_2,\ y_3 = c_3$. To convert a particular solution to a general solution one must be able to change all three initial values c_j through the complete allowed range. If all three constants may range over all the reals, then transformations with the generators $\exp(a_j\ \partial/\partial y_j)$ will have the effect of converting each c_j to $c_j' = c_j + a_j$, and thereby yield all other solutions. In this case one or more of the y_k might be an angular variable. If one of the constants, say c_2, is restricted to be nonnegative, the dilatation operator $\exp(\alpha y_2\partial/\partial y_2)$ will convert the solution $y_2 = c_2$ to $\exp(\alpha)x_2 = c_2$, i.e. to $y_2' = \exp(-\alpha)c_2$, giving all allowed nonzero solutions if c_2 is nonzero. Such a situation would arise if y_2 were to be a radial distance, r.

One can distinguish the various possible cases if one knows the transformation that puts the original differential equations into canonical form. One can avoid this by solving the determining equations that arise from the original differential equations themselves. The availability of computer programs that set up and solve determining equations often make this the most practical procedure. A basic mathematical discovery of Reid enables some of these programs to produce series approximations to generators.[4] This is particularly useful when one does not expect exact results in closed form.[5]

Jacobi's relation ensures that if U_1 and U_2 satisfy

$$[\mathrm{V}, \mathrm{U}_1] = c_1\mathrm{V}, \ [\mathrm{V}, \mathrm{U}_2] = c_2\mathrm{V}, \tag{5.9.18}$$

then

$$\mathrm{U}_3 = [\mathrm{U}_1, \mathrm{U}_2] \tag{5.9.19a}$$

satisfies

$$[\mathrm{V}, \mathrm{U}_3] = 0. \tag{5.9.19b}$$

As a consequence, if V (or V_t) is the evolution operator of a set of ODEs, knowing a U_1 and a U_2, which satisfy (5.9.18) and do not commute, allows one to know the generator of another one-parameter invariance group of the ODEs. This procedure may often be repeated. A set of operators $\mathrm{U}_1, \mathrm{U}_2, \mathrm{U}_3, \ldots$ obtained in this way generates groups that often provide a great deal of information about the physical system whose evolution is governed by V and U. These groups with multiple generators are the subject of the next chapter.

5.10 Conclusion

This chapter has developed basic methods for uncovering and using invariance properties of ordinary functions and equations, and of expressions involving infinitesimal changes — metrics, Pfaffians and derivatives. Particular attention has been paid to invariance properties of systems governed by ordinary differential equations.

Trajectories or solutions defined by a given set of ODEs may be converted into one another by one-parameter transformation groups if the trajectories or solutions are diffeomorphic. The groups are invariance groups of the differential equations; they define intrinsic symmetries possessed by the equations.

Ordinary differential equations are left invariant by an uncountable infinity of one-parameter transformation groups. These establish a host of relations between solutions. One may choose among the groups according to the purpose at hand. A common choice is to provide a set of one-parameter groups sufficient to convert any given particular solution into all other solutions that have diffeomorphic trajectories. Especially useful information comes to light if a set of one-parameter groups comprise a *many-parameter group*. Many-parameter groups are defined and explored in the next chapter. Their mathematical properties will be found to restrict geometric interpretations of the intrinsic symmetries of the differential equations.

For further information on the use of systematic group theoretic methods to solve ordinary differential equations the reader is referred to references listed at the end of this chapter. The recent monographs of Bluman and Kumei,[6] and Cantwell,[7] provide clear, useful, and extensive treatments of the invariance properties of many differential equations important in science and engineering. Further helpful monographs are also listed at the end of the chapter.

Appendix A. Relation between Symmetries and Intergrating Factors

As noted earlier, a first-order ODE $\mathrm{dy/dx = f(x,y)}$ may be written as

$$\mathrm{M(x,y)\ dx + N(x,y)\ dy = 0}, \tag{5.A.1}$$

and its family of solutions may be put in the form

$$\phi(\mathrm{x,y}) = \mathrm{c}, \tag{5.A.2}$$

where c is a parameter. Both expressions avoid a distinction between *independent* and *dependent* variables. From (5.A.2) it follows that

$$\mathrm{d}\phi = \partial\phi/\partial\mathrm{x}\ \mathrm{dx} + \partial\phi/\partial\mathrm{y}\ \mathrm{dy} = 0. \tag{5.A.3}$$

For both (5.A.1) and (5.A.3) to hold true it is necessary that

$$\partial\phi/\partial\mathrm{x} = \mathrm{A(x,y)\ M(x,y)}, \quad \partial\phi/\partial\mathrm{y} = \mathrm{A(x,y)\ N(x,y)}. \tag{5.A.4}$$

A(x,y) is an integrating factor for Eq. (5.A.1), for multiplying (5.A.1) by A(x,y) yields

$$\mathrm{A(x,y)(M(x,y)dx + N(x,y)dy)} \tag{5.A.5a}$$

$$= \partial\phi/\partial\mathrm{x}\ \mathrm{dx} + \partial\phi/\partial\mathrm{y}\ \mathrm{dy} = \mathrm{d}\phi. \tag{5.A.5b}$$

It was shown in (5.1.13) that, if $U = \xi(x,y)\partial/\partial x + \eta(x,y)\partial/\partial y$ is such that

$$U\phi(x,y) = \xi(x,y)\partial\phi/\partial x + \eta(x,y)\partial\phi/\partial y = 1, \qquad (5.A.6)$$

then $\exp(\alpha U)$ has the effect of converting a solution $\phi(x,y) = c$, into a different solution $\phi(x,y) = c'$ with $c' = c\exp(-\alpha)$. We next show that knowing such a U allows one to determine the integrating factor A(x,y).

Using (5.A.4) to eliminate the derivatives in (5.A.6) one obtains

$$A(x,y)(\xi(x,y)M(x,y) + \eta(x,y)N(x,y)) = 1. \qquad (5.A.7)$$

It follows that

$$A(x,y) = 1/(\xi M + \eta N). \qquad (5.A.8)$$

Thus, $1/(\xi M + \eta N)$ *is an integrating factor of differential equation* (5.A.1).

This connection between integrating factors for differential equations and transformations that interconvert solutions of differential equations is of considerable utility. Cohen's monograph, *The Lie Theory of One Parameter Groups,*[8] contains an extensive tabulation of the invariance groups of many types of first-order ordinary differential equations, together with their integrating factors.

Appendix B. Proof That the Commutator of Lie Generators is Invariant under Diffeomorphisms

Let T symbolize the change of variables defined by a diffeomorphism

$$x_j \to y_j = Tx_j = f_j(x), \qquad (5.B.1a)$$

for example,

$$x \to y = Tx = x^2, \quad 0 < x < \infty. \qquad (5.B.1b)$$

The inverse transformation, symbolized by T^{inv}, is defined by

$$y_j \to x_j = T^{inv}y_j = f_j^{inv}(y), \qquad (5.B.1c)$$

e.g.

$$y \to x = T^{ivn}y = y^{1/2}, \quad 0 < y < \infty. \qquad (5.B.1d)$$

When the transformations of a diffeomorphism act on a function of the x or y variables of (5.B.1), they, by definition, act on the variables x,y themselves. That is,

$$\mathrm{TG(x) = G(Tx) = G(y) = G(f(x)),} \tag{5.B.2a}$$
$$\mathrm{T^{inv}G(y) = G(T^{ivn}y) = G(x) = G(f^{inv}(y)).} \tag{5.B.2b}$$

Note that the transformations of a Lie group acting on an analytic function g(x) obey (5.B.2a) and (5.B.2b) because[a]

$$\mathrm{exp(\alpha U)g(x) = T(\alpha)g(x) = g(T(\alpha)x).} \tag{5.B.3}$$

In an expression such as $\mathrm{Tg_1(x)g_2(x)}$, T is construed to act on the compound expression, that is

$$\mathrm{Tg_1(x)g_2(x) = T(g_1(x)g_2(x)) = g_1(Tx)g_2(Tx).} \tag{5.B.4}$$

As a consequence one may use the identity $\mathrm{T^{inv}T = I}$ to write

$$\begin{aligned}\mathrm{Tg_1(x)g_2(x)} &= \mathrm{Tg_1(x)T^{inv}Tg_2(x) = (Tg_1(x)T^{inv})Tg_2(x),} \\ &= \mathrm{(Tg_1(x)T^{inv})g_2(Tx).}\end{aligned} \tag{5.B.5}$$

To determine which form

$$\mathrm{U = \Sigma\xi_j(x)\partial/\partial x_j} \tag{5.B.6}$$

takes when acting on the y variables of (5.B.1), we begin with the action of U on $\mathrm{x_k}$, *viz.* $\mathrm{Ux_k}$, and use T to transform it to an action on the corresponding y variable. Then

$$\mathrm{T(Ux_k) = (TUT^{inv})Tx_k} \tag{5.B.7a}$$
$$\mathrm{= (TUT^{inv})y_k = (TUT^{inv})f_k(x).} \tag{5.B.7b}$$

The first term in (5.B.7a) is evidently just $\mathrm{T\xi_k(x) = \xi_k(Tx)}$. The first term in (5.B.7b) represents the action of the transformed generator on $\mathrm{y_k}$. In Sec. 5.7, we investigated the effect of a change of variables on a group generator. Setting $\mathrm{y_k = x'_k}$ in Eqs. (4.7.3)–(4.7.5) gives

$$\mathrm{\xi'_k(y) = Uy_k|_{x=f^{inv}(y)} = (\Sigma_j\xi_j(x)f_{k,j}(x))|_{x=f^{inv}(y)},} \tag{5.B.8a}$$

[a] Some authors require (5.B.3) to be true, by definition, for any g(x). We have not done so because, in physical applications, doing so can easily lead to error.

where

$$f_{k,j}(x) = \partial f_k(x)/\partial x_j. \quad (5.B.8b)$$

This establishes that the diffeomorphism $x \to y = Tx$ *transforms* $U = \Sigma\, \xi_j(x)\partial/\partial x_j$ to

$$TUT^{inv} = \Sigma\, \xi'_k(y)\partial/\partial y_k. \quad (5.B.9)$$

We are now ready to prove that the commutator of two Lie generators is invariant under diffeomorphisms.

Theorem 5.B.1. *Let*

$$W = [U, V], \quad (5.B.10a)$$

with

$$U = \Sigma\, \xi_i(x)\partial/\partial x_i, \quad V = \Sigma\, \upsilon_j(x)\partial/\partial x_j, \quad W = \Sigma\, \omega_k(x)\partial/\partial x_k. \quad (5.B.10b)$$

Let

$$U' = TUT^{inv}, \quad V' = TVT^{inv}, \quad W' = TWT^{inv}. \quad (5.B.11)$$

Then

$$W' = [U', V']. \quad (5.B.12)$$

Proof.

$$\begin{aligned} TWT^{inv} &= T[U, V]T^{inv} = T(UV - VU)T^{inv} \\ &= TUT^{inv}TVT^{inv} - TVT^{inv}TUT^{inv} \\ &= [TUT^{in}, TVT^{in}]. \end{aligned} \quad (5.B.13a–c)$$

That is to say,

$$W' = [U', V']. \quad (5.B.14)$$

Q.E.D.

Appendix C. Isolating and Non-isolating Integrals of Motion

Roughly speaking, small errors in measurements lead to small errors in the value of isolating integrals, and small errors in estimates of the value of isolating integrals result in small errors in predicted motions. In contrast, small errors in calculating or estimating non-isolating integrals can lead to large errors in predicted motions. However, the difference in meaning of *small* and *large* in these statements is subtle. We can here only briefly address this subtlety. In Eqs. (5.9.4), fixing the value of c_1, and each of the constants of motion, $y_k = Y_k(x), k > 1$, at a time $t = t_0$ would completely fix the coordinates of the bodies in the space of the y's, and in the space of the x's, at t_0. As t varies, the point representing the state of the system traces out a line, which is the trajectory of the system in the space of the x's or y's. Each successive point on this line determines a state of the system at successive points of time. Because all the y_k are related to the x's by diffeomorphisms, the $Y_k(x)$ are single-valued functions. As a consequence trajectories evolving from different points cannot cross. Furthermore, except in the region of singular points of (5.9.1), infinitesimal changes δx_j in initial values $x_j|_{t=0}$, will produce infinitesimal changes in the constants of motion $Y_k(x)$. However, small finite changes $\Delta x_j|_{t=0}$ may produce changes $\Delta Y_k|_{t=0}$ that grow exponentially with the $\Delta x_j|_{t=0}$, and small finite changes $\Delta y_j|_{t=0}$ may produce changes in the $\Delta x_k|_{t=0}$ that grow exponentially with the $\Delta y_j|_{t=0}$. Under these circumstances, some trajectories initially well separated may closely approach one another, then recede — and in general wander about so much that they will pass through any finite hypervolume of the space allowed for them by conservation of energy, momentum, angular momentum, and any other isolating integrals. An example is provided by the Henon and Heiles system,[11] in which the potential function is $(x^2+y^2)/2+x^2y-y^3/3$. In these cases knowing a value of the corresponding constant of motion can tell little about the evolution of the system. The constants so affected are the ones that are said to be non-isolating.

The study of the stability and reliability of predictions based upon constants of motion of systems involving three or more gravitationally interacting bodies has a long history. For a discussion of the contributions made by Jacobi, Bruns, Painleve, and Poincare in the nineteenth and early twentieth century, the reader is referred to Whittaker's monograph.[12] These workers established fundamental mathematical results and showed that constants of motion other than those of energy, momentum, and angular momentum would generally have limited use in making long range predictions, and retrodictions, of planetary and lunar motions. In the mid twentieth century further fundamental advances were made by Kolmogorov,[13] Arnold,[14] and Moser.[15] Treating the sun and planets as point masses it was shown, for example, that for two planetary orbits whose planes intersect at low angles there are stable constants of motion which predict the constancy of the inclination.

Exercises

1. Is the equation dv − ydx = dc, with dc arbitrary, left invariant by a non-denumerable, or a denumerable, infinity of one-parameter groups?
2. What equations of the form (5.6.16) have path curves that are themselves circles?
3. Determine the generators of the transformations that leave invariant the space metrics obtained from (2.7.4) by setting dz = 0.
4. What is the action of each of the generators (5.7.12a–d) on the general solution of the equation of motion: $y - (A\sin(t) + B\cos(t)) = 0$? Which U's generate a transformation that leaves a particular solution invariant? For further information see Ref. 2.
5. Complete the solution of Eqs. (5.7.9a–d) and compare your result with (5.7.11).
6. Write down a set of first-order ODEs corresponding to the equation $d^2x/dt^2 = 0$. Obtain the determining equation for the point transformations that leave this equation invariant. Then use (5.9.16) to obtain the same determining equation. What choice of the function w must be made? Solve the determining equations. Show that the solution is a linear combination of eight linearly independent U's.

References

[1] S. Lie, *Christ. Forh. Aar.*, 1874 (1875) 242–254; Collected Works, Vol. III, No. XIII, p. 176.
[2] C. E. Wulfman and B. G. Wybourne, *J. Phys. A* **9** (1976) 507–518.
[3] S. Lie, *Vorlesungen uber Continuierliche Gruppen* (Chelsea, Bronx, N.Y., 1971).
[4] G. Reid, *J. Phys. A* **23** (1990) L853; G. J. Reid, *Eur. J. Appl. Math.* **2** (1991) 293; *loc. cit.*, 319.
[5] W. Hereman, *Euromath Bull.* **1** (1994) 45.
[6] G. W. Bluman and S. Kumei, *Symmetries and Differential Equations* (Springer, N.Y., 1989).
[7] B. J. Cantwell, *Introduction to Symmetry Analysis* (Cambridge University Press, Cambridge, 2002).
[8] A. Cohen, *An Introduction to The Lie Theory of One-parameter Groups* (D. C. Heath, 1911), reprinted by (G. E. Stechert, N.Y., 1931).
[9] P. J. Olver, *Application of Lie Groups to Differential Equations* (Springer-Verlag, N.Y., 1986).
[10] R. Sheshadri and T. Y. Na, *Group Invariance in Engineering Boundary Value Problems* (Springer-Verlag, N.Y., 1985).
[11] M. Henon and C. Heiles, *Astron. J.* **69** (1964) 73–9.

[12] E. T. Whittaker, *A Treatise on the Analytical Dynamics of Particles and Rigid Bodies*, 4th edn., Ch. XII - XV (Cambridge University Press, Cambridge 1959).
[13] A. N. Kolmogorov, *Proc. Int. Congress Math.* (Amsterdam, 1957); translated in Appendix D of R. Abraham's *Foundation of Mechanics* (Benjamin, N. Y., 1967).
[14] V. I. Arnold, *Russian Math. Surveys* **18** (1963) 5,6.
[15] J. Moser, *Stable and Random Motions in Dynamical Systems* (Princeton University Press, Princeton, N.J, 1973).

CHAPTER 6

Lie Transformation Groups and Lie Algebras

In this chapter many-parameter groups are defined, and the concepts of *Lie group* and Lie algebra are then developed along the lines initiated by Lie.[1] The further topics selected are those most often needed in applied work.

6.1 Relation of Many-Parameter Lie Transformation Groups to Lie Algebras

Suppose a succession of r different one-parameter transformation groups with group parameters α_k act on a set of variables $\mathbf{x} = (\mathbf{x}_1, \ldots, \mathbf{x}_n)$. This action may be expressed as

$$\mathbf{x} \to \mathbf{x}' = \exp(\alpha_r U_r)\exp(\alpha_{r-1}U_{r-1})\cdots\exp(\alpha_2 U_2)\exp(\alpha_1 U_1)\mathbf{x}.$$

$$= T(\alpha_r, \alpha_{r-1}, \ldots, \alpha_2, \alpha_1)\mathbf{x}. \tag{6.1.1a}$$

Here each

$$U_k = \Sigma_j\ \xi_{j,k}(\mathbf{x})\partial/\partial x_j \tag{6.1.1b}$$

is the generator of a one-parameter transformation group in which $T(\alpha_k)\mathbf{x} = \exp(\alpha_k U_k)\mathbf{x}$. Let us label the ordered set of group parameters by

$$\boldsymbol{\omega} = (\alpha_r, \alpha_{r-1}, \ldots, \alpha_2, \alpha_1), \tag{6.1.2a}$$

so $T(\alpha_r, \alpha_{r-1}, \ldots, \alpha_2, \alpha_1)$ becomes $T(\omega)$. Then

$$x' = T(\omega)x, \tag{6.1.2b}$$

and for $j = 1, \ldots, n$,

$$T(\omega)x_j = x_j'$$

with

$$x_j' = \Phi_j(x; \omega). \tag{6.1.2c}$$

We will write $\omega = 0$ iff all $\alpha_k = 0$. If the operation of $T(\omega)$ on x is that of an r-parameter group, then for each x_j the functions $\Phi_j(x;\omega)$ yield the identity transformation, T(0), when every group parameter α_k vanishes, and one must have

$$x_j = \Phi_j(x;\omega)|_{\omega=0}. \tag{6.1.2d}$$

For $T(\omega)$ to define a group, it is necessary that it has an inverse. Referring to (6.1.1) one sees that its inverse, $T^{inv}(\omega)$, is

$$T(\omega) = T(-\alpha_1, -\alpha_2, \ldots, -\alpha_{r-1}, -\alpha_r) \tag{6.1.3a}$$

(the parameters are listed in reverse order because the operations must be carried out in that order). Then

$$T(\omega^{in})T(\omega)x = T(\omega)T(\omega^{in})x = T(0)x = I\,x = x. \tag{6.1.3b}$$

For $T(\omega)$ to be the operator of a group, it is also necessary that the action of two successive operations can be expressed as the operation of the group, that is

$$T(\omega')T(\omega)x = T(\omega'')x. \tag{6.1.4a}$$

It requires that, for all $k = 1, \ldots, r$, one must have $\alpha''_k = \Omega_k(\omega', \omega)$, with

$$\alpha''_k \to \alpha'_k \text{ as } \omega \to 0, \quad \alpha'_k \to \alpha_k \text{ as } \omega' \to 0. \tag{6.1.4b}$$

Definition 6.1.1. A group is said to be a Lie group iff its composition functions $\Omega_k(\omega', \omega)$ are analytic functions in finite regions containing the values of ω' and ω that define identity transformations.

It follows that for Lie groups, the functions Ω_k can be expanded as a convergent power series in the group parameters α'_j and α_j within finite regions in which the parameters are sufficiently small.

To begin to develop an understanding of what is entailed when a group is a Lie group, consider the case of a two-parameter transformation group in which

$$x' = T(\alpha_2, \alpha_1)x = \exp(\alpha_2 U_2)\exp(\alpha_1 U_1)x, \tag{6.1.5}$$

with $x = (x_1, x_2)$. As the group-parameters approach zero,

$$T(\alpha_2, \alpha_1) \to T(\delta\alpha_2, \delta\alpha_1),$$

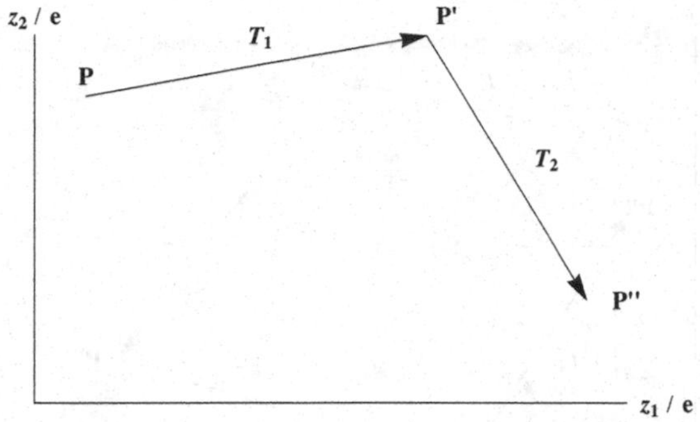

Fig. 6.1.1a Action of two successive infinitesimal transformations.

and

$$\begin{aligned} x' &\to (1 + \delta\alpha_2 U_2)(1 + \delta\alpha_1 U_1)x, \\ &= (1 + \delta\alpha_2 U_2 + \delta\alpha_1 U_1 + \delta\alpha_1\delta\alpha_2 U_2 U_1)x. \end{aligned} \tag{6.1.6a,b}$$

In **Fig. 6.1.1a** this transformation moves a point P with coordinates $x = (z_1, z_2)$ to P′ and then to P″ with coordinates $\mathbf{x}' = (z_1 + \delta z_1, z_2 + \delta z_2)$. Note that (6.1.6c) contains all terms in the expansion of (6.1.6a) that are linear in both $\delta\alpha_1$ and $\delta\alpha_2$. In the same approximation, the inverse transformation is

$$\begin{aligned} T(-\delta\alpha_1, -\delta\alpha_2) &= (1 - \delta\alpha_1 U_1)(1 - \delta\alpha_2 U_2) \\ &= (1 - \delta\alpha_2 U_2 - \delta\alpha_1 U_1 + \delta\alpha_1\delta\alpha_2 U_1 U_2). \end{aligned} \tag{6.1.7}$$

In **Fig. 6.1.1a**, the inverse transformation carries P''_x to P'_x and then to P. Expanding $T(-\delta\alpha_1,\ -\delta\alpha_2)T(\delta\alpha_2,\ \delta\alpha_1)$ one obtains

$$1 + O(\delta\alpha_1^2,\ \delta\alpha_2^2). \tag{6.1.8}$$

If one carries out the general transformation $T(\beta_2,\ \beta_i)T(\alpha_2,\ \alpha_1)x$, the resulting operator must be able to be expressed as

$$T(\gamma_2, \gamma_1) = \exp(\gamma_2 U_2)\exp(\gamma_1 U_1), \tag{6.1.9a}$$

with

$$\gamma_1 = \Omega_1(\alpha_2, \alpha_1, \beta_2, \beta_i), \quad \gamma_2 = \Omega_2(\alpha_2, \alpha_1, \beta_2, \beta_i). \tag{6.1.9b}$$

A fundamental property of Lie groups is exposed if, instead of following the transformation (6.1.6) by its inverse, it is followed by

$$\begin{aligned} T(-\delta\alpha_2, -\delta\alpha_1) &\to (1 - \delta\alpha_2 U_2)(1 - \delta\alpha_1 U_1) \\ &= (1 - \delta\alpha_2 U_2 - \delta\alpha_1 U_1 + \delta\alpha_1\delta\alpha_2 U_2 U_1). \end{aligned} \tag{6.1.10}$$

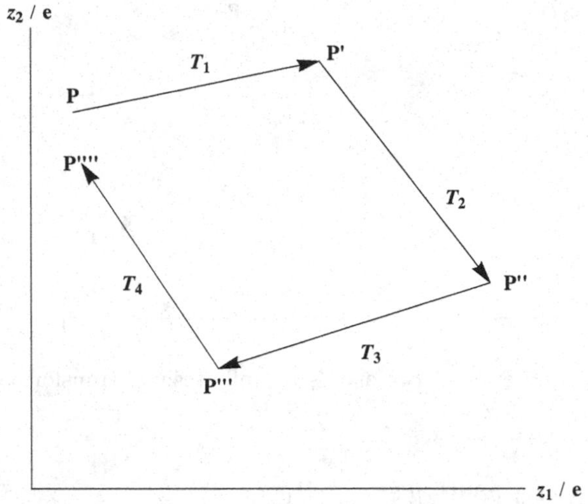

Fig. 6.1.1b Action of two successive infinitesimal transformations followed by the reverse of their inverse.

Figure 6.1.1b illustrates effect of the resulting transformation

$$\begin{aligned} &\mathrm{T}(-\delta\alpha_2, -\delta\alpha_1)\mathrm{T}(\delta\alpha_2, \delta\alpha_1)\mathrm{x} \\ &\quad \to (1 - \delta\alpha_2 \mathrm{U}_2)(1 - \delta\alpha_1 \mathrm{U}_1)(1 + \delta\alpha_2 \mathrm{U}_2)(1 + \delta\alpha_1 \mathrm{U}_1)\mathrm{x}. \end{aligned} \tag{6.1.11a}$$

Expanding the RHS of this, one finds that through terms first order in $\delta\alpha_1$ and/or $\delta\alpha_2$,

$$\mathrm{T}(-\delta\alpha_2, -\delta\alpha_1)\mathrm{T}(\delta\alpha_2, \delta\alpha_1) \to 1 + \delta\alpha_1\delta\alpha_2(\mathrm{U}_2\mathrm{U}_1 - \mathrm{U}_1\mathrm{U}_2). \tag{6.1.11b}$$

For this to be expressed as an approximation to $\exp(\gamma_2 \mathrm{U}_2)\exp(\gamma_1 \mathrm{U}_1)$, the commutator, $[\mathrm{U}_2, \mathrm{U}_1] = (\mathrm{U}_2\mathrm{U}_1 - \mathrm{U}_1\mathrm{U}_2)$, must be expressible as a linear combination of U_1 and U_2:

$$[\mathrm{U}_2, \mathrm{U}_1] = \mathrm{c}_{21}^1 \mathrm{U}_1 + \mathrm{c}_{21}^2 \mathrm{U}_2. \tag{6.1.12}$$

Then (6.1.11) becomes an approximation to

$$\exp(\gamma_2 \mathrm{U}_2)\exp(\gamma_1 \mathrm{U}_1), \tag{6.1.13}$$

in which

$$\gamma_2 \to \delta\alpha_1\ \delta\alpha_2\ \mathrm{c}_{21}^2, \quad \gamma_1 \to \delta\alpha_1\delta\alpha_2\ \mathrm{c}_{21}^1.$$

The result obtained in this example is a special case of the following

Theorem 6.1.1. *The generators*

$$U_\kappa = \Sigma_j \, \xi_{j\kappa}(x) \, \partial/\partial x_j, \quad \kappa = 1, \ldots, r, \tag{6.1.14a}$$

of r one-parameter groups are the generators of an r*-parameter Lie group iff they obey the commutation relations*

$$[U_\kappa, U_\lambda] = \Sigma_\mu C^\mu_{\kappa\lambda} U_\mu, \quad \kappa,\ \lambda,\ \mu = 1, \ldots, r, \tag{6.1.14b}$$

in which the $c^\mu_{\kappa\lambda}$ *are neither functions of the variables x nor of the group parameters ω.*

The generators satisfy a further condition

$$[[U_\kappa, U_\lambda], U_\mu] + [[U_\mu, U_\kappa], U_\lambda] + [[U_\lambda, U_\mu], U_\kappa] = 0. \tag{6.1.14c}$$

This condition, a Jacobi relation, is automatically satisfied because the Lie generators U are first-derivative operators, and for this reason it is also called the *derivative* condition.

Let us consider examples of operators that satisfy (6.1.14). In Euclidean three-space, the generators of rotations may be taken to be

$$R_x = y\,\partial/\partial z - z\,\partial/\partial y, \quad R_y = z\,\partial/\partial x - x\,\partial/\partial z, \quad R_z = x\,\partial/\partial y - y\,\partial/\partial x. \tag{6.1.15a}$$

They satisfy the commutation relations

$$[R_x, R_y] = -R_z, \quad [R_z, R_x] = -R_y, \ [R_y, R_z] = -R_x. \tag{6.1.15b}$$

The nonzero structure constants are

$$c^z_{xy} = -1, \quad c^y_{zx} = -1, \quad c^x_{yz} = -1 \tag{6.1.15c}$$

and

$$c^z_{yx} = -c^z_{xy} = 1, \quad c^y_{xz} = -c^y_{zx} = 1, \quad c^x_{zy} = -c^x_{yz} = 1.$$

We have previously seen that each of the generators R annihilates the function $x^2+y^2+z^2$, and that the operations of the corresponding one-parameter groups carry out rotations and leave invariant the Euclidean metric in three-space.

The generators of the group SO(2,1) which leaves invariant the value $x^2 + y^2 - z^2$ and the metric $(dx^2 + dy^2 - dz^2)$ are the R_z given above, and

$$S_x = y\,\partial/\partial z + z\,\partial/\partial y, \quad S_y = z\,\partial/\partial x + x\,\partial/\partial z. \tag{6.1.16a}$$

Their commutation relations are

$$[S_x, S_y] = -R_z, \quad [R_z, S_x] = S_y, \quad [S_y, R_z] = S_x. \tag{6.1.16b}$$

The nonzero structure constants are

$$c^z_{xy} = -1, \quad c^y_{zx} = 1, \quad c^x_{yz} = 1 \tag{6.1.16c}$$

and

$$c^z_{yx} = -c^z_{xy} = 1, \quad c^y_{xz} = -c^y_{zx} = -1, \quad c^x_{zy} = -c^x_{yz} = -1.$$

Note how these differ from the commutation relations obeyed by the generators of the rotation group.

Lie used the results stated in (6.1.14) to determine all possible many-parameter local Lie groups acting in spaces of one, two, and three variables.[2] Consider, for example, the following set of generators of one-parameter transformation groups acting on a single variable, x:

$$\partial/\partial x, \quad x\,\partial/\partial x, \quad x^2\,\partial/\partial x, \quad x^3\,\partial/\partial x. \tag{6.1.17}$$

By direct calculation,

$$[x^a\,\partial/\partial x, x^b\,\partial/\partial x]f(x) = (a-b)x^{(a+b-1)}\partial/\partial x f(x), \tag{6.1.18a}$$

i.e.

$$[x^a\,\partial/\partial x, x^b\,\partial/\partial x] = (a-b)x^{(a+b-1)}\partial/\partial x. \tag{6.1.18b}$$

Consequently

$$[x^2\partial/\partial x, x^3\partial/\partial x] = -x^4\partial/\partial x, \tag{6.1.18c}$$

and

$$[x\,\partial/\partial x, x^3\,\partial/\partial x] = -2x^3\,\partial/\partial x. \tag{6.1.18d}$$

It follows that $x\partial/\partial x$, $x^3\partial/\partial x$ are generators of a two-parameter group, but that $x^2\partial/\partial x$, $x^3\partial/\partial x$ are not. Relations (6.1.18) make it evident that there is no r-parameter group which contains amongst its generators both $x^2\partial/\partial x$ and $x^3\partial/\partial x$; this is simply because their commutator yields $x^4\partial/\partial x$ and this then gives rise to a commutator producing $x^5\partial/\partial x$, ad infinitum. However, the operators $\partial/\partial x$, and $x^n\partial/\partial x$ are generators of a two-parameter group. Note also that the operators $\partial/\partial x$, $x\partial/\partial x$, $x^2\partial/\partial x$ are generators of a three-parameter group.

These results have an immediate consequence. Suppose that the members of a set of generators are operators $U_{(a)} = \xi_a(x)\partial/\partial x$ whose $\xi_a(x)$ have power series expansions beginning with x^a. The previous discussion establishes that there is no r-parameter transformation group which contains as generators both $U_{(2)}$ and $U_{(3)}$, and so forth. Lie used extensions from this observation to construct a wide variety of many-parameter groups transforming many variables.[3]

In the previous examples, the commutator of two group generators is itself a group generator, rather than the more general linear combination of generators allowed by the theorem. However, it is easily shown that if a set of r generators U is closed under commutation, then any set of r generators U′ obtained by taking independent linear combinations of U, is also closed under commutation. As a consequence, a set of linearly independent linear combinations of generators of a group may themselves be considered generators of a group.

6.2 The Differential Equations That Define Many-Parameter Groups

In Chapter 4 it was shown that a one-parameter transformation group acting on n variables x_i is defined by a set of n first-order ODEs,

$$\partial x_i/\partial \alpha = \xi_i(x), \quad i = 1, 2, \ldots, n, \tag{6.2.1a}$$

and that the $\xi_i(x)$ define a group generator

$$U = \Sigma_i\ \xi_i(x)\partial/\partial x_i. \tag{6.2.1b}$$

At first glance, one might suspect that an analogous set

$$\partial x_i/\partial \alpha_j = \xi_{ij}(x), \quad i = 1, 2, \ldots, n, \quad j = 1, 2, \ldots \tag{6.2.1c}$$

would define a many-parameter group with generators

$$U_j = \Sigma_i\ \xi_{ij}(x)\ \partial/\partial x_i. \tag{6.2.1d}$$

However, these equations are not sufficiently restrictive; they impose no connection between the one-parameter groups defined for each value of j. Many-parameter groups can be defined by coupled first-order equations

$$\partial x_i/\partial \alpha_j = \zeta_{i,j}(x, \alpha_1, \alpha_2, \ldots) \tag{6.2.2a}$$

As shown in the appendix, for such equations to define a many-parameter group they must be further restricted. First of all, one must have

$$\zeta_{i,j}(x,\omega) = \Sigma_k\, u_{ik}(x)\lambda_{kj}(\omega), \tag{6.2.2b}$$

so that the defining differential equations are of the form

$$\partial x_i/\partial\alpha_j = \Sigma_k\, u_{ik}(x)\lambda_{kj}(\omega). \tag{6.2.2c}$$

For this set of partial differential equations to be integrable, one must have

$$\partial^2 x_i/\partial\alpha_j\, \partial\alpha_m = \partial^2 x_i/\partial\alpha_m\, \partial\alpha_j. \tag{6.2.2d}$$

This requires that

$$\Sigma_k\, \partial(u_{ik}(x)\lambda_{kj}(\omega))/\partial\alpha_m = \Sigma_k\, \partial(u_{ik}(x)\, \lambda_{km}(\omega))/\partial\alpha_j. \tag{6.2.2e}$$

These requirements are certainly satisfied if the $\zeta_{i,j}(x,\omega)$ are analytic functions of their arguments. Using them Lie established his First Fundamental Theorem[4]:

Theorem 6.2.1. *If the* $u_{ik}(x)$ *are linearly independent analytic functions, the equations*

$$\partial x_i/\partial\alpha_j = \Sigma_k\, u_{ik}(x)\lambda_{kj}(\omega), \quad i = 1,\ldots,n, \quad j,k = 1,\ldots,r, \tag{6.2.3a}$$

and

$$\Sigma_k\, \partial(u_{ik}(x)\, \lambda_{kj}(\omega))/\partial\alpha_m = \Sigma_k\, \partial(u_{ik}(x)\, \lambda_{km}(\omega))/\partial\alpha_j, \tag{6.2.3b}$$

determine an r-parameter group whose generators are

$$U_k = \Sigma_j\, u_{jk}(x)\, \partial/\partial x_j. \tag{6.2.3c}$$

The transformations of the group $x_j \to x'_j$ are defined by $x'_j = \Phi_j(x;\alpha)$ *in which the* $\Phi_j(x;\alpha)$ *are analytic functions of their arguments in regions surrounding* $(x;0)$.

We have previously dealt with a case in which the region does not include all of the space: the transformation $x' = x/(1-\alpha x)$ which is not defined when $\alpha x = 1$. Its generator is $U = u(x)\, \partial/\partial x$, where $u(x)$ is the analytic function x^2.

Because the functions $\xi(x)$ defining the generators of Lie groups are assumed to be analytic functions they may always be approximated by power series. The commutation relations of the generators may be determined from a knowledge of terms in the series that are of low order. This property of Lie transformation groups is, as expected, a great help in applied work.[5]

The operator of an r-parameter Lie group with generators U_κ can be put in the form

$$S(\alpha_1, \alpha_r) = \exp(\Sigma_\kappa\ \alpha_\kappa U_\kappa) \tag{6.2.4a}$$

as well in the form

$$\begin{aligned} T(\alpha_1, \alpha_r) &= \exp(\alpha_1 U_1)\exp(\alpha_2 U_2)\cdots\exp(\alpha_r U_r) \\ &= \Pi_\kappa \exp(\alpha_\kappa U_\kappa), \end{aligned} \tag{6.2.4b}$$

not to mention a variety of other forms. The first form is particularly useful when we deal with minuscule values of the parameters.

If the generators of a group do not commute, different orderings of the operations in (6.2.4b) will have different actions. A number of formulae, known as Baker–Campbell–Hausdorff relations, use commutation relations among the group generators to interrelate these.[6–8] Commutation relations of the group generators may also, in some cases, be used to express the operators $S(\alpha_1, \alpha_r)$, or $T(\alpha_1, \alpha_r)$, as products of one-parameter group operators $\exp(\alpha_\kappa U_\kappa)$, in which not all the U_κ are different. Operators of the rotation group are, for example, commonly put in the form

$$T(\gamma, \beta, \alpha) = \exp(\gamma U_3)\exp(\beta U_2)\exp(\alpha U_3), \tag{6.2.5}$$

where the γ, β, α are Euler's angles of rotation.[9]

Group operators of the form $\exp(\Sigma_\kappa\ \alpha_\kappa U_\kappa)$ avoid complications produced by different orderings in the form $\Pi_\kappa \exp(\alpha_\kappa U_\kappa)$.

6.3 Real Lie Algebras

In this section we investigate properties of group generators that make it possible to simplify and systematize much of the study of Lie groups that is important in physical applications.

We begin with a set of generators U_κ that satisfy the condition

$$[U_\kappa, U_\lambda] = \Sigma_\mu\ C^\mu_{\kappa\lambda} U_\mu, \quad \kappa, \lambda, \mu = 1, \ldots, r, \tag{6.3.1a}$$

and the Jacobi relation

$$[[U_\kappa, U_\lambda], U_\mu] + [[U_\mu, U_\kappa], U_\lambda] + [[U_\lambda, U_\mu], U_\kappa] = 0. \tag{6.3.1b}$$

As before, the $c^\mu_{\lambda\kappa}$ are neither functions of the variables x, nor of the group parameters ω. The r different U's are required to be linearly independent, that is, they do not satisfy any condition of the form

$$c_1U_1 + c_2U_2 + \cdots + c_rU_r = 0. \tag{6.3.1c}$$

One may imagine the U_κ to be represented by *basis vectors* u_κ of a linear algebra with, e.g.

$$U_1 \sim \mathbf{u}_1\ (1,0,0,\ldots),\ U_2\ \sim\ \mathbf{u}_2 = (0,1,0,\ldots),\quad etc. \tag{6.3.2a}$$

The general element of the algebra would then be $\mathbf{U} = \Sigma a_\kappa \mathbf{u}_\kappa \sim \Sigma a_\kappa U_\kappa$, with the a_κ being the scalars of the algebra. Commutation is the compositon operator of the algebra. The scalars $c^\mu_{\kappa\lambda}$ are restricted by the requirement

$$c^\mu_{\lambda\kappa} = -c^\mu_{\kappa\lambda}, \tag{6.3.2b}$$

which is a direct consequence of the fact that

$$[U_\lambda, U_\kappa] = -[U_\kappa, U_\lambda]. \tag{6.3.2c}$$

The reader is invited to show that the Jacobi relation implies that the $c^\mu_{\kappa\lambda}$ obey the further restrictions

$$\Sigma_\mu(c^\mu_{\alpha\beta}\ c^\nu_{\mu\lambda} + c^\mu_{\beta\lambda}\ c^\nu_{\mu\alpha} + c^\mu_{\lambda\alpha}\ c^\nu_{\mu\beta}) = 0. \tag{6.3.2d}$$

Matrix multiplication with a square matrix $\mathbf{M}$ will convert a vector $\mathbf{U} = \Sigma a_\kappa \mathbf{u}_\kappa$ to

$$\mathbf{U}' = \mathbf{MU}, \tag{6.3.3}$$

which will be a vector of the Lie algebra if the elements m_{ij} of $\mathbf{M}$ are scalars. The vector $\mathbf{U}'$ will represent a linearly independent set of generators U'_κ if the determinant of $\mathbf{M}$ is nonzero. It is then easy to show that (6.3.1a) implies

$$[U'_\kappa, U'_\lambda] = \Sigma_\mu\ (c^\mu_{\kappa\lambda})'\ U'_\mu,\quad \kappa,\lambda,\mu = 1,\ldots,r. \tag{6.3.4}$$

The structure constants are not, in general, identical to those in (6.3.1a).

The U's satisfy all the laws of a linear algebra — if one considers the **0** *vector* to be a U as well; henceforth it will be denoted U_0. We shall use bold face type to denote Lie generators when we wish to emphasize their linear-algebraic, vectorial, properties.

A linear algebra whose elements satisfy (6.3.1) and (6.3.2) is said to be a *Lie algebra*. It is a *real Lie algebra* if no element of the algebra is allowed to take on complex values. The $c^\mu_{\kappa\lambda}$ are termed the *structure constants* of the Lie algebra.

The *rank* of a Lie algebra is the maximum number of its generators that commute amongst themselves. If no generator other than U_0 commute with another, the rank is 1. A Lie algebra, all generators of which commute with one another, is termed *Abelian.* All the structure constants of an Abelian Lie algebra are zero.

Lie algebras with identical structure constants are said to be *isomorphic.* The freedom represented by the transformation from (6.3.1a) to (6.3.4) indicates that two Lie algebras of the same dimension, r, may be isomorphic though their structure constants are different. A method for determining whether Lie algebras are isomorphic, devised by Killing, is discussed in Sec. 6.6. It is extensively developed by Cartan and Dynkin and used to classify Lie algebras. The methodology is outlined in the Sec. 6.7, and the classification of Lie algebras and groups is outlined in the appendix.

6.4 Relations between Commutation Relations and the Action of Transformation Groups: Some Examples

We begin this section with a brief study of the effect of transformations of variables upon the action of a Lie transformation group. Suppose one has a set of generators

$$\mathrm{U_k} = \Sigma_{\mathrm{j}},\xi_{\mathrm{j,k}}(\mathrm{z})\ \partial/\partial \mathrm{z_j}, \quad \mathrm{k} = 1, \ldots, \mathrm{r}, \tag{6.4.1}$$

in which the $\xi_{\mathrm{j,k}}$ are $\mathrm{C^k}$ functions. If one subjects the $\mathrm{U_k}$ to a $\mathrm{C^k}$ diffeomorphism

$$\mathrm{z_j} \to \mathrm{z'_j} = \mathrm{f_j(z)}, \quad \mathrm{z'_j} \to \mathrm{z_j} = \mathrm{f_j^{inv}(z')}, \tag{6.4.2a,b}$$

then

$$\mathrm{U_k} \to \mathrm{U'_k} = \Sigma_{\mathrm{j}}\ \zeta'_{\mathrm{j,k}}(\mathrm{z'})\ \partial/\partial \mathrm{z'_j}. \tag{6.4.3}$$

In this, as usual, the $\zeta'_{\mathrm{j,k}}(\mathrm{z'})$ are obtained by using the inverse transformation, (6.4.2b), to replace z by z′ in $\mathrm{U_k z'_j} = \mathrm{U_k f_j(z)}$. Now let us consider the transformation from z to z′ passively, that is let us consider it to define a change from one local coordinate system z to another z′. Such a passive change in coordinates affects only the coordinate labels in **Fig. 6.4.1**. Consequently, one expects that

$$\text{if } [\mathrm{U}_\kappa, \mathrm{U}_\lambda] = \Sigma_\mu\ \mathrm{C}^\mu_{\kappa\lambda}\ \mathrm{U}_\mu, \text{ then } [\mathrm{U}'_\kappa, \mathrm{U}'_\lambda] = \Sigma_\mu\ \mathrm{C}^\mu_{\kappa\lambda}\ \mathrm{U}'_\mu, \tag{6.4.4}$$

then and the structure constants remaining unchanged. In short, diffeomorphisms do not alter commutation relations.

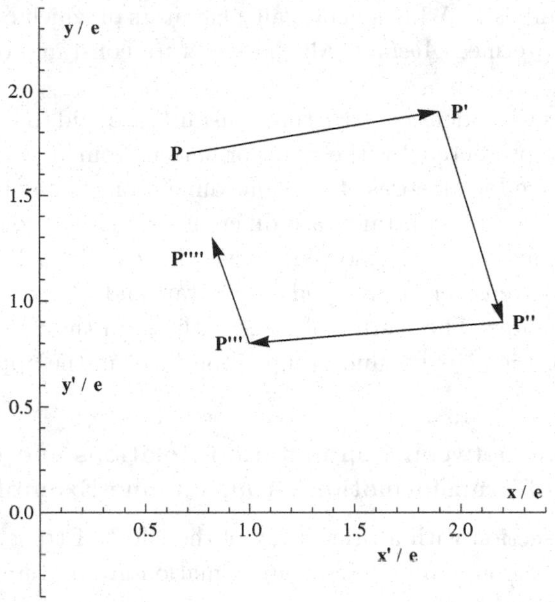

Fig. 6.4.1 Commutation diagram with second local coordinate.

This fact profoundly influences the physical interpretation of symmetry groups. Consider, for example, the commutation relations

$$[R_x, R_y] = -R_z, \quad [R_z, R_x] = -R_y, \quad [R_y, R_z] = -R_x \tag{6.4.5}$$

which are satisfied by the generators (6.1.15a) of the rotation group in Euclidean three-space. Let

$$x \to x/a, \quad y \to y/b, \quad z \to z/c. \tag{6.4.6}$$

This transformation converts spheres defined by

$$x^2 + y^2 + z^2 = r^2 \tag{6.4.7}$$

to ellipsoids defined by

$$(x/a)^2 + (y/b)^2 + (z/c)^2 = r^2. \tag{6.4.8}$$

It converts the generators R to generators such as

$$R'_x = (y/b)\ \partial/\partial(z/c) - (z/c)\ \partial/\partial(y/b) = (c/b)y\partial/\partial z - (b/c)z\partial/\partial y. \tag{6.4.9}$$

The reader can verify that if a, b, c are all nonzero, the commutation relations (6.4.5) are also satisfied by the R′. The R′ are generators of a group whose operator carries points on each ellipsoid into points on the ellipsoid, in exactly the same manner that the operator generated by the R carries points on a sphere into points on a sphere.

Lie transformation groups generated by two different sets of generators with isomorphic commutation relations are said to be *locally isomorphic* because, for some (finite) range of the group parameters, the actions $T(\alpha_1, \alpha_r)x$ and $T'(\alpha_1, \alpha_r)y$ of the two groups can be put in one-to-one correspondence. Locally isomorphic groups are *globally isomorphic* if the one-to-one correspondence persists for all allowed values of the group parameters. SO(3) and SU(2) are locally isomorphic, but not globally isomorphic. Different *realizations* of a group are produced when different group generators produce globally isomorphic groups. The R and the R′ of the previous paragraph are generators of different realizations of SO(3). Both SO(3) and SU(2) groups appear in a variety of realizations in physics and chemistry, some quite surprising.

All group parameters are allowed, in principle, to range through all the reals, but, as is the case for one-parameter groups, it may happen that, for any one-parameter subgroup, a group parameter and the same parameter mod some number, e.g. 2π or 4π, will give rise to the same transformation. One may, for convenience, then restrict the range of the group parameter so that the effect of the group operation is not repeated. The group parameter, say α, can then be contained within a compact region, e.g. $-\pi \leq \alpha \leq \pi$. Such a one-parameter group is termed a *compact one-parameter group*. A one-parameter group that is not compact is termed *noncompact*. The group parameter of a noncompact group ranges over an interval which may or may not contain all the real numbers, examples being $-\infty < \alpha < \infty$ and $-\pi < \alpha < \pi$. The group SO(2,1) contains a compact one-parameter subgroup, and two noncompact one-parameter subgroups. In the realization given above, $\exp(\alpha R_z)$ is the operator of a compact subgroup whose operation is the same for $\alpha = \pi$ as it is for $\alpha = -\pi$. On the other hand, the operators $\exp(\beta S_x)$ and $\exp(\gamma S_z)$ do not repeat their action as their group parameters range through all the real numbers. The group SO(3) contains only compact one-parameter subgroups.

If all one-parameter groups contained in a many-parameter group are compact, the many-parameter group is termed compact. If a many-parameter group contains one or more noncompact one-parameter groups, the many-parameter group is noncompact. The groups SO(n) are, for all values of n, compact. The groups SO(p, q) are noncompact for nonzero p and q.

The structure constants of the Lie algebra of SO(3) are different from the structure constants of the Lie algebra of SO(2,1). This is an example of a general truth: *the commutation relations of a real Lie algebra determine whether the corresponding Lie group is compact or noncompact.* Thus, rather surprisingly, the compactness or noncompactness of a Lie group is determined by relationships among its infinitesimal transformations. Subtleties may however be involved, as the following example demonstrates.

The projective group in one variable, say t, has generators

$$\partial/\partial t, \quad t\partial/\partial t, \quad t^2\partial/\partial t. \tag{6.4.10}$$

The general transformation of the group can be given the form

$$t \to t' = (t + a_1)/(a_2 t + a_3). \tag{6.4.11}$$

The identity transformation occurs when $\alpha_1 = -a_1/a_3$, $\alpha_2 = -a_2/a_3$, and $\alpha_3 = 1/a_3$. The linear combinations of the generators (6.4.10),

$$U_1 = (1/2)(1 - t^2)\,\partial/\partial t, \quad U_3 = (1/2)(1 + t^2)\,\partial/\partial t, \tag{6.4.12}$$

together with $U_2 = t\partial/\partial t$, obey the commutation relations

$$[U_1, U_2] = U_3, \quad [U_2, U_3] = -U_1, \quad [U_3, U_1] = -U_2. \tag{6.4.13}$$

The group is therefore locally isomorphic to SO(2,1), and the generator of the compact subgroup is U_3. As the real line $-\infty < t < \infty$ is an open interval it is rather surprising to find that a group that carries it into itself contains a compact subgroup, which, as we shall see, can carry any point on the line to any other point on the line.

The transformation (6.4.11), like that generated by $t^2\partial/\partial t$, undergoes a sudden change in sign from $\pm\infty$ to $\pm(-\infty)$ when $a_2 t + a_3$ passes through zero. In Chapter 2 we saw that a stereographic projection from the plane onto the sphere projects onto the South pole of the sphere all points whose (x, y) coordinates in the plane are $\pm\infty$. Let us consider the corresponding projection of the line of t onto a unit circle. Set

$$x = 2t/(t^2 + 1), \quad y = (1 - t^2)/(t^2 + 1). \tag{6.4.14}$$

Then $x^2 + y^2 = 1$, and one may set $x = \sin(\theta)$, $y = \cos(\theta)$. Consequently

$$t = \tan(\theta/2), \quad dt = (d\theta/2)\sec^2(\theta/2). \tag{6.4.15}$$

Figure 6.4.2 depicts the relation between t and θ. For $-\infty < t < \infty$ one has a diffeomorphism that maps t, dt to θ, $d\theta$ respectively with θ lying in

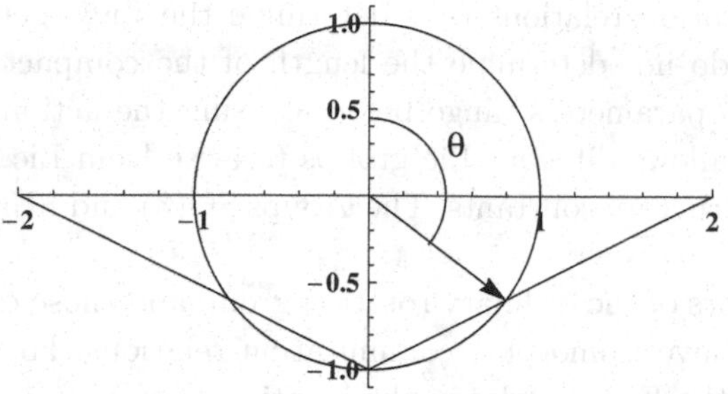

Fig. 6.4.2 Stereographic projection of t from the real line onto a unit circle.

the open interval $-\pi \leq \theta \leq \pi$. However θ also remains real over the closed interval $-\pi \leq \theta \leq \pi$ with $\theta = -\pi$ representing the same point as $\theta = \pi$.

The projection converts the U_j to the V_j defined below:

$$\begin{aligned} U_1 &\rightarrow y(y\partial/\partial x - x\partial/\partial y) \rightarrow \cos(\theta)\partial/\partial\theta = V_1, \\ U_2 &\rightarrow -x(y\partial/\partial x - x\partial/\partial y) \rightarrow -\sin(\theta)\partial/\partial\theta = V_2, \\ U_3 &\rightarrow -(y\partial/\partial x - x\partial/\partial y) \rightarrow -\partial/\partial\theta = V_3. \end{aligned} \tag{6.4.16}$$

V_1 vanishes when $\theta = \pm\pi/2$, and V_2 vanishes when $\theta = 0, \pm\pi$. This has the consequence (c.f. the exercises) that the finite transformations generated by V_1 and V_2 cannot carry θ out of the open interval $-\pi < \theta < \pi$. However, V_3 has no singular point, and $\exp(\gamma V_3)$ carries θ to $\theta - \gamma$, which can become $\pm\pi$.

The V's obey commutation relations isomorphic to those of the U's, but they obey them for all real values of θ, a range which includes values of θ in the compact interval $\pi \leq \theta \leq \pi$. This interval corresponds to the interval $-\infty \leq t \leq \infty$, with $-\infty$ being identified with $+\infty$. (Extending the interval by adding the point at $\theta = \pm\pi$ is termed *a one-point compactification of the interval*). Commutation relations express relations between infinitesimal transformations that move points over infinitesimal regions. Crudely speaking, an infinitesimal interval contains a continuum of points, and removing or adding a point is *unnoticeable* in so far as an infinitesimal transformation is concerned. In this case, the point added at $\theta = \pm\pi$ corresponds to $t = \pm\infty$. As the commutation relations are insensitive to such *trickery*, the three-parameter group generated by $\partial/\partial t$, $t\partial/\partial t$, $t^2\partial/\partial t$. contains a compact subgroup, isomorphic to that in the group generated by U_1, U_2, U_3.

The commutation relations that determine the Lie algebra of a compact Lie group do not determine the length of the compact interval over which the group parameters range before causing the action of the group to repeat. This allows different Lie groups to arise from Lie algebras that have the same structure constants. The groups SO(3) and SU(2) provide an example.

The generators of the ordinary rotation group and those of the spin 1/2 rotation group have isomorphic commutation relations. For the ordinary rotation group, the Euler angles in the rotation operator[9]

$$\mathrm{T}(\gamma, \beta, \alpha) = \exp(\gamma \mathrm{U}_3)\exp(\beta \mathrm{U}_2)\exp(\alpha \mathrm{U}_3) \tag{6.4.17}$$

range over the intervals $0 \le \alpha < 2\pi$, $0 \le \beta < \pi$, $0 \le \gamma < 2\pi$ without bringing a point back to itself. The group is a realization of the abstract Lie group SO(3). However, if half-integer spin vectors are being rotated, the action of $\exp(\gamma \mathrm{U}_3)$ only repeats when the group parameter γ moves over a range of 4π or greater. The standard range of the parameter is $0 \le \gamma < 4\pi$. The spin 1/2 rotation group is a realization of the abstract group SU(2), rather than the abstract group SO(3). SO(3) and SU(2) are locally isomorphic, but not globally isomorphic Lie groups. In short, *isomorphic Lie algebras yield locally isomorphic Lie groups, but do not necessarily yield globally isomorphic Lie groups.*

We conclude this section with several examples of physical interest. The linear symplectic group Sp(2R) has a realization with generators

$$\begin{aligned} \mathrm{U}_1 &= (\mathrm{q}\ \partial/\partial \mathrm{q} - \mathrm{p}\ \partial/\partial \mathrm{p})/2, \\ \mathrm{U}_2 &= (\mathrm{q}\ \partial/\partial \mathrm{p} + \mathrm{p}\ \partial/\partial \mathrm{q})/2, \\ \mathrm{U}_3 &= (\mathrm{p}\ \partial/\partial \mathrm{q} - \mathrm{q}\ \partial/\partial \mathrm{p})/2. \end{aligned} \tag{6.4.18}$$

The three operators $\exp(\alpha_j \mathrm{U}_j)$ carry out canonical transformations in the phase space of (p, q). The generators satisfy the commutation relations

$$[\mathrm{U}_1, \mathrm{U}_2] = -\mathrm{U}_3, \quad [\mathrm{U}_2, \mathrm{U}_3] = \mathrm{U}_1, \quad [\mathrm{U}_3, \mathrm{U}_1] = \mathrm{U}_2 \tag{6.4.19}$$

Now refer back to Eqs. (6.1.16a) and set up the correspondences $\mathrm{U}_1 \sim \mathrm{S}_x$, $\mathrm{U}_2 \sim \mathrm{S}_y$, $\mathrm{U}_3 \sim \mathrm{R}_z$. On comparing the commutation relations (6.1.16b) with (6.4.19) one sees that the structure constants of the two Lie algebras are identical. The Lie groups are consequently locally isomorphic. However

$$\mathrm{U}_3 = (\mathrm{p}\ \partial/\partial \mathrm{q} - \mathrm{q}\ \partial/\partial \mathrm{p})/2, \tag{6.4.20a}$$

while

$$R_z = (x\,\partial/\partial y - y\,\partial/\partial x), \tag{6.4.20b}$$

so

$$\exp(\alpha U_3) = \exp(\alpha' U_3), \quad \alpha' = \alpha + 4n\,\pi, \tag{6.4.20c}$$

while

$$\exp(\alpha R_z) = \exp(\alpha' R_z), \quad \alpha' = \alpha + 2n\,\pi, \tag{6.4.20d}$$

where n is an integer. Thus the groups are not globally isomorphic.

It is usually assumed that physical time runs continuously from an indefinite past through the present to an indefinite future: $-\infty < t < \infty$, or even $-\infty \leq t \leq \infty$. One expects that when $\exp(\alpha\,\partial/\partial t)$ acts on t to give $t' = t + \alpha$, one must allow α the full range $-\infty < \alpha < \infty$. However, even in Newtonian mechanics, if one is not careful this view can lead to inconsistencies. In Chapter 5 we found eight generators of one-parameter groups that leave invariant Newton's equation of motion for a harmonic oscillator. The commutation relations of three of these generators,

$$U_1 = \partial/\partial t, \tag{6.4.21a}$$

$$U_2 = (1 + y^2)\cos(t)\,\partial/\partial y + y\,\sin(t)\,\partial/\partial t, \tag{6.4.21b}$$

$$U_3 = (1 + y^2)\sin(t)\,\partial/\partial y - y\,\cos(t)\,\partial/\partial t, \tag{6.4.21c}$$

are

$$[U_3, U_1] = -U_2, \quad [U_2, U_3] = -U_1, \quad [U_1, U_2] = -U_3. \tag{6.4.21d}$$

These are the commutation relations of the compact groups SO(3) and SU(2). Each solution $x = A\sin(t) + B\cos(t)$ is converted into itself by the transformation $t \to t' = t + 2\pi$. Thus, $\partial/\partial t$ is the generator of a compact group, which is in fact, SO(3).[10] In so far as the oscillator is concerned, t and $t + 2\pi$ are indistinguishable.

In quantum mechanics one deals with the operations of matrices $\mathbf{M}$ that act on the coefficients c_j of a linear combination of wave functions, $c_j\,\psi_j(x)$, where the x are electronic coordinates. The matrix elements m_{jk} may be functions of parameters, $\mathbf{R}$, such as nuclear coordinates. One may, for example, be dealing with an eigenvalue problem $H_{op}\psi = E\psi$, $\psi = \Sigma_j c_j \psi_j(x)$. If $\mathbf{H}$ is a matrix whose elements, $H_{ij}(\mathbf{R}) = <\psi_i(x)|H_{op}|\psi_j(x)>$, are functions of a set of parameters $\mathbf{R}$, one obtains an eigenvalue equation of the form

$$(\mathbf{H}(\mathbf{R}) - E)C = 0, \quad C = (c_1, c_2, c_n). \tag{6.4.22}$$

As one varies the parameters, the coefficients c_j vary. It is a general property of such linear equations, that E may be left invariant by transformations of the general real linear group $Gl(n, R)$ of real transformations, if the c_j and the matrix elements $H_{jk}(\mathbf{R})$ are real variables. This group contains both SO(n) and SU(n) subgroups. If complex variables are involved the group is $GL(n, C)$, which also has SO(n) and SU(n)subgroups. In both cases the SU(n) subgroups have operators that vary some c_j's at half the rate the $H_{jk}(\mathbf{R})$ are varied.[11] This has the consequence that varying the $\mathbf{R}$ through a cycle that returns the matrix elements to their initial values can cause the c's to return to the *negative* of their initial values! Aharonov and Bohm were the first to notice this phenomenon in a physical context.[12] In chemical systems with degenerate molecular wavefunctions, similar phase shifts are found to be produced by cyclic nuclear motions. These can produce surprising effects when potential energy curves or surfaces $E(\mathbf{R})$ and $E'(\mathbf{R})$ become degenerate at crossings or intersections.[13] Berry[14] discovered that similar, but more varied, phase shifts occur in systems governed by the time-dependent analog of (6.4.22),

$$(\mathbf{H}(\mathbf{R}) - i\, \partial/\partial t)C = 0. \tag{6.4.23}$$

6.5 Transitivity

A transformation group acts *transitively* in a space or on a manifold if every point of the space or manifold can be carried into every other point. A group that fails this criterion, acts *intransitively* in the space or on the manifold. Transitivity is a property that is not determined by commutation relations alone, it depends upon the realization of the group, the functional form of the $\xi_{j,k}(z)$ in the group generators $U_k = \Sigma_j\, \xi_{j,k}(z)\, \partial/\partial z_j$. The translation subgroup of the Euclidean group E(3) in three-dimensional Euclidean space acts transitively in the space because it can move any point in the space to any other point in the space. The SO(3) group that carries out rotations about the origin acts transitively on the surface of a sphere centered at the origin. However this group acts intransitively in three-dimensional Euclidean space because it cannot move points from one sphere centered at the origin to points on a concentric sphere of different radius. This exemplifies a general observation: a group that acts transitively in a space S cannot simply move points about in a subspace of S. If S is a space whose points are determined by assigning values to n independent variables, x_j, a group that acts transitively in the space is consequently able to change the value of any function $f(x_1, x_2, \ldots, x_n)$ defined at each point P in the space. Such a group cannot leave invariant any function or manifold defined within S by such functions. It can leave metrics and coordinate differences invariant. As an example, the group of translations $x \to x + a, y \to y + b, z \to z + c$ acts transitively in the space of (x, y, z), and it leaves invariant the Euclidean distance between two points P_1 and P_2 or any function depending only upon $x_1 - x_2$, $y_1 - y_2$, $z_1 - z_2$.

The most useful dynamical groups and Lie algebras contain operations that act transitively on the space of solutions of dynamical equations. They often act intransitively on the space of variables in which the solutions are defined. Thus, for example, solutions of Hamilton's equations are defined in the phase space of the variables p, q, E and t. Transformations that carry solutions of the equations into solutions of the same energy leave invariant the expression $H(p, q) - E$. This expression defines a submanifold in the space of p, q, E and t. In Chapters 7–9 many-parameter groups that transform solutions of ODEs into solutions will be used to uncover many properties of the equations of classical dynamics. This will provide the foundation for our subsequent investigations of Schrödinger's and Maxwell's equations. The following sections of this chapter are primarily devoted to providing mathematical tools that help uncover consequences of group structure.

6.6 Complex Lie Algebras

In quantum mechanics one deals with complex extensions of real Lie algebras. The generators of the complex extension of a real Lie algebra are allowed to be functions of complex variables. In the linear combinations $\Sigma\, C_\kappa U_k$, the c_κ are allowed to be complex numbers, as are structure constants. In the Lie groups generated by the operators of a complex Lie algebra, the group parameters are also allowed to be complex variables.

As a first example of the consequences of the complexification of a Lie algebra, consider the effect on the rotation generators, (6.1.15a), by replacing z with i t. The operators then become

$$R'_x = -i(y\partial/\partial t + t\partial/\partial y), \quad R'_y = i(t\partial/\partial x + x\,\partial/\partial t), \quad R_z = x\,\partial/\partial y - y\,\partial/\partial x. \tag{6.6.1}$$

Note the similarity to the SO(2,1) generators of (6.1.16), *viz*

$$S_x = y\,\partial/\partial z + z\,\partial/\partial y, \quad S_y = z\,\partial/\partial x + x\,\partial/\partial z, \quad R_z = x\,\partial/\partial y - y\,\partial/\partial x. \tag{6.6.2}$$

The operators of (6.6.1) generate one-parameter groups which leave invariant $x^2 + y^2 - t^2$; the operators of (6.6.2) generate one-parameter groups which leave $x^2 + y^2 - z^2$ invariant.

The key commutation relations among the generators in (6.6.1) are

$$[R'_x, R'_y] = -R_z, \quad [R_z, R'_x] = -R'_y, \quad [R'_y, R_z] = -R'_x. \tag{6.6.3}$$

These are isomorphic to the commutation relations obeyed by the original rotation generators, and all the structure constants associated with (6.1.15) are the same as those associated with (6.6.3). They are not the same as the structure constants associated with the SO(2,1) generators in (6.1.16). When complexification is allowed, the commutation relations of Lie algebra of SO(2,1) and that of SO(3) can be made isomorphic, to have the same structure constants. The same may be said of the commutation relations of SO(p, q) and SO(p + q). The corresponding inhomogenous groups are similarly affected.

6.7 The Cartan–Killing Form; Labeling and Shift Operators

Because both complexification and changing the basis of a Lie algebra can change its commutation relations, it is helpful to have a way of determining properties of the algebra that are basis independent. A tool that does this was invented by Killing.[15] The *Cartan–Killing form* is a matrix (and second rank tensor) with elements

$$g_{jk} = g_{kj} = \Sigma_\mu \Sigma_\nu C^\mu_{j\nu}\, c^\nu_{k\mu}. \tag{6.7.1a}$$

Using the Einstein summation convention this is

$$g_{jk} = g_{kj} = c^\mu_{j\nu}\, c^\nu_{k\mu}. \tag{6.7.1b}$$

Cartan[16] proved that *if the determinant of this matrix is nonzero, the algebra contains operators that satisfy commutation relations of the form*

$$[U_0, U_+] = k\, U_+, \quad [U_0, U_-] = -k\, U_-, \tag{6.7.2}$$

where k is a constant. If the labeling operator U_0 *has eigenfunctions, the eigenvalues of the labeling operator may be used to label the eigenfunctions. The shift operators then shift this eigenvalue by* $\pm$k.

As a first example of these observations, consider the Lie algebra of the groups SO(3) and SU(2). We have seen that their commutation relations may be chosen to be

$$[R_x, R_y] = -R_z, \quad [R_z, R_x] = -R_y, \quad [R_y, R_z] = -R_x. \tag{6.7.3a}$$

Renaming R_x, R_y, R_z as U_1, U_2, U_3, respectively, the nonzero structure constants are

$$\begin{aligned} &c_{12}^3 = -1, \quad c_{31}^2 = -1, \quad c_{23}^1 = -1, \\ &c_{21}^3 = -c_{12}^3 = 1, \quad c_{13}^2 = -c_{31}^2 = 1, \quad c_{32}^1 = -c_{23}^1 = 1. \end{aligned} \tag{6.7.3b}$$

and Consequently,

$$\begin{aligned} g_{11} = c_{1\nu}^{\mu}\, c_{1\mu}^{\nu} = c_{12}^3\, c_{13}^2 + c_{13}^2\, c_{12}^3 = (-1)(1) + (1)(-1) = -2 \\ g_{12} = c_{1\nu}^{\mu}\, c_{2\mu}^{\nu} = c_{13}^2\, c_{22}^3 + c_{12}^3\, c_{23}^2 = 0, \\ g_{22} = c_{2\nu}^{\mu}\, c_{2\mu}^{\nu} = c_{21}^3\, c_{23}^1 + c_{23}^1\, c_{21}^3 = -2. \end{aligned} \tag{6.7.4}$$

Similarly, one finds that $g_{33} = -2, g_{13} = g_{23} = 0$. The Killing form is thus

$$\begin{vmatrix} -2 & 0 & 0 \\ 0 & -2 & 0 \\ 0 & 0 & -2 \end{vmatrix}. \tag{6.7.5}$$

The corresponding determinant has value -8, so the three generators U may be used to construct a labeling operator and a pair of shift operators.

The inhomogeneous group ISO(3) has in addition to the rotation generators U_1, U_2, U_3, the translation generators $U_4 = \partial x, U_5 = \partial y, U_6 = \partial z$, which commute among themselves. Their commutation relations with the rotation generators give rise to the nonzero structure constants:

$$\begin{aligned} &c_{24}^6 = 1 = -c_{42}^6, \quad c_{34}^5 = -1 = -c_{43}^5, \quad c_{15}^6 = -1 = -c_{51}^6, \\ &c_{35}^4 = 1 = -c_{53}^4, \quad c_{16}^5 = 1 = -c_{61}^5, \quad c_{26}^4 = -1 = -c_{62}^4. \end{aligned} \tag{6.7.6}$$

Using these one finds that the only nonzero elements of the 6×6 dimensional Killing form for ISO(3) are those of its SO(3) subgroup, the diagonal elements in (6.7.5). The determinant of the form is thus zero.

The class of groups termed *semi-simple* all have Killing forms whose determinant is nonzero. This class, defined in Appendix B, includes the groups SO(p,q), SO(p+q), and groups locally isomorphic to them, but not ISO(p, q), ISO(p+q).

The structure constants of the Lie algebra of the angular momentum operators L_j are $-ih/2\pi$ times those of the generators R_j of the rotation group SO(3). The determinant of the Killing form is consequently $8(h/2\pi)^3$. If we let L_z be the labeling operator, the shift operators are

$$\begin{aligned} L_+ &= L_x + iL_y = -i(R_x + iR_y), \\ L_- &= L_x - iL_y = -i(R_x - iR_y), \end{aligned} \tag{6.7.7a,b}$$

and

$$[L_z, L_+] = L_+, \quad [L_z, L_-] = -L_-. \tag{6.7.7c,d}$$

It follows that

$$L_z L_+ = L_+(L_z + 1), \quad L_z L_- = L_-(L_z - 1). \tag{6.7.7e,f}$$

The spherical harmonics $Y_{lm}(\theta, \phi)$ satisfy

$$L_z Y_{lm} = m(h/2\pi) Y_{lm}, \tag{6.7.8}$$

with l a nonnegative integer, and m an integer. Equations (6.6.7c,d) imply that

$$\begin{aligned} L_z(L_+ Y_m) &= (m + 1)(h/2\pi)(L_+ Y_{lm}), \\ L_z(L_- Y_{lm}) &= (m - 1)(h/2\pi)(L_- Y_{lm}). \end{aligned} \tag{6.7.9}$$

Consequently, if $L_+ Y_{lm}$ does not vanish, it has eigenvalue $(m+1)(h/2\pi)$; on the other hand, if $L_- Y_{lm}$ does not vanish, it has eigenvalue $(m-1)(h/2\pi)$. If the Y_{lm} are those defined by Condon and Shortley,[17] an elegant argument[18] establishes

$$L_+ Y_{lm} = \nu_{+lm} Y_{lm+1} = (l(l + 1) - m(m + 1))^{1/2}(h/2\pi) Y_{lm+1}, \tag{6.7.10a}$$

$$L_- Y_{lm} = \nu_{-lm} Y_{lm-1} = (l(l + 1) - m(m - 1))^{1/2}(h/2\pi) Y_{lm-1}. \tag{6.7.10b}$$

These relations imply that

$$L_+ Y_{l,l} = 0, \quad L_- Y_{l,l} = 0. \tag{6.7.10c}$$

As a consequence, m can only take on the $2l + 1$ integer values that begin with $-l$ and run through to $+l$.

In the general case, the allowed range of the eigenvalues of a labeling operator may be finite or infinite, and the eigenvalues need not be integers. The generators of the group SU(2) obey the same commutation relations as those of SO(3), but in contrast to SO(3) the labeling operator of SU(2) may have half-integer, as well as integer eigenvalues.

6.8 Casimir Operators

Though no generator of a group may commute with all generators of the group, there may be polynomial functions of the generators that commute with all of them. Since the importance of this fact was first recognized by the Dutch physicist Casimir,[19] these polynomial functions of generators are called Casimir operators. The best example known is the Casimir operator of SO(3), the total angular momentum operator, L^2, of quantum mechanics, whose eigenvalues are $l(l+1)(h/2\pi)$.[2]

The rotation group in four-space has a quadratic Casimir operator and a quartic one. Let the six generators of the group be U_{12}, U_{23}, U_{31}, U_{14}, U_{24}, and U_{34} with

$$U_{\kappa\lambda} = x_\kappa \partial/\partial x_\lambda - x_\lambda \partial/\partial x_\kappa. \qquad (6.8.1a)$$

Define

$$\mathbf{R} = (U_{23}, U_{31}, U_{12}) = (R_x, R_y, R_z), \qquad (6.8.1b)$$

$$\mathbf{A} = (U_{14}, U_{24}, U_{34}) = (A_x, A_y, A_z). \qquad (6.8.1c)$$

The Casimir operators may then be taken to be

$$C_2 = -(\mathbf{A}\cdot\mathbf{A} + \mathbf{R}\cdot\mathbf{R}), \quad C_4 = (-\mathbf{A}\cdot\mathbf{R})^2. \qquad (6.8.2)$$

These Casimir operators of SO(4) play prominent roles in later chapters.

If one lets the generators of the Lorentz group be components of $\mathbf{R}$ and

$$\mathbf{T} = (V_{14}, V_{24}, V_{34}) = (T_x, T_y, T_z), \quad V_{\kappa\lambda} = x_\kappa \partial/\partial x_\lambda + x_\lambda \partial/\partial x_\kappa, \quad x_4 = ct, \qquad (6.8.3)$$

then the quadratic and quartic Casimir operators are

$$C_2 = (\mathbf{T}\cdot\mathbf{T} - \mathbf{R}\cdot\mathbf{R}), \quad C_4 = (-\mathbf{T}\cdot\mathbf{R})^2. \qquad (6.8.4)$$

A semi-simple group of rank r *possesses* r *independent Casimir operators.*[20]

The spherical harmonics are perhaps the most used among the *special functions* of mathematical physics. If a Lie group, or its subgroups, has labeling and shift operators as well as Casimir operators, these operators may be used to greatly simplify investigations of any function that can be made an eigenfunction of the labeling and Casimir operators. Sharp[21] and Wigner[22] established that the theory of a wide variety of special functions of mathematical physics can be simplifed in this manner. The economy of thought that results is quite marvelous. In this connection the reader is invited to compare Watson's *Theory of Bessel Functions*[23] with Sharp's discussion of Bessel functions.[21] Vilenkin's monograph on the special functions[24] approaches the subject from a similar group theoretic standpoint.

The generators of a Lie group need not be first-order differential operators. They may be differential operators of order greater than one, operators of integration, matrices, *etc.* All continuous groups whose composition functions satisfy Definition 6.1 are Lie groups.

6.9 Groups That Vary the Parameters of Transformation Groups

The part of the theory of Lie groups that deals only with the action of the operators upon themselves is known as the theory of *abstract Lie groups.* During these actions of the abstract group only the group parameters α vary. Operators of these abstract groups may be considered to act on the group parameters. If α_1, α_2 are group parameters, and b is not contained in the space of the α's, then $\exp(\mathrm{b}(\alpha_1\, \partial/\partial\alpha_2 - \alpha_2\, \partial/\partial\alpha_1))$ is a such operator.

Many properties of a transformation group depend only upon properties of its parameter space, the space the parameters sweep through as they vary. There are several different ways of defining such groups.[25] The theory of topological groups is based on an analysis of parameter groups.[26] If a group has no generators which commute with all its generators, the *adjoint group* can be specially revealing. Adjoint groups also play a central role in the classification of Lie algebras. The generators of the adjoint group are defined to be

$$\Gamma_\mathrm{s} = \Sigma\, \Sigma\ \mathrm{a_{jk}}\ \alpha_\mathrm{j}\partial/\partial\alpha_\mathrm{k}, \quad \mathrm{a_{jk}} = \mathrm{c^k_{js}}. \tag{6.9.1}$$

As an example, consider the transformations of the rotation group SO(3) with operator of the form

$$\exp(\alpha_3 \mathrm{R}_3) \exp(\alpha_2 \mathrm{R}_2) \exp(\alpha_1 \mathrm{R}_1) \tag{6.9.2}$$

the R's being those defined in Eqs. (6.7.3). Using the structure constants in (6.7.3) to evaluate (6.9.1), one finds that the generators of the adjoint group are given by

$$\Gamma_1 = \alpha_2\partial/\partial\alpha_3 - \alpha_3\partial/\partial\alpha_2, \quad \Gamma_2 = \alpha_3\partial/\partial\alpha_1 - \alpha_1\partial/\partial\alpha_3,$$
$$\Gamma_3 = \alpha_1\partial/\partial\alpha_2 - \alpha_2\partial/\partial\alpha_1. \tag{6.9.3}$$

Though the adjoint group of the rotation group acts in a space of three dimensions, in most cases adjoint groups act in spaces of higher dimensions than does the group upon which they are based. Rotations in a four-dimensional space with coordinates $(\mathrm{x}_1, \mathrm{x}_2, \mathrm{x}_3, \mathrm{x}_4)$ lead to a six-dimensional space of group parameters: the group SO(4) is a six-parameter group. The adjoint group of SO(4) has the generators (6.9.3), and three additional generators:

$$\begin{aligned}\Gamma_4 &= \alpha_5 \partial/\partial\alpha_6 - \alpha_6 \partial/\partial\alpha_5,\\ \Gamma_5 &= \alpha_6 \partial/\partial\alpha_4 - \alpha_4 \partial/\partial\alpha_6,\\ \Gamma_6 &= \alpha_4 \partial/\partial\alpha_5 - \alpha_5 \partial/\partial\alpha_4.\end{aligned} \tag{6.9.4}$$

The SO(3,1) Lorentz group of spacetime also has a six-dimensional parameter space and an adjoint group with six generators. The generators of the Poincare group, an ISO(3) group, are those of the Lorentz group supplemented with the generators of four translations in spacetime. The parameter space of the adjoint group is therefore ten-dimensional.

6.10 Lie Symmetries Induced from Observations

We have seen that differential equations define Lie groups of one or more parameters, that Lie groups interconvert solutions of the differential equations of the natural and physical sciences, and that Lie groups define intrinsic symmetries of these differential equations. Relationships uncovered by Lie theory are found to be expressed in corresponding relationships between physical observations. Intrinsic symmetries thereby become symmetries in the natural world. These symmetries may not even be possible in three-dimensional Euclidean space. Their geometric interpretation involves further considerations, and is not necessarily implied by a differential equation itself or its physical interpretation.

What can be said about the reverse process of inducing mathematical connections from observations which establish group theoretic connections between observables? One must face the fact that the operations of a Lie transformation group, for example, SO(3), have an infinite number of realizations as differential operators. The space of continuous variables upon which they act is far from fully determined by the knowledge that observations are connected by operations of the group.

The operations of a Lie group may be operations of the group upon a variety of variables, or operations on a variety of differential (and integral) operators that operate on such variables. The operations of a Lie group can also be represented by matrices acting on vectors — vectors that may be conceived of in a wide variety of ways. And, as we have seen, associated with each r-parameter Lie transformation group, T(ω)x are groups whose operators, like those of the adjoint group, act on the group parameters of the transformation group. The dimension of the parameter space of an r-parameter group is r, and r is commonly greater than the dimension of the physical space in which the transformation group is conceived to act. Thus, e.g. the Poincare group acts on the four variables of spacetime, or the eight variables of the corresponding phase space, but its parameter group is ten-dimensional.

Russell once, and famously, said, "Mathematicians do not know what they are talking about, nor whether what they are saying is true."[27] As coauthor with Whitehead of *Principia Mathematica*,[28] Russell presumably knew what he was talking about, but was not explicitly talking about the theory of Lie groups, so Eddington added the following, "and a group theoretician is a supermathematician who does not know what he is doing."[29]

Natural scientists are supposed to *know what they are doing* — but presuppositions, sometimes unrecognized, occur beneath every theory. Newton supposed that mass is an intrinsic property of the object, independent of its velocity or acceleration. He presupposed that physical space was continuous as well as three-dimensional. And he presupposed that time flowed continuously from a distant past, through the present, and into the future. These presuppositions allowed him to define arbitrarily small changes in spacetime, and *instantaneous* velocities $d\mathbf{x}/dt$, accelerations $d(d\mathbf{x}/dt)/dt$, momenta $\mathbf{p} = md\mathbf{x}/dt$, and their rate of change $d\mathbf{p}/dt$. Observers equipped with clocks and measuring sticks were able to approximately measure each of these and give physical meaning. Classical mechanics developed as a theory based on the physical existence of continuous chains of causes $\mathbf{f}(\mathbf{x}, t)dt$, and effects $d\mathbf{q}, d\mathbf{p}$, evolving in spacetime. Several centuries after Newton, Noether realized that these presuppositions implied the conservation laws mentioned in Chapter 1, and their connection to Lie symmetries became evident.

Measurement errors always swamp any attempt to measure arbitrarily small changes such as dx_j and dt, so the continuity of space and time remains a hypothesis. And even in classical mechanics, simple, even separable, systems can smoothly evolve into unstable states, whose further evolution can be greatly altered by small perturbations. A tossed claw hammer, or any asymmetric top rotating about its intermediate axis, provides an intriguing example of such behavior. It is now well known that the classical equations of motion for systems of three or more bodies can develop a variety of instabilities, some of which lead to effectively unpredictable motions.

Electromagnetic theory, relativity theory, and quantum theory, all retain the presupposition of continuous spacetime: the dynamical equations of all are differential equations involving operators such as $\partial/\partial x_j$, and $\partial/\partial t$. However, in quantum theory the continuous evolution of causal chains in a continuous spacetime is replaced by the conception of the continuous spacetime evolution of wave functions, ψ, ψ^*, which determine the probabilities that an observable will be found to have a given eigenvalue.

There are times in the evolution of any science when presuppositions are challenged for good reason, and conjecture is required. After Gell-Mann proposed SU(3) as a symmetry group interrelating the properties of baryons, there was a period when theoretical physicists conjectured that properties of the (infinitely) stable elementary particles, the short lived elementary particles observed in high-energy collisions, as well as the properties assumed for quarks are the physical expressions of structure found in a great variety of groups. It is however one thing to find group structure, and another to find the underlying space of variables upon which the group acts, or to distinguish transformation groups from their parameter groups, which as we have seen often have higher dimensionality. Current attempts to provide unifying theories of physical phenomena have led to proposals that the observed relationships are consequences of symmetries expressed in multidimensional spaces with discontinuities such as *strings* and *wormholes*.[30]

6.11 Conclusion

Lie established the result that every many-parameter transformation group whose generators are first-order differential operators is defined by a restricted class of first-order partial differential equations. He showed that first-order differential operators are those of a many-parameter transformation group *iff* they are closed under commutation.

The commutator of two operators is the noncommutative multiplication operation used to define Lie algebras: operators that are closed under commutation are, together with scalars, the elementary objects of a Lie algebra.

Individual Lie groups are characterized by Lie-algebraic properties of their generators, and the intervals over which the group parameters range — the topology of their parameter space. If one is only concerned with the interaction of the operators upon each other, their characterization as abstract topological groups is complete.

Each differential equation of the sciences is invariant under the action of a Lie transformation group; its solutions are interconverted by the group. It is now known that partial differential equations can be invariant under the action of continuous transformation groups whose generators are differential operators of order greater than one.[a] Each particular transformation group can also be realized in a great number of ways. These truths have the consequence that Lie groups have greater relevance to the natural sciences than expected. When relations between observations are expressions of Lie symmetries, this freedom of interpretation also allows one to consistently conceive the Lie symmetries as geometric symmetries in a variety of spaces — Euclidean three-space, Newtonian spacetime, Hamiltonian phase spaces, Lorentzian spacetime, Einsteinian spacetimes with mass dependent metrics, the spaces of string theories, *etc.*

Appendix A. Definition of Lie Groups by Partial Differential Equations

In this appendix we give proofs of two of Lie's basic theorems. The starting point is more general than that in Sec. 6.1, because it is not assumed that a many-parameter group can be expressed as a product of one-parameter groups.

We first establish the conditions that a set of first-order partial differential equations

$$\partial \mathrm{x}_\mathrm{i}'/\partial \alpha_\kappa = \xi_{\mathrm{i}\kappa}(\mathbf{x}', \alpha), \tag{6.A.1}$$

must satisfy, to define a many-parameter Lie transformation group of variables $\mathbf{x} = (\mathrm{x}_1, \mathrm{x}_2, \ldots, \mathrm{x}_\mathrm{n})$. Let

$$\mathrm{x}_\mathrm{i} \to \mathrm{x}_\mathrm{i}' = \mathrm{f}_\mathrm{i}(\mathbf{x}; \alpha), \quad \mathrm{i} = 1, \ldots, \mathrm{n} \tag{6.A.2a}$$

define a transformation $\mathrm{T}(\alpha)\mathbf{x}$ depending upon r independent, continuously variable, parameters

$$\alpha = (\alpha_\mathrm{r}, \alpha_{\mathrm{r}-1}, \ldots, \alpha_2, \alpha_1). \tag{6.A.2b}$$

Let the identity transformation be given, by setting $\alpha = 0$. Denoting the succession of two transformations by $\mathrm{T}(\beta)\ \mathrm{T}(\alpha)$, we wish to establish the form of the differential Eqs. (6.A.1) that will ensure

$$\mathrm{T}(\beta)\ \mathrm{T}(\alpha) = \mathrm{T}(\gamma), \tag{6.A.3a}$$

with $\boldsymbol{\gamma} = \gamma(\alpha, \beta)$, i.e.

$$\gamma_\kappa = \gamma_\kappa(\alpha, \beta), \quad \kappa = 1, \ldots, \mathrm{r}. \tag{6.A.3b}$$

---

[a]These are introduced in Chapter 9.

If the first of two transformations is $\mathbf{x}' = f(\mathbf{x};\alpha)$, and the second is $\mathbf{x}'' = f(\mathbf{x}';\beta)$, then

$$\mathbf{x}'' = f(f(\mathbf{x};\alpha),\beta) = f(\mathbf{x},\gamma). \tag{6.A.3c}$$

The first key question is What restrictions do these requirements impose on differential equations that define infinitesimal transformations and group generators? In his paper of 1880,[1] Lie showed that this question could be answered by considering one of the transformations to be infinitesimal. The reader will find it rewarding to consult this paper: in it Lie first considers a number of examples involving transformations of one or two variables. He then derives his general theorems by explicitly considering the effects of variations of every pair of the group parameters. In the derivation below, this restriction is removed.[25]

Let

$$T(a)\mathbf{x} = \mathbf{x}' = f(\mathbf{x};\alpha), \tag{6.A.4a}$$

and

$$T(\delta a)T(a)\mathbf{x} = T(a + da)\mathbf{x}. \tag{6.A.4b}$$

Define

$$T(\delta a)T(a)\mathbf{x} - T(a)\mathbf{x} = \delta\mathbf{x}', \tag{6.A.4c}$$

and

$$T(a + da)\mathbf{x} - T(a)\mathbf{x} = d\mathbf{x}'. \tag{6.A.4d}$$

The composition stated in (6.A.4b) requires that $\delta\mathbf{x}' = d\mathbf{x}'$. Given (6.A.3b), this can only occur if

$$da = \boldsymbol{\gamma}(\alpha, \delta\alpha), \tag{6.A.5}$$

i.e.

$$da_\kappa = \gamma_\kappa(\alpha, \delta\alpha).$$

Now Eq. (6.A.4a) requires that

$$dx'_i = \Sigma_\kappa(\partial f_i(\mathbf{x}';\alpha)/\partial\alpha_\kappa)\, d\alpha_\kappa, \quad \text{and} \tag{6.A.6a}$$

Equation (6.A.5) requires that

$$\delta x'_i = \Sigma_\lambda((\partial f_i(\mathbf{x}';\alpha)/\partial\alpha_\lambda)|_{\alpha=0})\delta\alpha_\lambda = \Sigma_\lambda u_{i\lambda}(\mathbf{x}')\delta\alpha_\lambda. \tag{6.A.6b}$$

Because $\delta x_i' = dx_i'$, it must also be true that

$$\Sigma_\lambda u_{i\lambda}(\mathbf{x}')\delta\alpha_\lambda = \Sigma_\kappa(\partial f_i(\mathbf{x}';\alpha)/\partial\alpha_\kappa)\, d\alpha_\kappa, \tag{6.A.7}$$

a relation in which the $\delta\alpha_\lambda$ and $d\alpha_\kappa$ are related by (6.A.5). The latter requires that

$$\alpha_\kappa = \gamma_\kappa(\alpha, 0), \tag{6.A.8a}$$

and hence, that

$$\alpha_\kappa + d\alpha_\kappa = \gamma_\kappa(\alpha, \delta\alpha). \tag{6.A.8b}$$

Let

$$\nu_{\kappa\lambda}(\alpha) = (\partial\gamma_\kappa(\alpha,\beta)/\partial\beta_\lambda)|_{\beta=0}. \tag{6.A.9a}$$

Then

$$\alpha_\kappa + d\alpha_\kappa = \alpha_\kappa + \Sigma_\lambda\, \nu_{\kappa\lambda}(\alpha)\, \delta\alpha_\lambda, \tag{6.A.9b}$$

and consequently

$$d\alpha_\kappa = \Sigma_\lambda\, \nu_{\kappa\lambda}(\alpha)\, \delta\alpha_\lambda. \tag{6.A.10}$$

This linear relation between the $d\alpha$'s and the $\delta\alpha$'s can be solved for the latter, if the determinant of the $\nu_{\kappa\lambda}$ does not vanish. If this were not true the r group parameters would not all be independent. Thus *the requirement that the group parameters are independent requires that the determinant,* $\det(\nu_{\kappa\lambda}(\alpha))$, *be nonzero.* One then has

$$\delta\alpha_\lambda = \Sigma_\kappa\, \omega_{\lambda\kappa}(\alpha) d\alpha_\kappa, \tag{6.A.11a}$$

where the $\omega_{\lambda\kappa}$ satisfy

$$\Sigma_\kappa\, \omega_{\lambda\kappa}\, \nu_{\kappa\lambda'} = \delta_{\lambda,\lambda'}. \tag{6.A.11b}$$

Here $\delta_{\lambda,\lambda'}$ is the Kronecker delta. If the $\omega_{\lambda\kappa}$ and the $\nu_{\kappa\lambda}$ are considered to be elements of $r \times r$ matrices, $\boldsymbol{\omega}$ and $\boldsymbol{\nu}$, then (6.A.11b) states that $\boldsymbol{\omega\nu} = \mathbf{I}$. The existence of $\boldsymbol{\omega}$ implies the existence of its inverse, $\boldsymbol{\nu}$, and vice versa.

Using (6.A.11) one can express (6.A.6a) as

$$dx_i' = \Sigma_\lambda u_{i\lambda}(\mathbf{x})\, \omega_{\lambda\kappa}(\alpha)\, d\alpha_\kappa, \tag{6.A.12a}$$

so

$$\partial x_i'/\partial\alpha_\kappa = \Sigma_\lambda u_{i\lambda}(\mathbf{x})\, \omega_{\lambda\kappa}(\alpha). \tag{6.A.12b}$$

Consequently, in the differential Eqs. (6.A.1) one must have

$$\xi_{i\kappa}(\mathbf{x}', \alpha) = \Sigma_\lambda u_{i\lambda}(\mathbf{x}')\, \omega_{\lambda\kappa}(\alpha). \tag{6.A.13a}$$

This implies:

The differential equations defining a many-parameter group must be of the form

$$\partial x_i'/\partial\alpha_\kappa = \Sigma_\lambda u_{i\lambda}(\mathbf{x}')\ \omega_{\lambda\kappa}(\alpha), \quad i = 1, \ldots, n, \quad \lambda,\ \kappa = 1, \ldots, r. \quad (6.A.13b)$$

This result states a necessary, but not sufficient, condition that must be satisfied if the differential equations are to define a many-parameter group. The problem is that it may or may not be possible to integrate them, and so, for each value of κ, determine the transformations of a one-parameter group with parameter α_κ. For Eqs. (6.A.13b) to be integrable it is necessary and sufficient that

$$\partial^2 x_i'/\partial\alpha_\kappa\ \partial\alpha_\mu = \partial^2 x_i'/\partial\alpha_\mu\partial\alpha_\kappa. \quad (6.A.14)$$

This requires that

$$\partial\ \xi_{i\kappa}(\mathbf{x}', \alpha)/\partial\alpha_\mu - \partial\ \xi_{i\mu}(\mathbf{x}', \alpha)/\partial\alpha_\kappa = 0, \quad (6.A.15a)$$

with

$$\xi_{i\nu}(\mathbf{x}', \alpha) = \Sigma_\lambda u_{i\lambda}(\mathbf{x}')\ \omega_{\lambda\nu}\ (\alpha). \quad (6.A.15b)$$

When these conditions are satisfied, one obtains:

Lie's First Fundamental Theorem. *A set of integrable differential equations of the form*

$$\partial x_i/\partial\alpha_\kappa = \Sigma_\rho u_{i\rho}(\mathbf{x})\omega_{\rho\kappa}(\alpha), \quad i = 1, \ldots, n, \quad \rho, \kappa = 1, \ldots, r, \quad (6.A.16a)$$

defines an r*-parameter continuous group with generators*

$$U_\rho = \Sigma_i u_{i\rho}(\mathbf{x})x_i, \quad (6.A.16b)$$

where the $\omega_{\rho\kappa}(\alpha)$ are *elements of an invertible* $r \times r$ *matrix.*

The following paragraphs show that this theorem implies a further theorem which states that the generators of a many-parameter transformation group are closed under commutation. It is derived by applying the integrabilty conditions to the right-hand side of (6.A.16a).

For each pair of values of μ and κ, the integrabily conditions yield the equation

$$\Sigma_\rho(\partial u_{i\rho}(\mathbf{x}')/\partial\alpha_\mu\ \omega_{\rho\kappa}(\alpha) + u_{i\rho}(\mathbf{x}')\partial\omega_{\rho\kappa}(\alpha)/\partial\alpha_\mu)$$
$$- \Sigma_\rho(\partial u_{i\rho}(\mathbf{x}')/\partial\alpha_\kappa\ \omega_{\rho\mu}(\alpha) + u_{i\rho}(\mathbf{x}')\partial\omega_{\rho\mu}(\alpha)/\partial\alpha_\kappa) = 0. \quad (6.A.17a)$$

This may be rearranged to read

$$\Sigma_\rho \partial \mathrm{u}_{i\rho}(\mathbf{x}')/\partial\alpha_\mu\ \omega_{\rho\kappa}(\alpha) - \Sigma_\rho \partial \mathrm{u}_{i\rho}(\mathbf{x}')/\partial\alpha_\kappa \omega_{\rho\mu}(\alpha)$$
$$= \Sigma_\rho \mathrm{u}_{i\rho}(\mathbf{x}')(\partial\omega_{\rho\mu}(\alpha)/\partial\alpha_\kappa - \partial\omega_{\rho\kappa}(\alpha)/\partial\alpha_\mu). \qquad (6.A.17b)$$

Let us first deal with the derivatives in the first line of (6.A.17b). As the α's vary and produce variations in the $\mathbf{x}$'s, the resulting local changes in the $\mathrm{u}_{i\rho}(\mathbf{x}')$ are given by

$$\partial \mathrm{u}_{i\rho}(\mathbf{x}')/\partial\alpha_\tau = \Sigma_j\ \partial \mathrm{u}_{i\rho}(\mathbf{x}')/\partial \mathrm{x}'_j\ \partial \mathrm{x}'_j/\partial\alpha_\tau. \qquad (6.A.18a)$$

As (6.A.16a) requires

$$\partial \mathrm{x}'_j/\partial\alpha_\tau = \Sigma_\sigma \mathrm{u}_{j\sigma}(\mathbf{x}')\omega_{\sigma\tau}(\alpha),$$

one has

$$\partial \mathrm{u}_{i\rho}(\mathbf{x}')/\partial\alpha_\tau = \Sigma_j\ \Sigma_\sigma\ \partial \mathrm{u}_{i\rho}(\mathbf{x}')/\partial \mathrm{x}'_j\ \mathrm{u}_{j\sigma}(\mathbf{x}')\ \omega_{\sigma\tau}(\alpha). \qquad (6.A.18b)$$

The appearance of both $\mathrm{u}_{i\rho}(\mathbf{x}')$ and $\mathrm{u}_{j\sigma}(\mathbf{x}')$ suggests that (6.A.18b) may be used to establish relations between several generators. Applying (6.A.18b) to the first sum in (6.A.17b), by setting $\tau = \mu$, yields

$$\Sigma_\rho \partial \mathrm{u}_{i\rho}(\mathbf{x}')/\partial\alpha_\mu\ \omega_{\rho\kappa}(\alpha) = \Sigma_\rho\ \Sigma_\sigma\ \Sigma_j\ \mathrm{u}_{j\sigma}(\mathbf{x}')\ \partial \mathrm{u}_{i\rho}(\mathbf{x}')/\partial \mathrm{x}'_j\ \omega_{\sigma\mu}(\alpha)\ \omega_{\rho\kappa}(\alpha). \qquad (6.A.19a)$$

Applying (6.A.18b) to the second sum in (6.A.17b), by setting $\tau = \kappa$, yields

$$\Sigma_\rho \partial \mathrm{u}_{i\rho}(\mathbf{x}')/\partial\alpha_\kappa\ \omega_{\rho\mu}(\alpha) = \Sigma_\rho\Sigma_\sigma\ \Sigma_j\ \mathrm{u}_{j\sigma}(\mathbf{x}')\ \partial \mathrm{u}_{i\rho}(\mathbf{x}')/\partial \mathrm{x}'_j\ \omega_{\sigma\kappa}(\alpha)\ \omega_{\rho\mu}(\alpha). \qquad (6.A.19b)$$

Much of the following analysis involves reorganizing terms in multiple sums. This may be accomplished by altering summation indices. We begin this process by interchanging the summation indices ρ and σ in (6.A.19b), obtaining

$$\Sigma_\sigma\ \Sigma_\rho\{(\Sigma_j\ \mathrm{u}_{j\rho}(\mathbf{x}')\ \partial \mathrm{u}_{i\sigma}(\mathbf{x}')/\partial \mathrm{x}'_j)\ \omega_{\rho\kappa}(\alpha)\ \omega_{\sigma\mu}(\alpha)\}. \qquad (6.A.19c)$$

Subtracting it from (6.A.19a) yields the following expression for the first line in (6.A.17b):

$$\Sigma_\rho\Sigma_\sigma\{\Sigma_j(\mathrm{u}_{j\sigma}(\mathbf{x}')\ \partial \mathrm{u}_{i\rho}(\mathbf{x}')/\partial \mathrm{x}'_j - \mathrm{u}_{j\rho}(\mathbf{x}')\ \partial \mathrm{u}_{i\sigma}(\mathbf{x}')/\partial \mathrm{x}'_j)\}\omega_{\sigma\mu}(\alpha)\ \omega_{\rho\kappa}(\alpha). \qquad (6.A.19d)$$

Referring to Eq. (5.8.9) of Sec. 5.8, the reader will find that the expression enclosed in curly brackets is the coefficient of x'_i defined by the commutator of two generators $U_\sigma = \Sigma_j u_{j\sigma}(\mathbf{x}')x'_j$, and $U_\rho = \Sigma_i u_{i\rho}(\mathbf{x}')x'_i$. That is, one can write

$$[U_\sigma, U_\rho] = \{\Sigma_j(u_{j\sigma}(\mathbf{x}')\ \partial u_{i\rho}(\mathbf{x}')/\partial x'_j - u_{j\rho}(\mathbf{x}')\ \partial u_{i\sigma}(\mathbf{x}')/\partial x'_j)\}x'_i. \quad (6.A.20)$$

With this observation in mind, let us proceed.

We first rename the summation indices ρ and σ in (6.A.19d) by attaching primes to them. In the second line of (6.A.17b) we also change the summation index, from ρ to τ. Equation (6.A.17b) then becomes

$$\begin{aligned}\Sigma_{\rho'}\Sigma_{\sigma'}\{\Sigma_j(u_{j\sigma'}(\mathbf{x}')\ \partial u_{i\rho'}(\mathbf{x}')/\partial x'_j - u_{j\rho'}(\mathbf{x}')\ \partial u_{i\sigma'}(\mathbf{x}')/\partial x'_j)\ \omega_{\sigma'\mu}(\alpha)\ \omega_{\rho'\kappa}(\alpha)\}\\ = \Sigma_\tau u_{i\tau}(\mathbf{x}')(\partial\omega_{\tau\mu}(\alpha)/\partial\alpha_\kappa - \partial\omega_{\tau\kappa}(\alpha)/\partial\alpha_\mu). \quad (6.A.21)\end{aligned}$$

The terms $\omega_{\sigma'\mu}\ (\alpha)\omega_{\rho'\kappa}(\alpha)$ on the left-hand side of this expression may be removed with the aid of the general relation (6.A.11b): multiplying (6.A.21) by $\nu_{\mu\sigma}$ and $\nu_{\kappa\rho}$, and then summing over μ and κ converts (6.A.21) to

$$\Sigma_j(u_{j\sigma}(\mathbf{x}')x'_j\ u_{i\rho}(\mathbf{x}') - u_{j\rho}(\mathbf{x}')x'_j\ u_{i\sigma}(\mathbf{x}')) = \Sigma_\tau u_{i\tau}(\mathbf{x}')\ C^\tau_{\sigma\rho}(\alpha), \quad (6.A.22a)$$

where

$$C^\tau_{\sigma\rho}(\alpha) = \Sigma_\mu\Sigma_\kappa(\partial\omega_{\tau\mu}(\alpha)/\partial\alpha_\kappa - \partial\omega_{\tau\kappa}(\alpha)/\partial\alpha_\mu)\nu_{\mu\sigma}(\alpha)\nu_{\kappa\rho}(\alpha). \quad (6.A.22b)$$

In (6.A.22a) *we have a relation in which the* **x**'s *and the* α's *can in principle be varied independently.* The reader is invited to use this freedom and a power series expansion of the $C^\tau_{\sigma\rho}(\alpha)$, to show that for (6.A.22a) to hold true for any choice of the **x**'s and α's, each $C^\tau_{\sigma\rho}(\alpha)$ must be independent of the α's, and may be replaced by a constant, $c^\tau_{\sigma\rho}$.

Having established this, one may replace (6.A.22) by

$$\Sigma_j(u_{j\sigma}(\mathbf{x}')x'_j\ u_{i\rho}(\mathbf{x}') - u_{j\rho}(\mathbf{x}')x'_j\ u_{i\sigma}(\mathbf{x}')) = \Sigma_\tau\ u_{i\tau}(\mathbf{x}')c^\tau_{\sigma\rho}, \quad (6.A.23a)$$

with each

$$c^\tau_{\sigma\rho} = \Sigma_\mu\Sigma_\kappa(\partial\omega_{\tau\mu}(\alpha)/\partial\alpha_\kappa - \partial\omega_{\tau\kappa}(\alpha)/\partial\alpha_\mu)\nu_{\mu\sigma}(\alpha)\ \nu_{\kappa\rho}(\alpha). \quad (6.A.23b)$$

a constant.

As the right-hand side of (6.A.23a) is a sum of the coefficient functions $u_{i\tau}(\mathbf{x}')$ of generators $U_\tau = \Sigma_i u_{i\tau}(\mathbf{x}')x'_i$, and the left-hand side is the corresponding coefficient of the x_i defined by the commutator of

$U_\sigma = \Sigma_j u_{j\sigma}(\mathbf{x}')x'_j$, and $U_\rho = \Sigma_i u_{i\rho}(\mathbf{x}')x'_i$, we have obtained:

Lie's Second Fundamental Theorem. *The generators* U_γ *of an* r-*parameter group satisfy the relations*

$$[U_\sigma, U_\rho] = \Sigma_\tau c^\tau_{\sigma\rho}\, U_\tau, \quad \sigma,\ \rho,\ \tau = 1, \ldots, r. \tag{6.A.24}$$

Lie's Third Fundamental Theorem, the proof of which is suggested in Sec. 6.3, states that:

The structure constants $c^\tau_{\sigma\rho}$ *satisfy the relation*

$$\Sigma_\mu(c^\mu_{\alpha\beta}\, c^\nu_{\mu\gamma} + c^\mu_{\beta\gamma}\, c^\nu_{\mu\alpha} + c^\mu_{\gamma\alpha}\, c^\nu_{\mu\beta}) = 0. \tag{6.A.25}$$

The converse of the Second Fundamental Theorem was stated in Theorem 6.1.1 above. For derivations of the converse of each of the theorems, the reader is referred to Eisenhart's book.[31]

Appendix B. Classification of Lie Algebras and Lie Groups[31]

Two Lie algebras with the same set of structure constants are, from an abstract algebraic standpoint, identical, and are classified as such. Lie algebras may be classified in a variety of ways, and each classification has a counterpart in the classification of Lie groups. *It is customary in classifying Lie algebras to eliminate the zero operator*, U_0, *from consideration. This should be kept in mind in reading the material in this section.*

A Lie algebra and its corresponding group are said to be *Abelian* if all its generators commute, i.e. $[U_\kappa, U_\lambda] = 0$ for the generators U in the algebra. If a subset of the generators of an algebra commute among themselves they constitute a *Abelian gruop, subalgebra.* The generators of an Abelian subalgebra constitute the *center* of the whole algebra.

We move on to a consideration of further properties of Lie algebras important in their classification. Each of the properties defined will be illustrated in the exercises.

If the operator A of a Lie algebra can be separated into two subalgebras B, C such that none of the generators in B are contained in C and *vice versa*, and that the generators in B commute with those in C, then the Lie algebra A is a *direct sum* of the ideals B and C. This is written

$$A = B \oplus C. \tag{6.B.1}$$

With some abuse of notation one may write the conditions as

$$[B, B] = B, \quad [C, C] = C, \quad [B, C] = 0. \tag{6.B.2}$$

The corresponding group operator, T_A, is then a direct product of the subgroup operators T_B and T_C. This relationship is expressed as

$$T_A = T_B \otimes T_C. \tag{6.B.3}$$

An example of two such commuting algebras is provided by the operators $y_1\partial/\partial z_1 - z_1\partial/\partial y_1$ and $y_2\partial/\partial z_2 - z_2\partial/\partial y_2$, which may be considered to be generators of a two-parameter group that rotates two different particles in the y–z plane.

A Lie algebra A is a *semidirect sum* of the subalgebras B and C if

$$[B, B] = B, \quad [C, C] = C, \quad [B, C] = B. \tag{6.B.4}$$

This may be written

$$A = B \oplus_s C, \tag{6.B.5}$$

with the subalgebra B placed to the left of $\oplus_s$. The corresponding group operator, T_A, is termed a semidirect product of the subgroup operators T_B and T_C. This can be expressed as

$$T_A = T_B \otimes_S T_C. \tag{6.B.6}$$

The subalgebra B is termed a *ideal* or *normal subalgebra.* If an ideal does not contain U_0, the ideal is a *proper ideal.*

A Lie algebra is said to be *simple* if it contains no proper ideals. A Lie algebra is *semi-simple* if it contains no Abelian ideals other than U_0. A Lie algebra is semi-simple *iff* the determinant of its Killing form is nonzero.

A Lie algebra A^r of dimension r is a *solvable Lie algebra* if it possesses a chain of subgroups A^k such that

$$A^1 \supset A^2 \supset A^{r-1} \supset A^r. \tag{6.B.7}$$

The term *solvable* was introduced by Lie. In his investigations of ordinary differential equations he established that if an ordinary differential equation of order R is left invariant by a solvable group of order r whose generators are first-order differential operators U, then the order of the differential equation may be reduced to R − r without carrying out any integrations.[33]

The Levi–Malcev theorem shows that every Lie algebra A may be expressed as a semidirect sum of a solvable Lie algebra S_o and a semi-simple Lie algebra S_e

$$A = S_o \oplus_s Se. \tag{6.B.8}$$

Every Lie group may be expressed as the corresponding semidirect product.

Further information on the classification of Lie algebras and groups will be found in Gilmore's monograph, which contains extensive tables of groups and their subgroups.[34]

Exercises.

1. Let R_x, R_y, R_z be generators of rotations on the sphere. Find the commutation relations of the three operators $R_p = (R_x + R_y)$, $R_m = (R_x - R_y)$, R_z. Determine the Killing form.
2. Let $U = t^2\partial/\partial t$, which, as we have seen, generates finite transformations that have singularities. Determine the action of U on $\theta = 2\arctan(t)$, the form of the corresponding $V = v\partial/\partial\theta$, and the finite transformations of θ that it generates. Is V the generator of a compact group?
3. Let $R_{jk} = x_j\ \partial/\partial x_k - x_k\ \partial/\partial x_j$, and define $U_1 = (R_{23} + R_{14})/2$, $U_2 = (R_{31} + R_{24})/2$, $U_3 = (R_{12} + R_{34})/2$.
   (a) Show that the U's obey the commutation relations of SO(3) and SU(2).
   (b) Prove that the U's generate a transformation group whoes operations leave invariant the hyperspheres defined by $x_1^2 + x_2^2 + x_3^2 + x_4^2 = r^2$.
   (c) Show that, acting on this sphere, the group is SU(2).
   (d) Change the R_{j4} to $S_{j4} = x_j\ \partial/\partial x_4 + x_4\ \partial/\partial x_j$, then show that operators of the form $V_1 = (R_{23} + R_{14})/2$, $V_2 = (R_{31} + R_{24})/2$, $V_3 = (R_{12} + R_{34})/2$ do not generate SU(2) groups.
4. Use the invariant ds^2 of the Lorentz realization of SO(3,1) to establish the form of the manifolds in spacetime that are left invariant by transformations of the group.
5. Construct the generators of an SO(3,1) group that leaves invariant a unit sphere. Hint: cf. Eqs. (5.1.9).
6. Let $\mathbf{L} = (L_x, L_y, L_z)$, $\mathbf{p} = -i\boldsymbol{\nabla}$. Do $\mathbf{p}$ and $\mathbf{L}$ commute with $\mathbf{L}^2 + p^2$? With $\mathbf{L}\cdot\mathbf{p}$?
7. Let

$$U_1 = \partial/\partial x, \quad U_2 = x\partial/\partial x, \quad U_3 = (x^2/2)\ \partial/\partial x.$$

Show that

$$[U_1, U_2] = U_1, \quad [U_2,\ U_3] = 2U_3, \quad [U_3, U_1] = -U_2.$$

Then establish that

$$c_{12}^1 = 1 = -c_{21}^1, \quad c_{23}^3 = 2 = -c_{32}^3, \quad c_{31}^2 = -1 = -c_{13}^2,$$

and that all other c's vanish. Show that in the Killing form,

$$g_{11} = 0, \quad g_{12} = 0, \quad g_{13} = -3, \quad g_{22} = 5, \quad g_{23} = 0, \quad g_{33} = 0.$$

Evaluate the determinant of the form, and determine whether the group is semi-simple.

8. Acting in two-dimensional Euclidean space, the Euclidean group E(2) has generators

$$U_1 = \partial/\partial x, \quad U_2 = \partial/\partial y, \quad U_3 = x\partial/\partial y - y\partial/\partial x.$$

Show that

$$c^3_{12} = 0, \quad c^1_{23} = -1 = -c^1_{32}, \quad c^2_{31} = -1 = -c^2_{13},$$

and that all other c's vanish. Show that the Killing form is

$$\begin{vmatrix} 0 & 0 & 0 \\ 0 & 0 & 0 \\ 0 & 0 & -2 \end{vmatrix}.$$

9. Let

$$dx/da = Ax, \quad dx/db = Bx.$$

Let A, B respectively be linear and quadratic functions of a, b. In what cases can the equations define a two-parameter group?

References

[1] (a) S. Lie, *Math. Ann.* **16** (1880) 451–528; Coll. Works, Vol. 6, pp. 1–94, (Johnson Reprint, NY, 1973).
(b) *Sophus Lie's 1880 Transformation Group Paper*, translated by M. Ackerman and R. Hermann (Math. Sci. Press, Broookline. Massachusetts).

[2] Ref. 1(b) above, pp. 161–169; S. Lie, *Vorlesungen uber Continuirliche Gruppen*, reprint (Chelsea, Bronx, N.Y. (1971)).

[3] S. Lie, *Transformations Gruppen III*, reprint (Chelsea, Bronx, N.Y, 1970).

[4] For a proof of the theorem, see S. Pontriagin, *Topological Groups* (Princeton, 1946).

[5] G. J. Reid, *Eur. J. Appl. Math.* **2** (1991) 293, 319.

[6] H. F. Baker, *Proc. Lond. Math. Soc.* **34**(1) (1903) 347–360; **35**(1) (1903) 333–374; **2**(2) (1904) 293–296.

[7] J. E. Campbell, *Proc. Lond. Math. Soc. (1)* **29** (1898) 14–32.

[8] F. Hausdorff, *Ber. Sach. Akad. Wiss. (Math. Phys. Klasse) Leipzig* **58** (1906) 19–48.

[9] A. R. Edmonds, *Angular Momentum in Quantum Mechanics* (Princeton, 1957), pp. 5–7.

[10] C. E. Wulfman and B. G. Wybourne, *J. Phys. A* **9** (1976) 507–518.

[11] C. Wulfman, *Int. J. Quantum Chem.* **49** (1994) 185–195.
[12] Y. Aharonov and D. Bohm, *Phys. Rev.* **115** (1959) 485–491.
[13] (a) G. Herzberg and H. C. Longuet-Higgins, *Discuss. Faraday Soc.* **35** (1963) 77–82;
(b) H. C. Longuett-Higgins, *Proc. Roy. Soc. (London), A* **344** (1975) 147–156;
(c) C. A. Mead, *J. Chem. Phys.* **49** (1980); 23–32. 33–38.
(d) C. A. Mead and D. G. Truhlar, *J. Chem. Phys.* **70** (1979) 2284–2296.
[14] M. V. Berry, *Proc. Roy. Soc. (London), A* **392** (1984) 45–57.
[15] W. Killing, *Math. Ann.* **31** (1888) 252; **33** (1889) 1; **34** (1889) 55; **36** (1890) 161.
[16] E. Cartan, *Oeuvres Completes I*, pt. 1,2 (Gauthier-Villers, Paris, 1952).
[17] E. U. Condon and G. H. Shortley, *Theory of Atomic Spectra* (Cambridge, 1935).
[18] A. R. Edmonds, *Angular Momentum in Quantum Mechanics* (Princeton, 1957).
[19] H. B. G. Casimir, *Rotation of a Rigid Body in Quantum Mechanics*, thesis, (University of Leiden Wolters, Gronigen, 1931).
[20] B. G. Wybourne, *Classical Groups for Physicists*, Chap. XV (Wiley-Interscience, NY, 1974).
[21] W. T. Sharp, Thesis (Princeton University, Princeton, N.J., 1957).
[22] J. D. Talman, *Special Functions. A Group Theoretic Approach, Based on Lectures by Eugene P. Wigner* (Benjamin, NY 1968).
[23] G. N. Watson, *Theory of Bessel Functions*, 2nd edn. (Cambridge, 1966).
[24] N. J. Vilenkin, *Special Functions and the Theory of Group Representations, Am. Math. Soc.*, translation by V. N. Singh (Providence, R.I, 1968).
[25] G. Racah, *Group Theory and Spectroscopy*, Ergeb. Ex. Naturwiss, Vol. 37, (Springer-Verlag, Berlin, 1965), pp. 28–84.
[26] S. Pontriagin, *Topological Groups* (Princeton, 1942).
[27] B. Russell, quoted in A.S. Eddington, *New Pathways in Science* (Cambridge, 1935), p. 257.
[28] B. Russell and A. N. Whitehead, Principia Mathematica, I–III (Cambridge 1910–1913).
[29] A. S. Eddington, *loc. cit.*
[30] For a popular account, see B. Greene, *The Elegant Universe* (Vintage Books, N.Y., 2000).
[31] L. P. Eisenhart, *Continuous Groups of Transformations* (Princeton University Press, Princeton, N. J. 1933). Note: there is considerable variation in the literature as to the nameing of Lie's theorems and their converses.
[32] B. G. Wybourne, *Lie Groups for Physicists* (Wiley, NY, 1974), pp. 43–55.
[33] For an excellent exposition of the use of solvable Lie groups see G. W. Bluman and S. Kumei, *Symmetries and Differential Equations* (Springer-Verlag, NY, 1989).
[34] R. Gilmore, *Lie Groups, Lie Algebras, and Some of Their Applications* (Wiley, NY, 1974).

CHAPTER 7

Dynamical Symmetry in Hamiltonian Mechanics

This chapter generalizes and extends the discussion of the Hamiltonian mechanics introduced in Chapter 3. Because Hamiltonian mechanics inherits certain general invariance properties from Newtonian mechanics, we begin with a brief discussion of the latter.

7.1 General Invariance Properties of Newtonian Mechanics

Let a set of *point masses* $m_j, j = 1,\ldots,n$, have position vectors $\mathbf{r}_j = (x_j, y_j, z_j)$ in a Cartesian frame of reference, X,Y,Z. If their velocity vectors are $\mathbf{v}_j = (v_{xj}, v_{yj}, v_z)$, then their momentum vectors are $\mathbf{p}_j = (p_{xj}, p_{yj}, p_{zj}) = (m_j\, v_{xj}, m_j\, v_{yj}, m_j\, v_{zj})$. Newton's first law states that if no force is acting on the masses, then for each mass, m_j,

$$d\mathbf{p}_j/dt = 0. \tag{7.1.1}$$

If the masses are constant, (7.1.1) reduces to

$$m_j\, dv_j/dt = 0. \tag{7.1.2}$$

Newton's second law requires the change of momenta, $\mathbf{p}_j$, to be due the action of quantifiable *forces*, $\mathbf{F}_j = F_{xj}, F_{yj}, F_{zj}$, such that

$$d\mathbf{p}_j/dt = d^2(m_j\mathbf{r}_j)/dt^2 = \mathbf{F}_j. \tag{7.1.3}$$

Newton's third law states that if particle i exerts a force $\mathbf{F}_{ij}$ on particle j, then particle j exerts a force $\mathbf{F}_{ji}$ on particle i, with

$$\mathbf{F}_{ji} = -\mathbf{F}_{ij}. \tag{7.1.4}$$

In an isolated system of interacting particles, the total force, $\mathbf{F}_j$, acting on particle j is $\Sigma_i\, \mathbf{F}_{ij}$, so the total force, $\Sigma_i\Sigma_j\, \mathbf{F}_{ij}$, acting between all particles is zero. From this it follows that the total momentum of the system is conserved:

$$\Sigma_j \, d\mathbf{p}_j/dt = d^2(\Sigma_j m_j \mathbf{r}_j)/dt^2 = 0. \tag{7.1.5}$$

Newton proved that spherically symmetric masses of total mass m_j behave as point masses of mass m_j when acted on by external inverse square law forces, such as his force of gravity.

Equations (7.1.3) are second-order ordinary differential equations. A knowledge of the $\mathbf{F}_j$, and initial position and initial velocity vectors, is therefore sufficient to predict the motions in a classical mechanical system. This quite amazing feature of Newtonian mechanics withstood all experimental tests and challenges until the advent of quantum theory.

Newton's equations of motion for a system of interacting particles, are left invariant by a group of transformations that relates measurements made by two observers moving with constant relative velocity using clocks that agree on time intervals. The equations are invariant under the passive transformation $t \to t' = t - (\pm)\, t_0$. They are invariant under passive coordinate transformations that convert one Cartesian system with axes (X, Y, Z) to another with axes (X', Y', Z'), if the two origins are moving with relative velocity $-(\pm)\, (v_{x0}, v_{y0}, v_{z0})$, and at time $t = 0$, have origins whose position vectors differ by $-(\pm)\, x_0, y_0, z_0$. The two systems of axes may differ in orientation. Measurements of positions and velocities in the two systems are related by the active transformations

$$t \to t' = t \pm t_0, \tag{7.1.6a}$$

$$\mathbf{r}_j \to \mathbf{r}'_j = \mathbf{r}_j \pm (x_0, y_0, z_0), \tag{7.1.6b}$$

$$\mathbf{r}'_j = \pm \mathbf{R}(x_j, y_j, z_j)^T, \tag{7.1.6c}$$

where $\mathbf{R}$ is a rotation matrix. The transformation

$$\mathbf{v}_j \to \mathbf{v}'_j = \mathbf{v}_j \pm (v_{x0}, v_{y0}, v_{z0}), \tag{7.1.6d}$$

also converts $\mathbf{r}_j$ to

$$\mathbf{r}'_j = \mathbf{r}_j \pm (t v_{x0}, v_{y0}, v_{z0}). \tag{7.1.6e}$$

The transformations (7.1.4a–e) are those of the Lie transformation group termed the *Galilei group.*[1] From a purely *mathematical* standpoint, the transformations of the group may be active or passive ones. From a *physical* standpoint, the transformations must relate measurements made in one reference system to those in another by passive transformations of coordinates. If one were to use an active physical interpretation of the group's action, one would need to apply the transformations to physical particles moving with respect to a single set of position and time coordinates. This would require the application of forces to accelerate and/or decelerate the particles. The particles, together with the required measuring sticks and clocks, would no longer constitute an isolated system. To turn the system into an isolated one requires making the experimenter part of the system — a problem which still vexes quantum theory. However, because of conceptual simplicity, one commonly uses active transformations to mathematically interconvert solutions of a given equation of motion.

The Galilei group is a ten-parameter group whose passive operations on a (t, r) coordinate system may be summarized by

$$\mathrm{t} \rightarrow \mathrm{t}' = \mathrm{t} + \tau, \quad \mathbf{r} \rightarrow \mathbf{r}' = \mathbf{R}(\alpha, \beta, \gamma)\mathbf{r} + \mathbf{v}_0 \mathrm{t} + \mathbf{r}_0, \tag{7.1.7}$$

where $\mathbf{R}(\alpha, \beta, \gamma)$ is a rotation operator. The group parameters may be τ, α, β, γ, and the six components of $\mathbf{v}_0$ and $\mathbf{r}_0$. The generators of the subgroup that does not depend on v_0 may be

$$\mathrm{T}_0 = \partial/\partial\mathrm{t}, \quad \mathrm{T}_1 = \partial/\partial\mathrm{x}, \quad \mathrm{T}_2 = \partial/\partial\mathrm{y}, \quad \mathrm{T}_3 = \partial/\partial\mathrm{z}, \tag{7.1.8}$$

$$\mathrm{R}_1 = \mathrm{y}\partial/\partial\mathrm{z} - \mathrm{z}\partial/\partial\mathrm{y}, \quad \mathrm{R}_2 = \mathrm{z}\partial/\partial\mathrm{x} - \mathrm{x}\partial/\partial\mathrm{z}, \quad \mathrm{R}_3 = \mathrm{x}\partial/\partial\mathrm{y} - \mathrm{y}\partial/\partial\mathrm{x}.$$

R_1, R_2, R_3, and the T's generate an E(3) group of point transformations that acts in the Euclidean space of (X, Y, Z). T_0 generates an E(1) group that shifts the value of t, the *independent variable* in Newton's equations. Together, the seven operators generate $\mathrm{E}(3) \times \mathrm{E}(1)$. The generators of the passive transformations corresponding to (7.1.6d,e), act on both $\mathbf{r}$ and $\mathbf{v} = d\mathbf{r}/dt$, and are

$$\mathrm{K}_1 = \mathrm{t}\partial/\partial\mathrm{x} + \partial/\partial\mathrm{v_x}, \quad \mathrm{K}_2 = \mathrm{t}\partial/\partial\mathrm{y} + \partial/\partial\mathrm{v_y}, \quad \mathrm{K}_3 = \mathrm{t}\partial/\partial\mathrm{z} + \partial/\partial\mathrm{v_z}. \tag{7.1.9}$$

These generate one-parameter groups whose parameters are, respectively, the $\mathrm{v_{xo}}$, $\mathrm{v_{yo}}$, $\mathrm{v_{yo}}$, components of $\mathbf{v}_0$.

A space whose points are carried into one another by the group of translations and rotations is continuous, and is said to be *homogeneous* and *isotropic*. Newtonian time is continuous, homogeneous, and isotropic. These continuity properties make the calculus physically relevant, and make it possible for Newtonian physical causality to be continuous.

There are also discrete transformations that leave Newton's equations invariant. For isolated systems, Eqs. (7.1.1) and (7.1.2), are invariant under the combined passive time-reversal and space-inversion whose active form is

$$t \to -t, \quad \mathbf{r}_j \to -\mathbf{r}_j. \tag{7.1.10}$$

If the forces in Eq. (7.1.3) are time-independent, the equation is time-reversal invariant. If $\mathbf{F}_j$ is an odd function of the $\mathbf{r}$'s, Eq. (7.1.3) is also space-inversion invariant. Transformations such as

$$X \to Y, \quad Y \to X, \quad Z \to -Z, \tag{7.1.11}$$

which convert a right-handed coordinate system into a left-handed one, leave all of Newton's equations invariant if the forces $\mathbf{F}_j(\mathbf{r}_1, \ldots, \mathbf{r}_n)$ are invariant under the corresponding active transformations of the $\mathbf{r}_i$.

The conversion of the space, time, and velocity transformations of the Galilei group to transformations in phase space will be discussed in Sec. 7.6 below.

7.2 Relationship of Phase Space to Abstract Symplectic Space

This section further develops and generalizes Sec. 3.5, which introduced a geometric view of the Hamiltonian mechanics of systems with one q and one p. Such systems are, for historical reasons, said to have *one degree of freedom*, which they do have in position space.

We begin the treatment of Hamiltonian systems with n degrees of freedom by classifying the coordinates into two sets of n variables

$$\mathbf{q} = (q_1, \ldots, q_n), \quad \mathbf{p} = (p_1, \ldots, p_n). \tag{7.2.1}$$

At this point, the q's and p's are considered to be mathematical symbols whose physical interpretations have yet to be supplied. To generalize the discussion of Hamiltonian mechanics in Chapter 3, we consider transformations of the combined sets of variables (q, p) to sets (q', p') in which

$$q_i \to q_i' = Q_i(\mathbf{q}, \mathbf{p}), \quad p_j \to p_j' = P_j(q, p). \tag{7.2.2}$$

These give rise to the active local transformations

$$\begin{aligned} dq_i \to dq_i' &= \Sigma_k(\partial Q_i/\partial q_k dq_k + \partial Q_i/\partial p_k dp_k) = \Sigma_k(a_{ik} dq_k + b_{ik} dp_k), \\ dp_j \to dp_j' &= \Sigma_k(\partial P_j/\partial q_k\, dq_k + \partial P_j/\partial p_k\, dp_k) = \Sigma_k(c_{jk}\, dq_k + e_{jk}\, dp_k), \end{aligned} \tag{7.2.3}$$

with

$$\partial Q_i/\partial q_k = a_{ik}, \quad \partial Q_i/\partial p_k = b_{ik}, \quad \partial P_j/\partial q_k = c_{jk}, \quad \partial P_j/\partial p_k = e_{jk}.$$

A point $\mathbf{S}_s$ is considered to lie at the terminus of a Cartesian vector

$$\mathbf{s} = (q_1, \ldots, q_n, p_1, \ldots, p_n), \tag{7.2.4a}$$

the vector sum of

$$\mathbf{q} = (q_1, \ldots, q_n, 0, \ldots, 0) \tag{7.2.4b}$$

and

$$\mathbf{p} = (0, \ldots, 0, p_1, \ldots, p_n).$$

The original dq's and dp's are components of 2n-dimensional displacement vectors,

$$\mathbf{ds} = (dq_1, \ldots, dq_n, dp_1, \ldots, dp_n) = \mathbf{dq} + \mathbf{dp}. \tag{7.2.4c}$$

The pair of orthogonal unit vectors of Sec. 3.5 may generalized to a set of 2n orthogonal unit vectors which we denote $\mathbf{i}_1,\ldots,\mathbf{i}_n$ and $\mathbf{j}_1,\ldots,\mathbf{j}_n$. Eq. (7.2.3) then state

$$\begin{aligned} \mathbf{dq}_i \to \mathbf{dq}_i' &= \Sigma_k(a_{ik}\, \mathbf{dq}_k + b_{ik}\, \mathbf{dp}_k) = \mathbf{ds}_i', \\ \mathbf{dp}_j \to \mathbf{dp}_j' &= \Sigma_k(c_{jk}\, \mathbf{dq}_k + e_{jk}\, \mathbf{dp}_k) = \mathbf{ds}_j'. \end{aligned} \tag{7.2.5a}$$

The coefficients a_{ik} and b_{ik} represent the components of the local displacement vectors on the unit vectors of the reference system:

$$\begin{aligned} a_{ik} &= \mathbf{dq}_i' \cdot \mathbf{i}_k, \quad & b_{ik} &= \mathbf{dp}_i' \cdot \mathbf{j}_k, \\ c_{jk} &= \mathbf{dq}_j' \cdot \mathbf{i}_k, \quad & e_{jk} &= \mathbf{dp}_j' \cdot \mathbf{j}_k. \end{aligned} \tag{7.2.6}$$

The vectors $\mathbf{dq}$ and $\mathbf{dp}$ have a common origin at the point $\mathbf{S}_s$ and together define a two-plane. **Figure 7.2.1a** is an attempt to depict two such planes that are produced as $\mathbf{S}_s$ moves in a space of dimension greater than two.

In Chapter 3, transformations which convert Hamilton's equations of motion into Hamilton's equations of motion were characterized analytically with the aid of Poisson brackets, and geometrically by oriented areas defined with the aid of the wedge product analog of the vector cross product. To generalize these concepts to 2n dimensions, one first of all defines

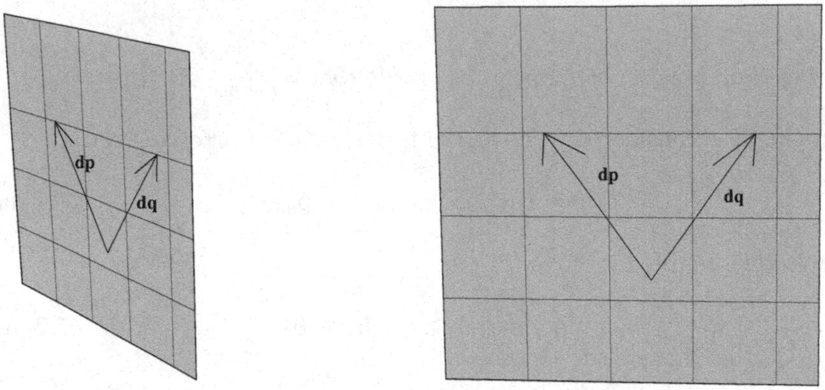

Fig. 7.2.1 Motion of the plane containing dq and dp.

the Poisson bracket $\{A(q,p), B(q,p)\}$ of functions A(q,p) and B(q,p) by

$$\{A, B\} = \Sigma_j(\partial A/\partial q_j\ \partial B/\partial p_j - \partial B/\partial q_j\ \partial A/\partial p_j). \tag{7.2.7}$$

As will be explained in Sec. 7.4 below, it is necessary to evaluate multiple Poisson brackets involving three functions, e.g. $\{\{A(q,p), B(q,p)\}, C(q,p)\}$. This requires that the functions possess second derivatives. As it will be necessary that the resulting derivatives are themselves continuous functions, *we henceforth require all functions of the q's and p's to be* C^2.

The transformations (7.2.2) are defined to be symplectic if

$$\Sigma_k(\partial Q_i/\partial q_k\ \partial P_j/\partial p_k - \partial P_j/\partial q_k\ \partial Q_i/\partial p_k) = \delta_{ij},$$

$$\Sigma_k(\partial Q_i/\partial q_k\ \partial Q_j/\partial p_k - \partial Q_j/\partial q_k\ \partial Q_i/\partial p_k) = 0, \tag{7.2.8}$$

$$\Sigma_k(\partial P_i/\partial q_k\ \partial P_j/\partial p_k - \partial P_j/\partial q_k\ \partial P_i/\partial p_k) = 0.$$

In the Poisson bracket notation these equations become

$$\{q_i', p_j'\} = \delta_{ij}, \quad \{q_i', q_j'\} = 0, \quad \{p_i', p_j'\} = 0. \tag{7.2.9}$$

One gives a geometric interpretation to the equations with the aid of the wedge product defined in Chapter 3. At the points $\mathbf{S}_s$ and $\mathbf{S}_s'$ the vectors $\mathbf{dq}$ and $\mathbf{dp}$, and the vectors $\mathbf{dq}'$ and $\mathbf{dp}'$ define local wedge products whose magnitude is twice the area of the plane triangle obtained by connecting their termini. The transformation of q and p moves $\mathbf{S}_s$ to $\mathbf{S}'_s$, and in the process

$$\mathbf{dq} \wedge \mathbf{dp} \rightarrow \mathbf{dq}' \wedge \mathbf{dp}'. \tag{7.2.10}$$

If $\mathbf{dq}_k \wedge \mathbf{dp}_k$ is the wedge product in the local q_k–p_k subplane at $\mathbf{s}$, and $\mathbf{dq}_k \wedge \mathbf{dp}_k$ is the wedge product in the local q'_k–p'_k subplane at $\mathbf{s}'$, then (cf. the exercises), Eq. (7.2.9) imply that

$$\mathbf{dq} \wedge \mathbf{dp} = \sum_k (\mathbf{dq}_k \wedge \mathbf{dp}_k), \tag{7.2.11a}$$

and that at $\mathbf{s}'$,

$$\mathbf{dq}' \wedge \mathbf{dp}' = \sum_k (\mathbf{dq}_k \wedge \mathbf{dp}_k). \tag{7.2.11b}$$

In the coordinate system at, say $\mathbf{s}$, Eq. (7.2.9) imply

$$\mathbf{dq} \wedge \mathbf{dp} = \mathbf{dq}' \wedge \mathbf{dp}'. \tag{7.2.11c}$$

The most direct physical interpretations of symplectic transformations develop when one is dealing with Cartesian variables x_j, and momentum variables defined by $p_{xj} = m_j dx_j/dt = m_j v_j$. The Poisson bracket of two functions A, B is then defined using Eqs. (7.2.7) and (7.2.9), with $q_j = x_j$ and $p_j = p_{xj}$ as the independent variables in (7.2.7). When generalized coordinates are used, one begins with the Cartesian position coordinates, makes transformations to new coordinates q_j, and then determines their conjugate momenta, p_j. If the system has potential energy function V(q), the determination of the new conjugate momenta is most simply done if one uses the transformation from x's to q's to transform the kinetic energy, T, of the system from $T(m_j, dx_j/dt)$ to $T(m_j, q_j, q'_j)$, where $q'_j = dq_j/dt$. The Lagrangian, L, is then $T - V(q)$, and Lagrange's equations are

$$\partial L(m_j, q_j, q'_j)/\partial q_j = q'_j, \tag{7.2.12a}$$

$$p_j \equiv \partial L(m_j, q_j, q'_j)/\partial q'_j. \tag{7.2.12b}$$

For the assumed form of L, Eq. (7.2.12b) determines the momenta p_j corresponding to the q_j. In Appendix A it is proved that the p_j so defined are canonical conjugates of the q_j. The proof is prefaced by, and depends upon, Thomson and Tait's direct derivation of Lagrange's equations of motion from Newton's.[2]

In the discussion of the Hamiltonian mechanics in Sec. 3.7, the phase space of Hamiltonian mechanics differs from symplectic space, because in physics one must consider scale changes $\mathbf{q} \to a\mathbf{q}$, $\mathbf{p} \to b\mathbf{p}$, that need not be the scale changes $\mathbf{q} \to a\mathbf{q}$, $\mathbf{p} \to a^{-1}\mathbf{p}$, which leave the wedge product $d\mathbf{q} \wedge d\mathbf{p}$ invariant. The same arguments that were applied to shifts from clockwise directed, to counterclockwise directed, two-dimensional wedge products, also apply to (7.2.11a). Thus, we adopt the following.

Definition of the Hamiltonian Phase Space

The phase space of 2n canonically conjugate variables q, *and* p, *is the space containing all points that are carried into one another by diffeomorphisms that leave invariant*

$$|d\mathbf{q} \wedge d\mathbf{p}|, \tag{7.2.13}$$

or multiply it by a positive constant.

7.3 Hamilton's Equations in PQ Space. Constants of Motion

Hamilton's equations of motion determine the evolution of congugate position and momentum variables in a system with a known Hamiltonian function. In this section it is assumed that the Hamiltonian, H(p,q), is not a function of time. Systems with time dependent Hamiltonians will be considered in Sec. 7.5 below.

We begin by deriving Hamilton's equations from Lagrange's equations of the second kind (cf. Appendix A). For fixed values of the masses, m_i, L is assumed to be a function only of the q_j and q_j' and not an explicit function of t. Consequently

$$dL = \sum_j (\partial L/\partial q_j \, dq_j + \partial L/\partial q_j' \, dq_j'). \tag{7.3.1}$$

Using Eq. (7.2.12) one obtains

$$dL = \sum_j (p_j' \, dq_j + p_j \, dq_j'). \tag{7.3.2}$$

The second term on the rhs of this relation may be expressed as

$$d\left(\sum_j p_j \, q_j'\right) - \sum_j q_j' \, dp_j. \tag{7.3.3}$$

Using this, one converts (7.3.2) to

$$d\left(L - \sum_j p_j q_j'\right) = \sum_j (p_j' \, dq_j - q_j' \, dp_j). \tag{7.3.4}$$

On defining

$$H = -\left(L - \sum_j p_j\, q_j'\right), \qquad (7.3.5)$$

Equation (7.3.4) can be written as

$$dH = \sum_j (q_j'\, dp_j - p_j'\, dq_j). \qquad (7.3.6)$$

The fact that dH is a function of the dp's and dq's suggests that it might be useful to convert H in (7.3.5) to a function of q's and p's. This is done by defining Hamilton's function, H, by

$$H(q, p) \equiv \left(\sum p_j v_j - L\right)|_{v=g(q,p)}. \qquad (7.3.7)$$

It has

$$dH = \sum_j (\partial H(q, p)/\partial p_j\, dp_j + \partial H(q, p)/\partial q_j\, dq_j). \qquad (7.3.8)$$

Comparing this with (7.3.6) one observes that dH and dH become identical if

$$\partial H(q, p)/\partial p_j = q_j', \quad \partial H(q, p)/\partial q_j = -p_j'. \qquad (7.3.9a)$$

Choosing this to be so, one obtains Hamilton's equations of motion

$$dq_j/dt = \partial H(q, p)/\partial p_j, \quad dp_j/dt = -\partial H(q, p)/\partial q_j. \qquad (7.3.9b)$$

As H is a function only of q,p,

$$dH = \sum(\partial H(q, p)/\partial p_j dp_j + \partial H(q, p)/\partial q_j dq_j), \qquad (7.3.10a)$$

and so

$$dH/dt = \sum(dq_j/dt\, dp_j/dt - dq_j/dt\, dp_j/dt) = 0. \qquad (7.3.10b)$$

Now (7.3.7) can be expressed as

$$H(q, p) = (2T - (T - V))|_{v=g(q,p)} = T(p, q) + V(q). \qquad (7.3.11)$$

The value of H(q,p) is thus the mechanical energy, E, of the system. Equation (7.3.10b) then states that the energy of the system with the Hamiltonian of (7.3.11) does not change with time.

Using Poisson bracket notation, Hamilton's equations (7.3.9b) become

$$\mathrm{d}q_j/\mathrm{d}t = \{q_j, H\}, \quad \mathrm{d}p_j/\mathrm{d}t = \{p_j, H\}, \tag{7.3.12}$$

or, equivalently

$$\mathrm{d}q_j/\mathrm{d}t = -\{H, q_j\}, \quad \mathrm{d}p_j/\mathrm{d}t = -\{H, p_j\}.$$

These last relations imply that one can identify $\{-H, \cdot\}$, a first-order differential operator, as the generator of a one-parameter Lie group whose group parameter is t. It follows directly that

$$q_j' = \exp(t\{-H, \cdot\})q_j = Q_j'(t), \quad p_j' = \exp(t\{-H, \cdot\})p_j = P_j'(t). \tag{7.3.13}$$

Differentiating q_j' and p_j' with respect to t, one checks that they satisfy Hamilton's equations, that is, $Q_j'(t)$ and $P_j'(t)$ are the solutions of the equations of motion that develop as q_j and p_j evolve. One may term $\exp(t\{-H, \cdot\})$ the *evolution operator* of the system with Hamiltonian H, and call $\{-H, \cdot\}$ its *evolution generator.* Then t is the group parameter of the *evolution group.*

Equations (7.3.12) may be considered to be special cases of the relation

$$\mathrm{d}f(q, p, t)/\mathrm{d}t = \{f, H\} + \partial f/\partial t, \tag{7.3.14}$$

which applies when the evolution of q and p is governed by Hamilton's equations with Hamiltonian H. If f(q,p) is not an explicit function of time, and

$$\{H, f(q, p)\} = 0, \tag{7.3.15a}$$

it follows that

$$\mathrm{d}f/\mathrm{d}t = 0. \tag{7.3.15b}$$

The function f(q,p) is then said to be a *constant of the motion* of the system with Hamiltonian H. The constants of motion of a Hamiltonian may be different for different values of E, or different ranges of E. Stated definitively, f(q, p) is a constant of motion when $H(q, p) = E$, iff

$$0 = \{-H, \cdot\}f|_{H=E}. \tag{7.3.15c}$$

It follows that, if f(q,p) is a constant of motion, it is invariant under the one-parameter evolution group generated by $\{-H, \cdot\}$. Because (7.4.15c) can also be expressed as

$$0 = (\{f(q, p), \cdot\}H(q, p))|_{H=E}, \tag{7.3.15d}$$

it also follows that $\{f(q, p), \cdot\}$ generates a one-parameter Lie group which leaves H(q,p) invariant for this value of E.

The function $q_1(H\text{–}E)$ provides an example of a trivial constant of motion. For this,

$$\{f, H\} = q_1\{(H\text{–}E), H\} + \{q_1, H\}(H\text{–}E) = p_1(H\text{–}E). \qquad (7.3.16)$$

This vanishes for all values of E, simply because H–E is a factor of f. This is of little import. The physically interesting cases arise when f does not directly contain H–E, but contains a factor that substitutions convert to H–E. The situation is then similar to that illustrated in **Fig. 5.1.1**.

7.4 Poisson Bracket Operators

From the previous remarks it is apparent that, because Poisson bracket operators generate one-parameter Lie groups, they play a fundamental role in associating Lie transformation groups with symmetries in Hamiltonian mechanics. In fact, because they will be so ubiquitous throughout the book, we will often abbreviate *Poisson bracket* as PB. This section will deal, first of all, with the connection of PB operators to operators in position space. It will then be shown that PB operators are Lie generators of one-parameter groups of canonical transformations in phase space. Thereafter, *PB Lie algebras* will be defined and related to Lie algebras of PB operators.

To begin, we note that one has the operator relations

$$\partial/\partial q_k = -\{p_k, \cdot\}, \quad \partial/\partial p_k = \{q_k, \cdot\}. \qquad (7.4.1)$$

As an example of the application of these correspondences, consider the generator of rotations in the q_i–q_j plane of position space

$$R_{ij} = (q_i\partial/\partial q_j - q_j\partial/\partial q_i). \qquad (7.4.2a)$$

On replacing each $\partial/\partial q_k$ in R_{ij} by $-p_k$, one finds

$$R_{ij} \rightarrow -(q_i p_j - q_j p_i) = -L_{ij}, \qquad (7.4.2b)$$

where $L_{ij} = q_i p_j - q_j p_i$ represents the angular momentum of a particle moving in the q_i–q_j plane. The full PB operator obtained from $-L_{ij}$ is

$$\{-L_{ij}, \cdot\} = (q_i\partial/\partial q_j - q_j\partial/\partial q_i) + (p_i\partial/\partial p_j - p_j\partial/\partial p_i). \qquad (7.4.2c)$$

It is a Lie generator of rotations in the two orthogonal two-planes q_i–q_j and p_i–p_j in phase space. On identifying p_k with mv_k, one sees that (7.4.2c) is also the first extension of $(q_i\ \partial/\partial q_j - q_j\ \partial/\partial q_i)$.

In the general case, to convert a Lie generator, $\sum \xi_k(q)\partial/\partial q_k$ of point transformations in position space, Q, to a PB operator that acts in the phase space PQ, one first associates the position space operator with a function of q,p by letting

$$\sum \xi_k(q)\partial/\partial q_k \rightarrow \sum \xi_k(q)(-p_k) = -\sum \xi_k(q)p_k. \tag{7.4.3a}$$

One then associates this function with a Lie operator in phase space by letting

$$-\sum \xi_k(q)\ p_k \rightarrow \{-\sum \xi_k(q)p_k, \cdot\}, \tag{7.4.3b}$$

The correspondences setup in (7.4.3) are also two-way correspondences in the following sense: a function f(q,p) of the form $-\sum \xi_k(q)\ p_k$ gives rise to the PB operator $\{-\sum \xi_k(q)p_k, \cdot\}$ and to a position space operator $\sum \xi_k(q)\partial/\partial q_k$.

Only Lie generators of point transformations $q \rightarrow q' = g(q)$ yield Poisson bracket operators whose position space part is of the form $\sum \xi_k(q)\partial/\partial q_k$; they correspond to functions f(q, p) that are linear in the momenta p_k. If f(q, p) is not linear in the p's, then $\{f(q,p), \cdot\}$ will contain terms whose action on the q's is a function of the p's. As will be seen in Chapter 8, some of the generators of transformations that convert Kepler orbits into Kepler orbits have this property. The following paragraphs deal with properties of PB operators of the general form $\{f(q,p), \cdot\}$. Each function of positions and momenta determines a PB operator that has both physical and geometric interpretations.

One may show by direct calculation that if A, B and X are C^2 functions of the q's and p's, then Poisson brackets obey the Jacobi relation

$$\{A, \{B, X\}\} + \{X, \{A, B\}\} + \{B, \{X, A\}\} = 0. \tag{7.4.4}$$

This enables one to quickly establish that PB operators are generators of canonical transformations. Let

$$q_i \rightarrow q_i' = \exp(\alpha\{A, \cdot\})q_i, \quad p_j \rightarrow p_j' = \exp(\alpha\{A, \cdot\})p_j. \tag{7.4.5a}$$

Expanding $\{q_i', p_j'\}$ in powers of α one obtains

$$\{q_i', p_j'\} = \{q_i, p_j\} + \alpha\{\{A, q_i\}, p_j\} + \alpha\{q_i, \{A, p_j\}\} + O(\alpha^2). \tag{7.4.5b}$$

The sum of the terms multiplying α can be rewriten as

$$\{q_i, \{A, p_j\}\} - \{p_j, \{A, q_i\}\} = \{q_i, \{A, p_j\}\} + \{p_j, \{q_i, A\}\}. \tag{7.4.5c}$$

From Jacobi's relation one sees that adding $\{A, \{p_j, q_i\}\}$ to this gives zero. As $\{p_j, q_i\}$ is a number, $\{A, \{p_j, q_i\}\}$ has the value zero. Consequently

$$\{q_i, \{A, p_j\}\} + \{p_j, \{q_i, A\}\} = 0, \tag{7.4.5d}$$

and

$$\{q_i', p_j'\} = \{q_i, p_j\} = \delta_{ij}. \tag{7.4.5e}$$

Using the properties of Lie transformation groups established in previous chapters, one sees that (7.4.5e) holds to all orders in α. Applying the same argument to $\{q_i', q_j'\}$ and $\{p_i', p_j'\}$, one finds that they vanish. *Thus, if* $A(q, p)$ *is a* C^2 *function,* $\{A, \cdot\}$ *generates canonical transformations.*

We next prove the following relation between the PB's of C^2 functions and PB operators:

If

$$\{A, B\} = C, \tag{7.4.6a}$$

then

$$[\{A, \cdot\}, \{B, \cdot\}]X(q, p) = \{C, \cdot\}X(q, p). \tag{7.4.6b}$$

This is a direct consequence of Jacobi's relation amongst A,B and X, since the relation can be rewritten as

$$\{A, \{B, X\}\} - \{B, \{A, X\}\} = -\{X, \{A, B\}\} = \{\{A, B\}, X\}. \tag{7.4.7}$$

As the lhs is $[\{A, \cdot\}, \{B, \cdot\}]X$, and the rhs is $\{\{A, B\}, \cdot\}X$, the relationship of (7.4.6) follows.

The reader may prove that if A, B and C are functions that satisfy the Jacobi relation

$$\{A, \{B, C\}\} + \{C, \{A, B\}\} + \{B, \{C, A\}\} = 0, \tag{7.4.8a}$$

then $\{A, \cdot\}$, $\{B, \cdot\}$ and $\{C, \cdot\}$ satisfy the Jacobi relation

$$\begin{aligned}&[[\{A, \cdot\}, \{B, \cdot\}], \{C, \cdot\}] + [[\{C, \cdot\}, \{A, \cdot\}], \{B, \cdot\}]\\ &\quad + [[\{B, \cdot\}, \{C, \cdot\}], \{A, \cdot\}] = 0.\end{aligned} \tag{7.4.8b}$$

If a finite set of C^2 functions $f_i(q,p)$ satisfies (7.4.8a) and

$$\{f_i, f_j\} = \sum c_{ij}^k f_k, \tag{7.4.8c}$$

then

$$[\{f_i, \cdot\}, \{f_j, \cdot\}] = \sum c_{ij}^k \{f_k, \cdot\}. \tag{7.4.8d}$$

It follows from the general definition of Lie algebras (cf. Chapter 6, Appendix C), that

1. If r linearly independent operators, $\{f_i, \cdot\}$, satisfy (7.4.8d), and the f_i are C^2 functions, then the $\{f_i, \cdot\}$ are elements of a Lie algebra with composition operation [,], and they generate an r-parameter Lie transformation group.
2. The f_i are elements of a *Poisson bracket* Lie algebra whose composition operation is { , }, each giving rise to a function $\exp(\alpha_i f_i)$. The group which the f_i generate is termed a *function group.*
3. The commutator and PB Lie algebras are formally isomorphic.

However, if any f_k is a constant, then $\{f_i, \cdot\} = 0$, and $c_{ij}^k \{f_k, \cdot\} = 0$, a relation that also results if $c_{ij}^k = 0$. An example can be found in Sec. 7.7 below. It will be further investigated in the discussion of the quantum mechanical Correspondence Principle in Chapter 9.

In this and the following chapters we will make use of Lie algebras defined by (7.4.8c), as well as those defined by (7.4.8d).

7.5 Hamiltonian Dynamical Symmetries in PQ Space

For a system with n Cartesian q's and n conjugate p's, the equation

$$H(p, q) - E = 0, \tag{7.5.1}$$

defines a manifold in a 2n-dimensional phase space, *PQ space.* For each value of E it requires 2n − 1 numbers to fix a point in the manifold. Hamilton's equations of motion define evolution trajectories on these (2n − 1)-dimensional hypersurfaces. The manifold defined by (7.5.1) possesses Euclidean geometric symmetries. Furthermore it possesses symplectic geometric symmetries expressed by a dynamical group. The operations of the latter leave Hamilton's equations invariant, and define dynamical symmetries.

We have seen that each constant of motion u(q, p) determines a Poisson bracket operator $U = \{u, \cdot\}$ that satisfies

$$0 = UH(q, p)|_{H=E}, \tag{7.5.2a}$$

and that, if UH can not contain H − E as a factor, then U also satisfies

$$UH(q, p) = 0. \tag{7.5.2b}$$

Each U generates a one-parameter Lie group that transforms evolution trajectories into one another or into themselves. All such transformations carry points on the hypersurface of fixed energy into points on the same hypersurface.

If one knows two constants of the motion, u_1 and u_2, then $U_1 = \{u_1, \cdot\}$, and $U_2 = \{u_2, \cdot\}$ satisfy

$$U_1 H(q,p)|_{H=E} = 0, \quad U_2 H(q,p)|_{H=E} = 0. \tag{7.5.3}$$

Then (cf. the exercises) $[U_1, U_2]$ satisfies

$$[U_1, U_2] H(q,p)|_{H=E} = 0. \tag{7.5.4}$$

If $[U_1, U_2]$ is not identically zero, then $u_3 = \{u_1, u_2\}$ is a further constant of the motion. If a set of r linearly independent constants of motion, u_j, is closed under the PB operation, they generate an r-parameter Lie function group.

Each u, defines a PB operator U. If a set of r constants of motion, u_j, provides a set of PB operators, U_j, that are linearly independent and closed under commutation, they generate an r-parameter Lie transformation group. For the given value of E, the operations of this dynamical group convert solutions of Hamilton's equations into solutions. The group is the classical mechanical analog of a quantum mechanical degeneracy group, whose operations interconvert wave functions of the same energy.

As illustrated below, knowing the constants of motion, u, of a system enables one to solve problems involving systems with Hamiltonians $H' = H + f(u)$. Constants of motion and the groups they determine can also imply a variety of revealing connections between the dynamics of a system and its geometrical properties in phase space.

Let

$$\mathbf{s} = (s_1, \ldots, s_n, s_{n+1}, \ldots, s_{2n}) = (q_1, \ldots, q_n, p_1, \ldots, p_n), \tag{7.5.5}$$

be a vector whose components are conjugate q,p variables in this phase–space. The components may be, but are not necessarily, Cartesian. Suppose that $\mathbf{s}$ is a vector from the origin to a point $\boldsymbol{\rho}_s$ in the Hamiltonian hypersurface defined by $H(p, q) = E$. If $u_k = \mu_k(\mathbf{s})$ is a constant of motion, then at $\boldsymbol{\rho}_s$ each group generator, $U_k = \{u_k, \cdot\}$, and infinitesimal parameter, $\delta\alpha_k$, determines an infinitesimal displacement, $\delta \mathbf{s}_k = \delta\alpha_k U_k \mathbf{s}$, defined by

$$\mathbf{s} \to \mathbf{s} + \delta\alpha_k U_k \mathbf{s} + O(\delta\alpha_k^2) \tag{7.5.6}$$

This may be imagined to lie in the hypersurface defined by H(q,p) =E. If one lets $\boldsymbol{\Delta}\mathbf{s}_k = U_k \mathbf{s}$, the components of $\boldsymbol{\Delta}\mathbf{s}_k$ are just the functions $\xi_{km}(\mathbf{s})$ in the generator $U_k = \sum_m \xi_{km}(\mathbf{s})\, \partial/\partial s_m$:

$$\boldsymbol{\Delta}\mathbf{s}_k = (\xi_{k1}, \ldots, \xi_{kn}, \xi_{kn+1}, \ldots, \xi_{k2n}). \tag{7.5.7}$$

The finite vector $\boldsymbol{\Delta}\mathbf{s}_k$ is tangent to the hypersurface at $\boldsymbol{\rho}_s$.

Because the generators are linearly independent, the $\mathbf{\Delta s}_k$ are linearly independent. They may be used to establish a local coordinate system at $\boldsymbol{\rho}_s$. If, for a fixed value of $\mathbf{s}$, the tangent vectors $\mathbf{\Delta s}_k$ associated with each generator, including $\mathbf{s}$ itself, all transform as Cartesian vectors, then the $\mathbf{\Delta s}_k$ may be transferred to the origin of s and used to set up a Cartesian coordinate system there. Evaluating these vectors at a particular time, t_0, provides a set of fixed vectors to which motions of $\boldsymbol{\rho}_s$ may be referred.

When the constants of motion u_j transform as components of a Cartesian vector, they can also provide a fixed Cartesian coordinate system. When the coordinate system provided by these fixed vectors, or by the tangent vectors, is Cartesian, one is best able to use one's geometric intuition to guide investigations of the dynamics governed by the equations of motion.

Harmonic oscillators with Hamiltonian

$$H = (p_1^2 + q_1^2 + p_2^2 + q_2^2)/2 = H_1(q_1, p_1) + H_2(q_2, p_2) \tag{7.5.8a}$$

provide examples of the foregoing general remarks. The physical system with this Hamiltonian may be considered to be a mass moving in the q_1q_2 Euclidean plane, or two masses moving on a line. In either case, Hamilton's equations of motion are

$$dq_1/dt = p_1, \quad dp_1/dt = -q_1, \tag{7.5.8b}$$

$$dq_2/dt = p_2, \quad dp_2/dt = -q_2. \tag{7.5.8c}$$

One may assume q_1, q_2, p_1, p_2 to be coordinates in a four-dimensional phase space, and define Cartesian vectors

$$\mathbf{s} = (q_1, q_2, p_1, p_2) = (s_1, s_2, s_3, s_4) \tag{7.5.9a}$$

in the space. If s is a radius vector that stretches from the origin to a point defined by H = E, then, for the Hamiltonian (7.5.8), all such points satisfy the relation

$$(s_1, s_2, s_3, s_4)\cdot(s_1, s_2, s_3, s_4) = 2E, \tag{7.5.9b}$$

and the length of s is $\sqrt{2E}$. For each value of E, the equation H = E then defines an S(3) hypersphere of radius $\sqrt{2E}$. The surface of each such hypersphere is a three-dimensional manifold, a space allowing three orthonormal local Cartesian coordinates at each of its points. Though the manifold defined by $H - E = 0$ has O(4) symmetry, we have yet to determine the dynamical symmetries of the system.

The solutions of (7.5.8b,c) depend upon the four initial values of the q's and p's at $t_0 = 0$; $q_1(0)$, $p_1(0)$, $q_2(0)$, and $p_2(0)$. They are

$$\begin{aligned} q_1 &= q_1(0)\cos(t) + p_1(0)\sin(t), \quad & p_1 &= p_1(0)\cos(t) - q_1(0)\sin(t), \\ q_2 &= q_2(0)\cos(t) + p_2(0)\sin(t), \quad & p_2 &= p_2(0)\cos(t) - q_2(0)\sin(t). \end{aligned} \tag{7.5.10}$$

The simple relations of (7.5.8) and (7.5.10) imply a host of geometrical and dynamical connections. These are the subject of the remainder of this section.

In the four-dimensional Euclidean space of (q_1, q_2, p_1, p_2), there are six mutually orthogonal two-planes that pass through the origin — the planes of (q_1, q_2), (p_1, p_2), (q_1, p_1), (q_2, p_2), (p_1, q_2), and (p_2, q_1). For given initial values, Eqs. (7.5.10) determine the motion of points in each of these two-planes. One may project the motion in (q_1, p_1) and (q_2, p_2) onto a common q–p plane. **Figure 7.5.1** depicts the result if $q_1(0) = 1$, $p_1(0) = 4$, $q_2(0) = 2$, and $p_2(0) = 3$.

To determine the constants of motion u(p, q) of the oscillator and their implications, we seek functions u(p, q) annihilated by the PB operator of the Hamiltonian (7.5.8a). In this case, one procedure is to solve the equations that result when u(q, p) is expanded in power series in the p's and q's.

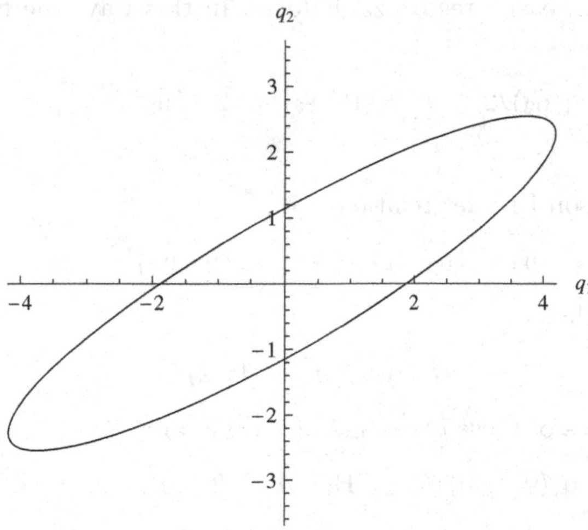

Fig. 7.5.1 Projection to trajectories of isotropic oscillator onto position space.

One lets

$$u = \sum c_{m_1 n_1 m_2 n_2} q_1^{m_1} p_1^{n_1} q_2^{m_2} p_2^{n_2}, \tag{7.5.11a}$$

and requires

$$\sum c_{m_1 n_1 m_2 n_2} \{H, q_1^{m_1} p_1^{n_1} q_2^{m_2} p_2^{n_2}\} = 0. \tag{7.5.11b}$$

The solving of (7.5.11b) is investigated in the exercises. One finds it has no solutions that are linear combinations of the p's and q's. It has three independent solutions that are quadratic functions of the positions and momenta, functions that, together with H, are constants of motion. Letting $\mathbf{S}_1 = (q_1, p_1)$, and $\mathbf{S}_2 = (q_2, p_2)$, these constants of the motion may be chosen to be

$$(H_1 - H_2) = (\mathbf{S}_1, \cdot\mathbf{S}_1 - \mathbf{S}_2, \cdot\mathbf{S}_2)/2, \tag{7.5.12a}$$

$$(p_1 p_2 + q_1 q_2) = \mathbf{S}_1 \cdot \mathbf{S}_2, \tag{7.5.12b}$$

and the vector $\mathbf{S}_1 \times \mathbf{S}_2$, with value $(q_1 p_2 - p_1 q_2)$. If the coordinates are those of a single particle, then

$$\mathbf{S}_1 \times \mathbf{S}_2 = \mathbf{L}_{12}, \tag{7.5.12c}$$

i.e. its angular momentum.

In determining the Lie algebras defined by the constants of motion, it is convenient to begin by multiplying the functions in (7.5.12) by constants a, b and c respectively. The constants can then be chosen so as to put the Lie algebra in an easily recognizable form. In this way, one finds that the three functions

$$u_1 = (p_1 p_2 + q_1 q_2)/2, \quad u_2 = (H_1 - H_2)/2, \quad u_3 = (q_1 p_2 - q_2 p_1)/2, \tag{7.5.13a}$$

satisfy the Poisson Bracket relations

$$\{u_1, u_2\} = -u_3, \quad \{u_3, u_1\} = -u_2, \quad \{u_2, u_3\} = -u_1. \tag{7.5.13b}$$

One also finds that

$$u_1^2 + u_2^2 + u_3^2 = (H/2)^2. \tag{7.5.13c}$$

The initial values of these constants of motion are

$$(p_1(0)p_2(0) + q_1(0)q_2(0))/2, \quad (E_1 - E_2)/2, \quad L_{12}(0)/2, \;\; E^2/4. \tag{7.5.14}$$

Adding any of the constants of motion to H gives a new Hamiltonian function H'. The motions determined by H' may be obtained from those determined by the original Hamiltonian. In the simplest case,

$$\mathrm{H}' = \mathrm{H} + \mathrm{cu}_1 = \mathrm{c}_1\mathrm{H}_1 + \mathrm{c}_2\mathrm{H}_2, \tag{7.5.15a}$$

with

$$\mathrm{c}_1 = 1 + \mathrm{c}/2, \quad \mathrm{c}_2 = 1 - \mathrm{c}/2. \tag{7.5.15b}$$

This eliminates the energy degeneracy $E_1 = E_2$, without altering the form of the solutions of (7.5.11). One may (see below) determine the motions when H' is $H + cu_2$ or $H + cu_3$ by making use of transformations generated by the PB operators of u_1, u_2 and u_3. These are the linearly independent operators

$$\begin{aligned}
\mathrm{U}_1 &= \{\mathrm{u}_1, \cdot\} = (-1/2)(\mathrm{p}_2\, \partial/\partial \mathrm{q}_1 + \mathrm{p}_1\, \partial/\partial \mathrm{q}_2 - \mathrm{q}_2\, \partial/\partial \mathrm{p}_1 - \mathrm{q}_1\, \partial/\partial \mathrm{p}_2),\\
\mathrm{U}_2 &= \{\mathrm{u}_2, \cdot\} = (-1/2)(\mathrm{p}_1\, \partial/\partial \mathrm{q}_1 - \mathrm{p}_2\, \partial/\partial \mathrm{q}_2 - \mathrm{q}_1\, \partial/\partial \mathrm{p}_1 + \mathrm{q}_2\, \partial/\partial \mathrm{p}_2),\\
\mathrm{U}_3 &= \{\mathrm{u}_3, \cdot\} = (-1/2)(\mathrm{q}_1\, \partial/\partial \mathrm{q}_2 - \mathrm{q}_2\, \partial/\partial \mathrm{q}_1 + \mathrm{p}_1\, \partial/\partial \mathrm{p}_2 - \mathrm{p}_2\, \partial/\partial \mathrm{p}_1).
\end{aligned} \tag{7.5.16}$$

They satisfy the commutation relations of the commutator Lie algebra

$$[\mathrm{U}_1, \mathrm{U}_2] = -\mathrm{U}_3, \quad [\mathrm{U}_3, \mathrm{U}_1] = -\mathrm{U}_2, \quad [\mathrm{U}_2, \mathrm{U}_3] = -\mathrm{U}_1, \tag{7.5.17a}$$

with Casimir invariant

$$\mathrm{U}_1^2 + \mathrm{U}_2^2 + \mathrm{U}_3^2. \tag{7.5.17b}$$

The operators U_1, U_2, U_3 each generate a one-parameter group of transformations, $\exp(\alpha_k U_k)$, which produces invariance transformations of the equations of motion, each necessarily a one-parameter group of symplectic transformations. The commutation relations establish that the resulting three-parameter group is a linear symplectic group Sp(2).

If an isotropic oscillator with Hamiltonian H is affected by an outside source that gives it angular momentum, c $(q_1p_2 - q_2p_1)$, then H' can be expressed as

$$\mathrm{H}' = \mathrm{H} + 2\mathrm{cu}_3. \tag{7.5.18a}$$

The resulting trajectories in the q_1p_1 and q_2p_2 subspaces are no longer independent. However, acting on H' with the group operator $\exp(\alpha_2 U_2)$ one finds

$$\mathrm{H}' = \mathrm{H} + 2\mathrm{c}(\mathrm{u}_3 \cos(\alpha_2) + \mathrm{u}_1 \sin(\alpha_2)). \tag{7.5.18b}$$

These become Hamiltonians of the form (7.5.15a) if $\alpha_2 = \pi$.

All of the $\{U, \cdot\}$'s necessarily commute with the evolution generator

$$\{-H, \cdot\} = \{2u_4, \cdot\} = p_1\partial/\partial q_1 + p_2\partial/\partial q_2 - q_1\partial/\partial p_1 - q_2\partial/\partial p_2. \quad (7.5.19)$$

The group generated by $\{-H, \cdot\}$ could, in the abstract, be any one-parameter Lie group, but as we know, it generates symplectic transformations, the group is always a realization of Sp(1). In analogy to U_1, U_2 and U_3, we defined U_4 to be $\{-H, \cdot\}/2$. The group generated by the four U's is then Sp(2) × Sp(1). The quadratic Casimir invariant of the Sp(2) groups is equal to the square of the generator of the Sp(1) group. It is the symplectic group Sp(2) × Sp(1) that defines dynamical symmetries of the oscillator.

As noted earlier, the Hamiltonian (7.5.8a) has evident four-dimensional rotational symmetry. Let us investigate the relationship between its SO(4) symmetry group and the Sp(2) × Sp(1) symplectic group uncovered to this point. To this end, we let

$$J_{ij} = s_i\partial/\partial s_j - s_j\partial/\partial s_i, \quad (7.5.20)$$

be the generator of rotations in the i–j plane of the four-space in which the Hamiltonian is defined. Six of the J_{ij} are linearly independent. Together they generate rotations of the SO(4) group that leaves invariant the Hamiltonian (7.5.8a). Inspecting (7.5.16) and (7.5.18) one finds that

$$U_1 = -(J_{14} + J_{23})/2, \quad U_2 = -(J_{24} + J_{31})/2, \quad U_3 = (J_{34} + J_{12})/2, \quad (7.5.21a)$$

and

$$U_4 = (J_{31} - J_{24})/2. \quad (7.5.21b)$$

Note that each U is composed of two rotation generators that commute. Equations (7.5.21) allow one to connect the operations of the symplectic group Sp(2) × Sp(1) generated by the U's, to operations of the group that leaves H invariant.

In addition to these U's, there are two linear combinations of the rotation generators (7.5.20), which annihilate H but do not generate canonical transformations. These are

$$V_2 = (J_{14} - J_{23})/2, \quad V_3 = (J_{34} - J_{12})/2. \quad (7.5.22)$$

Let us re-label U_4, which equals $\{-H, \cdot\}/2$, and call it V_1. Then

$$[V_1, V_2] = -V_3, \quad [V_3, V_1] = -V_2, \quad [V_2, V_3] = -V_1. \quad (7.5.23)$$

Though $V_2\ H = 0$ and $V_3\ H = 0$, the operators V_2, V_3 generate neither symplectic transformations nor transformations that convert solutions of the equations of motion (7.5.8b,c) into solutions. However, they do commute with the three U's in (7.5.21a) that generate the invariance group of these equations!

As each U is a linear combination of two commuting rotation generators, each generates finite transformations that are the product of two rotation transformations, and one has

$$\exp(\alpha_1 U_1) = \exp((-\alpha_1/2)J_{23})\exp(-(\alpha_1/2)J_{14}), \qquad (7.5.24a)$$

$$\exp(\alpha_2 U_2) = \exp(-(\alpha_2/2)J_{31})\exp(-(\alpha_2/2)J_{24}), \qquad (7.5.24b)$$

and

$$\exp(\alpha_3 U_3) = \exp((\alpha_3/2)J_{12})\exp((\alpha_3/2)J_{34}), \qquad (7.5.24c)$$

where each α_k is a group parameter. The evolution operator is

$$\exp(t\{-H,\cdot\}) = \exp(2tU_4) = \exp(tJ_{31})\exp(-tJ_{24}). \qquad (7.5.24d)$$

In SO(4), the operator of each one-parameter rotation subgroup with generator J_{ij} and group parameter γ_{ij} is expressed as $\exp(\gamma_{ij}\ J_{ij})$, and in each case the parameter ranges over an interval of 2π before the transformation repeats. In contrast, the finite transformations generated by the Sp(2) generators U_1, U_2, and U_3 return a point in the phase space to itself only after the group parameters α_k range through 4π. For the same reason, t in the operator $\exp(tU_4)$ must range through 4π to return a point in the phase space to itself.

The four generators U have the following action on $\mathbf{s}$:

$$U_1\mathbf{S} = (-p_2/2, -p_1/2, q_2/2, q_1/2) = (-s_4/2, -s_3/2, s_2/2, s_1/2) \equiv \mathbf{\Delta s}_2,$$

$$U_2\mathbf{S} = (-p_1/2, p_2/2, q_1/2, -q_2/2) = (-s_3/2, s_4/2, s_1/2, -s_2/2) \equiv \mathbf{\Delta s}_1,$$

$$U_3\mathbf{S} = (q_2/2, -q_1/2, p_2/2, -p_1/2) = (s_2/2, -s_1/2, s_4/2, -s_3/2) \equiv \mathbf{\Delta s}_3,$$

$$U_4\mathbf{S} = (-p_1/2, -p_2/2, q_1/2, q_2/2) = (-s_3/2, -s_4/2, s_1/2, s_2/2) \equiv \mathbf{\Delta s}_4.$$

(7.5.25a–d)

At a point $\boldsymbol{\rho}_s$ on the three-dimensional hypersurface of the hypersphere, the linearly independent vectors $\boldsymbol{\delta}\mathbf{s}_k = \varepsilon U_k \mathbf{s}$ can provide axes for a local coordinate system determined by the constants of motion. The corresponding finite tangent vectors $\mathbf{\Delta s}_k$, k = 1, 2, 3, provide local systems of coordinates in the four-dimensional space. The infinitesimal version of $\mathbf{\Delta s}_4$ points in the direction in which $\boldsymbol{\rho}_s$ will evolve.

The length of each of the the vectors $\mathbf{\Delta s}_1$, $\mathbf{\Delta s}_2$ and $\mathbf{\Delta s}_3$ is $\sqrt{E}/\sqrt{2}$. The vector dot products with each other are all zero, so they are mutually orthogonal. The rotation generator in the phase space is $\mathrm{L}_{12} = \{\mathrm{L}_{12}, \cdot\} = \{2\mathrm{u}_3, \cdot\}$. As $\mathrm{L}_{12}\boldsymbol{\Delta}\mathbf{s}_1 = \boldsymbol{\Delta}\mathbf{s}_2$ and $\mathrm{L}_{12}\boldsymbol{\Delta}\mathbf{s}_2 = -\boldsymbol{\Delta}\mathbf{s}_1$, the operator $\exp(\alpha_3\mathrm{L}_{12})$ carries out rotations in the plane defined by $\boldsymbol{\Delta}\mathbf{s}_1$, $\boldsymbol{\Delta}\mathbf{s}_2$, rotations about an axis along $\boldsymbol{\Delta}\mathbf{s}_3$. Consequently, at every point $\boldsymbol{\rho}_\mathrm{s}$ the unit tangent vectors $\mathrm{i}(\mathbf{s}) = (\sqrt{2}/\sqrt{\mathrm{E}})\boldsymbol{\Delta}\mathbf{s}_1$, $\mathrm{j}(\mathbf{s}) = (\sqrt{2}/\sqrt{\mathrm{E}})\boldsymbol{\Delta}\mathbf{s}_2$, and $\mathrm{k}(\mathbf{s}) = (\sqrt{2}/\sqrt{\mathrm{E}})\boldsymbol{\Delta}\mathbf{s}_3$, provide the basis for a local Cartesian coordinate system which can be used to describe infinitesimal displacements in the hypersurface. The vector $\boldsymbol{\Delta}\mathbf{s}_4$, has the following components on these axes:

$$\text{on } \mathbf{i} : \mathbf{S}_1 \cdot \mathbf{S}_2/(\sqrt{2}\mathrm{E}), \quad \text{on } \mathbf{j} : (\mathrm{E}_1 - \mathrm{E}_2)/(\sqrt{2}\mathrm{E}), \quad \text{on } \mathbf{k} : \mathrm{L}_{12}/(\sqrt{2}\mathrm{E}). \tag{7.5.26}$$

It follows that the orientation of trajectories on the hypersurface is determined by the three constants of motion whose value is fixed by the initial values of the positions and momenta as per (7.5.6). For a given value of $\mathbf{s}$, say $\mathbf{s}_0$, unit vectors $\mathbf{I}_0$, $\mathbf{J}_0$ and $\mathbf{K}_0$ that define a set of axes parallel to $\mathbf{i}(\mathbf{s}_0)$, $\mathbf{j}(\mathbf{s}_0)$ and $\mathbf{k}(\mathbf{s}_0)$ can be set up at the center of the hypersphere. The initial position vector, $\mathbf{s}_0$, when divided by its length, $\sqrt{2}\mathrm{E}$, provides a fourth unit vector, $\mathbf{h}_0$, orthogonal to all the others. To sum up, the three constants of motion have determined PB operators whose tangent vectors, together with $\mathbf{s}_0$, define a *natural* global Cartesian coordinate system in phase space.

Figure 7.5.2a depicts the unit vectors **i**, **j** of a local axis system tangent to the surface of an ordinary sphere at $\mathbf{s}_0$ the terminus of the radius vector $\mathbf{s}_0$. This set of axes is analogous to the axes at the surface of our hypersphere. The unit vector **h** lies along the extension of the radius vector to the point. In **Fig. 7.5.2b**, **I**, **J** and **h** axes that are parallel to the **i**, **j** and **h** axes of **Fig. 7.5.2a** have been set up at the center of the sphere. The components of any radius vector **s** on the **I**, **J** and **h** axes are clearly identical with its components on the **h**, **i** and **j** axes.

The operator $\exp(\alpha_1\mathrm{U}_1)$ moves a point $\boldsymbol{\rho}_{\mathrm{so}}$ on the hypersphere defined by (7.5.9b) to a one-parameter family of points, each of which may be considered a new initial point. Evolving these with $\exp(\alpha_4\mathrm{U}_4)$ one obtains a one-parameter family of trajectories in phase space, trajectories that depict solutions to the equations of motion, (7.5.8b,c). Because $[\mathrm{U}_4,\mathrm{U}_1] = 0$, the result is the same as that obtained if the initial point is first evolved, then transformed by $\exp(\alpha_1\mathrm{U}_1)$. The operators $\exp(\alpha_2\mathrm{U}_2)$ and $\exp(\alpha_3\mathrm{U}_3)$ act analogously, altogether generating a three-parameter family of trajectories and solutions. Because the three U_k commute with H, all of the solutions have the same energy. It can be proved that this three-parameter family

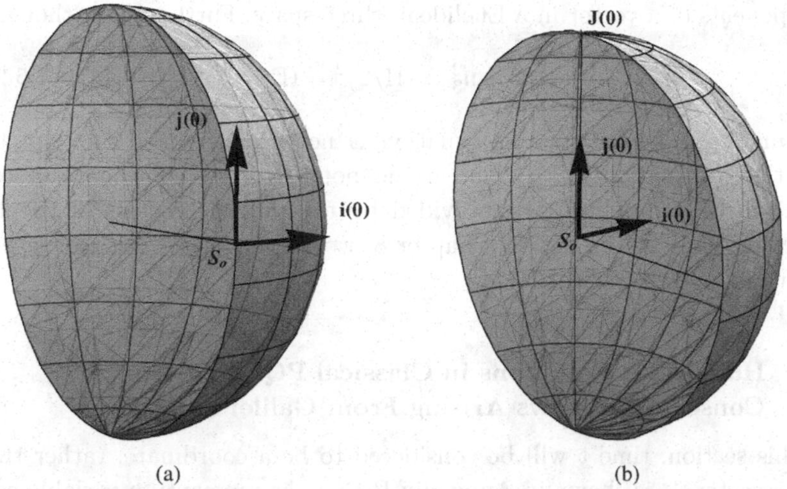

Fig. 7.5.2 (a) Radius vector S_a and unit tangent vectors. (b) Tangent vectors translated to origin of S_a.

contains all the solutions of the equations of motion which have a common energy.

In the foregoing we have used the Lie generators $\{u_j, \cdot\}$ to define coordinate systems and motions in the four-dimensional phase space. An alternative approach is suggested by the fact that $\{u_3, \cdot\}$ is a rotation generator divided by two. Relations (7.5.8) imply that

$$\{u_3, u_1\} = u_2, \quad \{u_3, u_2\} = -u_1, \tag{7.5.27}$$

as would be expected if $\{u_3, \cdot\}$ generates rotations of a Cartesian two-vector $[u_1, u_2]$. However,

$$\begin{aligned} \exp(\alpha_3\{u_3, \cdot\})u_1 &= \cos(\alpha_3/2)u_1 + \sin(\alpha_3/2)u_2, \\ \exp(\alpha_3\{u_3, \cdot\})u_2 &= \cos(\alpha_3/2)u_2 - \sin(\alpha_3/2)u_1. \end{aligned} \tag{7.5.28}$$

Because the group parameter α_3 must range through 4π in order to cycle u_1 and u_2 back to their initial point, the operation of $\exp(\alpha_3\{u_3, \cdot\})$ may be considered to move a point with coordinates u_1, u_2, around the girdle of a Möbius band. The same argument can be applied to the action of $\exp(\alpha_2\{u_2, \cdot\})$ on u_3, u_1, and that of $\exp(\alpha_1\{u_1, \cdot\})$ on u_2, u_3. The space of u_1, u_2, u_3 is not Euclidean, and u_1, u_2, u_3 cannot be considered to be components of a vector in a Euclidean three-space. Furthermore, though

$$u_1^2 + u_2^2 + u_3^2 = (H/2)^2 = (E/2)^2, \qquad (7.5.29)$$

the manifold defined by this equation is not a sphere. To sum up: the constants of motion u_1, u_2, and u_3 do not provide a coordinate system as intuitively useful as that provided by the tangent vectors of the PB operators they determine. In Chapter 8, we will find that just the reverse is true in Kepler systems.

7.6 Hamilton's Equations in Classical PQET Space; Conservation Laws Arising From Galilei Invariance

In this section, time t will be considered to be a coordinate, rather than a parameter. It is shown in Appendix B that the momentum variable conjugate to t is −E, the negative of an energy. The resulting phase space is termed *PQET space*. When Hamiltonian mechanics is extended to this space one replaces t by an *evolution parameter*, which we shall denote by τ. It is an independent variable that ranges through all the real numbers, and is used to define the Lagrangian function that determines Hamiltonian functions in PQET space. In nonrelativistic Hamiltonian mechanics in PQET space it is presupposed that, though different particles have different P, Q coordinates, the coordinates p_j and q_j, may all be observed simultaneously at time t. The corresponding variable E is represents the total energy of a system. With these presuppositions, there is only one t coordinate and one E coordinate. The Poisson bracket, $\{A, B\}$, of two functions A(P,Q,E,t), B(P,Q,E,t) is therefore defined by

$$\textstyle\sum_1^n(\partial A/\partial q_j\, \partial B/\partial p_j - \partial B/\partial q_j\, \partial A/\partial p_j) - (\partial A/\partial t\, \partial B/\partial E - \partial B/\partial t\, \partial A/\partial E). \qquad (7.6.1a)$$

It is often convenient to set $t = q_0$ and $-E = p_0$. When this is done,

$$\{A, B\} = \textstyle\sum_0^n (\partial A/\partial q_j\, \partial B/\partial p_j - \partial B/\partial q_j\, \partial A/\partial p_j). \qquad (7.6.1b)$$

For a Hamiltonian system defined in PQET space, the Hamiltonian function, H, may, as in Appendix B, be derived from a time-dependent Lagrangian. Let

$$W(P, Q, E, t) \equiv H(P, Q, t) - E. \qquad (7.6.2a)$$

Then Hamilton's equations of motion in PQET space are

$$\begin{aligned} dq_j/d\tau &= \partial W/\partial p_j, \quad dp_j/d\tau = -\partial W/\partial q_j, \\ dt/d\tau &= \partial W/\partial(-E), \quad d(-E)/d\tau = -\partial W/\partial t. \end{aligned} \tag{7.6.2b–e}$$

If f = F(P,Q,E,t), one has the general relation

$$df/d\tau = \{f, W\}. \tag{7.6.3}$$

Applying this to a system with a time-independent Hamiltonian, one obtains

$$\begin{aligned} dq_j/d\tau &= \partial W/\partial p_j, \quad dp_j/d\tau = -\partial W/\partial q_j \\ dt/d\tau &= 1, \quad dE/d\tau = 0. \end{aligned} \tag{7.6.4a–d}$$

It follows that $\Delta t = \Delta\tau$, so if the motion is not periodic

$$t = \tau + \text{const} \tag{7.6.4e}$$

and the motion is periodic,

$$t - \text{const} = (\tau)_{\text{modT}}, \tag{7.6.4f}$$

where T is the period. This case was studied in Sec. 3.3.

Lagrange's equations of motion inherit the Galilei invariance of Newton's equations of motion for isolated systems of particles. Let us investigate the manner in which this invariance is passed on to Hamilton's equations of motion in PQET space. Using (7.4.3), one finds that the generators of the passive group given in (7.1.8) become the following PB operators in PQET space:

$$T_0 = \partial/\partial t \to \{-E, \cdot\}, \tag{7.6.5a}$$

$$T_x = \partial/\partial q_x \to \{-p_x, \cdot\}, \textit{etc.}, \tag{7.6.5b}$$

and

$$\begin{aligned} R_{xy} &= q_x\partial/\partial q_y - q_y\partial/\partial q_x \\ &\to \{-(q_x p_y - q_y p_x), \cdot\} = (q_x\partial/\partial q_y - q_y\partial/\partial q_x) \\ &\qquad + (p_x\partial/\partial p_y - p_y\partial/\partial p_x), \textit{etc.} \end{aligned} \tag{7.6.5c}$$

To express the active transformations determined by these operators for a system with n q's, and n p's, define the position and momentum vectors of particle i by

$$\mathbf{q}_i = (q_{xi}, q_{yi}, q_{zi}), \quad \mathbf{p}_i = (p_{xi}, p_{yi}, p_{zi}). \tag{7.6.6}$$

The total momentum vector is $\mathbf{p} = \sum \mathbf{p}_i$. The operators

$$\boldsymbol{\nabla}_{\mathbf{q}i} = (\partial/\partial q_{xi}, \partial/\partial q_{yi}, \partial/\partial q_{zi}), \quad \boldsymbol{\nabla}_{\mathbf{p}i} = (\partial/\partial p_{xi}/\partial p_{yi}/\partial p_{zi}), \tag{7.6.7}$$

generate active translations of the position and momentum of particle i in PQET space. The corresponding many-particle operators are

$$\boldsymbol{\nabla}_{\mathbf{q}} = \sum \boldsymbol{\nabla}_{\mathbf{q}i}, \quad \boldsymbol{\nabla}_{\mathbf{p}} = \sum \boldsymbol{\nabla}_{\mathbf{p}i}. \tag{7.6.8}$$

These, together with **p**, enable one to express the effects of active Galilei transformations on a many-particle system.

a. It follows immediately from (7.6.5a) that

$$\{H - E, E\} = 0 \quad \text{if } \partial H/\partial t = 0, \tag{7.6.9a}$$

the energy of a system is conserved if its Hamiltonian is time-translation invariant.

b. Equation (7.6.5b) implies that

$$\{H, \mathbf{p}\} = 0 \quad \text{if } \boldsymbol{\nabla}_q H = 0, \tag{7.6.9b}$$

i.e. *the total momentum of a system is conserved if its Hamiltonian is space-translation invariant.*

c. The total angular momentum of the system is $\mathbf{L} = \sum \mathbf{q}_i \,\mathbf{x}\, \mathbf{p}_i$, and the vector operator that generates the active rotations that result from rotations of the coordinate system is $\mathbf{R} = \{\mathbf{L}, \cdot\}$. Equation (7.6.5c) requires that

$$\{H, \mathbf{L}\} = 0 \quad \text{if } \mathbf{R}H = 0, \tag{7.6.9c}$$

i.e. *the angular momentum of a system is conserved if its Hamiltonian is rotation invariant.*

Though Newton's equations are invariant under the passive transformation $\mathbf{v} \to \mathbf{v} + \mathbf{v}_0$, the same is not true of Hamilton's equations in PQ space because kinetic energies are not invariant under the transformation. As the operators K_x, K_y, K_z depend upon t, they determine operators that act in PQET space. For this reason one must formulate the Hamiltonian analog of Newtonian Galilei transformations as transformations in PQET space. For

$$K_x = t\partial/\partial q_x + \partial/\partial v_x,$$

one finds

$$K_x \to \{(-tp_x + mq_x), \cdot\} = t\,\partial/\partial q_x + p_x\,\partial/\partial E + m\,\partial/\partial p_x, \quad (7.6.10a)$$

etc. These components of the vector operator $\mathbf{K}$ do not annihilate $\mathbf{p}\cdot\mathbf{p}/2m$. However, H – E is invariant under transformations generated by an operator of active transformations closely related to $\mathbf{K}$. To see this, let

$$\mathbf{K}_i = -t\mathbf{p}_i + m_i\mathbf{q}_i, \quad (7.6.10b)$$

and

$$\mathbf{u} = -\sum K_i. \quad (7.6.10c)$$

Now, let $m = \sum m_i$, then $\boldsymbol{\rho} \equiv \sum m_i\mathbf{q}_i/m$ is the vector from the origin to the center of mass of the system. Also let $\sum t\mathbf{p}_i = t\mathbf{P}$, where $\mathbf{P}$ is the total momentum vector of the system. Then

$$\mathbf{u} = -m\boldsymbol{\rho} + t\mathbf{P}. \quad (7.6.10d)$$

The PB operator determined by $\mathbf{u}$ is

$$\mathbf{U} = \{\mathbf{u}, \cdot\} = \sum(\mathbf{p}_i\partial/\partial E + t\boldsymbol{\nabla}_{qi} + m_i\boldsymbol{\nabla}_{pi}). \quad (7.6.10e)$$

The components of $\mathbf{U}$, PB operators that one may denote by U_x, U_y and U_z, are the active forms of the generators K_x, K_y and K_z, of the Galilei group. $\mathbf{U}$ generates the transformations

$$\exp(\mathbf{v}_0\cdot\mathbf{U})\mathbf{q}_i = \mathbf{q}_i + v_0\,t = \exp(\mathbf{v}_0\cdot\mathbf{K})\mathbf{q}_i, \quad (7.6.11a)$$

and

$$\exp(\mathbf{v}_0\cdot\mathbf{U})\mathbf{p}_i = \mathbf{p}_i + \mathbf{v}_0 m_i. \quad (7.6.11b)$$

Thus $\exp(\mathbf{v}_0\cdot\mathbf{U})$ has the same effect as $\exp(\mathbf{v}_0\cdot\mathbf{K})$ on $\mathbf{q}$ and $\mathbf{v}$. One finds that

$$\exp(\mathbf{v}_0\cdot\mathbf{U})E = E + \sum\mathbf{v}_0\cdot\mathbf{p}_i + (1/2)\sum m_i\mathbf{v}_0\cdot\mathbf{v}_0, \quad (7.6.11c)$$

and

$$\exp(\mathbf{v}_0\cdot\mathbf{U})\sum\mathbf{p}_i\cdot\mathbf{p}_i/2m_i = \sum\mathbf{v}_0\cdot\mathbf{p}_i + (1/2)\sum m_i\mathbf{v}_0\cdot\mathbf{v}_0. \quad (7.6.11d)$$

Consequently, if

$$H = \sum \mathbf{p}_i \cdot \mathbf{p}_i / 2m_i + \sum \sum v(|\mathbf{q}_i - \mathbf{q}_j|), \tag{7.6.12a}$$

then

$$\mathbf{U}(H - E) = 0. \tag{7.6.12b}$$

Thus, though observers in systems moving at constant velocity with respect to each other will assign different values to E and H, they will agree that H = E.

7.7 Time-dependent Constants of Motion; Dynamical Groups That Act Transitively

A function G(p,q,E,t) defined in PQET space, is a time-dependent constant of motion in a system with Hamiltonian H(p,q) if

$$0 = \{H - E, G\}_{H-E=0}. \tag{7.7.1a}$$

Now

$$\{E, \cdot\} = \partial/\partial t, \tag{7.7.1b}$$

so (7.7.1a) is equivalent to requiring that

$$\{H, \cdot\}G - \partial G/\partial t = 0, \tag{7.7.1c}$$

on the manifold in PQET space defined by $H(q, p) - E = 0$. The function **u** of the previous paragraph is an example of a time-dependent vector constant of motion. If H–E in (7.7.1a) is replaced by (H–E)f, with f any C_2 function of the dynamical variables, then a function G that satisfies (7.7.1a) will also satisfy

$$0 = \{(H - E)f, G\}|_{H-E=0}. \tag{7.7.1d}$$

This turns out to be a useful property of Hamiltonian systems and constants of motion in PQET space.

Harmonic oscillators provide one of the simplest examples of physically important time-dependent constants of motion. To investigate them for a one-dimensional oscillator, suppose that $H = (p^2 + q^2)/2$. Then

$$\{H, \cdot\} = q\partial/\partial p - p\partial/\partial q. \tag{7.7.2}$$

Let u(p,q,E,t) be a prospective constant of motion, and expand it in a power series. Setting

$$u = \sum q^a \, p^b \, g_{ab}(t, E), \tag{7.7.3}$$

one requires that

$$\{H - E, u\} = (q\partial/\partial p - p\partial/\partial q)u + \partial u/\partial t = 0 \tag{7.7.4}$$

when H = E. The first term in the expansion of u that gives a nontrivial result is $u_1 = q\, g_{10}(t,E) + p\, g_{01}(t,E)$. The equation that u_1 must satisfy is

$$(q\partial/\partial p - p\partial/\partial q)u_1 - \partial u_1/\partial t$$
$$= (g_{01} - \partial g_{10}/\partial t)q - (g_{10} + \partial g_{01}/\partial t)p = 0. \tag{7.7.5a}$$

Because q, p are independently variable, (7.7.5a) requires that

$$g_{01} = \partial g_{10}/\partial t, \quad \partial g_{01}/\partial t = -g_{10}. \tag{7.7.5b}$$

These equations have two linearly independent solutions. To simplify the subsequent transformation of our results to quantum mechanics, we allow these solutions to be complex, and choose them to be

$$g_{10+} = \exp(it), \quad g_{10-} = \exp(-it). \tag{7.7.6a}$$

Then

$$g_{01+} = i\exp(it), \quad g_{01-} = -i\exp(-it). \tag{7.7.6b}$$

These yield the constants of motion

$$(q + ip)\exp(it) \equiv u_+ \tag{7.7.7a}$$

and

$$(q - ip)\exp(-it) \equiv u_-. \tag{7.7.7b}$$

One has

$$\{H, \cdot\}u_+ = \partial u_+/\partial t = iu_+, \tag{7.7.8a}$$

$$\{H, \cdot\}u_- = \partial u_-/\partial t = -iu_-. \tag{7.7.8b}$$

The two time-dependent constants of motion satisfy

$$\{u_+, u_-\} = 2i. \tag{7.7.9}$$

The PB operators obtained from them are

$$U_+ = \{u_+, \cdot\} = \exp(it)(\partial/\partial p - i\partial/\partial q - i(q + ip)\partial/\partial E), \tag{7.7.10a}$$

and

$$U_- = \{u_-, \cdot\} = \exp(-it)\partial/\partial p + i\partial/\partial q + i(q - ip)\partial/\partial E). \tag{7.7.10b}$$

When acting only in PQ space,

$$U_+ \to \exp(it)(\partial/\partial p - i\partial/\partial q), U_- \to \exp(-it)(\partial/\partial p + i\partial/\partial q). \tag{7.7.11a,b}$$

In both PQ and PQET space,

$$[\mathrm{U}_+, \mathrm{U}_-] = 0. \tag{7.7.12}$$

Thus U_+ and U_- commute. Since $\{1, \cdot\} = 0$, this result is compatible with relations (7.4.8).

Integrating the infinitesimal transformations defined by the generators U_+ and U_-, (cf. Sec. 4.5), one finds the finite transformations produced by the operators $\exp(\alpha_\pm \mathrm{U}_\pm)$:

$$\begin{gathered} \mathrm{p} \to \mathrm{p} + \alpha_+ \exp(\mathrm{it}), \quad \mathrm{q} \to \mathrm{q} - \mathrm{i}\alpha_+ \exp(\mathrm{it}), \\ \mathrm{E} \to \mathrm{E} - \mathrm{i}\alpha_+ \mathrm{u}_+, \quad \mathrm{H} \to \mathrm{H} - \mathrm{i}\alpha_+ \mathrm{u}_+; \end{gathered} \tag{7.7.13a}$$

$$\begin{gathered} \mathrm{p} \to \mathrm{p} + \alpha_- \exp(-\mathrm{it}), \quad \mathrm{q} \to \mathrm{q} + \mathrm{i}\alpha_- \exp(-\mathrm{it}), \\ \mathrm{E} \to \mathrm{E} + \mathrm{i}\alpha_- \mathrm{u}_-, \quad \mathrm{H} \to \mathrm{H} + \mathrm{i}\alpha_- \mathrm{u}_-. \end{gathered} \tag{7.7.13b}$$

At $\mathrm{t} = 0$, the transformations of p, q and E become

$$\mathrm{p}_0 \to \mathrm{p}_0 + \alpha_+, \quad \mathrm{q}_0 \to \mathrm{q}_0 - \mathrm{i}\alpha_+, \quad \mathrm{E}_0 \to \mathrm{E}_0 - \mathrm{i}\alpha_+(\mathrm{q}_0 + \mathrm{ip}_0); \tag{7.7.14a}$$

$$\mathrm{p}_0 \to \mathrm{p}_0 + \alpha_-, \quad \mathrm{q}_0 \to \mathrm{q}_0 + \mathrm{i}\alpha_-, \quad \mathrm{E}_0 \to \mathrm{E}_0 + \mathrm{i}\alpha_-(\mathrm{q}_0 - \mathrm{ip}_0). \tag{7.7.14b}$$

Equation (7.7.12) states that U_+ and U_- commute, so

$$\exp(\alpha_+ \mathrm{U}_+)\exp(\alpha_- \mathrm{U}_-) = \exp(\alpha_+ \mathrm{U}_+ + \alpha_- \mathrm{U}_-). \tag{7.7.15}$$

To express the right-hand side of (7.7.15) in terms of real variables we let the linear combination of PB generators be re-expressed as

$$\alpha_+\{(\mathrm{q} + \mathrm{ip})\exp(\mathrm{it}), \cdot\} + \alpha_-\{(\mathrm{q} - \mathrm{ip})\exp(-\mathrm{it}), \cdot\}. \tag{7.7.16}$$

Define

$$\alpha = \alpha_+ + \alpha_-, \quad \beta = \mathrm{i}(\alpha_+ - \alpha_-), \tag{7.7.17a}$$

$$\mathrm{u}_\alpha = \mathrm{q}\cos(\mathrm{t}) - \mathrm{p}\sin(\mathrm{t}), \quad \mathrm{u}_\beta = \mathrm{q}\sin(\mathrm{t}) + \mathrm{p}\cos(\mathrm{t}), \tag{7.7.17b}$$

and

$$\mathrm{U}_\alpha = \{\mathrm{u}_\alpha, \cdot\}, \mathrm{U}_\beta = \{\mathrm{u}_\beta, \cdot\}. \tag{7.7.17c}$$

Then

$$\begin{gathered} \exp(\alpha_+ \mathrm{U}_+ + \alpha_- \mathrm{U}_-) = \exp(\alpha \mathrm{U}_\alpha + \beta \mathrm{U}_\beta) \\ = \exp(\alpha \mathrm{U}_\alpha)\exp(\beta \mathrm{U}_\beta) = \mathrm{T}(\alpha, \beta). \end{gathered} \tag{7.7.18}$$

$T(\alpha,\beta)$ carries out real transformations that may be obtained from the composition of the two transformations (7.7.13). Evaluating the action of $T(\alpha,\beta)$ at $t = 0$, one finds that it carries out the conversions

$$p_0 \to p_0 + \alpha, \quad q_0 \to q_0 + \beta, \quad E_0 \to E_0 + \alpha p_0 + \beta q_0 + (\alpha^2\ \beta^2)/2. \quad (7.7.18)$$

Using $T(\alpha,0)$ and $T(0,\beta)$, one is able to independently shift the two initial values q_0 and p_0, so the operations of the group can convert any solution of the equations of motion into any other solution. Using the definition of transitivity in Chapter 5, the group is seen to act *transitively* on the space of solutions.

The generators U_α and U_β satisfy the commutation relations

$$[\{H,\cdot\},U_\alpha] = U_\beta, \quad [\{H,\cdot\},U_\beta] = U_\alpha, \quad [U_\alpha,U_\beta] = 0. \qquad (7.7.19)$$

The corresponding group, denoted N(3), is called the Heisenberg group.

To apply the results of this section to the oscillator of Sec. 7.6, with $H = H_1 + H_2$, as defined in (7.5.8a), one uses operators $T(\alpha_1, \beta_1)$ and $T(\alpha_2,\beta_2)$ to carry out the operations of (7.7.19) in the subspaces of p_1, q_1, E_1 and p_2, q_2, E_2. This enables one to independently change each of the four initial values $p_1(0)$, $q_1(0)$, $p_2(0)$, $q_2(0)$, and thereby change E_1, E_2 and E. The operations of the two Heisenberg groups carried out by $T(\alpha_1,\beta_1)$,$T(\alpha_2,\beta_2)$, combined with the operations of the degeneracy group, produce a subgroup of Sp(3) that acts transitively on the space of solutions of the oscillator equations of motion (7.5.8b,c).

For classical harmonic oscillators, knowing a dynamical group that acts transitively is of little practical value. For more complex systems, knowing analogous groups can be very revealing, and of great utility — even if the required group generators are only approximately known. For this reason, as well as conceptual reasons previously set forth, we will often be working with Hamiltonian dynamics and symmetries in PQET space, or with their quantum mechanical analogs.

Subsequent chapters will also make much use of the PB operators defined in this chapter, the commutator Lie algebras they define, and the corresponding Poisson bracket Lie algebras. Chapter 9 establishes quantum mechanical versions of a number of results contained in this chapter.

7.8 The Symplectic Groups Sp(2n,r)

The PB operators determined by quadratic functions of conjugate q's and p's generate symplectic transformation groups that act linearly in phase space. If there are n real q's and n real p's, these groups are realizations of the group Sp(2n,r). The following Lie generators provide a basis for the Lie algebra of Sp(2n,r) when i and j run from 1 to n:

$$\begin{gathered}\{q_i^2/2,\cdot\} = q_i\partial/\partial p_i, \quad \{p_i^2/2,\cdot\} = -p_i\partial/\partial q_i,\\ \{q_ip_i,\cdot\} = p_i\partial/\partial p_i - q_i\partial/\partial q_i, \quad \{q_ip_j,\cdot\} = p_j\partial/\partial p_i - q_i\partial/\partial q_j,\\ \{q_jp_i,\cdot\} = p_i\partial/\partial p_j - q_j\partial/\partial q_i, \quad \{q_iq_j,\cdot\} = q_i\partial/\partial p_j + q_j\partial/\partial p_i,\\ \{p_ip_j,\cdot\} = -(p_i\partial/\partial q_j + p_j\partial/\partial q_i).\end{gathered} \tag{7.8.1}$$

The commutator and PB Lie algebras obtained from the functions above are isomorphic. Table 7.1 below gives Poisson bracket relations that are sufficient to determine the Lie algebras of both types of symplectic groups.

Sp(2n,r) has n(2n + 1) generators and possesses the following subgroup chains:[2]

$$\begin{gathered}\mathrm{Sp}(2n,r) \supset \mathrm{U}(n) \supset \mathrm{SU}(n) \supset \mathrm{SO}(n),\\ \mathrm{Sp}(2n,r) \supset \mathrm{SL}(n,r)\times \mathrm{R} \supset \mathrm{SL}(n,r) \supset \mathrm{SO}(n).\end{gathered} \tag{7.8.2}$$

Writing 2n = 2a + 2b, with a and b integers, one also has

$$\begin{gathered}\mathrm{Sp}(2a+2b,r) \supset \mathrm{U}(a,b) \supset \mathrm{SU}(a,b) \supset \mathrm{SO}(a,b),\\ \mathrm{Sp}(2a+2b,r) \supset \mathrm{Sp}(2a,r)\times \mathrm{Sp}(2b,r).\end{gathered} \tag{7.8.3}$$

Numerous group–subgroup chains of SO(n) and SO(a,b) are tabulated in Ref. 2.

The group N(3) of the previous section is a symplectic group though the functions determining that its PB generators are not quadratic in p,q,E,t.

Nonlinear transformations can convert the transformations of an Sp(2n,r) group that acts linearly in phase space into those of Sp(2n,r) groups that act nonlinearly. This enables Sp(2n,r) to take on a particularly important role in Hamiltonian dynamics. An example of a nonlinear realization of Sp(2n,r) will be found in Chapter 8.

Table 7.1. The Poisson Brackets {A, B}

A\B	$q_i^2/2$	$p_i^2/2$	q_ip_i	q_ip_j	q_jp_i	q_iq_j	p_ip_j
$q_i^2/2$	0	q_ip_i	q_i^2	0	q_iq_j	0	q_ip_j
$p_i^2/2$		0	$-p_i^2$	$-p_ip_j$	0	$-q_jp_i$	0
q_ip_i			0	$-q_ip_j$	q_jp_i	$-q_iq_j$	p_ip_j
q_ip_j				0	$q_jp_j - q_ip_i$	$-q_i^2$	p_j^2
q_jp_i					0	$-q_j^2$	p_i^2
q_iq_j						0	$q_jp_j+q_ip_i$

7.9 Generalizations of Symplectic Groups That Have an Infinite Number of One-parameter Groups

If one generalizes the concept of closure used in defining Lie groups so as to allow an infinite set of generators, two different generalizations may be used. One may allow either a countable or uncountable infinite set.

For transformation groups of the first type one requires

$$[U_i, U_j] = \sum_k^{\infty} c_{ij}^k U_k. \tag{7.9.1}$$

An example is provided when one begins to compose transformations from one-parameter groups generated by PB operators of polynomial functions of the q's and p's of order greater than three.

Generalizations of the second type arise when generators U of one-parameter groups depend upon a continuous parameter, say α. Varying the parameter develops a one-parameter family of transformation groups. If

$$[U(\alpha), U(\alpha')] = \Phi(\alpha, \alpha', \alpha'')U(\alpha''), \quad \alpha'' = A(\alpha, \alpha'), \tag{7.9.2}$$

the transformations are said to belong to an *infinite group*. If the $U(\alpha)$ are Poisson bracket operators, the resulting group is a symplectic one. As an example of this generalization, let $u(\alpha) = p\,\exp(\alpha q)$. Then

$$U(\alpha) = \exp(\alpha q)(\alpha p\partial/\partial p - \partial/\partial q), \tag{7.9.3}$$

and

$$\begin{aligned}[U(\alpha), U(\alpha')] &= (\alpha - \alpha')\exp((\alpha\ \alpha')q)((\alpha + \alpha')p\partial/\partial p - \partial/\partial q)\\ &= (\alpha - \alpha')U(\alpha\ \alpha').\end{aligned}$$

In this case, the function A is $(\alpha + \alpha')$ and the function Φ is $(\alpha - \alpha')$.

The commutation relations (7.9.2) must be generalized when one is composing one-parameter groups whose generators differ by more than a single continuous parameter. This is necessary, for example, when one wishes to compose the one-parameter groups generated by (7.9.3a) with those generated by $\{-p, \cdot\} = \partial/\partial q$.

Though infinite dimensional symplectic groups will not be utilized in subsequent chapters, infinite dimensional groups in general, play an important role in the study of partial differential equations whose solutions depend upon arbitrary functions rather than arbitrary constants.

Appendix A. Lagrange's Equations and the Definition of Phase Space

The equations of Lagrange and of Hamilton directly formulate the evolution of kinetic energies, potential energies and work, rather than the evolution of positions and momenta alone. Both treat Cartesian coordinates x_i on the same footing as generalized coordinates q_j, related to the Cartesian coordinates by C^2 transformations

$$x_i = X_i(q), \quad q = (q_1, q_2, \ldots, q_n). \tag{7.A.1}$$

Using Cartesian coordinates, (x_1, x_2, x_3), and velocity components $v_i = dx_i/dt$, the kinetic energy of a mass point in an inertial Cartesian coordinate system is $m(v_1^2 + v_2^2 + v_3^2)/2$. For a system with n particles and 3n, or 2n, Cartesian coordinates, it is convenient to associate a mass m_i with each degree of freedom, keeping in mind that some of the m_i have the same value. If the masses are constant, and f_i is the component of the force acting on m_i, Newton's equations of motion are then

$$m_i(dv_i/dt) = f_i, \quad i = 1, 2 \ldots. \tag{7.A.2}$$

The total kinetic energy of the system is

$$T = \sum m_i v_i^2/2. \tag{7.A.3a}$$

To investigate the evolution of the kinetic energy under the action of forces, one considers dT/dt. As

$$dT/dt = \sum m_i v_i \, dv_i/dt, \tag{7.A.3b}$$

Newton's equations imply

$$dT/dt = \sum f_i v_i. \tag{7.A.3c}$$

One then has the well known relation,

$$dT = \sum f_i dx_i = dW, \tag{7.A.3d}$$

where dW is the element of work done by all the forces during the displacements dx.

Lagrange's equations are obtained by determining the way in which relations (7.A.3b–d) are altered when one replaces Cartesian coordinates and velocities by *generalized coordinates* and their time derivatives.

The simplest derivation of Lagrange's equations from Newton's equations, appears to be that provided by Thompson and Tait.[3] It begins with the transformation of coordinates defined by (7.A.1), which implies that

$$dx_i = \sum_j (\partial X_i/\partial q_j)dq_j = \sum_j (dx_i/dq_j)\, dq_j. \tag{7.A.4a}$$

This is used to express the element of work in (7A.3d) as

$$dW = \sum_i f_i \sum_j (\partial X_i/\partial q_j)dq_j = \sum_j F_j\, dq_j, \tag{7.A.4b}$$

where

$$F_j = \sum_i f_i\, \partial X_i/\partial q_j = \sum_i f_i dx_i/dq_j. \tag{7.A.4c}$$

Here, and hereafter, each of the generalized coordinates q_j are considered independently variable, so each F_j becomes the local component of the force that acts in the direction of dq_j.

The connection between the F_j and the f_i is put to use in converting the left-hand side of (7.A.2) to an expression analogous to (7.A.3c), one in which the q_j and their time derivatives replace the x_i and their time derivatives. To begin, for each value of i in Newton's equations (7.A.2), multiply both sides by dx_i/dq_j, to obtain

$$m_i(dv_i/dt)dx_i/dq_j = f_i dx_i/dq_j. \tag{7.A.5}$$

The term multiplying m_i on the left-hand side of this may be re-expressed as

$$d/dt(v_i\, dx_i/dq_j) - v_i(d/dt)dx_i/dq_j. \tag{7.A.6a}$$

Now (cf. the exercises) it can be shown that

$$dx_i/dq_j = (dx_i/dt)/(dq_j/dt). \tag{7.A.6b}$$

On writing q_j' for dq_j/dt, (7A.6a) can thus be re-expressed as

$$(d/dt)(v_i dv_i/dq_j') - v_i dv_i/dq_j, \tag{7.A.6c}$$

and as

$$(d/dt)(d(v_i^2/2)/dq_j') - (d(v_i^2/2)/dq_j). \tag{7.A.6d}$$

Using (7.A.6d) to replace the coefficient of m_i in the left-hand side of (7.A.5), Eq. (7.A.5) becomes

$$m_i\left[(d/dt)\left(\frac{d(v_i^2/2)}{dq_j'}\right) - d(v_i^2/2)/dq_j\right] = f_i dx_i/dq_j. \tag{7.A.7}$$

Summing this over i yields

$$\sum_i \left(\frac{(d/dt)(d(m_i v_i^2/2))}{dq_j'} - \frac{d(m_i v_i^2/2)}{dq_j} \right) = F_j. \tag{7.A.8a}$$

Thus,

$$(d/dt)\partial T/\partial q_j' - \partial T/\partial q_j = F_j. \tag{7.A.8b}$$

These equations, *Lagrange's equations of the first kind,* are valid for all coordinate systems that can be defined by Eqs. (7.A.1). If the forces are derivable from a potential, V(q), then

$$F_j = -\partial V(q)/\partial q_j, \tag{7.A.9a}$$

and, on defining

$$L(q, q') = T(q, q') - V(q), \tag{7.A.9b}$$

(7.A.8b) becomes

$$d(\partial L/\partial q_j')/dt - \partial L/\partial q_j = 0. \tag{7.A.9c}$$

These are *Lagrange's equations of the second kind.*

To utilize Lagrange's equations it is necessary to express T as a function of the q's and their time derivatives. Using (7.A.1) one finds

$$T = (1/2)\sum_j \sum_k g_{jk}(q) q_j' q_k', \tag{7.A.10}$$

with

$$g_{jk}(q) = \sum_i m_i\ \partial X_i/\partial q_j\ \partial X_i/\partial q_k.$$

The variables,

$$p_j \equiv \partial T/\partial q_j' = \partial L/\partial q_j', \tag{7.A.11}$$

are the *canonical momenta* of the system.

We next show that the p_j and q_j are, in fact, members of canonically conjugate sets of variables. Let the Cartesian positions and momenta be denoted by χ_k and π_k, with $\pi_k = m_k v_k$. Using Poisson brackets defined using derivatives with respect to χ_k and π_k, one has

$$\{q_i, p_j\} = \{q_i, \partial T/\partial q_j'\}, \tag{7.A.12a}$$

$$= \{q_i, \partial \sum (m_k v_k^2/2)/\partial q_j'\} = \{q_i, \sum m_k v_k dv_k/dq_j'\}. \tag{7.A.12b}$$

Hence

$$\{q_i, p_j\} = \{q_i, \sum \pi_k \, dv_k/dq'_j\}. \quad (7.A.12c)$$

As in (7.A.6b), one has

$$dv_k/dq'_j = d\chi_k/dq_j. \quad (7.A.12d)$$

Consequently,

$$\{q_i, p_j\} = \{q_i, \sum \pi_k \, d\chi_k/dq_j\}. \quad (7.A.12e)$$

Because $d\chi_k/dq_j$ can be expressed as a function of the χ's only, for the Poisson bracket on the right-hand side of (7.A.12e) one has

$$\begin{aligned}\{q_i, \sum_k \pi_k d\chi_k/dq_j\} &= \sum_k \{q_i, \pi_k\} d\chi_k/dq_j \\ &= \sum_k (\sum_m \partial q_i/\partial \chi_m \; \partial \pi_k/\partial \pi_m) \; d\chi_k/dq_j. \quad (7.A.12f)\end{aligned}$$

Only the terms $\partial \pi_k/\partial \pi_m$ with $k = m$ are nonvanishing, so

$$\{q_i, p_j\} = \sum_m \partial q_i/\partial \chi_m d\chi_m/dq_j = \sum_m dq_i/d\chi_m d\chi_m/dq_j = dq_i/dq_j. \quad (7.A.12g)$$

Thus

$$\{q_i, p_j\} = \delta_{ij}. \quad (7.A.13a)$$

Similar, but shorter, analyses establish that

$$\{q_i, q_j\} = 0 = \{p_i, p_j\}. \quad (7.A.13b,c)$$

To sum up, for a transformation from Cartesian coordinates to generalized coordinates, q_j, the momenta p_j conjugate to the q_j are given by (7.A.11), as claimed.

Appendix B. The Variable Conjugate to Time in PQET Space

We wish to establish that in PQET space, the variable conjugate to t is −E. Following Lanczos,[4] we begin the argument with Hamilton's variational derivation of Lagrange's equations. Let $L(m_j, q_j, q'_j, t)$ be the Lagrangian function of a system of masses m_j, with position coordinates q_j, and velocity coordinates $q'_j = dq_j/dt$. The integral

$$S \equiv \int_{t_1}^{t_2} L(m_j, q_j, q', t)\, dt, \tag{7.B.1}$$

has the dimensions of an action, and represents a total action developed as the masses move between times t_1 and t_2. Hamilton established that Lagrange's equations result if one requires that, as t evolves, the point with coordinates q_j, q'_j, describes a path through the space of positions and velocities which makes S an extremum. Put more precisely, the motions are such that $\delta S = 0$ if the position functions $q'_j(t)$, and velocity functions $q'_j(t) = dq_j(t)/dt$, are subjected to infinitesimal variations, $\delta q_j(t)$, $\delta q'_j(t) = d(\delta q_j(t))/dt$, at all values of t between t_1 and t_2. The variations are restricted by the requirement that they vanish at t_1 and t_2:

$$\delta q_j(t_1) = \delta q_j(t_2) = 0 = \delta q'_j(t_1) = \delta q'_j(t_2). \tag{7.B.2}$$

Thus, one requires

$$\delta S = \int_{t_1}^{t_2} (L(m_j, q_j + \delta q'_j, q'_j + \delta q_j, t) - L(m_j, q_j, q', t))\, dt, \tag{7.B.3a}$$

$$= \int_{t_1}^{t_2} \sum((\partial L/\partial q_j)\delta q_j + (\partial L/\partial q'_j)\delta q'_j)\, dt = 0. \tag{7.B.3b}$$

Integrating the second term by parts enables one to express this as

$$\delta S = \sum(\partial L/\partial q'_j)\delta q_j|_{t_1}^{t_2} + \int_{t1}^{t2} \sum((\partial L/\partial q_j) - (d/dt)(\partial L/\partial q'_j))\delta q_j dt = 0. \tag{7.B.4}$$

As (7B.2) requires that the variation vanishes at t_1 and t_2, the first term must vanish, one has

$$\delta S = \int_{t_1}^{t_2} \sum((\partial L/\partial q_j) - (d/dt)(\partial L/\partial q'_j))\delta q_j dt = 0. \tag{7.B.5}$$

The integral vanishes iff for each value of j, Lagrange's equations

$$(d/dt)(\partial L/\partial q'_j)\ \partial L/\partial q_j = 0 \tag{7.B.6}$$

are satisfied.

The momentum conjugate to each q_j is $p_j = \partial L/\partial q'_j$. If L is not a function of t, the Hamiltonian, H, corresponding to it is obtained by expressing the function

$$H = \sum q'_j p_j - L(m_j, q_j, q'_j, t), \tag{7.B.7}$$

as a function of the q's and p's.

To determine the applications of this derivation to motions in PQET space, we change the variable of integration in (7.B.1). The evolution parameter t is replaced by an evolution parameter τ, and t is replaced by t', a new independent dynamical variable. To this end we let $dt \to (dt/d\tau)d\tau \equiv t'd\tau$, and replace d/dt by $(1/t')d/d\tau$. If we write $q_j'' = dq_j/d\tau$, then $q_j' = q_j''/t'$. Supposing that L does not depend explicitly upon t, and writing

$$L'(m_j, q_j, q_j''/t') = t'L(m_j, q_j, q_j''/t'), \tag{7.B.8a}$$

(7.B.1) becomes

$$S' \equiv \int_{\tau_1}^{\tau_2} L'(m_j, q_j, q_j''/t')d\tau. \tag{7.B.8b}$$

The Lagrangian L' depends upon the generalized coordinates q_j t', and the generalized velocities q_j''. As L' is not a function of t, the variable conjugate to t is a constant of motion. Using the S' and L' of (7.B.8), one finds Eq. (7.B.6) replaced by

$$(d/d\tau)(\partial L'/\partial q_j'') - \partial L'/\partial q_j = 0. \tag{7.B.9}$$

The momenta conjugate to the q_j' are $p_j' \equiv \partial L'/\partial q_j''$, and the momentum conjugate to t' is

$$p_{t'}' \equiv \partial L'/\partial t'. \tag{7.B.10}$$

Thus

$$p_t' = L - t'\sum((q_j''/t'^2)\partial L/\partial q_j') \tag{7.B.11a}$$

$$= L - \sum((q_j''/t')\partial L/\partial q_j') \tag{7.B.11b}$$

$$= L - \sum q_j'\partial L/\partial q_j' \tag{7.B.11c}$$

$$= L - \sum q_j'p_j \tag{7.B.11d}$$

$$= -H \to -H(q, p). \tag{7.B.11e}$$

Consequently

$$p_t' \to -E, \tag{7.B.11f}$$

and $-E$ is the variable conjugate to t.

To define functions $W(p, q, E, t)$ that are generalizations of $H(p, q) - E$, one begins by defining a *generalized kinetic energy* that is allowed to be a function of generalized coordinates q, t, and *generalized velocities* $dq/d\tau$ and $dt/d\tau$. One then defines a generalized Lagrangian $T - V(q)$ and proceeds as in (7.B.10) to determine the generalized momenta conjugate to the coordinates q and t. W(p, q, E, t) is then defined by the same process that converts (7.B.11a) to (7.B.11e).

Exercises

1. In matrix notation, (7.2.3) may be expressed as $\mathbf{ds}' = \mathbf{M}\,\mathbf{ds}^T$, with **M** composed from n x n submatrices **A**, **B**, **C** and **E**, with the upper left sub-matrix, **A**, having elements a_{ij}, the upper right submatrix, **B**, having elements b_{ij}, and the lower left and right sub-matrices, **C**, **E** having elements c_{ij}, e_{ij} respectively. What relations between the matrix elements are required by Eqs. (7.2.8)?
2. a) Produce plots, analogous to **Fig. 7.5.1**, of trajectories in a common q–q plane and a common p–p plane. b) Plot motions implied by (7.5.10) in each of the six two-planes.
3. Show that u_1,u_2,u_3 satisfy Eq. (7.5.11b).
4. Expand (7.5.11b), collect terms that multiply the same powers of each q and p, and determine the relations that the coefficients c must satisfy when H is given by (7.5.8a).
5. Derive (7.2.11) from (7.2.9).
6. Derive (7.7.5a) from (7.7.4).

References

[1] J.-M. Levy-Leblond, Galilei Group and Galilei Invariance, in *Group Theory and it Applications, II*, ed. E. M. Loebl (Academic Press, N.Y., 1971).

[2] R. Gilmore, *Lie Groups, Lie Algebras, and Some of Their Applications* (Wiley-Interscience, N.Y., 1974).

[3] Sir W. Thomson and P. G. Tait, *Treatise on Natural Philosophy* (Cambridge University Press, Cambridge, 1879), p. 302.

[4] C. Lanczos, *The Variational Principles of Mechanics* (University Toronto Press, Toronto, 1949), p. 133.

CHAPTER 8

Symmetries of Classical Keplerian Motion

In recent years it has become possible to turn Kepler's intuitions of the "music of the spheres" into a mathematically and physically sound understanding of the dynamical symmetries of classical Keplerian motion. A few of their consequences were pointed out in Chapter 1. This chapter develops their conceptual underpinnings.

8.1 Newtonian Mechanics of Planetary Motion

Let GmM/r^2 be the magnitude of the gravitational force attracting a spherically symmetrical planet of mass m, and the sun of mass M, with G being the universal constant of gravitation, and r the distance between the centers of mass of the two bodies. The reduced mass of the planet is $\mu = Mm/(M + m)$. If one defines $\kappa = GmM/\mu = G(m + M)$, and $\mathbf{r}$ is the vector from the center of mass of the sun to the center of mass of the planet, then Newton's equation for the motion of the center of mass of the planet reduces to

$$d\mathbf{v}/dt = -f(r)\mathbf{r}, \tag{8.1.1}$$

with

$$\mathbf{v} = d\mathbf{r}/dt, \quad f(r) = \kappa/r^3.$$

Kepler's three laws of motion are direct consequences of dynamical symmetries of this equation.

We first deduce the scaling symmetry that underlies Kepler's law of periodic times. Let T represent a measure of time, and $\mathbf{R}$ a position vector of length R. Suppose $t \to T' = aT$, $\mathbf{R} \to \mathbf{R}' = b\mathbf{R}$, and $\kappa \to \kappa' = c\kappa$. Equation (8.1.1) with $T = t$, $\mathbf{R} = \mathbf{r}$, is invariant under such a transformation if $a^2c/b^3 = 1$. Because the masses of the planets are small compared to the mass of the sun, κ is nearly the same for all, so one may in good approximation set $c = 1$. For such systems, $a^2/b^3 = 1$, and the time and distance scalings must produce the approximate relation $(T'/T)^2 = (R'/R)^3$. As the planetary orbits are all ellipses of small eccentricity, they may be approximated by circles of radii R and R′. Letting T and T′ be the period of time it takes for two planets with orbital radii R and R′, to traverse such a circle, one obtains *Kepler's law of periodic times*:

$$(T'/T)^2 = (R'/R)^3. \tag{8.1.2}$$

A discussion of the accuracy of this relation will be found in Ref. 1 at the end of the chapter. The exact law will be derived below.

We next consider time-independent properties of planetary orbits that are consequences of two different vector constants of motion of Keplerian systems. Subsequent sections of the chapter will investigate the way in which these constants of motion are related to dynamical symmetries.

Taking the cross product of (8.1.1) with $\mathbf{r}$, one obtains

$$\mathbf{r} \times d\mathbf{v}/dt = -f(r)\, \mathbf{r} \times \mathbf{r} = \mathbf{0}. \tag{8.1.3}$$

Now

$$d(\mathbf{r} \times \mathbf{v})/dt = (\mathbf{v} \times \mathbf{v}) + (\mathbf{r} \times d\mathbf{v}/dt) = (\mathbf{r} \times d\mathbf{v}/dt). \tag{8.1.4a}$$

It then follows from (8.1.3) that

$$d(\mathbf{r} \times \mathbf{v})/dt = \mathbf{0}. \tag{8.1.4b}$$

Hence,

$$\mathbf{h} \equiv \mathbf{r} \times \mathbf{v} \tag{8.1.4c}$$

is a vector that remains constant as $\mathbf{r}$ and $\mathbf{v}$ evolve. It has a length, h, given by

$$h^2 = (\mathbf{h} \cdot \mathbf{h}) = r^2 v^2 - (\mathbf{r} \cdot \mathbf{v})^2. \tag{8.1.4d}$$

From (8.1.4c) it follows that

$$\mathbf{r} \times d\mathbf{r} = \mathbf{h}\, dt, \tag{8.1.5a}$$

so $\mathbf{r}$ and $d\mathbf{r}$ lie in a fixed plane perpendicular to $\mathbf{h}$. This plane contains the center of mass of the planet and that of the sun. In every time interval $\Delta t = \int dt$, the area swept out by the radius vector $\mathbf{r}$ has the same value,

$$\int (\mathbf{r}(t) \times d\mathbf{r}(t))/2 = \mathbf{h}\Delta t/2. \tag{8.1.5b}$$

This implies *Kepler's law of equal areas: a line from the sun to the planet sweeps out equal areas in equal times.* This law may also be considered to arise from the constancy of the angular momentum vector of the planet,

$$\mathbf{L} = \mathbf{r} \times \mathbf{p}, \quad \mathbf{p} = \mu\mathbf{v}. \tag{8.1.6}$$

Integrating (8.1.5b) over an entire period of length T, the area swept out is an ellipse of area $\pi\Lambda_a\Lambda_b$, where $2\Lambda_a$ is length of its major axis, and $2\Lambda_b$ is the length of its minor axis. Consequently, for an elliptical orbit,

$$\mathrm{T} = 2\pi\Lambda_a\Lambda_b/\mathrm{h}. \tag{8.1.7}$$

Kepler systems possess a further vector constant of motion.[2] *We shall see that it and* **L** *together determine the dynamical symmetry of every Keplerian manifold defined by* H = E, *and provide the initial conditions that determine motion upon it.* To uncover this vector, take the vector cross-product of **h** with the equation of motion, (8.1.1), to obtain

$$\mathbf{h} \times \mathrm{d}\mathbf{v}/\mathrm{dt} = -(\kappa/\mathrm{r}^3)\,\mathbf{h} \times \mathbf{r}. \tag{8.1.8a}$$

Because **h** is constant, this implies that

$$\mathrm{d}(\mathbf{h} \times \mathbf{v})/\mathrm{dt} = -(\kappa/\mathrm{r}^3)\,\mathbf{h} \times \mathbf{r}. \tag{8.1.8b}$$

Using the identity

$$(\mathbf{a} \times \mathbf{b}) \times \mathbf{c} = (\mathbf{c} \cdot \mathbf{a})\mathbf{b} - \mathbf{a}(\mathbf{c} \cdot \mathbf{b})$$

to expand

$$\mathbf{h} \times \mathbf{r} = (\mathbf{r} \times \mathrm{d}\mathbf{r}/\mathrm{dt}) \times \mathbf{r},$$

one obtains

$$\mathbf{h} \times \mathbf{r} = \mathrm{r}^2\mathrm{d}\mathbf{r}/\mathrm{dt} - \mathrm{r}(\mathbf{r} \cdot \mathrm{d}\mathbf{r}/\mathrm{dt}) = \mathrm{r}^3\mathrm{d}(\mathbf{r}/\mathrm{r})/\mathrm{dt}. \tag{8.1.8c}$$

Thus (8.1.8b) becomes

$$\mathrm{d}(\mathbf{h} \times \mathbf{v})/\mathrm{dt} = -\kappa\mathrm{d}(\mathbf{r}/\mathrm{r})/\mathrm{dt}. \tag{8.1.8d}$$

Consequently

$$\mathrm{d}((\mathbf{h} \times \mathbf{v}) + \kappa\mathbf{r}/\mathrm{r})/\mathrm{dt} = 0, \tag{8.1.8e}$$

and the vector

$$\mathbf{b} \equiv (\mathbf{h} \times \mathbf{v}) + \kappa\mathbf{r}/\mathrm{r}, \tag{8.1.9a}$$

is a constant of the motion.

By expanding $\mathbf{h} \times \mathbf{v}$ one may obtain an alternative expression for **b**.

Letting $v = |\mathbf{v}|$, one finds

$$\mathbf{b} = (\mathbf{r} \cdot \mathbf{v})\mathbf{v} + (\kappa/r - v^2)\mathbf{r}. \tag{8.1.9b}$$

This vector is perpendicular to $\mathbf{h}$ (and $\mathbf{L} = \mu\mathbf{h}$) because both $\mathbf{v} \cdot \mathbf{h}$ and $\mathbf{r} \cdot \mathbf{h}$ vanish. The constancy of its components was known to Laplace and to Hamilton.[3] Soon after Gibbs[4] introduced the concept of vector they were assembled into one. Today, $\mathbf{b}$ and vectors proportional to it, are commonly called Runge–Lenz vectors.[5]

From initial measurements of a planet's, or comet's, position and velocity vectors, observers can determine $\mathbf{h}$ and $\mathbf{b}$ in their frame of reference and make them the basis of a new right-handed frame of reference that is particularly adapted to describing the motion of the body. To this end, one translates and rotates one's original coordinate system, putting the origin of the new system at the center of mass, and supposing the plane of the orbit is the x–y plane. Letting the x axis and unit vector $\mathbf{i}$ lie on the major axis of the orbit, one may choose its positive direction to be that from the center of mass to aphelion. Then one may set the z axis along $\mathbf{h} = \mathbf{r} \times \mathbf{v}$. If the motion is counter-clockwise, one chooses $\mathbf{k} = \mathbf{h}/h$, if clockwise, $\mathbf{k} = -\mathbf{h}/h$. Then if $\mathbf{c} = \mathbf{h}\times\mathbf{b}$, one has $\mathbf{j} = \mathbf{c}/c$. It is also convenient to shift the origin of clock time so that at $t = 0$ the planet is at perihelion. This frame of reference is used in **Figs. 8.1.1**. The planet's initial coordinates are $x = x_0$, $y = 0$ and $z = 0$. Its initial velocities are $v_x = 0$, $v_y = v_0$ and $v_z = 0$. These values define the further initial values

$$r = |x_0|, \quad E = \mu(v_0^2/2 - \kappa/|x_0|); \tag{8.1.10a}$$

and

$$\mathbf{h} = (0, 0, h_z), \quad h_z = x_0\, v_0; \tag{8.1.10b}$$

$$\mathbf{b} = (b_x, 0, 0), \quad b_x = x_0(\kappa/|x_0| - v_0^2); \tag{8.1.10c}$$

$$\mathbf{c} = (0, c_y, 0), \quad c_y = b_x\, h_z. \tag{8.1.10d}$$

At perihelion, $v_0^2 > \kappa/|x_0|$, so b_x is positive, and $\mathbf{b}$ points toward aphelion. In the Exercises one finds that for an orbit with *eccentricity* ε, and *semilatus rectum* λ,

$$|\mathbf{b}| = \varepsilon\kappa, \tag{8.1.10e}$$

$$|\mathbf{h}| = (\lambda\kappa)^{1/2}. \tag{8.1.10f}$$

Figure 8.1.1 illustrate three orbits obtained when the eccentricity is nonzero. It is assumed that x_0 is negative and that the motion is counterclockwise, so v_0 is also negative. The perihelion distance is $|x_0|$.

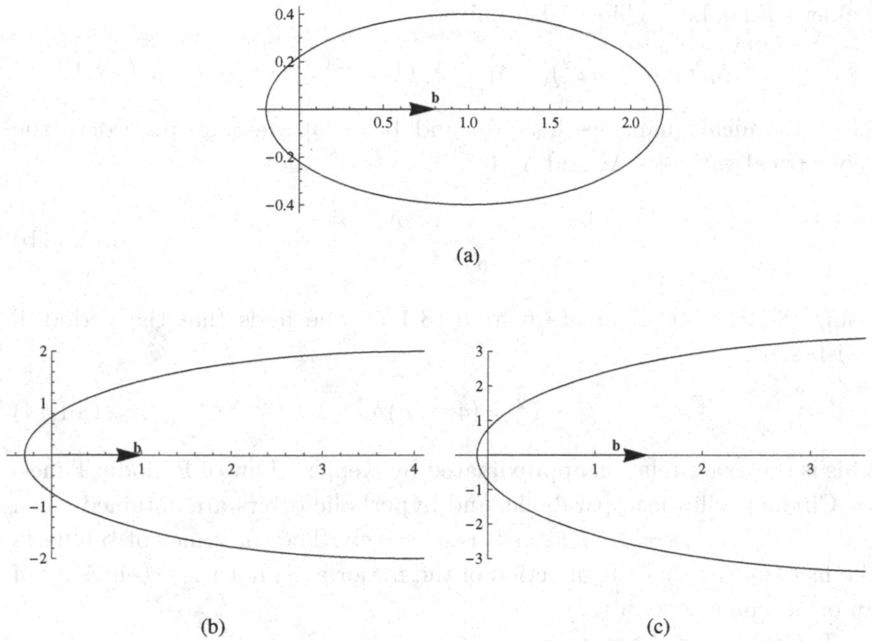

Fig. 8.1.1 Runge–Lenz vectors for elliptical, parabolic, and hyperbolic orbits, (a) $|\mathrm{b}| = 0.8, |\mathrm{b}| = 0.71$, (b) $|\mathrm{b}| = 1,0, |\mathrm{b}| = 0.71$, (c) $|\mathrm{b}| = 1.2, |\mathrm{b}| = 0.71$.

All the orbits available to bodies moving under an inverse square law of force can be determined by considering the scalar product of $\mathbf{r}$ with $\mathbf{b}$.[6] One has

$$\mathbf{b}\cdot\mathbf{r} \equiv (\mathbf{h}\times\mathbf{v})\cdot\mathbf{r} + \kappa \mathrm{r} = -\mathrm{h}^2 + \kappa\mathrm{r}. \tag{8.1.11a}$$

Letting θ be the angle between $\mathbf{b}$ and $\mathbf{r}$, one may express this equation as

$$\mathrm{b}\,\mathrm{r}\cos(\theta) = -\mathrm{h}^2 + \kappa\mathrm{r}, \tag{8.1.11b}$$

so the dependence of r upon θ is expressed by

$$\mathrm{r} = \frac{(\mathrm{h}^2/\kappa)}{(1 - (\mathrm{b}/\kappa)\cos(\theta)).} \tag{8.1.11c}$$

This is the equation of a conic section with *eccentricity* $\varepsilon = \mathrm{b}/\kappa$, and *semilatus rectum* $\lambda = \mathrm{h}^2/\kappa$. For planets the conic sections are, as stated in

Kepler's First Law, ellipses. For ellipses

$$\Lambda_a = \lambda/(1-\varepsilon^2), \quad \Lambda_b = \lambda/(1-\varepsilon^2)^{1/2}. \tag{8.1.12a,b}$$

The dynamical quantities $h = |\mathbf{h}|$ and $b = |\mathbf{b}|$ are then related to the geometrical variables Λ_a and Λ_b by

$$\begin{gathered} b/\kappa = (1-(\Lambda_b/\Lambda_a)^2)^{1/2}, \\ h(b/\kappa)^{1/2} = \Lambda_a. \end{gathered} \tag{8.1.13a,b}$$

Using (8.1.12a) to eliminate h from (8.1.7), one finds that the period T satisfies

$$T^2 = (4\pi^2/\kappa)\Lambda_a^3. \tag{8.1.14}$$

This is the exact relation approximated by Kepler's Law of Periodic Times.

Circular, elliptical, parabolic, and hyperbolic orbits are obtained when $\varepsilon = 0$, $0 < \varepsilon < 1$, $\varepsilon = 1$, and $\varepsilon > 1$, respectively. The constancy of **b** reflects the fact that neither the direction of the major axis, nor the eccentricity of an orbit, changes with time.

The energy of the system,

$$E = \mu v^2/2 - GMm/r = \mu(v^2/2 - \kappa/r), \tag{8.1.15}$$

is a scalar constant of the motion. It is shown in standard monographs on theoretical mechanics[7] that for $0 \le \varepsilon < 1$,

$$E = -\mu/2\Lambda_{a.}. \tag{8.1.16a}$$

For $\varepsilon > 1$,

$$E = \mu/2\Lambda_{a.}. \tag{8.1.16b}$$

As these energies do not depend upon the length of the minor axis, for given values of μ and a, varying Λ_b produces a family of *degenerate* orbits — orbits of the same energy. The dynamical symmetries responsible for this degeneracy are investigated in the upcoming sections of the chapter.

A development analogous to that of (8.1.10) may be used to obtain the relation among velocity components that defines the *hodograph* of Keplerian motion.[6] The component of **v** on the **j** axis is v_y, so (8.1.10) implies

$$v_y = (bh)^{-1}(\mathbf{h} \times \mathbf{b}) \cdot \mathbf{v} = -(bh)^{-1}h^2(v^2 - \kappa/r) = -(h/b)(v^2/2 + E/\mu). \tag{8.1.17a}$$

Because **v** has no component on the **k** axis, (8.1.17a) can be expressed as

$$\mathrm{v_x^2 + v_y^2 + 2v_y b/h} + 2E/\mu = 0, \tag{8.1.17b}$$

so

$$\mathrm{v_x^2 + (v_y + b/h)^2 = (b/h)^2} - 2E/\mu. \tag{8.1.17c}$$

As the right-hand side of this equation is only a function of constants of the motion, *for every Keplerian motion the hodograph is a circle*, of radius $(\mathrm{b/h})^2 - 2E/\mu$, with its center at a point $-\mathrm{b/h}$ on the **j** axis. **Figures 8.1.2**

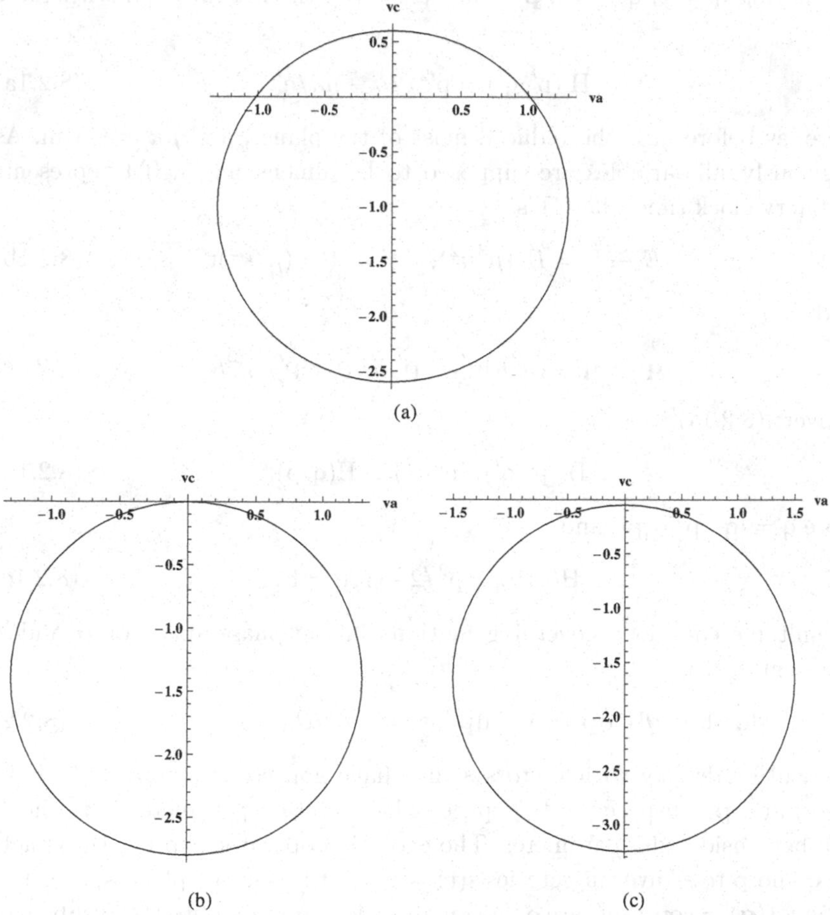

Fig. 8.1.2 Hodographs corresponding to orbits of Figs. 8.1.1, for (a) $|\mathrm{b}| = 0.8, |\mathrm{h}| = 0.71$, (b) $|\mathrm{b}| = 1.0, |\mathrm{h}| = 0.7$, and (c) $|\mathrm{b}| = 1.2, |\mathrm{h}| = 0.71$.

illustrate the hodographs corresponding to the orbits of **Figs. 8.1.1**. For them, the value of the radius defined by the right-hand side of (8.1.17c) is $\kappa^2/(x_0^2 v_0^2)$.

8.2 Hamiltonian Formulation of Keplerian Motions in Phase Space

Hamilton's mechanics describes how constants of motion determine dynamical symmetries in a phase space. We therefore move from a Newtonian to a Hamiltonian description of Keplerian motions.

Setting $\mathbf{q}' = \mathbf{r}$, $q' = |\mathbf{r}|$, $\mathbf{p}' = \mu\mathbf{v}$, $p' = |\mathbf{p}'|$, the Hamiltonian with value E is

$$H'(p', q') = p'^2/2\mu - \mu\kappa/q'. \tag{8.2.1a}$$

Here, as before, μ is the reduced mass of the planet, and $\mu\kappa = GMm$. As previously, all variables are supposed to be dimensionless. If t represents ordinary clock time, the CT's

$$E \to E = E/(\mu^3\kappa^2), \quad t' \to t = (\mu^3\kappa^2)t', \tag{8.2.1b}$$

and

$$\mathbf{q}' \to \mathbf{q} = (\mu^2\kappa)\mathbf{q}', \quad \mathbf{p}' \to \mathbf{p} = \mathbf{p}'/(\mu^2\kappa), \tag{8.2.1c}$$

convert (8.2.1a) to

$$H'(p', q')/(\mu^3\kappa^2) \equiv H(q, p). \tag{8.2.1d}$$

Here $q = |\mathbf{q}|$, $p = |\mathbf{p}|$, and

$$H(q, p) = p^2/2 - 1/q = E. \tag{8.2.1e}$$

Hamilton's equations governing motions in the phase space of $\mathbf{p}$ and $\mathbf{q}$ are then

$$d\mathbf{q}/dt = \partial H/\partial\mathbf{p} = \mathbf{p}, \quad d\mathbf{p}/dt = -\partial H/\partial\mathbf{q} = -\mathbf{q}/q^3. \tag{8.2.2}$$

We will be dealing with vectors $\mathbf{s}$ that have nonzero components in both the $\mathbf{q}$ and $\mathbf{p}$ subspaces of PQ space. The evolution parameter is t, and E can be considered a parameter. The evolution operator $\exp(t\{-H, \cdot\})$ acts on $\mathbf{q}$ and $\mathbf{p}$ to evolve trajectories $\mathbf{s}(t) = (\mathbf{q}(t), \mathbf{p}(t))$ in the phase space. The orbits, $f(\mathbf{q}) =$ constant, are orthogonal projections of $\mathbf{s}$ onto the $\mathbf{q}$ subspace of PQ space, and the hodographs, $g(\mathbf{p}) =$ constant, are projections onto its $\mathbf{p}$ subspace. If one considers planetary motions in a fixed plane, both q-space and p-space become two-dimensional, and the PQ space is reduced to a four-dimensional phase space.

For a function $X(\mathbf{q}, \mathbf{p}, E)$ to be a constant of motion in PQ space, it must satisfy

$$\{X, H\}|_{H=E} = 0. \tag{8.2.3}$$

Two vectors whose components satisfy this requirement are

$$\mathbf{L} = \mu\mathbf{h}, \tag{8.2.4a}$$
$$\mathbf{A} \equiv -(\mathbf{L} \times \mathbf{p} + \mathbf{q}/q) = -\mathbf{b}/\kappa \equiv \mathbf{a}/\kappa. \tag{8.2.4b}$$

The negative sign has been introduced in the definition of **A** and **a**, because doing so substantially simplifies the discussion in subsequent sections. In position space, the vectors **A** and **a** point from the center of mass toward perhelion. After expanding the $\mathbf{L} \times \mathbf{p}$ term in **A** to $(\mathbf{q} \cdot \mathbf{p})\mathbf{p} - \mathbf{q}p^2$, one can use (8.2.1d,e) to make the conversions

$$\mathbf{A} \to \mathbf{A}(H) = \mathbf{q}(p^2 + 2H)/2 - (\mathbf{q} \cdot \mathbf{p})\mathbf{p}, \tag{8.2.5a}$$

and

$$\mathbf{A} \to \mathbf{A}(E) = \mathbf{q}(p^2 + 2E)/2 - (\mathbf{q} \cdot \mathbf{p})\mathbf{p}, \tag{8.2.5b}$$

both forms of **A** satisfy (8.2.3) with $X = \mathbf{A}$. All the expressions for **A** change sign when $\mathbf{q} \to -\mathbf{q}$ and $\mathbf{p} \to -\mathbf{p}$, so **A**, unlike **L**, is a proper vector in PQ space.

From the components of **L** and **A** one obtains the components of the PB operators $\boldsymbol{L} \equiv \{\mathbf{L}, \cdot\}$ and $\boldsymbol{A} \equiv \{\mathbf{A}, \cdot\}$ that generate invariance transformations of the equation $H = E$. One has

$$\{L_{ij}, \cdot\} = -(q_i\partial/\partial q_j - q_j\partial/\partial q_i) - (p_i\partial/\partial p_j - p_j\partial/\partial p_i) \equiv L_{ij}. \tag{8.2.5}$$

The form of the components of $\boldsymbol{A}$ is more complicated. In Sec. 8.3 we will use a CT to greatly simplify them.

Here we wish to consider the Lie algebras generated by the these PB operators. In expressing the commutation relations it is simplest to make use of the notation in which, e.g. $L_x = L_1$ rather than L_{23}. We will also make use of the conventional symbol ε_{ijk}, which has the value +1 when the indices are in clockwise order, and −1 when they are not. Then one finds that the components of $\boldsymbol{L}$ and $\boldsymbol{A}$ obey the commutation relations:

$$[\mathrm{L_i}, \mathrm{L_j}] = \varepsilon_{\mathrm{ijk}}\mathrm{L_k}, \quad [\mathrm{L_i}, \mathrm{A_j}] = \varepsilon_{\mathrm{ijk}}A_{\mathrm{k}}, \tag{8.2.6a,b}$$

$$\begin{aligned} [\mathrm{A_i}, \mathrm{A_j}] &= (-2\mathrm{H})\varepsilon_{\mathrm{ijk}}\mathrm{L_k}, \quad \mathbf{A} = \mathbf{A}(\mathrm{H}), \\ [\mathrm{A_i}, \mathrm{A_j}] &= (-2\mathrm{E})\varepsilon_{\mathrm{ijk}}\mathrm{L_k}, \quad \mathbf{A} = \mathbf{A}(\mathrm{E}). \end{aligned} \tag{8.2.6c,d}$$

Because of the presence of H in (8.2.6c), Eqs. (8.2.6a–c) do not define a finite-dimensional Lie algebra. However, because $\boldsymbol{A}$ and $\boldsymbol{L}$ commute with H, equation c may be replaced by equation d, and E may be treated as a parameter.

For bound states E is negative. In the subspace of phase space that contains only bound trajectories

$$\mathbf{A}_<(\mathrm{E}) \equiv \mathbf{A}(\mathrm{E})/(-2\mathrm{E})^{1/2}, \tag{8.2.7a}$$

defines PB operators $\boldsymbol{A}_< = \{\mathbf{A}_<(\mathrm{E}), \cdot\}$, that given (8.2.6d), satisfy the relations

$$[\mathrm{L_i}, \mathrm{L_j}] = \varepsilon_{\mathrm{ijk}}\mathrm{L_k}, \tag{8.2.7b}$$

$$[\mathrm{L_i}, \mathrm{A_{<j}}] = \varepsilon_{\mathrm{ijk}}\mathrm{A_{<k}}, \tag{8.2.7c}$$

$$[\mathrm{A_{<i}}, \mathrm{A_{<j}}] = \varepsilon_{\mathrm{ijk}}\mathrm{L_k}. \tag{8.2.7d}$$

These are the commutation relations of an SO(4) Lie algebra. The quadratic Casimir operator is

$$\mathrm{A}_<^2 + \mathrm{L}^2 = \Sigma(\mathrm{A}_{<\mathrm{j}}^2 + \mathrm{L}_\mathrm{j}^2). \tag{8.2.8a}$$

Substituting H for E in (8.2.7a) one obtains $\mathbf{A}_<(\mathrm{H})$. Expanding $\mathbf{A}_<(\mathrm{H})\cdot\mathbf{A}_<(\mathrm{H})$ and $\mathbf{L}\cdot\mathbf{L}$ one finds that

$$(2\mathrm{H})(\mathrm{A}_<(\mathrm{H})^2 + \mathrm{L}^2) = -(\mathrm{q}(\mathrm{p}^2 - 2\mathrm{H})/2)^2. \tag{8.2.8b}$$

Now $\mathrm{q}(\mathrm{p}^2 - 2\mathrm{H})/2 = 1$, so the right-hand side of this equation is equal to -1. Consequently

$$\mathrm{H} = -(1/2)/(\mathrm{A}_<^2 + \mathrm{L}^2). \tag{8.2.9}$$

The Lie algebra is of rank 2, and consequently possesses a quartic Casimir operator, which is $(\boldsymbol{L}\cdot\boldsymbol{A}_<)^2$. Because $\mathbf{L}$ is perpendicular to $\mathbf{A}$, for all values of E, $(\mathbf{L}\cdot\mathbf{A}_>)^2$ vanishes.

The operators $\exp(\alpha_\mathrm{i}\mathrm{A}_{<\mathrm{i}})$ and $\exp(\beta_\mathrm{i}\mathrm{L_i})$ may be shown to be those of the group SO(4). They alter the components of $\mathrm{A}_{<\mathrm{j}}$, and $\mathrm{L_j}$, $\mathrm{j} \neq \mathrm{i}$, while leaving $(\mathrm{A}_<^2 + \mathrm{L}^2)$ unaltered. Because of this, *when E is negative, the equation* H = E *is SO(4) invariant and the dynamical symmetry of the system is that of a hypersphere S(3).*

For Keplerian systems with positive energy, one may proceed analogously. Let

$$\mathbf{A}_{>}(\mathrm{E}) \equiv \mathrm{A}(\mathrm{E})/(2\mathrm{E})^{1/2} = (\mathbf{q}(\mathrm{p}^2 + 2\mathrm{E})/2 - (\mathbf{q}\cdot\mathbf{p})\mathbf{p})/(2\mathrm{E})^{1/2}. \tag{8.2.10a}$$

The PB operators, $\boldsymbol{A}_{>} = \{\mathbf{A}_{>}(\mathrm{E}), \cdot\}$ satisfy the commutation relations

$$[\mathrm{L_i}, \mathrm{L_j}] = \varepsilon_{\mathrm{ijk}}\mathrm{L_k}, \tag{8.2.10b}$$

$$[\mathrm{L_i}, \mathrm{A_{>j}}] = \varepsilon_{\mathrm{ijk}}\mathrm{A_{>k}}, \tag{8.2.10c}$$

$$[\mathrm{A_{>i}}, \mathrm{A_{>j}}] = -\varepsilon_{\mathrm{ijk}}\mathrm{L_k}. \tag{8.2.10d}$$

These are the commutation relations of the Lie algebra of an SO(3, 1) group. The quadratic Casimir operator is $\mathrm{L}^2 - \mathrm{A}^2_{>}$. In this case one finds that

$$\mathrm{H} = (1/2)/(\mathrm{L}^2 - \mathrm{A}^2_{<}). \tag{8.2.11}$$

The dynamical symmetry is that of a hyperboloid of revolution in four-space.

When E = 0, H → 0, and

$$\boldsymbol{A} \to \boldsymbol{A}_0(\mathrm{E}) \equiv \{\mathbf{q}\mathrm{p}^2/2 - (\mathbf{q}\cdot\mathbf{p})\mathbf{p}, \cdot\}, \tag{8.2.12a}$$

the resulting commutation relations are those of the Lie algebra of E(3):

$$[\mathrm{L_i}, \mathrm{L_j}] = \varepsilon_{\mathrm{ijk}}\mathrm{L_k}, \tag{8.2.12b}$$

$$[\mathrm{L_i}, \mathrm{A_{>j}}] = \varepsilon_{\mathrm{ijk}}\mathrm{A_{>k}}, \tag{8.2.12c}$$

$$[\mathrm{A_{>i}}, \mathrm{A_{>j}}] = 0. \tag{8.2.12d}$$

The dynamical symmetry is that of a three-hyperplane in four-space.

The same question arises for all of these three dynamical systems: *what are the geometric objects which possess these symmetries in a phase space?* The next two sections are devoted to providing the answer to this question.

8.3 Symmetry Coordinates For Keplerian Motions

In this section we will be introducing a pair of canonical diffeomorphisms that produce coordinate systems in which the dynamical symmetries of Keplerian systems become self-evident geometric symmetries.

To this end, because E is a variable, we consider the motions to be in PQET space, letting $\mathrm{t} = \mathrm{q}_4$, and $-\mathrm{E} = \mathrm{p}_4$. The $1/\mathrm{q}$ singularity in H can be eliminated by multiplying the equation $H - E = 0$ through by q. This replaces it by the *regularized* equation

$$W = 1, \tag{8.3.1}$$

with

$$W = q(p^2 - 2E)/2.$$

For all values of the dynamical variables, this equation defines a six-dimensional manifold in a seven-dimensional PQE subspace of the PQET space. Let τ be an evolution parameter that ranges through all the real numbers, let $t = q_4$, $-E = p_4$, and define the PB of functions F and G to be

$$\{F, G\} = \Sigma_{j=1}^{4}(\partial F/\partial q_j \partial G/\partial p_j - \partial F/\partial p_j \partial G/\partial q_j). \tag{8.3.2}$$

Hamilton's equations of motion in PQET space are then special cases of the general equation

$$dF/d\tau = \{F, W\}. \tag{8.3.3}$$

One has

$$d\mathbf{L}/d\tau = \{\mathbf{L}, W\} = 0, \quad d\mathbf{A}/d\tau = \{\mathbf{A}, W\} = 0, \tag{8.3.4}$$

so **L** and **A** remain constants of motion.

We next take advantage of a useful property of Runge–Lenz vectors. Because

$$\mathbf{A}(E) = \mathbf{q}(p^2 + 2E)/2 - (\mathbf{q} \cdot \mathbf{p})\mathbf{p}, \tag{8.3.5a}$$

one has

$$(\mathbf{A}(E) - \mathbf{A}(-E))/2E = \mathbf{q}. \tag{8.3.5b}$$

When E is negative, $E = -|E|$, and

$$\mathbf{q} = |-2E|^{-1}(\mathbf{A}(-|E|) - \mathbf{A}(|E|)) = (-2E)^{-1/2}(\mathbf{A}_<(E) - \mathbf{A}_<(-E)). \tag{8.3.5c}$$

It follows that for each negative value of E, the motion of **r** can be described by the motion of $\mathbf{A}_<(-E)$ with respect to the constant vector $\mathbf{A}_<(E)$. **Figure 8.3.1** illustrates this relation.

When E is positive, one has

$$\mathbf{q} = (1/(2E))(\mathbf{A}(|E|) - \mathbf{A}(-|E|)) = (2E)^{-1/2}(\mathbf{A}_>(E) - \mathbf{A}_>(-E)). \tag{8.3.5d}$$

The motion of **r** can thus be described by the motion of $\mathbf{A}_>(-E)$ with respect to the stationary vector $\mathbf{A}_>(E)$.

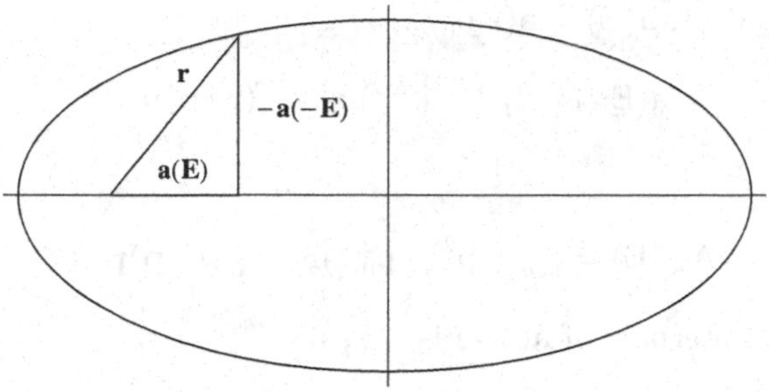

Fig. 8.3.1 Relation of position vector to Runge–Lenz vectors $\mathbf{a}(E), \mathbf{a}(-E)$ when $E = \frac{1}{2}$.

We now describe the first canonical diffeomorphism. Equation (8.3.3) implies that

$$dt/d\tau = \{t, W\} = r, \tag{8.3.6}$$

and so $dt = r\, d\tau$. For states with $E \geq 0$, both r and t range over an open interval, but for bound states, r is cyclic, so mathematically, t is cyclic, and confined within a compact interval. For these reasons we use different coordinates for different ranges of E in the $(-E, t)$ subspace of PQET space. A canonical transformation will be used to define these. For $E < 0$, $E > 0$, and $0_- \leq E \leq 0_+$, define coordinates $p_0' = (p_{0<}', p_{0>}', p_{00}')$ and $q_0' = (q_{0<}', q_{0>}', q_{00}')$ by

$$p_{0<}' \equiv (-2E)^{1/2}\cos(t), \quad q_{0<}' \equiv (-2E)^{1/2}\sin(t), \;\; E < 0; \tag{8.3.7a}$$

$$p_{0>}' \equiv -(2E)^{1/2}\cosh(t), \quad q_{0>}' \equiv (2E)^{1/2}\sinh(t), \;\; E > 0; \tag{8.3.7b}$$

$$p_{00}' \equiv -E, \quad q_{00}' \equiv t, \;\; 0_- \leq E \leq 0_+. \tag{8.3.7c}$$

The transformation (8.3.7a) is due to Poincare[8]. For all real values of E and t, the transformation $(-E, t) \to (p_0', q_0')$ is a one-to-one map from the full space of (E, t) to that of (p_0', q_0') and is C^2 in each interval. It converts $t = q_4$ and $-E = p_4$ in the of PB (8.3.2) to q_0' and p_0'. If one denotes negative values of E by $E_<$, and positive values of E by $E_>$, then when $t \to 0$, one has $q_0' \to 0$ in all three cases, and

$$p_{0<}' \to (-2E_<)^{1/2} \equiv p_{0<}, \quad E_< = -p_{0<}^2/2; \tag{8.3.8a}$$

$$p_{0>}' \to -(2E_>)^{1/2} \equiv p_{0>}, \quad E_> = p_{0>}^2/2; \tag{8.3.8b}$$

and

$$p_0' \to 0_\pm \equiv p_{00}. \tag{8.3.8c}$$

When E is negative, $\mathbf{a}(\mathrm{E}) = \mathbf{a}(-\mathrm{p}_0^2/2)$ is given by

$$\mathbf{a}(\mathrm{E}_<) = (\mathrm{p}^2 - \mathrm{p}_{0<}^2)\mathbf{r}/2 - (\mathbf{r}\cdot\mathbf{p})\mathbf{p}, \tag{8.3.9a}$$

and

$$\mathbf{A}_<(\mathrm{E}) = \mathrm{p}_{0<}^{-1}((\mathrm{p}^2 - \mathrm{p}_{0<}^2)\mathbf{r}/2 - (\mathbf{r}\cdot\mathbf{p})\mathbf{p}). \tag{8.3.9b}$$

The moving counterpart of $\mathbf{a}(\mathrm{E}_<)$ is

$$\mathbf{a}(-\mathrm{E}_<) = (\mathrm{p}^2 + \mathrm{p}_{0<}^2)\mathbf{r}/2 - (\mathbf{r}\cdot\mathbf{p})\mathbf{p}. \tag{8.3.9c}$$

For $\mathrm{E} > 0$, one has the stationary vectors

$$\mathbf{a}(\mathrm{E}_>) = (\mathrm{p}^2 + \mathrm{p}_{0>}^2)\mathbf{r}/2 - (\mathbf{r}\cdot\mathbf{p})\mathbf{p}, \tag{8.3.9d}$$

$$\mathbf{A}_>(\mathrm{E}) = \mathrm{p}_{0>}^{-1}((\mathrm{p}^2 + \mathrm{p}_{0>}^2)\mathbf{r}/2 - (\mathbf{r}\cdot\mathbf{p})\mathbf{p}), \tag{8.3.9e}$$

and the moving vector

$$\mathbf{a}(-\mathrm{E}_>) = (\mathrm{p}^2 - \mathrm{p}_{0>}^2)\mathbf{r}/2 - (\mathbf{r}\cdot\mathbf{p})\mathbf{p}. \tag{8.3.9f}$$

The second canonical transformation in PQET space is based on an inversion in the momentum subspace of $\mathrm{p}_0, \mathrm{p}_1, \mathrm{p}_2, \mathrm{p}_3$. Set

$$\mathbf{p}'^4 = (\mathrm{p}_0', \mathrm{p}_1, \mathrm{p}_2, \mathrm{p}_3), \tag{8.3.10a}$$

and define

$$(\mathbf{p}'^{(4)}\cdot\mathbf{p}'^{(4)})_\pm = \mathrm{p}^2 \pm \mathrm{p}_0'^2. \tag{8.3.10b}$$

Then, letting

$$\mathbf{p}'^{(4)} \to \mathbf{P}_\pm'^{(4)} = 2\mathbf{p}'^{(4)}/(\mathbf{p}'^{(4)}\cdot\mathbf{p}'^{(4)})_\pm, \tag{8.3.10c}$$

and setting

$$\mathbf{q}'^{(4)} = (\mathrm{q}_0', \mathrm{q}_1, \mathrm{q}_2, \mathrm{q}_3), \quad (\mathbf{q}'^{(4)}\cdot\mathbf{q}'^{(4)})_\pm = \mathrm{q}^2 \pm \mathrm{q}_0'^2, \tag{8.3.10d}$$

the canonically conjugate transformation is

$$\mathbf{q}'^{(4)} \to \mathbf{Q}_\pm'^{(4)} = (\mathbf{p}'^{(4)}\cdot\mathbf{p}'^{(4)})_\pm\mathbf{q}'^{(4)}/2 - (\mathbf{q}'^{(4)}\cdot\mathbf{p}'^{(4)})\mathbf{p}'^{(4)}. \tag{8.3.10e}$$

Here,

$$\mathbf{q}'^{(4)}\cdot\mathbf{p}'^{(4)} = \mathrm{q}_0'\mathrm{p}_0' + \mathbf{q}\cdot\mathbf{p}, \tag{8.3.10f}$$

is unaffected by the choice of ± signs. The transformations defined by (8.3.10) are generalizations of a transformation due to Levi-Civita.[9] The inverse transformations are obtained by interchanging the roles of the variables represented by lower case letters with those represented by capital letters. They are diffeomorphisms in every region not containing the origin in PE space. Under them

$$\mathbf{q}'^{(4)} \cdot \mathbf{p}'^{(4)} \to -\mathbf{Q}'^{(4)} \cdot \mathbf{P}'^{(4)}, \tag{8.3.11a}$$

thus orthogonal $\mathbf{q}'^{(4)}$ and $\mathbf{p}'^{(4)}$ produce orthogonal $\mathbf{Q}'^{(4)}$ and $\mathbf{P}'^{(4)}$.

When the positive signs in (8.3.10) are chosen,

$$L_{ij} = q_i p_j - q_j p_i = Q_i^4 P_j^4 - Q_j^4 P_i^4, \quad i < j = 0, 1, 2, 3. \tag{8.3.11b}$$

When the negative signs are chosen,

$$\begin{aligned} L_{ij} &= q_i p_j - q_j p_i \to Q_i^4 P_j^4 - Q_j^4 P_i^4, \quad i < j = 1, 2, 3; \\ K_{0j} &= q_0 p_j + q_j p_0 \to Q_0^4 P_i^4 + Q_i^4 P_0^4, \quad i = 1, 2, 3. \end{aligned} \tag{8.3.11c,d}$$

To use these coordinate systems as reference systems in PQE space, we set $t = q_0 = 0$. The resulting momentum vectors are

$$\mathbf{p}^{(4)} = (p_0, p_1, p_2, p_3) = (p_0, \mathbf{p}), \quad \mathbf{p} = (p_1, p_2, p_3); \tag{8.3.12a}$$

and

$$\mathbf{P}_\pm^{(4)} = 2\mathbf{p}^{(4)}/(\mathbf{p}^{(4)} \cdot \mathbf{p}^{(4)})_\pm = (P_{0\pm},\ \mathbf{P}_\pm); \tag{8.3.12b}$$

with

$$P_{0\pm} = 2p_0/(p^2 \pm p_0^2); \tag{8.3.12c}$$

$$\mathbf{P}_\pm = 2\mathbf{p}/(p^2 \pm p_0^2). \tag{8.3.12d}$$

The corresponding position vectors are

$$\mathbf{q}^{(4)} = (0, q_1, q_2, q_3) = (0, q) = (0, r); \tag{8.3.13a}$$

$$\mathbf{Q}_\pm^{(4)} = (\mathbf{p}^{(4)} \cdot \mathbf{p}^{(4)})_\pm \mathbf{q}^{(4)}/2 - (\mathbf{q}^{(4)} \cdot \mathbf{p}^{(4)})\mathbf{p}^{(4)} = (Q_{0\pm}, \mathbf{Q}_\pm); \tag{8.3.13b}$$

with

$$Q_{0\pm} = \mathbf{q} \cdot \mathbf{p} p_0; \tag{8.3.13c}$$

$$\mathbf{Q}_\pm = (p^2 \pm p_0^2)\mathbf{q}/2 - (\mathbf{q} \cdot \mathbf{p})\mathbf{p} = \mathbf{a}(\pm p_0^2/2). \tag{8.3.13d}$$

To summarize: the combined canonical transformations have converted ordinary position and momentum coordinates in PQET space into a new set of coordinates, diffeomorphic to the first in regions that do not contain the origin. Setting $t = 0$ has turned the original coordinates into a set of coordinates in PQE space: three of the resulting position coordinates are components of Runge–Lenz vectors, and three of the resulting momentum coordinates are proportional to Fock's projective momentum coordinates.[10] The role of the fourth position and momentum coordinates will appear in the following section.

8.4 Geometrical Symmetries of Bound Keplerian Systems in Phase Space

In this section the coordinate systems of the previous section are used to determine the manifolds in phase space upon which bound motions evolve, and the dependence of these motions on the evolution parameter.

We first consider bound motion and begin by renaming the moving Q_j as follows:

$$\rho_{j<} = (p^2 + p_{0<}^2)q_j/2 - (\mathbf{q}\cdot\mathbf{p})p_j, \quad j = 1, 2, 3, \tag{8.4.1a}$$

and

$$\rho_{4<} = (\mathbf{q}\cdot\mathbf{p})p_{0<}. \tag{8.4.1b}$$

One has

$$\mathbf{a}(E_<) = (p^2 - p_{0<}^2)\mathbf{r}/2 - (\mathbf{r}\cdot\mathbf{p})\mathbf{p}, \quad \mathbf{A}_< = \mathbf{a}(E_<)/p_{0<}. \tag{8.4.1c}$$

The four $\rho_{j<}$ are functionally independent in PQE space. The SO(4) group operators $\exp(\alpha_j\{A_{j<},\cdot\})$ and $\exp(\lambda_{ij}\{L_{ij},\cdot\})$ move points at the terminus of $\boldsymbol{\rho}_< = (\rho_{1<}, \rho_{2<}, \rho_{3<}, \rho_{4<})$. The action of $\{A_{i<},\cdot\}$ and $\{L_{ij},\cdot\}$ upon the $\rho_{j<}$ is as follows:

$$\begin{aligned}\{A_{i<},\cdot\}\rho_{j<} = \delta_{ij}\rho_{4<}, \quad \{A_{i<},\cdot\}\rho_{4<} = -\rho_{i<},\\ \{L_{ij},\cdot\}\rho_{j<} = -\rho_{j<}.\end{aligned} \tag{8.4.2}$$

Consequently, the SO(4) generators $\{A_{i<},\cdot\}$ and $\{L_{ij},\cdot\}$ generate rotations of the four-vector $\boldsymbol{\rho}_<$. One finds that

$$\rho_<^2 \equiv \boldsymbol{\rho}_<\cdot\boldsymbol{\rho}_< = \Sigma_1^4\rho_{j<}^2 = (r/2)(p^2 + p_{0<}^2)^2 = W(E_<)^2. \tag{8.4.3}$$

As W has unit value, the rotations move points on the surface defined by

$$\rho_< = 1. \tag{8.4.4}$$

Thus, in the coordinate system of the ρ_j, Eqs. (8.4.1a,b) define a unit hypersphere.

We next turn our attention to the $P_j(E)$ with $E = p_{0<}^2$. Define

$$\pi_{i<} = P_i(p_{0<}^2) = 2p_i/(p^2 + p_{0<}^2), \quad i = 1, 2, 3. \tag{8.4.5}$$

The action of the $\{A_{i<}, \cdot\}$ on these $\pi_{j<}$ produces a further momentum coordinate:

$$\{A_{i<}, \cdot\}\pi_{j<} = \delta_{ij}\pi_{4<}. \tag{8.4.6a}$$

$$\pi_{4<} \equiv (1/p_{0<})(p^2 - p_{0<}^2)/(p^2 + p_{0<}^2). \tag{8.4.6b}$$

Also

$$\{A_{i<}, \cdot\}\pi_{4<} = -\pi_{i<}, \quad \{L_{ij}, \cdot\}\rho_{j<} = -\rho_{j<}, \; i = 1, 2, 3. \tag{8.4.6c}$$

Equations (8.4.6) imply that the $\{A_{i<}, \cdot\}$ and $\{L_{ij}, \cdot\}$ are generators of rotations of the locally cartesian four-vectors $\boldsymbol{\pi}_< \equiv (\pi_{1<}, \pi_{2<}, \pi_{3<}, \pi_{4<})$, as well as rotations of the vectors $\boldsymbol{\rho}_< \equiv (\rho_{1<}, \rho_{2<}, \rho_{3<}, \rho_{4<})$. Equations (8.4.2) and (8.4.6) together imply that the generators have the realization

$$\{A_{i<}, \cdot\} = -(\rho_{i<}\partial/\partial\rho_{4<} - \rho_{4<}\partial/\partial\rho_{i<} + \pi_{i<}\partial/\partial\pi_{4<} - \pi_{4<}\partial/\partial\pi_{i<}),$$
$$i = 1, 2, 3; \tag{8.4.7a}$$

$$\{L_{ij}, \cdot\} = -(\rho_{i<}\partial/\partial\rho_{j<} - \rho_{j<}\partial/\partial\rho_{i<} + \pi_{i<}\partial/\partial\pi_{j<} - \pi_{j<}\partial/\partial\pi_{i<}),$$
$$i < j = 1, 2, 3. \tag{8.4.7b}$$

The variable $\pi_{4<}$ is related to the coordinate

$$P_{0+} = 2p_{0<}/(p^2 + p_{0<}^2), \tag{8.4.8a}$$

by

$$\pi_{4<} = P_{0+} - 1/p_{0<}. \tag{8.4.8b}$$

The four functions $\pi_{i<}$ are Fock's projective momentum coordinates divided by $p_{<0}$. They are linearly independent in the PE subspace of PQE space, but satisfy the identity

$$E_< = (-1/2)/\left(\pi_{1<}^2 + \pi_{2<}^2 + \pi_{3<}^2 + \pi_{4<}^2\right). \tag{8.4.9a}$$

Consequently

$$(p_{0<}\boldsymbol{\pi}_<) \cdot (p_{0<}\boldsymbol{\pi}_<) = 1. \tag{8.4.9b}$$

Direct calculation shows that vectors $\boldsymbol{\rho}_<$ and $\boldsymbol{\pi}_<$ are orthogonal:

$$\Sigma_4\rho_{j<}\pi_{j<} \equiv \boldsymbol{\rho}_< \cdot \boldsymbol{\pi}_< = 0. \tag{8.4.10}$$

Consequently, for each value of E, bound state Keplerian motions in ordinary six-dimensional phase space evolve on an SO(4)-invariant manifold defined by three equations

$$\boldsymbol{\rho}_< \cdot \boldsymbol{\rho}_< = 1, \quad (\mathrm{p}_{0<}\boldsymbol{\pi}_<) \cdot (\mathrm{p}_{0<}\boldsymbol{\pi}_<) = 1, \;\; \boldsymbol{\rho}_< \cdot (\mathrm{p}_{0<}\boldsymbol{\pi}_<) = 0. \tag{8.4.11}$$

For any fixed value of the energy, the vector $\mathbf{S}_< \equiv \boldsymbol{\rho}_< + \mathrm{p}_{0<}\boldsymbol{\pi}_<$ *terminates on an* $SO(4)$*-invariant manifold in six-dimensional PQ space. It has length* $2^{1/2}$*, and the vectors* $\boldsymbol{\rho}_<$ *and* $\mathrm{p}_{0<}\boldsymbol{\pi}_<$ *may be considered to be projections of* $\mathbf{S}_<$ *onto locally orthogonal subspaces.*

The geometric interpretation of these statements is most transparent for one-dimensional regularized Keplerian motion, which exhibits SO(2) symmetry. The generator of the SO(2) group is

$$\begin{aligned}\{\mathrm{A}_<, \cdot\} &= \mathrm{q}_1\mathrm{p}_1\partial/\partial\mathrm{q}_1 - ((\mathrm{p}_{0<}^2 + \mathrm{p}_1^2)/2)\partial/\partial\mathrm{p}_1 \\ &\to -(\rho_{1<}\partial/\partial\rho_{4<} - \rho_{4<}\partial/\partial\rho_{1<} + \pi_{1<}\partial/\partial\pi_{4<} - \pi_{4<}\partial/\partial\pi_{1<}),\end{aligned} \tag{8.4.12}$$

and the unit four-vectors of the three-dimensional system are replaced by unit two-vectors. The vector

$$\mathrm{p}_{0<}\boldsymbol{\pi}_< = (2\mathrm{p}_{0<}\mathrm{p}_1/(\mathrm{p}_{0<}^2 + \mathrm{p}_1^2), (\mathrm{p}_1^2 - \mathrm{p}_{0<}^2)/(\mathrm{p}_{0<}^2 + \mathrm{p}_1^2)) \tag{8.4.13a}$$

$$\equiv (\sin(\theta), \cos(\theta)), \tag{8.4.13b}$$

expresses the stereographic projection of p_1 onto a unit circle. In **Fig. 8.4.1a**, p_1 is depicted lying on the y axis, and as p_1 changes, $\mathrm{p}_{0<}\boldsymbol{\pi}_<$ rotates about the center of the unit circle in the y–z plane. In the figure p_1 is denoted by p, and the point at the terminus of $\mathrm{p}_{0<}\boldsymbol{\pi}_<$ is denoted by $\mathrm{p}_0\pi$. When the group operator $\exp(\alpha\{\mathrm{A}, \cdot\})$ rotates $\mathrm{p}_{0<}\boldsymbol{\pi}_<$, the angle θ is converted to $\theta + \alpha$. The vector

$$\boldsymbol{\rho}_< = ((\mathrm{p}_1^2 + \mathrm{p}_{0<}^2)(\mathrm{q}_1/2) - \mathrm{q}_1\mathrm{p}_1^2, \mathrm{q}_1\mathrm{p}_1\mathrm{p}_{0<}) \tag{8.4.14a}$$

$$\boldsymbol{\rho}_< = ((\mathrm{p}_{0<}^2 - \mathrm{p}_1^2)(\mathrm{q}_1/2), \mathrm{q}_1\mathrm{p}_1\mathrm{p}_{0<}), \tag{8.4.14b}$$

$$\boldsymbol{\rho}_< \equiv (\sin(\phi), \cos(\phi)), \tag{8.4.14c}$$

is a function of p_1 as well as q_1. When $\mathrm{p}_1 \to 0$,

$$\boldsymbol{\rho}_< \to \mathrm{q}_1(\mathrm{p}_{0<}^2/2, 0) = \mathrm{q}_1(-\mathrm{E}_1, 0), \tag{8.4.14d}$$

and when $\mathrm{p}_1 \to \pm\,\mathrm{p}_{0<}$,

$$\boldsymbol{\rho}_< \to \mathrm{q}_1(0, -\mathrm{E}). \tag{8.4.14e}$$

The variables q_1 and p_1 are related by the equation

$$\boldsymbol{\rho}_< \cdot \boldsymbol{\rho}_< = (\mathrm{q}_1/2)^2(\mathrm{p}_1^2 + \mathrm{p}_{0<}^2)^2 = \mathrm{W}^2 = 1. \tag{8.4.15a}$$

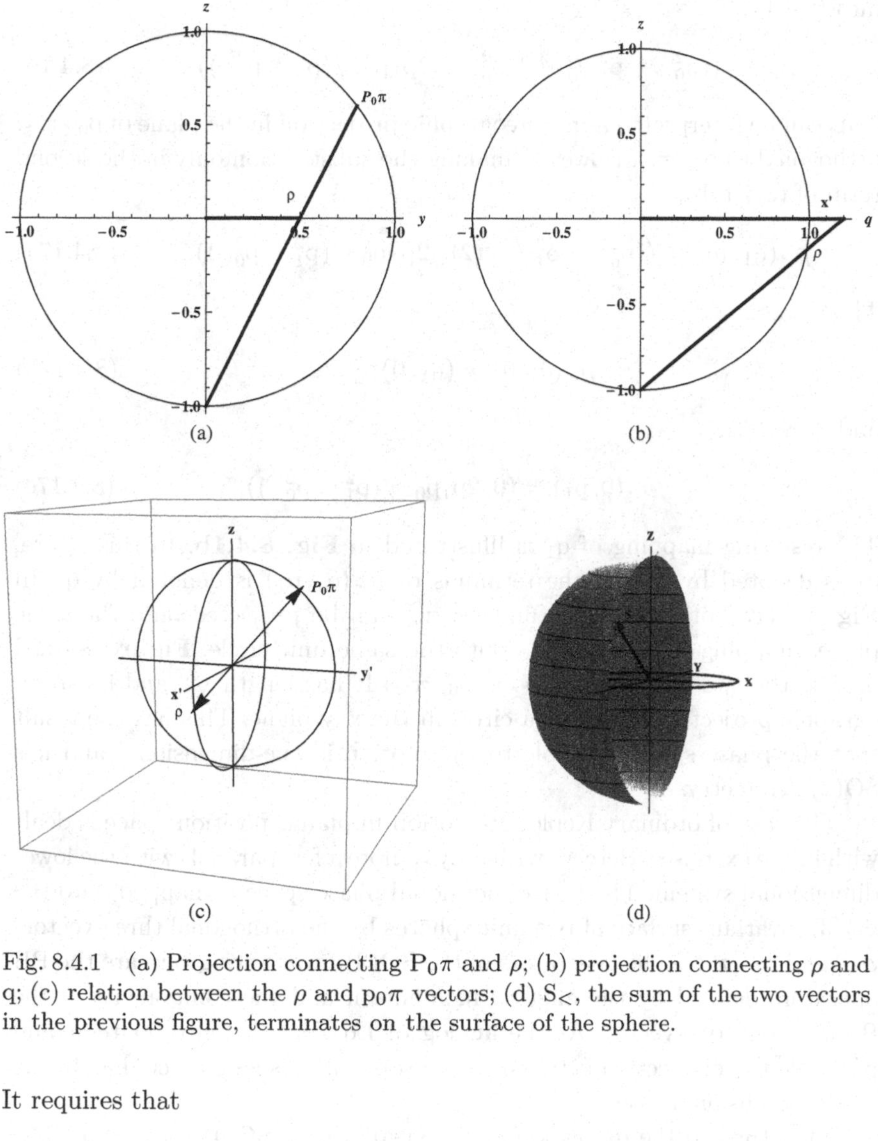

Fig. 8.4.1 (a) Projection connecting $P_0\pi$ and ρ; (b) projection connecting ρ and q; (c) relation between the ρ and $p_0\pi$ vectors; (d) $S_<$, the sum of the two vectors in the previous figure, terminates on the surface of the sphere.

It requires that

$$q_1 = \pm 2/(p_1^2 + p_{0<}^2). \tag{8.4.15b}$$

The potential function, $1/|q_1|$, becomes infinite at the origin, so for the kinetic energy to remain finite, q_1 cannot go to zero, and the sign on the right-hand side of (8.4.15b) cannot change during the motion. We will assume that the positive sign remains. Then, on substituting (8.4.15b) into (8.4.14b),

$$\boldsymbol{\rho}_< \rightarrow ((p_{0<}^2 - p_1^2)/(p_1^2 + p_{0<}^2),\ 2p_1p_{0<}/(p_1^2 + p_{0<}^2)). \tag{8.4.16}$$

This can be interpreted as a stereographic projection in the plane of $p_{0<}\boldsymbol{\pi}_<$, orthogonal to $p_{0<}\boldsymbol{\pi}_<$. However, making the substitution only in the second term of (8.4.14b),

$$\boldsymbol{\rho}_<(q_1, p_1) = ((p_{0<}^2 - p_1^2)(q_1/2),\ 2p_1p_{0<}/(p_1^2 + p_{0<}^2)). \tag{8.4.17a}$$

Then

$$\boldsymbol{\rho}_<(q_1, 0) = (q_1, 0)p_{0<}^2/2, \tag{8.4.17b}$$

and

$$\boldsymbol{\rho}_<(0, p_1) = (0, 2p_1p_{0<}/(p_1^2 + p_{0<}^2)). \tag{8.4.17c}$$

The resulting mapping of q_1 is illustrated in **Fig. 8.4.1b**. In this figure, q_1 is denoted by q, and the terminus of $\boldsymbol{\rho}_<(q_1, p_1)$ is denoted by q′. In **Fig. 8.4.1c** both $\boldsymbol{\rho}_<(q_1, p_1)$ and $p_{0<}\boldsymbol{\pi}_<$ can be projected onto the same plane, mapping $\boldsymbol{\rho}_<$ and $p_{0<}\boldsymbol{\pi}_<$ onto the same unit circle. **Figure 8.4.1d** depicts the two-vector $\mathbf{S}_< = \boldsymbol{\rho}_< + p_{0<}\boldsymbol{\pi}_<$. It has length $\sqrt{2}$ and its stereographic projection describes a circle in the x–y plane. This has the result that the phase space available to the motion is one-dimensional and has SO(2) symmetry.

The case of ordinary Keplerian motion in planar position-space is dealt with in the exercises. Here we wish only to note a few parallels with the lower dimensional system. The four-dimensional phase space is mapped onto the SO(3) invariant surface of two unit spheres by the orthogonal three-vectors $\boldsymbol{\rho}_<$ and $p_{0<}\boldsymbol{\pi}_<$. The generators of the SO(3) invariance group are the PB operators defined by the angular momentum and the components of the Runge–Lenz two-vector. As the analog of Eq. (8.4.15) only relates r and p^2, while the connection between $\boldsymbol{\rho}_<$ and $p_{0<}\boldsymbol{\pi}_<$ is less direct than in the lower dimensional case.

We return to the discussion of the system with SO(4) invariance. The evolution of $\boldsymbol{\rho}_<$ is determined by the equation

$$d\boldsymbol{\rho}_</d\tau = \{\boldsymbol{\rho}_<, \rho_<\} = p_{0<}^2 W\boldsymbol{\pi}_< = p_{0<}^2\boldsymbol{\pi}_<, \tag{8.4.18a}$$

and that of the $\boldsymbol{\pi}_<$ is determined by

$$d\boldsymbol{\pi}_</d\tau = \{\boldsymbol{\pi}_<, \rho_<\} = -\boldsymbol{\rho}_</W = -\boldsymbol{\rho}_<. \tag{8.4.18b}$$

Consequently,

$$d^2\boldsymbol{\rho}_</d\tau^2 = -p_{0<}^2\boldsymbol{\rho}_<, \tag{8.4.18c}$$

$$d^2\boldsymbol{\pi}_</d\tau^2 = -p_{0<}^2\boldsymbol{\pi}_<. \tag{8.4.18d}$$

If the initial values of $\rho_{i<}$ and $\pi_{i<}$ are $\rho_{i<}(0)$ and $\pi_{i<}(0)$ respectively, then integration of each component of these vector equation yields

$$\begin{aligned}\rho_{i<}(\tau) &= \rho_{i<}(0)\cos(p_{0<}\tau) + p_{0<}\pi_{i<}(0)\sin(p_{0<}\tau),\\ \pi_{i<}(\tau) &= \pi_{i<}(0)\cos(p_{0<}\tau) - (\rho_{i<}(0)/p_{0<})\sin(p_{0<}\tau).\end{aligned} \tag{8.4.19}$$

These describe uniform circular motions of the vectors with period $2\pi/p_{0<}$. The relation between τ and t is considered in the exercises.

8.5 Symmetries of Keplerian Systems with Non-negative Energies

We conclude this section with a brief discussion of Keplerian systems with fixed non-negative energies. When the energy is positive, $E = E_> = p_{0>}^2/2$. The constant reference three-vector, $\mathbf{A}_> \equiv \mathbf{a}_>/|p_{0>}|$ has components

$$A_{j>} = ((\mathbf{r}\cdot\mathbf{p})p_j - (p^2 + p_{0>}^2)q_j/2)/|p_{0>}|, \tag{8.5.1}$$

and the moving four-vector $\boldsymbol{\rho}_>$ has components

$$\rho_{j>} = (p^2 - p_{0>}^2)q_j/2 - (\mathbf{r}\cdot\mathbf{p})p_j,\ \ i = 1,2,3; \tag{8.5.2a}$$

$$\rho_{4>} \equiv (\mathbf{r}\cdot\mathbf{p})p_{0>}. \tag{8.5.2b}$$

The corresponding components of the vector $\boldsymbol{\pi}_>$ are

$$\pi_{j>} = 2p_j/(p^2 - p_{0>}^2),\ \ \pi_{4>} = (1/p_{0>})(p^2 + p_{0>}^2)/(p_{0<}^2 - p^2). \tag{8.5.3}$$

One finds

$$\rho_{1>}^2 + \rho_{2>}^2 + \rho_{3>}^2 - \rho_{4>}^2 = W^2, \tag{8.5.4a}$$

$$E_> = \frac{(1/2)}{(\pi_{4>}^2 - (\pi_{1>}^2 + \pi_{2>}^2 + \pi_{3>}^2))}, \tag{8.5.4b}$$

$$\Sigma_1^3\rho_{j>}\pi_{j>} + \rho_{4>}\pi_{4>} = 0. \tag{8.5.4c}$$

The action of the PB operators of $A_{i>}$ and L_{ij} on the coordinates $\rho_{j>}$ and π_j is expressed by

$$\begin{aligned}\{A_{i<},\cdot\} &= -(\rho_{i<}\,\partial/\partial\rho_{4<} + \rho_{4<}\,\partial/\partial\rho_{i<} + \pi_{i<}\,\partial/\partial\pi_{4<} + \pi_{4<}\,\partial/\partial\pi_{i<}),\\ &\quad i = 1,2,3,\\ \{L_{ij},\cdot\} &= -(\rho_{i<}\,\partial/\partial\rho_{j<} - \rho_{j<}\,\partial/\partial\rho_{i<} + \pi_{i<}\,\partial/\partial\pi_{j<} - \pi_{j<}\,\partial/\partial\pi_{i<}),\\ &\quad i < j = 1,2,3,\end{aligned} \tag{8.5.5}$$

and the evolution of $\boldsymbol{\rho}_>$ and $\boldsymbol{\pi}_>$ is governed by the equations

$$d\boldsymbol{\rho}_>/d\tau = \{\boldsymbol{\rho}_>, \rho_>\} = -2E_>\boldsymbol{\pi}_>, \tag{8.5.6a}$$

$$d\boldsymbol{\pi}_>/d\tau = \{\boldsymbol{\pi}_>, \rho_>\} = -\boldsymbol{\rho}_>, \tag{8.5.6b}$$

and

$$d^2\boldsymbol{\rho}_>/d\tau^2 = p_{0>}^2\boldsymbol{\rho}_>, \tag{8.5.6c}$$

$$d^2\boldsymbol{\pi}_>/dt^2 = p_{0>}^2\boldsymbol{\pi}_>. \tag{8.5.6d}$$

Equations (8.5.4) determine SO(3,1)-invariant manifolds in phase space upon which the the termini of the vectors $\boldsymbol{\rho}_>$ and $\boldsymbol{\pi}_>$ evolve.

When $E = 0$, we consider the vector $\boldsymbol{\pi}_0$ to move with respect to the fixed Runge–Lenz vector. For the latter, one has

$$a_{j0} = (\mathbf{r}\cdot\mathbf{p})p_j - p^2q_j/2, \ \ j = 1, 2, 3; \tag{8.5.7a}$$

$$a_{40} = (\mathbf{r}\cdot\mathbf{p})p_0. \tag{8.5.7b}$$

Consequently,

$$\mathbf{r}_{j0} = -a_{j0}, \tag{8.5.7c}$$

$$r_{40} = (\mathbf{r}\cdot\mathbf{p})p_{00}, \tag{8.5.7d}$$

with $0_- \le p_{00} \le 0_+$. The components of the vector $\boldsymbol{\pi}_0$ are

$$\pi_{j0} = 2p_j/p^2, \quad j = 1, 2, 3, \tag{8.5.8a}$$

while

$$p_0\pi_{40} = 1. \tag{8.5.8b}$$

It follows that

$$\Sigma_1^4\ \rho_{j0}^2 = (rp^2/2)^2 = W^2, \tag{8.5.9a}$$

$$\Sigma_1^4\ \pi_{j0}^2 = 4/p^2, \tag{8.5.9b}$$

$$\Sigma_1^4\ \rho_{j0}\pi_{j0} = 0. \tag{8.5.9c}$$

One finds

$$\{a_{j0}, \cdot\} = \partial/\partial\pi_{j0}, \quad i = 1, 2, 3; \tag{8.5.9d}$$

$$\{L_{ij}, \cdot\} = -(\rho_{i0}\ \partial/\partial\rho_{j0} - \rho_{j0}\ \partial/\partial\rho_{i0} + \pi_{i0}\ \partial/\partial\pi_{j0} - \pi_{j0}\ \partial/\partial\pi_{i0}), \quad i < j = 1, 2, 3. \tag{8.5.9e}$$

Because $\mathbf{a}(0) = \mathbf{a}(\mathrm{E}) + \mathbf{a}(-\mathrm{E})$, a Lie algebra that contains $\mathbf{a}(\mathrm{E})$ and $\mathbf{a}(-\mathrm{E})$ also contains $\mathbf{a}(0)$. The same can be said of the corresponding PB operators. The evolution of the vectors $\boldsymbol{\rho}_0$ and $\boldsymbol{\pi}_0$ is determined by the equations

$$\mathrm{d}\boldsymbol{\rho}_0/\mathrm{d}\tau = \{\boldsymbol{\rho}_0, \rho_0\} = 0, \tag{8.5.10a}$$

$$\mathrm{d}\boldsymbol{\pi}_0/\mathrm{d}\tau = \{\boldsymbol{\pi}_0, \rho_0\} = 0. \tag{8.5.10b}$$

Consequently,

$$\mathrm{d}^2\boldsymbol{\rho}_0/\mathrm{d}\tau^2 = 0, \tag{8.5.10c}$$

$$\mathrm{d}^2\boldsymbol{\pi}_0,/\mathrm{d}\tau^2 = 0. \tag{8.5.10d}$$

To summarize, as E varies, all $\mathbf{a}(\mathrm{E})$, $\mathbf{r}(\mathrm{E}) = \mathbf{a}(\mathrm{E}) - \mathbf{a}(-\mathrm{E})$, and $\mathbf{p}(\mathrm{E})$, adjust themselves so that the relation $\mathrm{r}(\mathrm{H} - \mathrm{E}) = 0$, i.e., $\mathrm{r}(\mathrm{p}^2/2 - \mathrm{E}) = 1$, determines manifolds in PQET phase space defined by

$$(\rho_1^2 + \rho_2^2 + \rho_3^2 + \mathrm{g}^{44}\rho_4^2)^{1/2} = \mathrm{K}, \tag{8.5.11a}$$

$$\mathrm{E}_> = (-1/2)/(\pi_1^2 + \pi_2^2 + \pi_3^2 + \mathrm{g}_{44}\,\pi_4^2), \tag{8.5.11b}$$

$$(\rho_1\pi_1 + \rho_2\pi_2 + \rho_3\pi_3 + \mathrm{g}_4^4\,\rho_4\pi_4) = 0, \tag{8.5.11c}$$

which for the case when $\mathrm{E} < 0$,

$$\mathrm{g}^{44} = \mathrm{g}_{44} = \mathrm{g}_4^4 = 1, \tag{8.5.11d}$$

and when $\mathrm{E} = 0$,

$$\mathrm{g}^{44} = \mathrm{g}_{44} = \mathrm{g}_4^4 = 0, \tag{8.5.11e}$$

and when $\mathrm{E} > 0$,

$$\mathrm{g}^{44} = \mathrm{g}_{44} = -1, \mathrm{g}_4^4 = 1. \tag{8.5.11f}$$

For each range of E, the points that represent the system evolve in phase space on a three-dimensional manifold determined by these equations. These manifolds have everywhere-local geometric symmetry determined by the Lie generators of the corresponding degeneracy group, all of which are first-order differential operators that are of degree one or zero in the variables ρ_j and π_j.

8.6 The SO(4,1) Dynamical Symmetry

In this section, the fourth component of the vectors $\boldsymbol{\rho}$ is used to provide the group generator

$$\mathrm{D} = \{(\mathbf{q}\cdot\mathbf{p}),\cdot\} = \mathbf{p}\cdot\boldsymbol{\nabla}_{\mathrm{p}} - \mathbf{q}\cdot\boldsymbol{\nabla}_{\mathrm{q}}. \tag{8.6.1a}$$

The operator,

$$\mathrm{S}(\alpha) = \exp(\alpha\mathrm{D}), \tag{8.6.1b}$$

of the corresponding one-parameter group carries out the dilatations

$$\mathrm{S}(\alpha)\mathbf{p} = \exp(\alpha)\mathbf{p}, \quad \mathrm{S}(\alpha)\mathbf{q} = \exp(-\alpha)\mathbf{q}. \tag{8.6.1c,d}$$

This property enables one to, indirectly change energies because

$$\begin{aligned}\mathrm{S}(\alpha)\mathrm{W}(\mathrm{E}) &= \mathrm{S}(\alpha)\mathrm{q}(\mathrm{p}^2 - 2\mathrm{E})/2\\ &= (\exp(\alpha)\mathrm{qp}^2 - \exp(-\alpha)\mathrm{q}\ 2\mathrm{E})/2\\ &= \exp(\alpha)(\mathrm{qp}^2 - 2\mathrm{E}'\mathrm{q})/2 = \exp(\alpha)\mathrm{W}(\mathrm{E}'),\end{aligned} \tag{8.6.2}$$

with

$$\mathrm{E}' = \exp(-2\alpha)\mathrm{E}.$$

In the two preceding sections dynamical symmetries were considered to be geometrical symmetries in PQ space, and the energy, E, was treated as a parameter. In this section E will be treated as a dynamical variable, and dynamical symmetries will be considered to be geometrical symmetries in PQE space.

We begin by using the finite dilatations to convert the vectors $\mathbf{a}(\mathrm{E})$ with $\mathrm{p}_0 = 1$ and $\mathrm{E} = \pm 1/2$, to the general vectors $\mathbf{A}(\mathrm{E})$. If $\exp(\alpha) = \mathrm{p}_0$, then

$$\begin{aligned}\mathrm{S}(-\alpha)\mathbf{a}(-1/2) &= \mathrm{S}(-\alpha)((\mathbf{q}\,\mathrm{p}^2/2 - \mathbf{q}\cdot\mathbf{p}\,\mathbf{p}) - \mathbf{q}/2)\\ &= \mathrm{p}_0^{-1}(\mathbf{q}\,\mathrm{p}^2/2 - \mathbf{q}\cdot\mathbf{pp}) - \mathrm{p}_0\mathbf{q}/2\\ &= \mathrm{p}_0^{-1}(\mathrm{q}(\mathbf{q}\,\mathrm{p}^2/2 - \mathbf{q}\cdot\mathbf{p}\ \mathbf{p}) - \mathrm{p}_0^2\mathbf{q}/2) = \mathbf{A}_<(-\mathrm{p}_0^2/2).\end{aligned} \tag{8.6.3a}$$

Similarly

$$\mathrm{S}(-\alpha)\mathbf{a}(1/2) = \mathbf{A}_>(\mathrm{p}_0^2/2). \tag{8.6.3b}$$

(Because $\mathbf{a}(0) = (\mathbf{a}(-1/2)+\mathbf{a}(1/2))/2$, it is not necessary to treat the $\mathrm{E} = 0$ case separately.) The general effect of $\mathrm{S}(\alpha)$ on $\mathbf{a}(-1/2)$ and $\mathbf{a}(1/2)$ is

$$\mathrm{S}(\alpha)\{\mathbf{a}(1/2),\cdot\} = \cosh(\alpha)\{\mathbf{a}(1/2),\cdot\} + \sinh(\alpha)\{\mathbf{a}(-1/2),\cdot\}, \tag{8.6.4a}$$

and

$$S(\alpha)\{\mathbf{a}(-1/2),\cdot\} = \cosh(\alpha)\{\mathbf{a}(-1/2),\cdot\} + \sinh(\alpha)\{\mathbf{a}(1/2),\cdot\}. \quad (8.6.4b)$$

These relations enable one to use D and the components of $\mathbf{a}(-1/2)$ and $\mathbf{a}(1/2)$ to provide PB operators that constitute a basis for the Lie algebras and groups that act in the $E < 0$, $0_- \leq E \leq 0_+$, and $E > 0$, subspaces of PQET space.[11] One may set

$$D = K_{45} \equiv K_{54}, \quad (8.6.5a)$$

and for $i < j$, $j = 1, 2, 3$, define the operators

$$J_{ij} = \{L_{ij},\cdot\} = -J_{ji}, \quad (8.6.5b)$$

$$J_{j4} = \{a_j(-1/2),\cdot\} \equiv -J_{4j}, \quad (8.6.5c)$$

$$K_{j5} = \{a_j(1/2),\cdot\} \equiv K_{5j}. \quad (8.6.5d)$$

These J_{ij}, J_{j4} satisfy the commutation relations of the SO(4) bound-state degeneracy group, and the J_{ij}, K_{j5} satisfy the commutation relations of the SO(3,1), $E > 0$, degeneracy group. For $E = 0$, the generators of the E(3) degeneracy group may be taken to be J_{ij} and $(J_{j4} + K_{j5})/2$. Together, the J_{ij}, J_{j4}, K_{j5}, and K_{45} generate an SO(4,1) invariance group of W(E) and the equations of motion.

8.7 Concluding Remarks

The following general features of the remarkable symmetries of Kepler systems are worth emphasizing:

1. All the dynamical symmetries of Keplerian systems can be realized as geometrical symmetries in PQET phase space or its subspaces. Kepler's laws, and many further unifying relations, may be considered to be consequences of these symmetries. The bound motions in unbounded PQE space map into motions on the compact surfaces of hyperspheres. Operations of the SO(4) degeneracy group that interconvert Keplerian trajectories and hodographs carry out rotations on these hyperspheres. The time-evolution of points on the trajectories and hodographs also become rotations on the hyperspheres.
2. The dynamical groups of all systems governed by Hamilton's equations of motion may be interpreted as geometrical symmetry groups in the appropriate phase spaces.
3. There are systematic ways to determine the generators of the dynamical groups of these systems.

4. However, currently, there is no known general method for finding a canonical transformation that will convert these generators, which may be of the generalized Lie form, to first-order differential operators.
5. The SO(4, 1) dynamical symmetry of Keplerian systems can be enlarged to an SO(4, 2) symmetry. Because it is particularly relevant to atomic and molecular physics, this symmetry will be considered in Chapter 11.
6. As is evident from the citations, the treatment of Keplerian systems given here is based on a long history of discoveries in classical mechanics, followed by the discoveries of Fock[10] and of Bargmann,[12] that exposed the *hidden symmetries* of hydrogen-like atoms in the 1930s. In the early 1960s, interest in the role of symmetries in elementary particle theory led to a great extension of the work of Fock and Bargmann, and to Barut's introduction of the concept of *dynamical symmetry.*[13] In addition to works already cited we would call attention to a review,[14] and to an article by Macintosh,[15] published during this period. The transformations that convert ordinary position and momentum variables to Runge–Lenz vectors and Fock variables are nonrelativistic versions of transformations discovered in 2008.[16]

Exercises:

1. Is $\mathbf{q} \cdot \mathbf{p}$ invariant under the rotations generated by $q_j \partial/\partial p_j - q_j \partial/\partial p_j$, or transformations generated by $q_j \partial/\partial p_j + q_j \partial/\partial p_j$?
2. Let the elliptical orbit of a planet lying in the q_1–q_2 plane have its Hamilton–Runge–Lenz vector $\mathbf{b}$ on the q_1 axis. Determine the effect on $\mathbf{q}$ of infinitesimal transformations with the PB generators determined by L_{ij}, A_i. Average the effect over a circular orbit. What do these infinitesimal transformations do to $\mathbf{p}$? Average their effect over the corresponding hodograph.
3. Let $s = p_0 \tau$, and use the equations of motion (8.3.3) to determine the relation between physical time, t, and s. Set both t and τ equal to zero when a planet or comet is closest to the sun. For the planetary case verify that your periodic time agrees with the exact form of Kepler's Law.
4. Determine the stereographic projection of Keplerian hodographs onto a unit sphere (cf. *Am. J. Phys.* **33** (1965) 570). Generalize your results by considering rotations of the plane of the hodograph, and compare with the work of Fock in Ref. 10.

5. Letting E = 0, carry through the developments analogous to those of sections three through six.
6. Verify that the PB relations of the components of the four-vectors Q^4, P^4 introduced in Sec. 8.3 are those of canonically conjugate variables.

References

[1] B. J. Cantwell, *Introduction to Symmetry Analysis* (Cambridge University Press, Cambridge, 2002).
[2] E. A. Milne, *Vectorial Mechanics* (Methuen, London, Interscience, N.Y., 1948), pp. 236–7.
[3] H. Goldstein, *Am. J. Phys.* **43** (1975) 737.
[4] E. B. Wilson, *Vector Analysis Founded upon the Lectures of J. Willard Gibbs* (Scribners, N.Y., 1901).
[5] W. Pauli, *Z. Phys.* **36** 336.
[6] E. A. Milne, *ibid.* p. 238.
[7] A. L. Fetter, J. D. Walecka, *Theoretical Mechanics of Particles and Continua* (McGraw-Hill, N.Y, 1980); H. Goldstein, *Classical Mechanics* (Addison-Wesley, Reading, Mass, 1980).
[8] H. Poincare, *New Methods of Celestial Mechanics I*, NASA Technical Translation TTF- 450 (Washington, D.C., 1967) pp. 14, 24.
[9] C. Caratheodory, *Calculus of Variations and Partial Differential Equations of the First Order*, translated by R. B. Dean, J. J. Brandstatter (Holden-Day, San Francisco, 1965).
[10] V. Fock, *Z. Phys.* **98** (1935) 145.
[11] The Lie generators in equations (8.6.5) are classical mechanical versions of operators introduced by M. Bednar in *Ann. Phys. (N.Y.)* **75** (1973) 305; see also A. O. Barut and H. Kleinert, *Phys. Rev.* **156** (1967) 1541, and **157** (1967) 1180.
[12] V. Bargmann, *Z. Physik* **99** (1936) 576.
[13] A. O. Barut, *Dynamical Groups and Generalized Symmetries in Quantum Theory* (University of Canterbury Press, Christchurch, N. Z. 1972).
[14] M. Bander, C. Itzykson, *Rev. Mod. Phys.* **38** (1966) 330, 346.
[15] H. V. Macintosh, *Symmetry and Degeneracy*, Group Theory and its Applications, Vol. 2, ed. E. M. Loebl (Academic Press, N.Y., 1971).
[16] C. E. Wulfman, *J. Phys. A* **42** (2009) 185301.

CHAPTER 9

Dynamical Symmetry in Schrödinger Quantum Mechanics

In this chapter we set out the fundamental principles needed to uncover the intrinsic symmetries of systems obeying Schrödinger equations, and to interpret these as geometric symmetries. The chapter begins with a treatment of Dirac's correspondence principle connecting classical and quantal systems. A Lie algebraic extension of the principle is then established. It is used to connect intrinsic local symmetries of Schrödinger equations to local geometric symmetries of the corresponding classical systems in phase space. These connections will be shown to exist, despite the fact that Heisenberg's uncertainty principle establishes that the positions and momenta of a particle cannot, even in principle, both be measured with arbitrary accuracy.

Once the connection between quantal and classical symmetries has been established, the method used to determine invariance transformations of Schrödinger equations is presented. Finally, dynamical symmetry algebras of two simple quantal systems are determined, and their physical interpretations and applications are noted.

9.1 Superposition Invariance

Schrödinger's equations and all the usual equations governing quantum mechanical systems are linear partial differential equations. If a and b are constants, and ψ and ψ' are solutions of a linear equation, then $a\psi + b\psi'$ are also solutions. As a consequence, every such equation is invariant under the one-parameter groups with generators $\psi\partial/\partial\psi$, $\psi'\partial/\partial\psi$, $\psi\partial/\partial\psi'$, $\psi'\partial/\partial\psi'$. If $\psi = \Sigma c_k\psi_k$ is a general solution of the Schrödinger equation of interest, one may choose to transfer the variations in Ψ to variations of the c_k carried out by one-parameter groups with generators $c_k\partial/\partial c_k$, $c_j\partial/\partial c_k \pm c_k\partial/\partial c_j$. When there are N linearly independent real ψ's, the resulting group of superposition transformations is GL(N,R). It the ψ's are complex variables, the corresponding group is GL(N,C). Such superposition groups, in and of themselves, carry no information that is dependent upon the Hamiltonian operator that defines a particular quantum mechanical system. Hamiltonian-dependent information is carried by groups whose generators are functions of position and momentum operators. It is these later groups which can correspond to the dynamical groups of classical Hamiltonian systems.

9.2 The Correspondence Principle

In their first paper on the new quantum mechanics, Born and Jordan showed that Heisenberg's recently published paper on the subject could be interpreted in terms of the operations of matrix algebra, with physical observables being represented by *Hermitian* matrices.[1,2] In the general case, if $\mathbf{M}$ is a matrix with elements M_{ab}, the corresponding elements of $\mathbf{M}^T$, the transpose of $\mathbf{M}$, are M_{ba}. The matrix whose elements are the complex conjugate of those of $\mathbf{M}$ is denoted by $\mathbf{M^*}$. The adjoint of $\mathbf{M}$, denoted $\mathbf{M}^\dagger$, is the complex conjugate of $\mathbf{M}^T$, and its elements have the value M^*_{ba}. Born and Jordan required that if $\mathbf{H}$ is the Heisenberg matrix representing a classical Hamiltonian, then $\mathbf{H}$ must be Hermitian (self-adjoint), that is must satisfy the relation $\mathbf{H} = \mathbf{H}^\dagger$.

After the publication of Schrödinger's first paper it was realized that Schrödinger's wave functions, ψ, could be considered to correspond to components of Heisenberg vectors. The overall correspondence is as follows: Let ψ_a and ψ_b be general components of a *wave-vector*, $\boldsymbol{\psi}$, and let $\boldsymbol{\psi}^\dagger$ represent the transpose of the complex conjugate of $\boldsymbol{\psi}$. Then the elements of the matrix M representing the effect of the Schrödinger operator μ on this vector are

$$M_{ab} = \int dv\ \psi^*_a(\mu\ \psi_b). \tag{9.2.1}$$

Here dv is the volume element in position space, and ψ^*_k is the complex conjugate of the Schrödinger wave function ψ_k, which is in general a function of t as well as the position variables x. Dirac proposed the *bra* and *ket* notation that is commonly used to represent these integrals, and matrix elements in quantum theory. Because Dirac's notation can lead to confusion when one is dealing with operators that are not self-adjoint, we prefer to use the mathematician's notation (f, g), or $\langle f, g\rangle$, to denote a general scalar product of f and g. Using it, one would express (9.2.1) as

$$M_{ab} = (\psi^*_a, \mu\ \psi_b), \quad \text{or} \quad M_{ab} = \langle\psi^*_a, \mu\ \psi_b\rangle. \tag{9.2.2}$$

The reader may wish to establish that if μ is self-adjoint and ψ_a and ψ_b are eigenfunctions of it with eigenvalues, m_a, m_b, then the eigenvalues must be real numbers.

Because adjoints of Schrödinger operators are defined by the scalar products, (9.2.1), in which there is no integration over t, in Schrödinger mechanics the adjoint of $i\hbar\partial/\partial t$ and of t, is not defined. However, in time dependent-Schrödinger mechanics, energies are eigenvalues of the self-adjoint operator $H(q, p_{op})$ and the operator $i\hbar\partial/\partial t$. If

$$H(q, p_{op})\phi_k(q) = E_k\, \phi_k(q), \tag{9.2.3a}$$

then

$$\psi_k(q, t) = \exp(-iE_k\, t/\hbar)\phi_k(q) \tag{9.2.3b}$$

is the time-dependent Schrödinger function that is a solution of (9.2.3a) and

$$H(q, p_{op})\psi_k(q, t) = i\,\hbar\partial\psi_k(q, t)/\partial t. \tag{9.2.4a}$$

Though the adjoint of $i\,\hbar\partial/\partial t$ is not defined, it is evident from (9.2.3) and (9.2.4a) that $\psi_k(q, t)^*$ satisfies

$$H(q, p_{op})\psi_k(q, t)^* = (i\,\hbar\partial/\partial t)^*\psi_k(q, t)^* = -i\,\hbar\partial\psi_k(q, t)/\partial t^*. \tag{9.2.4b}$$

The adjoint of the product of two operators, AB, is $(AB)^{*T} = B^\dagger A^\dagger$; so the adjoint of the commutator of two operators A and B is

$$[A, B]^\dagger = [B^\dagger, A^\dagger]. \tag{9.2.5a}$$

If A and B are self-adjoint, their commutator is *skew-adjoint*, that is

$$[A, B]^\dagger = [B, A] = -[A, B]. \tag{9.2.5b}$$

If A and B are self-adjoint, and c is a real constant, either of the equations

$$C = ic[A, B] \quad \text{or} \quad C = -ic[A, B] \tag{9.2.6}$$

must produce from A and B a self-adjoint operator C.

Born and Jordan observed that in Cartesian coordinates the commutation relations

$$[q_i, -i\,\hbar\partial/\partial q_j] = i\,\hbar\delta_{ij}, \quad [q_i, q_j] = 0, \quad [-i\,\hbar\partial/\partial q_i, -i\,\hbar\partial/\partial q_j] = 0, \tag{9.2.7a}$$

are in one-to-one correspondence with the Poisson bracket relations

$$\{q_i, p_j\} = \delta_{ij}, \quad \{q_i, q_j\} = 0, \quad \{p_i, p_j\} = 0. \tag{9.2.7b}$$

For Cartesian q's and p's, this sets up a one-to-one correspondence

$$[\mathrm{q}, \mathrm{p}_{\mathrm{op}}] \leftrightarrow \mathrm{i}\,\hbar\{\mathrm{q}, \mathrm{p}\} \tag{9.2.8}$$

$$\mathbf{q} = (\mathrm{q}_1, \mathrm{q}_2, \ldots),\ \mathbf{p} = (\mathrm{p}_1, \mathrm{p}_2, \ldots), \tag{9.2.9}$$
$$\mathrm{p}_{\mathrm{op}} = (-\mathrm{i}\,\hbar\partial/\partial \mathrm{q}_1, -\mathrm{i}\,\hbar\partial/\partial \mathrm{q}_2, \ldots).$$

Since we are interested in establishing connections between classical symmetries and quantal symmetries, we require operations with the quantum mechanical operators that represent observables to correspond to operations that are on real functions of real classical variables. We thus prefer to express the correspondence (9.2.8) as

$$(-\mathrm{i}/\,\hbar)[\mathrm{q}_{\mathrm{j}}, \mathrm{p}_{\mathrm{kop}}] \leftrightarrow \{\mathrm{q}_{\mathrm{j}}, \mathrm{p}_{\mathrm{k}}\} = \delta_{\mathrm{ij}}; \tag{9.2.10a}$$

the left-hand side is then self adjoint and the right-hand side is real. Defining E_{op} to be $\mathrm{i}\partial/\partial \mathrm{t}$ one has for the classically conjugate variables $\mathrm{q}_4 = \mathrm{t}$, $\mathrm{p}_4 = -\mathrm{E}$,

$$(-\mathrm{i}/\,\hbar)[\mathrm{t}, -\mathrm{E}_{\mathrm{op}}] \leftrightarrow \{\mathrm{t}, -\mathrm{E}\} = 1, \tag{9.2.10b}$$

a special case of (9.2.9a), though as mentioned previously the adjoint of $\mathrm{i}\partial/\partial \mathrm{t}$ is not defined in Schrödinger mechanics.

With these observations in mind, we note that for each value of j in (9.2.9a), one has a quantal three-parameter Lie algebra, in which the operation of composition is via the commutator, and

$$(-\mathrm{i}/\,\hbar)[\mathrm{q}_{\mathrm{j}}, \mathrm{p}_{\mathrm{jop}}] = 1,\quad (-\mathrm{i}/\,\hbar)[\mathrm{q}_{\mathrm{j}}, 1] = 0,\quad (-\mathrm{i}/\,\hbar)[\mathrm{p}_{\mathrm{jop}}, 1] = 0. \tag{9.2.11a}$$

In the corresponding classical Lie algebra, composition is via the Poisson bracket, and

$$\{\mathrm{q}_{\mathrm{j}}, \mathrm{p}_{\mathrm{j}}\} = 1,\quad \{\mathrm{q}_{\mathrm{j}}, 1\} = 0,\quad \{\mathrm{p}_{\mathrm{j}}, 1\} = 0. \tag{9.2.11b}$$

These two isomorphic Lie algebras are termed *Heisenberg algebras.*

The Born-Jordan correspondence produces a revealing correspondence between transformation groups. The commutation relation $-\mathrm{i}[\mathrm{q}_{\mathrm{k}}, \mathrm{p}_{\mathrm{kop}}] = \hbar$ implies Weyl's relation

$$\exp(\mathrm{iap}_{\mathrm{kop}})\exp(\mathrm{ibq}_{\mathrm{k}}) = \exp(\mathrm{iab}\,\hbar)\exp(\mathrm{ibq}_{\mathrm{k}})\exp(\mathrm{iap}_{\mathrm{kop}}). \tag{9.2.12a}$$

The corresponding classical relation is, of course,

$$\exp(\mathrm{iap}_{\mathrm{k}})\exp(\mathrm{ibq}_{\mathrm{k}}) = \exp(\mathrm{ibq}_{\mathrm{k}})\exp(\mathrm{iap}_{\mathrm{k}}). \tag{9.2.12b}$$

The relation between (9.2.11a) and (9.2.11b), may be summarized as

$$\exp(\mathrm{iab}\,\hbar) \leftrightarrow 1,\quad \mathrm{q}_{\mathrm{k}} \leftrightarrow \mathrm{q}_{\mathrm{k}},\quad \mathrm{p}_{\mathrm{kop}} = -\mathrm{i}\,\hbar\partial/\partial \mathrm{q}_{\mathrm{j}} \leftrightarrow \mathrm{p}_{\mathrm{k}}. \tag{9.2.12c}$$

The function $\exp(\mathrm{iab}\,\hbar)$ only takes on unit value if its argument is zero. For this reason the spaces related by the correspondence (9.2.11), when $\hbar \neq 0$, are not the spaces one needs to relate classical and quantum mechanical symmetries in which the group parameters a,b can vary.

There is another way of viewing group-theoretic consequences of the Born–Jordan correspondence. Let g(q) be an analytic function and consider the expression

$$\begin{aligned}(-i/\hbar)[\exp(iap_{op}), g(q)] &= (-i/\hbar)[\exp(a\hbar\partial/\partial q), g(q)] \\ &= (-i/\hbar)(g(q+a\hbar) - g(q))\exp(a\hbar\partial/\partial q).\end{aligned} \tag{9.2.13a}$$

Its Poisson bracket analog is

$$\begin{aligned}\{\exp(ap), g(q)\} &= -(\partial\exp(ap)/\partial p)\,\partial g(q)/\partial q \\ &= -a\,\partial g(q)/\partial q\,\exp(ap).\end{aligned} \tag{9.2.13b}$$

A Maclaurin series expansion of the term $g(q+a\hbar)-g(q)$ in (9.2.12a) yields

$$-i(a\partial g(q)/\partial q + (a^2\hbar)(\partial^2 g(q)/\partial q^2)/2 + \cdots)\exp(a\hbar\partial/\partial q). \tag{9.2.13c}$$

Comparing this with the corresponding term, $-a\,\partial g(q)/\partial q$, in (9.2.12b) one sees that relations (9.2.12a) and (9.2.12b) are in one-to-one correspondence only to first order in $\hbar$.

Both these observations indicate that a change in viewpoint is required if one is to understand why it is that some, but not all, classical and quantum mechanical systems have identical local symmetry groups.

Intrinsic local symmetries of Schrödinger equations are determined by Lie algebras whose generators are functions of q's and p_{op}'s, t and $-E_{op} = -i\hbar\partial/\partial t$. On the other hand, as we have seen, the local dynamical symmetries of classical Hamiltonian systems are determined by Lie algebras whose generators are Poisson bracket operators. In Schrödinger mechanics

$$[\exp(iap_{kop}), g(q)] = [\exp(a\hbar\partial/\partial q_k), g(q)]. \tag{9.2.13}$$

In classical mechanics

$$[\exp(-a\{p_k, \cdot\}), g(q)] = [\exp(a\partial/\partial q_k), g(q)]. \tag{9.2.14}$$

The group operators in these expressions are in one-to-one correspondence to all orders in $\hbar$ As a consequence, correspondences between the group actions of Poisson bracket operators in phase space and the group actions of quantum mechanical operators in position or momentum subspaces of phase space are valid to all orders in $\hbar$.

It will be shown in Secs. 9.3 and 9.4, below, that these observations lead to a change in viewpoint that does much to clarify the relation between quantal dynamical symmetries and classical dynamical symmetries. These sections develop the correspondence that relates Lie algebras of Schrödinger operators of a more general form than q, p_{op}, to Lie algebras of their corresponding Poisson bracket operators. As in Chapter 3, position and momentum variables will be assumed dimensionless. Because the relations that will be set forth are valid independent of the value of $\hbar$, it simplifies matters if we require $\hbar$ to be dimensionless, and use units that have been so chosen that its value is 1. This puts a constraint on the units of time being used. If one uses atomic units with the Bohr radius, $a_o = 0.5292 \times 10^{-8}$ cm, as the unit of length, and m_o, the mass of the electron (or m_{or}, its reduced mass in the H atom, 9.102×10^{-28} g), as the unit of mass, then $\hbar$ will have unit value if t is measured in units of $t_0 = 2.41776 \times 10^{-17}$ s (or $t_{or} = 2.41782 \times 10^{-17}$ s). Using these units the absolute value, e, of the charge of the electron has unit value, and c, the velocity of light, has the value 136.958 (or 136.962).

9.3 Correspondence Between Quantum Mechanical Operators and Functions of Classical Dynamical Variables

The connection between classical functions of positions and momenta and Schrödinger operators, or their momentum space analogs, is ambiguous if one does not specify an ordering of the non-commuting quantum mechanical representatives of the p's and q's. These *ordering ambiguities* have been termed *impediments to quantization* of classical mechanics, and are still a subject of interest to theoreticians.[3] The ambiguity is reduced by requiring that Schrödinger operators are either self-adjoint or skew-adjoint. To further reduce the ambiguity: *one may require either that*

i) *a quantum mechanical operator A which contains a product of* q_j*'s and* $\partial/\partial q_j$*'s must also contain the adjoint of this product,*

or

ii) *the self-adjoint or skew-adjoint operator, A′ can be obtained from an operator of form i) by a similarity transformation* $A \to A' = S^{-1}AS$, *with a corresponding classical analog that is a canonical transformation.*

Examples of Schrödinger operators of the first kind and their corresponding classical functions are:

$$A_1 = (-i\partial/\partial x)^2/2 - i\partial/\partial t = p_{op}^2/2 - E_{op} \leftrightarrow A_1 = p^2/2 - E \qquad (9.3.1a)$$

and

$$A_2 = -it(x\partial/\partial x + \partial/\partial xx)/2 - x^2/2 - i(t^2\partial/\partial t + \partial/\partial tt^2)/2 \\ \leftrightarrow tqp - q^2/2 - t^2E. \tag{9.3.1b}$$

The self-adjoint Schrödinger operator A_2 generates invariance transformations of the time-dependent free-particle Schrödinger equation $A_1\psi = 0$, which will then be investigated in Sec. 9.6. One might use commutation relations to replace the symmetrized operators in (9.3.1) by $-i(x\partial/\partial x + 1/2)$ and $i(t^2\partial/\partial t + t)$, but we will not do so.

The operators in (9.3.1) can be considered to be special cases of a more general class of operators each of whose members are sums of *symmetrized products* of the form

$$\Gamma = \sum(F_j(q,t)g_j(p_{op}, E_{op}) + (F_j(q,t)g_j(p_{op}, E_{op}))^\dagger). \tag{9.3.2}$$

Operators of this form will be used throughout this chapter. It will be assumed that these operators have a well-defined action on the space of wave functions that is of interest.

An example of the second type is provided by the inversion transformation of Eqs. (8.3.10). Classically, it interconverts two sets of canonically conjugate variables. Its quantum mechanical analog produces self-adjoint Schrödinger operators and momentum space operators that are not necessarily of the form (9.3.2), but satisfy the relations (9.2.10a). In any given case it may not be easy to recognize whether an operator which contains terms not of the symmetrized form can be obtained by a similarity transformation from operators that are of this form. Finally, it should be noted that the considerations of this section apply, *ipso facto*, to momentum space realizations of quantum mechanics, as well as to position space realizations.

9.4 Lie Algebraic Extension of the Correspondence Principle

In this section we investigate the circumstances under which one can have the isomorphic relations

$$(-i/\hbar)[A(q, p_{op}), B(q, p_{op})] = C(q, p_{op}) \tag{9.4.1a}$$

$$\{A(q,p), B(q,p)\} = C(q,p), \tag{9.4.1b}$$

$$[\{A(q,p), \cdot\}, \{B(q,p), \cdot\}] = \{C(q,p), \cdot\}. \tag{9.4.1c}$$

We first consider the classical correspondence between (9.4.1b) and (9.4.1c) when a is not a function of q, p, E nor t. Then

$$\{\mathrm{a}, \cdot\} = 0. \tag{9.4.3}$$

Thus

$$\{(\mathrm{A}(\mathrm{q},\mathrm{p},\mathrm{t},\mathrm{E}) + \mathrm{a}), \cdot\} = \{\mathrm{A}(\mathrm{q},\mathrm{p},\mathrm{t},\mathrm{E}), \cdot\} \tag{9.4.4a}$$

and the two-way correspondence

$$\{\mathrm{A_k}(\mathrm{q},\mathrm{p},\mathrm{t},\mathrm{E}), \cdot\} \leftrightarrow \mathrm{A_k}(\mathrm{q},\mathrm{p},\mathrm{t},\mathrm{E}) \tag{9.4.4b}$$

becomes ambiguous. This can have particularly confusing consequences if one does not adhere to the use of symmetrized self-adjoint, or skew-adjoint, quantum-mechanical operators introduced in the previous section.

In Chapter 7 it was shown that if a set of C^2 functions $G_m(q, p)$ satisfy

$$\{\mathrm{G_m}(\mathrm{q},\mathrm{p}),\ \mathrm{G_n}(\mathrm{q},\mathrm{p})\} = \Sigma_k c^k_{mn} \mathrm{G_k}(\mathrm{q},\mathrm{p}), \tag{9.4.5}$$

then the corresponding Poisson bracket operators satisfy

$$[\{\mathrm{G_m}(\mathrm{q},\mathrm{p}), \cdot\}, \{\mathrm{G_n}(\mathrm{q},\mathrm{p}), \cdot\}] = \Sigma_k c^k_{mn} \{\mathrm{G_k}(\mathrm{q},\mathrm{p}), \cdot\}. \tag{9.4.6}$$

It was also noted that the apparent isomorphism between (9.4.5) and (9.4.6) was misleading if any G_k(q,p) could be a constant. If this occurs, a $\{G_k, \cdot\}$ will vanish, and the Lie algebras determined by (9.4.5) and (9.4.6) will not be isomorphic, the vanishing of a $\{G_k, \cdot\}$ being equivalent to the vanishing of every c^k_{mn} with the same value of k.

This is illustrated by the Lie algebra that is put in correspondence with the Heisenberg algebras of (9.2.10) by using (9.4.1). In this algebra

$$\{\mathrm{q_j}, \cdot\} = \partial/\partial \mathrm{p_j}, \quad \{\mathrm{p_j}, \cdot\} = -\partial/\partial \mathrm{q_j}, \tag{9.4.7a}$$

and

$$[\{\mathrm{q_j}, \cdot\}, \{\mathrm{p_j}, \cdot\}] = \{1, \cdot\} = 0, \quad [\{\mathrm{q_j}, \cdot\}, \{1, \cdot\}] = 0, \quad [\{\mathrm{p_j}, \cdot\}, \{1, \cdot\}] = 0). \tag{9.4.7b}$$

The algebra is Abelian, but its quantum mechanical version is not, because

$$[\mathrm{q_j}, \mathrm{p_j}] = \mathrm{i}\,\hbar. \tag{9.4.8}$$

It is not difficult to determine the circumstances under which the general correspondences of (9.4.1) give rise to Poisson bracket Lie algebras that contain this Abelian algebra as a subalgebra. Let two functions F(q, p) and

G(q, p) each be expandable in a Maclaurin series with fixed coefficients, and a remainder term, R, so that

$$F = f_{00} + f_{01}q + f_{10}p + R(f_{ab}q^a p^b), \quad a + b > 1, \tag{9.4.9a}$$

$$G = g_{00} + g_{01}q + g_{10}p + R(g_{ab}q^c p^d), \quad c + d > 1. \tag{9.4.9b}$$

Then

$$\{F, G\} = (f_{01}g_{10} - g_{01}f_{10}) + \sum(h_{mn}q^m p^n), \quad m + n > 1. \tag{9.4.9c}$$

Consequently, $\{F,G\}$ can be a nonzero constant *iff* $(f_{01}g_{10} - g_{01}f_{10})$ is nonzero and R vanishes. This is the only case in which F and G can be members of a Heisenberg algebra, and hence produce an Abelian subalgebra containing $\{q_j, \cdot\}$ and $\{p_j, \cdot\}$.

These observations lead to the following statement[4]:

Let a set of operators $\Gamma_k(q, p_{op}, t, E_{op})$ *of the form* (9.3.2) *be generators of invariance transformations of a Schrödinger equation, and members of a Lie algebra whose basis cannot contain a Heisenberg subalgebra or a generator that is a constant. The two-way correspondences*

$$\Gamma_k(q, p_{op}, t, E_{op}) \leftrightarrow G_k(q, p, t, E) \leftrightarrow \{G_k(q, p, t, E), \cdot\}, \tag{9.4.10a}$$

then produce the three isomorphic Lie algebras

$$-i[\Gamma_j, \Gamma_k] = \hbar c_{jk}^m \Gamma_m, \tag{9.4.10b}$$

$$\{G_j, G_k\} = c_{jk}^m G_m, \tag{9.4.10c}$$

$$[\{G_j, \cdot\}, \{G_k, \cdot\}] = c_{jk}^m \{G_m, \cdot\}. \tag{9.4.10d}$$

When the Γ*'s are generators of invariance transformations of a Schrödinger equation, the isomorphisms establish a one-to-one correspondence between local intrinsic symmetries of Schrödinger equations, and local geometrical symmetries in classical phase space.*

It should be noted that if $[A_j, A_k] = \hbar c_{jk}^R A_R$, and $K\{A_j, A_k\} = \hbar c_{jk}^R A_R$, with K a nonzero constant, then K can be absorbed into the structure constants without altering the Killing–Cartan classification of the Lie algebra.

When these conditions are satisfied one may say that Eqs. (9.4.10) define a *faithful Lie-algebraic extension of the correspondence principle.* In this case, the Lie algebra of quantum mechanical operators and the Lie algebra of PB operators define locally isomorphic dynamical groups. Heisenberg's uncertainty principle, a consequence of the commutation relation

$[q_{op}, p_{op}] = i\hbar$, does not affect this correspondence. The SO(4,2) local Lie groups of quantum-mechanical and classical Kepler systems are an example of this isomorphism which may be said to exist *despite the uncertainty principle.*

Harmonic oscillators provide a contrasting example. The Lie algebra of their Dirac energy shift operators is a Heisenberg algebra. Because of this, there is no one-to-one correspondence between the consequent quantum mechanical and classical dynamical groups of harmonic oscillators.

When a Poisson bracket Lie algebra contains a pair of operators $-\partial/\partial q_j$, and $\partial/\partial p_j$, they generate translations in the q and p subspaces of classical phase space. Thus translation invariance in phase space produces systems with quantum mechanical and classical symmetries that need not be isomorphic. In Schrödinger mechanics, $-\partial/\partial q_j$ innocuously becomes $-i\partial/\partial q_j$. On the other hand, $\partial/\partial p_j$ becomes $-iq_j$ and generates *gauge transformations* which multiply $\psi(q)$ by $\exp(iaq_j)$.

To determine whether corresponding quantal and classical systems with *locally* isomorphic symmetry groups have *globally* isomorphic groups, one must determine the interval over which their group parameters can range without producing an identity operation. In the classical mechanical cases the group moves a point in a phase space, so one determines the range which the group parameters can span before restoring points in the phase space to their initial positions. In Schrödinger quantum mechanics the wave function Ψ (q, t) defines points with coordinates (ψ, q, t), with ψ having the value Ψ(q, t). One determines the range which group parameters can span before restoring q,t to their original values, and ψ to its original value. When the correspondence between classical and quantum mechanical global Lie groups exists, intrinsic quantum mechanical global symmetries correspond to global geometrical symmetries in phase space.

An example is provided by the correspondence between quantum angular momentum operations and rotational symmetry. For L_{12}, the component of the angular momentum operator on the q_3 axis, one has

$$L_{12} = -i(q_1\partial/\partial q_2 - q_2\partial/\partial q_1) = (q_1 p_{2op} - q_2 p_{1op})$$

and

$$q_1p_2 - q_2p_1 \leftrightarrow \{(q_1p_2 - q_2p_1), \cdot\} = -R_{12}. \qquad (9.4.11)$$

Note that

$$R_{12} = (q_1\partial/\partial q_2 - q_2\partial/\partial q_1) + (p_1\partial/\partial p_2 - p_2\partial/\partial p_1).$$

Applying the correspondence to the other angular momentum components, $-i(q_2\partial/\partial q_3 - q_3\partial/\partial q_2)$ and $-i(q_3\partial/\partial q_1 - q_1\partial/\partial q_3)$, one obtains three R_{ij} corresponding to three L_{ij}. The L's obey the commutation relations

$$\sqrt{-1}[L_{ij}, L_{jk}] = L_{ki}, \quad i, j, k \text{ cyclic.} \tag{9.4.12a}$$

The $-$R's obey the isomorphic commutation relations

$$[-R_{ij}, -R_{jk}] = -R_{ki}, \quad i, j, k \text{ cyclic.} \tag{9.4.12b}$$

Allowing the operators $\exp(i\alpha_{jk}L_{jk})$ to act upon the spherical harmonics $Y_{lm}(\theta, \phi)$ in order to reproduce the identity operation, the group parameters must have the same ranges that are required of $\exp(-\alpha_{jk}R_{jk})$ when it acts in phase space — the group in both cases is SO(3).

9.5 Some Properties of Invariance Transformations of Partial Differential Equations Relevant to Quantum Mechanics

In this section we lay the ground work required for determining the intrinsic symmetries of Schrödinger equations. For the time being, it will not be assumed that group generators are self-adjoint.

Let $G_{op} = G(q, p_{op}, t, E_{op}) = G(x, -i\partial/\partial x, t, i\partial/\partial t)$ be a Schrödinger operator that acts on $\psi(x, t)$ and suppose that $\psi(x, t)$ satisfies the Schrödinger equation

$$(H(x, -i\partial/\partial x) - i\partial/\partial t)\psi(x, t) = 0. \tag{9.5.1}$$

Suppose also that, for arbitrary infinitesimal values of ϵ, the infinitesimal transformation $\psi(x, t) \rightarrow \psi'(x, t) = (1 + \epsilon G_{op})\psi(x, t)$ yields a $\psi'(x, t)$ that is a solution of (9.5.1). This requires that

$$(H - i\partial/\partial t)G_{op}\psi(x, t) = 0 \tag{9.5.2}$$

when

$$(H - i\partial/\partial t)\psi(x, t) = 0.$$

Note that $\psi'(x, t)$ may or may not be a solution of the equation

$$(H - E')G_{op}\psi(x, t) = 0, \tag{9.5.3a}$$

when

$$(H - E)\psi(q) = 0. \tag{9.5.3b}$$

Also, if it is, E′ may, or may not, equal E. We shall, for example, see in Sec. 9.6 that if

$$H = -(1/2)\partial^2/\partial x^2, \tag{9.5.4a}$$

then,

$$G_{op} = -i(x\partial/\partial x + 2t\partial/\partial t) \tag{9.5.4b}$$

satisfies (9.5.2). It converts solutions of the time-dependent equation into solutions, but does not convert an eigenfunction of H into an eigenfunction. Similarly, if

$$H = -(1/2)\partial^2/\partial x^2 + (1/2)kx^2, \tag{9.5.5a}$$

the shift operator

$$\exp(-ikt)(\partial/\partial x - ikx) \tag{9.5.5b}$$

satisfies (9.5.2), and converts eigenfunctions of energy E into eigenfunctions with E′ not equal to E.

If $G(\mathbf{q};\mathbf{p}) = G(\mathbf{x}, t; \mathbf{p}_x, -E)$ is an analytic function of the components of $\mathbf{p}$ in a region about $\mathbf{p} = 0$, we may expand it in a Maclaurin series in the components of p, E. We suppose this is the case, and let

$$G(q, p) = \Sigma g^{ijk\cdots}(p_1)^i(p_2)^j(p_3)^k\cdots, \tag{9.5.6}$$

where the g's are functions of the q's or constants. We use the form (9.5.6) in determining Lie generators, then subsequently convert them to self-adjoint operators $A(G + G^\dagger)/2$ or their symmetrized counterparts.

Before proceeding further let us make a change in notation to conform to usage in much literature on invariance transformations of Schrödinger equations. Let us express the Schrödinger operator corresponding to the power series expansion (9.5.6) as

$$G_{op} = Q(\mathbf{q}, \partial/\partial\mathbf{q}) = \Sigma g^{ijk\cdots}(\partial/\partial q_1)^i(\partial/\partial q_2)^j(\partial/\partial q_3)^k\cdots, \tag{9.5.7}$$

the $g^{ijk\cdots}$ being functions of the position variables, q = x, and/or t. The operators may have a term with all indices equal to zero, which contains no derivative operator, as well as terms containing derivatives of arbitrary order. If the coefficient of a term linear in a $\partial/\partial x$ is real, one may wish to multiply Q by i. Call the result Q_c. Then, if Q generates invariance transformations of a Schrödinger equation, $(Q_c \pm Q_c^\dagger)/2$ can be made self-adjoint or skew-adjoint. Doing this for each of the generators Q_k of interest, one can construct a set of operators with definite adjointness and definite commutation relations.

The Q of (9.5.7) appear different from the generators introduced by Lie, all of which were first-order differential operators. The Q's may have operators with terms free of derivatives, as well as terms with higher-order derivatives. To understand how to put the derivative-free terms in Lie form it is sufficient to consider the action of a simple Q of the form

$$\mathrm{Q} = \mathrm{q}^0(\mathrm{x}) + \mathrm{q}^1(\mathrm{x})\partial/\partial\mathrm{x}. \tag{9.5.8}$$

This acts upon solutions $\psi(\mathrm{x})$ to give

$$\mathrm{Q}\psi(\mathrm{x}) = \mathrm{q}^0(\mathrm{x})\psi(\mathrm{x}) + \mathrm{q}^1(\mathrm{x})\partial\psi(\mathrm{x})/\partial\mathrm{x}. \tag{9.5.9}$$

We have seen in Chapter 4 that it is always possible to vary solutions of differential equations without varying what are usually termed their *independent variables.* In this case $\psi(\mathrm{x})$ can be varied without varying x, so $\partial/\partial\psi$ can act upon it to give a nonzero result even if x is fixed. Thus Q has the same effect as the Lie generator

$$\mathrm{U} = \mathrm{q}^0(\mathrm{x})\psi\partial/\partial\psi + \mathrm{q}^1(\mathrm{x})\partial/\partial\mathrm{x}. \tag{9.5.10}$$

This relation between U and Q can be maintained as long as the wavefunction $\psi(\mathrm{x})$ is not completely fixed. If a Schrödinger equation had only one solution, fixed in phase and normalization, (9.5.10) could not replace (9.5.9). This never happens.

In the light of these remarks, let us consider Q's that involve second- and higher-order derivatives, operators of the form

$$\mathrm{Q} = \mathrm{q}^1(\mathrm{x})\partial/\partial\mathrm{x} + \mathrm{q}^2(\mathrm{x})\partial^2/\partial\mathrm{x}^2 + \mathrm{q}^3(\mathrm{x})\partial^3/\partial\mathrm{x}^3 + \cdots \tag{9.5.11}$$

As ψ and x can be independently varied, this Q has the same effect on $\Psi(\mathrm{x})$ as the action of the operator

$$\mathrm{U} = \mathrm{q}^1(\mathrm{x})\partial/\partial\mathrm{x} + (\mathrm{q}^2(\mathrm{x})\partial^2\Psi(\mathrm{x})/\partial\mathrm{x}^2 + \mathrm{q}^3(\mathrm{x})\partial^3\Psi(\mathrm{x})/\partial\mathrm{x}^3 + \cdots)\psi\partial/\partial\psi \tag{9.5.12}$$

on ψ. This U is that portion of the Lie generator of a contact transformation that acts in the space of x, $\Psi(\mathrm{x})$, and these derivatives. These considerations apply to the more general Q of (9.5.11), which can be expressed as a first-order differential operator U by converting all of its terms that are not those of a first-order differential operator to an analogous coefficent of $\psi\partial/\partial\psi$.

In applying (9.5.7) one finds a fundamental distinction between invariance transformations of ordinary differential equations and invariance transformations of partial differential equations: PDEs can be invariant under transformations with generators that depend upon derivatives of order higher than those in the PDE.

We now turn to a discussion of this difference. Consider an ordinary differential equation such as the Schrödinger equation for which the kinetic energy operator T is $-(1/2)\partial^2\Psi(x)/\partial x^2$ and the potential function is v(x). Let us rearrange the equation to isolate T, and write it as

$$\partial^2\Psi(x)/\partial x^2 = 2(v(x) - E)\Psi(x). \tag{9.5.13}$$

Taking the derivative of both sides of this equation one obtains the identity

$$\partial^3\Psi/\partial x^3 = 2(v\partial\Psi/\partial x + (\partial v/\partial x - E)\Psi). \tag{9.5.14}$$

This equation is said to be a *differential consequence* of (9.5.13). Inspecting it, one observes that at any point with coordinate x, $\partial^3\Psi/\partial x^3$ cannot be varied independently of Ψ and $\partial\Psi/\partial x$. A further differentiation then establishes that $\partial^4\Psi/\partial x^4$ is not independent of $\Psi, \partial\Psi/\partial x$, and $\partial^2\Psi(x)/\partial x^2$, and so forth. In short, because (9.5.13) is a second-order ODE, only Ψ and $\partial\Psi/\partial x$ can be independently varied at x. This has the consequence that the generators of invariance transformations of (9.5.13) need only depend upon, x, Ψ and $\partial\Psi/\partial x$. The result is not affected by the presence of the potential v(x), and a similar statement holds for all second-order ODEs of a form such that the highest derivative can be isolated, or more generally, expressed as a function of lower-order derivatives. An analogous argument establishes that for an n-th order ODE with isolable highest derivative, all derivatives of order n or greater, are functionally dependent on those of lower order. In Lie's theory of differential equations the dependencies on these derivatives are treated as contact transformations, and expressed by the use of extended generators. (cf., Chapter 5.)

Now consider the time-dependent free-particle Schrödinger equation

$$i\partial\Psi(x,t)/\partial t = -(1/2)\partial^2\Psi(x,t)/\partial x^2, \tag{9.5.15a}$$

and write it as

$$2i\Psi_t = -\Psi_{xx}. \tag{9.5.15b}$$

It has the differential consequences:

$$2i\Psi_{tt} = -\Psi_{xxt},\ 2i\Psi_{tx} = -\Psi_{xxx}, \ldots. \tag{9.5.16}$$

One may consider all the functions on one side of these equations to be independently variable. As a result generators of invariance transformations of (9.5.15) can depend upon derivatives of arbitrarily large order. *In contrast to ordinary differential equations, partial differential equations have invariance transformations whose generators* U *and* Q *may depend upon derivatives of arbitrary order.*[5]

Thus, when one is dealing with partial differential equations, Q must, in principle, be allowed to be a differential operator of arbitrarily high order. However, as will be seen in the following pages, for many purposes one may only require generators of point transformations, or generators Q that depend only on a few higher order derivatives.

From the standpoint of the correspondence principle, the fact that a Schrödinger operator can depend upon derivatives of arbitrary order should be expected, because the corresponding functions in a classical Poisson bracket operator can depend upon arbitrary powers of p. If one expands the functions in PB operators $\{G(q,p),\cdot\}$ in a power series in p, one confronts the classical analog of the discussion of this section. The reader may suspect that, because kinetic energies involve only first and second powers of p's, differential consequences of Schrödinger equations may not prove overly troublesome. This suspicion will be verified in subsequent pages, which provide a number of illustrations of the concepts introduced in this section.

9.6 Determination of Generators and Lie Algebra of Invariance Transformations of $((-1/2)\partial^2/\partial x^2 - i\partial/\partial t)\psi(x,t) = 0$

The equation may be considered as the time-dependent Schrödinger equation of a free-particle of unit mass moving in one dimension. To simplify matters we multiply it by −2, set $\psi = y$ and obtain the equivalent equation

$$S_{op}y(x,t) = 0, \quad S_{op} = \partial^2/\partial x^2 + 2i\partial/\partial t. \tag{9.6.1a}$$

Among the transformations which leave this equation invariant, some leave invariant the corresponding time-independent equation

$$(\partial^2/\partial x^2 + 2E)y(x,t) = 0. \tag{9.6.1b}$$

We will seek invariance generators of (9.6.1) that have the simple form

$$Q = q^{00}(x,t) + q^{10}(x,t)\partial/\partial x + q^{01}(x,t)\partial/\partial t. \tag{9.6.2}$$

This requires that

$$(\partial^2/\partial x^2 + 2i\partial/\partial t)Qy = 0 \tag{9.6.3}$$

for all y satisfying (9.6.1a) and all of its relevant differential consequences. Given the assumed form of Q, (9.6.3) contains a third-order derivative, and these differential consequences involve derivatives up to third-order. One has

$$\begin{aligned}
\partial Q/\partial x &= Q_x = q_x^{00}(x,t) + q_x^{10}(x,t)\partial/\partial x + q_x^{01}(x,t)\partial/\partial t,\\
\partial Q/\partial t &= Q_t = q_t^{00}(x,t) + q_t^{10}(x,t)\partial/\partial x + q_t^{01}(x,t)\partial/\partial t, \qquad (9.6.4)\\
\partial^2 Q/\partial x^2 &= Q_{xx} = q_{xx}^{00}(x,t) + q_{xx}^{10}(x,t)\partial/\partial x + q_{xx}^{01}(x,t)\partial/\partial t.
\end{aligned}$$

Using the same convention to denote derivatives of y, one has

$$\begin{aligned}
\partial Qy/\partial t = Q_t y + Qy_t, \quad \partial Qy/\partial x = Q_x y + Qy_x,\\
\partial^2 Qy/\partial x^2 = Q_{xx}y + 2Q_x y_x + Qy_{xx}.
\end{aligned} \qquad (9.6.5)$$

The invariance condition (9.6.3) requires

$$Q_{xx}y + 2Q_x y_x + 2iQ_t y = 0, \qquad (9.6.6a)$$

and is subject to the differential consequences of (9.6.1a), *viz.*

$$y_{xx} + 2iy_t = 0. \qquad (9.6.6b)$$

These are

$$y_{xxx} + 2iy_{tx} = 0, \quad y_{xxx} + 2iy_{tt} = 0. \qquad (9.6.6c,d)$$

Expanding (9.6.6a) yields

$$\begin{aligned}
&(q_{xx}^{00}y + q_{xx}^{10}y_x + q_{xx}^{01}y_t) + 2(q_x^{00}y_x + q_x^{10}y_{xx} + q_x^{01}y_{xt})\\
&\quad + 2i(q_t^{00}y + q_t^{10}y_x + q_t^{01}y_t) = 0
\end{aligned} \qquad (9.6.7)$$

We choose y, y_x, y_t, y_{tx} to be independently variables, and use (9.6.6b) to eliminate y_{xx} from (9.6.7). This converts (9.6.7) to

$$\begin{aligned}
&(q_{xx}^{00} + 2i\,q_t^{00})y + (q_{xx}^{10} + 2\,q_x^{00} + 2i\,q_t^{10})y_x\\
&\quad + q_{xx}^{01}y_t + 2i(q_t^{01} - 2\,q_x^{10})y_t + 2q_x^{01}y_{xt} = 0.
\end{aligned} \qquad (9.6.8)$$

As y, y_x, y_t, y_{tx} can be assigned arbitrary values, for (9.6.8) to hold, the coefficients of these functions in the equation must themselves vanish. We thus obtain the following determining equations for Q:

$$q_{xx}^{00} + 2iq_t^{00} = 0, \qquad (9.6.9a)$$

$$q_{xx}^{10} + 2q_x^{00} + 2iq_t^{10} = 0, \qquad (9.6.9b)$$

$$q_{xx}^{01} + 2i(q_t^{01} - 2q_x^{10}) = 0, \qquad (9.6.9c)$$

$$q_x^{01} = 0. \qquad (9.6.9d)$$

The last equation is the simplest, so we start out with it and work our way upward. For q^{01}_x to vanish, q^{01} can not be a function of x, so one may set

$$q^{01} = A(t). \tag{9.6.10}$$

This causes the first term in (9.6.9c) to vanish — with the consequence that

$$q^{10}_x = (1/2)q^{01}_t = (1/2)\partial A(t)/\partial t. \tag{9.6.11}$$

Hence

$$q^{10} = (x/2)\partial A(t)/\partial t + B(t). \tag{9.6.12}$$

It follows that in (9.6.9b), $q^{10}_{xx} = 0$, and

$$q^{10}_t = (x/2)\partial^2 A(t)/\partial t^2 + \partial B(t)/\partial t. \tag{9.6.13}$$

Thus (9.6.9b) requires

$$q^{00}_x = -iq^{10}_t = -i(x/2)\partial^2 A(t)/\partial t^2 - i\partial B(t)/\partial t, \tag{9.6.14}$$

so

$$q^{00} = -i(x^2/4)\partial^2 A(t)/\partial t^2 - ix\partial B(t)/\partial t + C(t). \tag{9.6.15}$$

Inserting (9.6.15) into (9.6.9a) yields

$$(i/2)\partial^2 A(t)/\partial t^2 + (x^2/2)\partial^3 A(t)/\partial t^3 + 2x\partial^2 B(t)/\partial t^2 + 2i\partial C(t)/\partial t = 0. \tag{9.6.16}$$

As x and x^2 are linearly independent this requires

$$\partial^3 A(t)/\partial t^3 = 0, \tag{9.6.17a}$$

$$\partial^2 B(t)/\partial t^2 = 0, \tag{9.6.17b}$$

and

$$-(i/2)\partial^2 A(t)/\partial t^2 + 2i\partial C(t)/\partial t = 0. \tag{9.6.17c}$$

The first two equations imply that

$$A(t) = c_1 t^2 + c_2 t + c_3, \quad B(t) = c_4 t + c_5. \tag{9.6.18a,b}$$

Inserting these results into (9.6.17c) gives

$$\partial C(t)/\partial t = c_1/2, \quad C(t) = c_1 t/2 + c_6. \tag{9.6.18c,d}$$

This completes the determination of q^{01}, q^{10}, and q^{00}. Inserting (9.6.18a,b,d) into (9.6.10), (9.6.12) and (9.6.15) one obtains

$$q^{01} = A(t) = c_1 t^2 + c_2 t + c_3, \quad (9.6.19a)$$

$$q^{10} = (x/2)\partial A(t)/\partial t + B(t) = (x/2)(2c_1 t + c_2) + c_4 t + c_5, \quad (9.6.19b)$$

$$q^{00} = -i(x^2/4)\partial^2 A(t)/\partial t^2 - ix\,\partial B(t)/\partial t + C(t), \quad (9.6.19c)$$

so

$$q^{00} = -i(x^2/2)c_1 - ix\,c_4 + c_1 t/2 + c_6. \quad (9.6.19d)$$

Inserting these results in (9.6.2), that is in $Q = q^{00} + q^{10}\partial/\partial x + q^{01}\partial/\partial t$, yields

$$Q = -ix^2 c_1/2 - ix\,c_4 + c_1 t/2 + c_6 + (c_1 xt + c_2 x/2 + c_4 t + c_5)\partial/\partial x + (c_1 t^2 + c_2 t + c_3)\partial/\partial t. \quad (9.6.20)$$

Collecting terms that are multiplied by the same c_j, one may express this as

$$Q = c_1 Q_1 + c_2 Q_2 + c_3 Q_3 + c_4 Q_4 + c_5 Q_5 + c_6 Q_6. \quad (9.6.21)$$

The coefficients c_j are arbitrary constants of integration.

The six Q's, their corresponding U's, and the transformations the U's generate are as follows:

$$\begin{aligned} &Q_1 = xt\partial/\partial x + t^2\partial/\partial t + 1/2(t - ix^2),\\ &U_1 = xt\partial/\partial x + t^2\partial/\partial t + 1/2(t - ix^2)u\partial/\partial u,\\ &x \to x/(1 - a_1 t), \quad t \to t' = t/(1 - a_1 t),\\ &u \to u' = u(1 - a_1 t)^{-1/2}\exp(-i(a_1/2)x^2/(1 - a_1 t)). \end{aligned} \quad (9.6.22a)$$

$$Q_2 = x\partial/\partial x + 2t\partial/\partial t: \quad x \to x' = \exp(a_2)x, \quad t \to t' = \exp(2a_2)t. \quad (9.6.22b)$$

$$Q_3 = \partial/\partial t: \quad t \to t' = t + a_3. \quad (9.6.22c)$$

$$\begin{aligned} &Q_4 = t\partial/\partial x - ix, \quad U_4 = t\partial/\partial x - ixu\partial/\partial u:\\ &x \to x' = x + ta_4, \quad u \to u' = u\exp(-ia_4(x + a_4 t/2)). \end{aligned} \quad (9.6.22d)$$

$$Q_5 = \partial/\partial x: \quad x \to x' = x + a_5. \quad (9.6.22e)$$

$$Q_6 = 1, \quad U_6 = u\partial/\partial u: \quad u \to u' = \exp(a_6)u. \quad (9.6.22f)$$

The operators so far obtained do not correspond to quantum mechanical observables. However, as the c's of (9.6.18)–(9.6.20) are arbitrary constants of integration, the Q's are determined only up to a multiplicative constant. (We used this freedom when defining Q_2 as a standard generator of dilatations.) The constants c may be chosen to be real, complex, or pure imaginary. Let us multiply the Q's by $-i$, so that wherever it appears, $\partial/\partial x$ is converted into the self-adjoint operator $-i\partial/\partial x$ that represents p_{op}. Symmetrizing products of x and $-i\partial/\partial x$ one converts the Q operators to self-adjoint operators, Q'. The resulting Q' and their classical analogs, are:

$$Q'_1 = -it(x\partial/\partial x + \partial/\partial x\, x)/2 - it^2\partial/\partial t - x^2/2 \to t\, x\, p_x - t^2 E - x^2/2; \tag{9.6.23a}$$

$$Q'_2 = -i(x\partial/\partial x + \partial/\partial x\, x)/2 - 2it\partial/\partial t \to x\, p_x - 2tE; \tag{9.6.23b}$$

$$Q'_3 = i\partial/\partial t \to E; \tag{9.6.23c}$$

$$Q'_4 = -it\partial/\partial x - x \to tp_x - x; \tag{9.6.23d}$$

$$Q'_5 = -i\partial/\partial x \to p_x; \tag{9.6.23e}$$

$$Q'_6 = -i \to 1. \tag{9.6.23f}$$

The commutation relations of the Q' are given in Table 9.1 below. Q'_1, Q'_2, Q'_3, Q'_4, and Q'_5 are the quantal analogs of classical operators which generate a five-parameter subgroup of the projective group in the two variables x, t. On inspecting the table one observes that Q'_4 and Q'_5 are members of a Heisenberg subalgebra, so the classical and quantal subgroups are not isomorphic. In this subgroup, Q'_6 generates transformations which multiply a wave function by the phase factor $e^{i\alpha}$. The generators Q'_1, Q'_2, Q'_3, are members of a Lie algebra that has no Abelian or Heisenberg subalgebra.

Table 9.1. Commutators of self-adjoint invariance generators of Eq. (9.6.1).

$[Q'_A, Q'_B]$						
$Q'_A \backslash Q'_B$	Q'_1,	Q'_2,	Q'_3,	Q'_4,	Q'_5,	Q'_6
Q'_1	0	$2iQ'_1$	$-iQ'_2$,	0	iQ'_4	0
Q'_2		0	$2iQ'_3$	$-iQ'_4$	iQ'_5	0
Q'_3			0	iQ'_5	0	0
Q'_4				0	$-i$	0
Q'_5					0	0
Q'_6						0

The corresponding classical and quantal three-parameter groups are locally isomorphic. The transformations of x and t generated by Q'_3, Q'_4, Q'_5 are also those of a subgroup. This is the Galilei group if, in the case of Q'_4 one considers the group parameter to be the relative velocity of two classical frames of reference used to measure time and position coordinates. Q'_3, Q'_5 commute with H_{op}, but not Q'_4 which commutes with S_{op}. Q'_1, Q'_2 commute with S_{op} only on the surface defined by $y(x,t) - u = 0$ when $S_{op}y(x,t) = 0$.

9.7 Eigenfunctions of the Constants of Motion of a Free-Particle

The eigenfunctions of the constants of motion of a quantal system are analogs of classical trajectories labeled by the values of constants of motion. The eigenvalues represent the values of these constants of motion. The classical trajectories of Hamiltonian systems may have definite values of every constant of motion of the system, but the wave functions of the corresponding quantal system can, in general, only have eigenvalues for commuting constants of motion. In the quantal, as in the classical case, transformations that leave invariant the equations of motion may be used to convert solutions of unpertubed equations of motion to solutions of equations perturbed by functions of the group generators.

The solutions of the time-dependent Schrödinger equation, (9.6.1a), that are eigenfunctions of the commuting operators $p_{op} = -i\partial/\partial x = Q'_5$ and $i\partial/\partial t = Q'_3$, are necessarily eigenfunctions of H. They may be chosen as

$$y_k(x,t) = (1/\sqrt{2\pi})\exp(ikx)\exp(-ik^2t/2). \tag{9.7.1}$$

One has

$$Q'_3 y_k(x,t) = (k^2/2)y_k(x,t) = E\, y_k(x,t) = H\, y_k(x,t), \tag{9.7.2}$$

and

$$p_{op}y_k(x,t) = p\, y_k(x,t), \; p = k.$$

Using the Schrödinger scalar product, these y_k are orthonormal in the generalized sense of Dirac's delta function:

$$\langle y_k, y_{k'}\rangle = \int_{-\infty}^{\infty} dx\, y_k^*(x,t)y_{k'}(x,t) = \delta(k' - k). \tag{9.7.3}$$

In this and subsequent relations, we use the standard notation where y^* represents the complex conjugate of y.

As k may take on a continuous range of values, the general solution y(x,t) of (9.6.1) is a continuous linear combination of energy eigenfunctions:

$$y(x,t) = \int_{-\infty}^{\infty} dkf(k)y_k(x,t), \tag{9.7.4}$$

a Fourier transform of y(k).

If Q is the self-adjoint Schrödinger operator corresponding to a classical constant of motion q(x,t), and if y(x,t) is the normalized wave function of a system, the mean value of measurements of the value of q(x, t) is

$$\begin{aligned} q_y \langle y, Qy\rangle &= \int_{-\infty}^{\infty} dx\ y^*(x,t)Qy(x,t) \\ &= \int_{\infty}^{\infty} dx(Qy(x,t))^* Qy(x,t) = \langle Qy, y\rangle. \end{aligned} \tag{9.7.5}$$

The time translation operator $\exp(iaQ'_3) = \exp(-a\partial/\partial t)$ multiplies y by a phase factor:

$$\exp(-a\partial/\partial t)y(x,t) = y'(x,t) = \exp(iak^2/2)y(x,t). \tag{9.7.6}$$

The space translation operator $\exp(b\partial/\partial x)$ multiplies y(x,t) by the phase factor $\exp(i\beta k)$. In either case, the other Schrödinger operators Q' will not alter these factors of the general form exp(iq). Writing $y' = \exp(iq_y)y$, one has

$$Q'y' = Q'\exp(iq_y)y = \exp(iq_y)Q'y. \tag{9.7.7a}$$

If y is an eigenfunction of Q', then so also is y', and both have the same eigenvalues. Whether y is or not an eigenfunction of Q',

$$y'^*Q'y' = y^*\exp(-iq_y)Q'\exp(iq_y)Q'\exp(iq_y)y = y^*Q'y. \tag{9.7.7b}$$

If y is normalizable then one has

$$\langle y', Q'y'\rangle = \langle y, Q'y\rangle. \tag{9.7.7c}$$

For the one-dimensional motion of a single particle, the classical Galilei transformations that relate observations made in coordinate systems moving at constant relative velocity have, as their generator, the PB operator of the constant of motion $tp_x - x$. The corresponding Schrödinger operator is Q'_4. As Q'_4 does not commute with either Q'_3 or Q'_5, one cannot expect to find sets of common eigenfunctions of all three generators. One concludes that, in general, it is the expectation values of Schrödinger operators, rather than their eigenvalues, that are invariant under the totality of operations of the Galilei group. The previous statement allows exceptions. For example, the time and space translation invariant eigenfunction $y_0(x,t) = (1)$ satisfies $Q'_4 y_0 = 0$.

The eigenfunctions of Q'_4 with eigenvalue λ are of the general form

$$\Phi_\lambda(x,t) = N(t)\exp((i/t)(x^2/2 + \lambda x)), \tag{9.7.8}$$

where N(t) is an arbitrary normalizable function of t. Thus one may choose $N(t) = (1/\sqrt{2\pi})\exp(if(t))$, with f being an arbitrary function of t. Observers whose clocks and meter sticks agree at $t = 0$, will agree on the value of λ if they are moving with uniform relative velocity.

Now, because Q'_1 commutes with Q'_4, one can replace N(t) with a function of t which makes Φ_λ(x,t) an eigenfuntion of Q'_1. Letting C be a constant, one finds the resulting functions

$$\begin{aligned}\Phi_{nl}(x,t) &= C\,t^{-1/2}\exp(-in/t)\exp((i/t)(x^2/2 + \lambda x))\\ &= C\,t^{-1/2}\exp((i/t)(x^2/2 + \lambda x - n)).\end{aligned} \tag{9.7.9}$$

One finds that these functions satisfy Schrödinger's equation if $n = -\lambda^2/2$. The resulting functions,

$$y_\lambda(x,t) = C\,t^{-1/2}\exp((i/t)(x+\lambda)^2/2), \tag{9.7.9}$$

also satisfy the relations

$$Q'_1 y_\lambda(x,t) = -(\lambda^2/2)y_\lambda(x,t), \quad Q'_4 y_\lambda(x,t) = \lambda y_\lambda(x,t). \tag{9.7.10}$$

The probability function is

$$y_\lambda(x,t)^* y_\lambda(x,t) = C^*C/t, \tag{9.7.11}$$

so

$$\langle y_{\lambda'}, y_\lambda\rangle = t^{-1}\int dxC^*C = t^{-1}C^*C\delta(\lambda' - \lambda).$$

Consequently, y_λ is not normalizable for all times. The $t^{-1/2}$ factor ensures that the resulting wave functions can only be normalized for fixed nonzero values of t. These observations lead to the conclusions that:

a. *Time-dependent Schrödinger equations may be left invariant by transformations that do not leave the Schrödinger scalar product invariant.*
b. *Time-dependent Schrödinger equations may have solutions, $y_n(x,t)$, with constents of motion whose eigenvalues, n, do not change with time, but the probability of finding a system in the eigenstate y_n(x,t) can vary with time, even though the eigenstate remains the same. This can be true even when the eigenvalues correspond to values of classical constants of motion.*

In the case involved here, it will be noticed that this unpleasantness did not appear for wavefunctions that are eigenstates of Q_4', but it did appear when they became eigenstates of Q_1'. Using the analysis and methods of Sec. 5.3, the reader will find that transformations with generator U_1 do not leave dx invariant, and that those with generator U_2 also do not leave dx invariant. As a result, the Schrödinger scalar product is not left invariant by transformations with either generator. On the other hand, the generators U_3, U_4, U_5, generate transformations that leave invariant the Schrödinger scalar product, and Q_3', Q_4', Q_5' leave invariant the time-dependent Schrödinger equation. These generators correspond to the classical generators of the Galilei group. One concludes that the Schrödinger Eq. (9.6.1a) and the Schrödinger scalar product (9.7.3) are both invariant under the transformations of the Galilei group.

This concludes our discussion of the quantal free-particle carried out to illustrate some of the general connections established in Secs. 9.1–7, their physical interpretations and consequences. The next section illustrates further connections.

9.8 Dynamical Symmetries of the Schrödinger Equations of a Harmonic Oscillator

For a system with potential energy function v(x), the determining equations corresponding to (9.6.9) are

$$q_{xx}^{00} + 2i\, q_t^{00} + 4v\, q_x^{00} + 2\partial v/\partial x q_x^{10} = 0, \tag{9.8.1a}$$

$$q_{xx}^{10} + 2q_x^{00} + 2i\, q_t^{10} = 0, \tag{9.8.1b}$$

$$q_{xx}^{01} + 2i\, q_t^{01} - 4i\, q_x^{10} = 0, \tag{9.8.1c}$$

$$q_x^{01} = 0. \tag{9.8.1d}$$

All of these equations, except the top one, are the same as the corresponding equations in (9.6.9); the strategy for solving both sets is the same. For a one-dimensional harmonic oscillator with

$$H = (p_{op}^2 + kx^2)/2, \quad p_{op} = -i\partial/\partial x, \tag{9.8.2}$$

one finds the following solutions of these determining equations:

$$\begin{aligned} Q_1 = 1, \quad Q_3 = \exp(i\sqrt{k}t)(\partial/\partial x + \sqrt{k}x),\\ Q_2 = i\partial/\partial t, \quad Q_4 = \exp(-i\sqrt{k}t)(\partial/\partial x - \sqrt{k}x). \end{aligned} \tag{9.8.3}$$

If one allows second-order differential operators, the resulting determining equations also produce Q_3^2, Q_4^2, Q_3Q_4, and of course, H_{op} itself.

Setting $\omega = \sqrt{k}$, define

$$\begin{aligned} Q_3 &\to b_t = (2\omega)^{-1/2} \exp(i\omega t)(\omega x + \partial/\partial x), \\ Q_4 &\to b_t^+ = (\omega)^{-1/2} \exp(-i\omega t)(\omega x - \partial/\partial x). \end{aligned} \tag{9.8.4}$$

Here b_t^+ is the adjoint of b_t, and the operators are those introduced by Dirac. They obey the following commutation relations:

$$\begin{gathered} [i\partial/\partial t, b_t] = -\omega b_t, \quad [i\partial/\partial t, b_t^+] = \omega\, b_t^+, \\ [b_t, b_t^+] = 1. \end{gathered} \tag{9.8.5a–c}$$

The first two equations imply that

$$[H, b_t] = -\omega\, b_t, \quad [H, b_t^+] = \omega\, b_t^+. \tag{9.8.5d,e}$$

One also finds

$$H = (\omega/2)(b_t^+ b_t + b_t b_t^+). \tag{9.8.6}$$

Much of the material described in this section will be found in many textbooks on quantum mechanics. It is not as well known that the Lie-algebraic properties of the oscillator are closely related to those of a variety of systems. In Chapter 11, for example, it will be shown that they are but a special case of Lie-algebraic properties of Gegenbauer's equation, which plays a central role in the theory of the orthogonal groups O(n).

Let us continue with this example to elucidate a general pattern of thought. If y_E has energy eigenvalue E, then (9.8.5e) implies

$$H\, b_t^+ y_E = b_t^+ (H + \omega) y_E = b_t^+ (E + 1) y_E = (E + \omega) b_t^+ y_E. \tag{9.8.7}$$

Thus $b_t^+ y_E$ is an eigenfunction of H with eigenvalue $E + \omega$. This shows that H (and hence $i\partial/\partial t$) has a discrete energy spectrum.

The lowering operator b_t has an important invariant function; that is, there is a function y'_0 such that

$$(2\omega)^{-1/2} \exp(i\omega t)(\partial/\partial x + \omega x) y'_0 = 0. \tag{9.8.8a}$$

Solving this ODE yields

$$y'_0 = A' \exp(-\omega x^2/2). \tag{9.8.8b}$$

Operating on this with H one finds

$$H y'_0 = E_0 y'_0, \quad E_0 = (\omega/2). \tag{9.8.9}$$

As the lowering operator b_t annihilates y'_0 there is no solution of the time independent Schrödinger equation with eigenvalue E less than E_0: the Lie algebraic properties of the system establish that its energy spectrum has a bottom level.

An integration gives the result that the normalized wave function of this bottom level is

$$\mathrm{y}_0 = (\omega^{1/2}/\mathrm{p}^{1/4})\exp(-\omega \mathrm{x}^2/2)\exp(-\mathrm{i}\omega \mathrm{t}/2). \tag{9.8.12}$$

The raising operator b_t^+ acts on y_0 to give

$$\mathrm{y}_1' = 2^{1/2}\omega \mathrm{x}\,\mathrm{y}_0. \tag{9.8.13}$$

The commutation relations (9.8.3) require that this is a solution of the Schrödinger equations with eigenvalue $3\omega/2$. They furthermore require

$$(\mathrm{b}_\mathrm{t}^+)^n \mathrm{y}_0 = \mathrm{y}_n' \tag{9.8.14}$$

to be a solution with eigenvalue $(n + 1/2)\omega$. These y_n' are not normalized. Using the relations obeyed by b_t^+ and b_t one can show that

$$\int \mathrm{dx}\ \mathrm{y}_n'^{*}\mathrm{y}_n' = n! \int \mathrm{dx}\,\mathrm{y}_0^{*}\mathrm{y}_0 = n!. \tag{9.8.15}$$

Consequently the functions

$$\mathrm{y}_n = (1/(n!)^{1/2})(\mathrm{b}_\mathrm{t}^+)^n \mathrm{y}_0. \tag{9.8.16}$$

are normalized. These energy eigenfunctions can be conveniently constructed using a computer math program to evaluate (9.8.16). The operator b_t^+ has been so defined that the functions obtained do not introduce any phase factors into the usual harmonic oscillator functions,

$$\mathrm{y}_n = \mathrm{N}_n \mathrm{H}_n(\omega^{1/2}\mathrm{x})\exp(-\omega \mathrm{x}^2/2)\exp(-\mathrm{i}\omega \mathrm{t}/2). \tag{9.8.17}$$

The raising and lowering operators act upon the normalized functions as follows:

$$\begin{aligned} \mathrm{b}_\mathrm{t}^+ \mathrm{y}_n &= (n+1)^{1/2}\mathrm{y}_{n+1}, \\ \mathrm{b}_\mathrm{t} \mathrm{y}_n &= n^{1/2}\mathrm{y}_{n-1}. \end{aligned} \tag{9.8.18}$$

If we replace x by q, and $\partial/\partial \mathrm{x}$ by ip, in b_t and b_t^+ we obtain the classical functions whose Poisson bracket operators are standard Lie generators corresponding to b_t and b_t^+. One finds

$$\begin{aligned} b_t \rightarrow B_t &= \exp(i\omega t)(i\partial/\partial p - \partial/\partial q + (q - ip)\partial/\partial E)/2^{1/2}, \\ b_t^+ \rightarrow B_t^+ &= \exp(-i\omega t)(-i\partial/\partial p - \partial/\partial q + (q + ip)\partial/\partial E)/2^{1/2}. \end{aligned} \tag{9.8.19}$$

Evaluating the commutator of these operators, one obtains, as expected from Sec. 9.4, that the Poisson bracket operator analog of the quantal relation $[b_t, b_t^+] = 1$ is

$$[B_t, B_t^+] = 0. \tag{9.8.20}$$

The quantal group and the group generated by the PB operators are not isomorphic. The group generated by b_t, b_t^+ and 1 has no isomorphic analog in the phase space PQET. In contrast, the Lie algebra of the quantum mechanical *double-jump* operators $Q_m = (b_t)^2$, $Q_p = (b_t^+)^2$, and the labeling operator $Q_{pm} = b_t^+ b_t + b_t b_t^+$ *is* isomorphic to the corresponding classical Lie algebra of PB operators.

9.9 Use of the Oscillator Group in Pertubation Calculations

When one can express a Hamiltonian in the form $H = H_0 + v$, where $H_0 = T + V$, and v is in some sense "small", much can be learned if v can be expressed as a function of operators of the invariance group of the Schrödinger equations involving H. The situation is analogous to that occuring in classical Hamiltonian mechanics, and once again the simplest example is provided by the harmonic oscillator. On setting $t = 0$ in the oscillator b_t, b_t^+,

$$\begin{aligned} b_t \rightarrow b &= (2\omega)^{-1/2}(\omega x + \partial/\partial x), \\ b_t^+ \rightarrow b^+ &= (2\omega)^{-1/2}(\omega x - \partial/\partial x). \end{aligned} \tag{9.9.1a,b}$$

Hence

$$x = (2/\omega)^{1/2}(b^+ + b), \quad \partial/\partial x = (2\omega)^{1/2}(b^+ - b). \tag{9.9.2a,b}$$

If $v = v(x)$ can be expanded in a power series in x, it can be expressed as a power series in $(b^+ + b)$. One can then use the expressions to determine the levels that interact under ther influence of the perturbation, and evaluate matrix elements of v. If v is velocity dependent, one can make similar use of $(b^+ - b)$. Examples can be found in the exercises.

9.10 Concluding Observations

In this chapter we have dealt with fundamental principles needed to uncover and interpret the intrinsic symmetries of systems obeying Schrödinger equations or their momentum-space analogs. This required a treatment of correspondences between classical and quantal systems, and the development of a Lie algebraic extension of the correspondence principle. The extension enables one to determine whether the intrinsic symmetries of the quantum mechanical equations correspond to geometric symmetries in classical phase space.

A method for determining the generators of invariance transformations of Schrödinger equations has been described. Several important and straightforward, but less than obvious, aspects of the mathematics were noted. The resulting *determining equations* were used to determine the generators of transformations which leave invariant the time-dependent Schrödinger equation governing a free-particle moving in one spatial dimension, and the generators of transformations that leave invariant the time-dependent Schrödinger equation for a one-dimensional harmonic oscillator.

The free-particle equation was found to admit a six-parameter dynamical group isomorphic to the corresponding group of Hamilton's equations in PQET space. The intrinsic symmetry of the Schrödinger equation does not, however, correspond to the geometric symmetry of the classical free particle in phase space. The six-parameter group contains a three-parameter subgroup of Galilei transformations. The Schrödinger scalar product is invariant under the transformations of this subgroup. The generators of its time translation and space translation groups commute. The remaining generator of the Galilei group does not commute with these. The Schrödinger scalar product is not invariant under all transformations of the six-parameter group that leave invariant the time-dependent Schrödinger equation of the free-particle. It follows from these observations that the quantum mechanical consequences of the classical dynamical symmetries are more surprising than one might expect.

The time-dependent Schrödinger equation for the oscillator admits a three-parameter invariance group. Two of its generators are the well known shift operators that generate the energy spectrum of the system, and convert eigenfunctions Ψ_n to Ψ_{n+1} and Ψ_{n-1}. The form of its lowering operator requires that the system has a lowest energy state, and determines the functional form of this ground state. Together with the energy operator, the operators generate a three-parameter Lie group. This group has no isomorphic classical analog: it defines a dynamical symmetry that is not isomorphic to the geometric symmetry in phase space defined by corresponding PB operators.

Historical Note

Lie did not develop a general theory of the invariance transformations of partial differential equations of order greater than one. Noether, in her treatment of the variational principles and conservation laws of mechanics, introduced group generators that were second-order partial derivative operators.[5a] Johnson was the first to generalize Lie's theory of ODEs to PDEs of arbitrary order.[5b] The lack of a theory of integration for the resulting infinite set of differential equations defining finite transformations stymied further developments of Johnson's theory (personal communication). In Ref. 5c it is argued that the generators of invariance transformations of second-order PDEs could not, in general, be second-order differential operators. Unaware of these mathematical developments, but knowing that second-order PDE's of physics admitted invariance generators that are second-order differential operators, several physicists in the early 70's independently began to use second- and higher-order differential operators as group generators.[5d,5e] The theory of invariance transformations of Schrödinger equations was first elaborated in Ref. 5e by Kumei, who then extended the theory to nonlinear PDEs in Ref. 5f. In 1978, Ibragimov and Anderson[5g] found the error in Ref. 5c, the implicit assumption of a finite dimensional space of derivatives, a "finite dimensional jet space". They introduced the name *Lie–Backlund* transformations for the group operations generated by differential operators of arbitrary order. (The transformations connect conceptions originating with Lie and Backlund, but because neither Lie nor Backlund used such transformations, some authors prefer to call them generalized Lie transformations.) There are now many articles that exploit these transformations, and by changing the definition of the base variables, integral-dependent transformations have been developed.[5h]

Exercises

1. Compute the commutator of $\mathrm{QP}_j^{(1,2)}$ with q_j and with $\mathrm{p}_{j\mathrm{op}}^2$.
2. Establish the condition under which the Q_1 and Q_2 of Sec. 9.6 can be expressed in terms of products of Q_3, Q_4, Q_5.

3. Does using the symmetrized self-adjoint form of $\mathrm{it}^2\partial/\partial\mathrm{t}$ in Q_1' produce an operator that leaves invariant the surface defined by $\mathrm{S_{op}y(x,t)} = 0$ in Eq. (9.6.1a)?
4. Solve the determining Eqs. (9.8.1) when the potential $\mathrm{v} = \mathrm{kx}$ or $\mathrm{k/x^2}$, or simply a constant, k.
5. Using the $\mathrm{B_t}$, $\mathrm{B_t^+}$ given in Eqs. (9.8.19), determine the finite transformations of p, q, E, t carried out by the groups generated by $(\mathrm{B_t} + \mathrm{B_t^+})/2$ and $\mathrm{i}(\mathrm{B_t} - \mathrm{B_t^+})/2$.
6. Obtain real linear combinations of the oscillator shift operators and compute their commutation relations. Do the same with the corresponding Poisson bracket operators. Determine the finite transformations of p, q, E, t generated by the Poisson bracket operators.
7. Use the scaling properties of the harmonic oscillator Hamiltonian to prove that

$$\langle \mathrm{Y(x)}, \mathrm{T_{op}Y(x)} \rangle = \langle \mathrm{Y(x)}, \mathrm{VY(x)} \rangle.$$

References

[1] W. Heisenberg, *Z. Physik*, **33** (1925) 879.

[2] M. Born and P. Jordan, *Z. Physik* **34** (1925) 858. For translations, of this and the previous reference, see *Sources of Quantum Mechanics*, ed. B. L. Van Der Warden (North-Holland, Amsterdam, 1967).

[3] a) H. J. Groenewald, *Physica* **12** (1946) 405–460.
b) L. Van Hove, *Acad Roy. Belgique Bull. Cl. Sci. (5)* **37** (1951) 610–620.
c) For a recent mathematical review see, Mark J. Gotay, in Mechanics: From theory to computation, *J. Nonlinear Science* (2000) 271–316.

[4] C. Wulfman, *J. Phys. Chem. A* **102** (1998) 9542-8; *J. Phys. A* **42** (2009) 185301.

[5] a) E. Noether, Nachr. Konig. Gesell. Wissen. Gottingen, Math.-Phys. p. 23; English translation: Transport Theory Stat. Phys. I, Vol. 186 (1971).
b) H. H. Johnson, *Proc. Am. Math. Soc.* **15** (1964) 432, 675 .
c) E. A. Muller and K. Matschat, *Miszellaneen der Angewandten Mechanik* (1962) 190.
d) Z. Khukhunashvili, *Izvestiy Fizika Vyss. Ucheb. Zavadenii* **3** 1971 95–103.
e) S. Kumei, M. Sc. Thesis, University of the Pacific (1972); R. L. Anderson, S. Kumei, C. Wulfman, *Phys. Rev. Lett.* **28** (1972) 988.
f) S. Kumei, *J. Math. Phys.* **16** (1975) 2461; **18** (1977) 256.
g) N. Ibragimov and R. L. Anderson, *Dokl. Akad. Nauk. SSSR* **227** (1976) 539.
h) G. W. Bluman and S. Kumei, *Symmetries and Differential Equations* (Springer, N.Y., 1989).

CHAPTER 10

Spectrum-Generating Lie Algebras and Groups Admitted by Schrödinger Equations

Introduction

The generators of invariance transformations of time-dependent Schrödinger equations provide operators that can change the eigenstates and eigenenergies determined by the corresponding time-independent equations. The previous chapter considered two examples of time-dependent Schrödinger equations whose invariance groups provide operators that commuted with H, along with those that did not commute with H. The eigenvalues of the group generators that commuted with H were used to label the energy eigenstates, and the generators that did not commute with H were used to change eigenstates with a given energy into states with a different energy. Both group generators and group operators were used for the latter purpose. Groups whose generators and/or operators convert an energy eigenstate into such sets of eigenstates are said to be spectrum generating.

Infeld and Hull developed a method which uncovered spectrum generating groups,[1] long before the extension of Lie's theory of differential equations to Schrödinger equations was available.[2] Their method, the *factorization method*, depends upon factoring a Hamiltonian into a product of raising and lowering operators, and determines the generators of a three-parameter group. A generalization produces Hamiltonians termed *supersymmetric*, along with the Lie algebra of their three-parameter groups.[3] In this chapter we develop a characterization of spectrum-generating Lie algebras and use the generalized Lie theory of differential equations to determine generators of the dynamical groups of a number of quantum mechanical systems. Appendix 10.A lists references in which the dynamical groups of other Schrödinger equations are displayed.

As a Hamiltonian operator, H, commutes with all the generators of its invariance group, it commutes with the Casimir operator(s) of its invariance group. Consequently, the eigenfunctions of H can be chosen to be eigenfunctions of both Casimir operators and a commuting set of generators of the group. An example is provided by the wave functions ψ_{nlm} of a hydrogen atom. The quantum number n fixes the eigenvalue of the quadratic Casimir operator of the O(4) invariance group of H, the quantum number l fixes the eigenvalue of the Casimir operator L^2 of its SO(3) subgroup, and m is the eigenvalue of a generator of this SO(3). Operators of SO(4) can alter the value of l and m, but not n. Operators of the invariance group of the time-dependent Schrödinger equation can alter n, as well as l, m. Hydrogen atoms, and most quantal systems, have both discrete and continuous spectra. The forthcoming sections show how the operators that generate these two types of spectra are characterized by their commutation relations.

10.1 Lie Algebras That Generate Continuous Spectra

In Chapter 9 it was shown that the time-dependent Schrödinger equation governing a free particle moving along the x axis admits a Lie generator $Q_2 = x\partial/\partial x + 2t\partial/\partial t$. Q_2 generated transformations that convert a given solution of the Schrödinger equations into a continuous family of solutions, each member of the family having a different energy. This implied the free particle possessed a continuous energy spectrum. The generator Q_2 satisfies the commutation relation

$$[H, Q_2] = -2\,H, \tag{10.1.1a}$$

and hence, also,

$$[i\partial/\partial t, Q_2] = -2\,i\partial/\partial t. \tag{10.1.1b}$$

In the next two paragraphs we establish that commutation relations of this form imply that H and $i\partial/\partial t$ have a continuous spectrum.[4]

Let Q be the generator of an invariance transformation of a time-dependent Schrödinger equation in any number of spatial variables q_j. Suppose Q satisfies the relations

$$[Q, H] = b\,H, \;\; [Q, i\partial/\partial t] = b\,i\partial/\partial t, \quad -\infty < b < \infty. \tag{10.1.2a,b}$$

Let $\Psi(q, t)$, $q = (q_1, q_2, \ldots)$, be any eigenstate of H, and $i\partial/\partial t$, satisfying the equations

$$H\Psi = i\partial\Psi/\partial t, \quad H\Psi = E\Psi, \tag{10.1.3a,b}$$

and physically appropriate boundary conditions. Let Ψ_0 represent the eigenstate with energy E_0. We shall shortly prove that if

$$\Psi_{\alpha b,0} = \exp(-\alpha Q)\Psi_0, \tag{10.1.4a}$$

then

$$H\Psi_{\alpha b,0} = i\partial\Psi_{\alpha b,0}/\partial t = E(\alpha b, 0)\Psi_{\alpha b,0}, \tag{10.1.4b}$$

with

$$E(\alpha b, 0) = \exp(\alpha b)E_0. \tag{10.1.4c}$$

Thus if E_0 is the energy eigenvalue for Ψ_0, then $E(\alpha b, 0)$, as a continuous function of the group parameter α, is the corresponding eigenvalue for $\Psi_{\alpha b,0} = \exp(\alpha Q)\Psi_0$. It follows, that if $\Psi_{\alpha b,0}$ satisfies the required boundary conditions, H must possess a continuous energy spectrum, and $\Psi_{\alpha b,0}$ will represent the eigenstate with energy $E(\alpha b, 0)$.

To prove this claim, we make use of the general expansion:

$$\exp(\alpha Q)B\exp(-\alpha Q) = B + [\alpha Q, B] + (1/2!)[\alpha Q, [\alpha Q, B]] + \cdots \tag{10.1.5}$$

in which the n-th term is 1/n times the commutator of αQ with the previous term. If $[Q, B] = bB$, then (10.1.5) implies

$$\begin{aligned} \exp(\alpha Q)B\exp(-\alpha Q) &= B + \alpha bB + ((\alpha b)^2/2!)B + \cdots \\ &= \exp(\alpha b)B. \end{aligned} \tag{10.1.6}$$

Applying this result to (10.1.3a), and using (10.1.2a), one has

$$\exp(-\alpha Q)H\Psi_0 = \exp(-\alpha Q)E\Psi_0 = E_0\exp(-\alpha Q)\Psi_0,$$

and

$$\exp(-\alpha Q)\,H\exp(\alpha Q)\exp(-\alpha Q)\Psi_0 = E_0\exp(-\alpha Q)\Psi_0, \tag{10.1.7}$$

so

$$\exp(-\alpha b)\,H\exp(\alpha Q)\Psi_0 = E_0\exp(-\alpha Q)\Psi_0.$$

Hence

$$H\Psi_{\alpha b,0} = \exp(\alpha b)E_0\Psi_{\alpha b,0} = E(\alpha b, 0)\Psi_{\alpha b,0}.$$

Applying the same process with H replaced by $i\partial/\partial t$ one obtains the analog of Eq. (9.6.29) in which H is replaced by $i\partial/\partial t$. It follows that *if the Schrödinger equation* $(H - i\partial/\partial t)\Psi = 0$ *has an invariance generator Q which satisfies Eq.* (10.1.2), and $\Psi_{\alpha b,0}$ *satisfies the proper boundary conditions, then* H *and* $i\partial/\partial t$ *have a continuous spectrum.*

10.2 Lie Algebras That Generate Discrete Spectra

Suppose that a time-dependent Schrödinger equation admits a set of generators N_j and $Q_{j\pm}$ which on the solution space $\Psi(q,t)$ satisfy

$$[(H - i\partial/\partial t), N_j] = 0, \quad [(H - i\partial/\partial t), Q_{j\pm}] = 0, \qquad (10.2.1a,b)$$

and which have the following commutation relations

$$[H, N_j] = 0 = [i\partial/\partial t, N_j], \quad [N_j, Q_{j\pm}] = \pm b_j Q_j, \quad j = 1, 2, \ldots, \qquad (10.2.2a,b)$$

where the b_j are finite constants. Equation (10.2.2a) indicates that the eigenfunctions of H may be chosen to be eigenfunctions of the operators N_j. Equation (10.2.2b) implies that if Ψ_{n_j}, is an eigenfunction of N_j with eigenvalue n_j, then

$$N_j Q_{j\pm} \Psi_{n_j} = Q_{j\pm}(N_j \pm b_j)\Psi_{n_j} = (n_j \pm b_j) Q_{j\pm} \Psi_{n_j}. \qquad (10.2.3)$$

Thus $Q_{j\pm}\Psi_{n_j}$ is an eigenfunction of N_j with eigenvalue $(n_j \pm b_j)$, and N_j has a discrete spectrum.

In the general case, one may choose a set of commuting N_j, and label energy eigenfunctions, Ψ, with the eigenvalues n_j of the N_j, so

$$N_j \Psi_{n_j} = n_j \Psi_{n_j}, \qquad (10.2.4a)$$

and

$$H\Psi_n = E(n)\Psi_n, \quad H\Psi_n = i\partial\Psi_n/\partial t, \quad n = (n_1, n_2, \ldots, n_j, \ldots, n_m). \qquad (10.2.4b,c)$$

Then

$$Q_{j\pm}\Psi_n = \Psi_{n'}, \quad n' = (n_1, n_2, \ldots, n_j \pm b_j, \ldots, n_m), \qquad (10.2.4d)$$

where $\Psi_{n'}$ is also a solution of both Schrödinger's equations. As the operators need not convert normalized solutions of Schrödinger equations into normalized solutions, it is convenient to replace (10.2.4d) by

$$Q_{j\pm}\Psi_n = c_\pm(n_j)\Psi_{n'} \qquad (10.2.5)$$

in which Ψ_n and $\Psi_{n'}$ are normalized, and the $c_\pm(n_j)$ are normalizing factors.

If for some value of n_j, $c_+(n_j) = 0$, the spectrum of N_j has an upper bound; if for some value of n_j, $c_-(n_j) = 0$, its spectrum has a lower bound. If the $Q_{j\pm}$ do not commute with H, then different values of n_j produce different values of E. If the $Q_{j\pm}$ commute with H, then E is not changed by the shifting of the values of n_j, and H has a spectrum that is degenerate.

Equations (10.2.1c) are equations of the type used by Cartan to classify the Lie algebras of semisimple groups. The symplectic groups Sp(2N) are semisimple groups. The Lie-Algebraic Correspondence Principle, and the fact that the invariance groups of Hamilton's equations are symplectic groups, gives (10.2.1c) a wide range of applications in quantum mechanics.

One often wishes to label eigenfunctions with the eigenvalues of Casimir operators rather than group generators. To alter the eigenvalues of the Casimir operator of a group, one must use operators that do not commute with it — operators not contained in the group. The next two sections deal with examples in which the label l in functions ψ_{lm} can be changed by operators which do not commute with the generators of SO(3).

10.3 Dynamical Groups of N-Dimensional Harmonic Oscillators

Let the Schrödinger Hamiltonian of a one-dimensional oscillator be expressed in terms of Cartesian coordinates by

$$\mathrm{H_j} = -(\partial/\partial \mathrm{q_j})^2/2 + (\mathrm{k_j}/2)\mathrm{q_j^2}. \tag{10.3.1a}$$

Setting $\omega_\mathrm{j} = \sqrt{\mathrm{k_j}}$, and substituting $\mathrm{y_j} = \mathrm{q_j}\sqrt{\omega_\mathrm{j}}$ into (10.3.1a), converts it to

$$\mathrm{H_j} = (\omega_\mathrm{j}/2)(-\partial^2/\partial \mathrm{y_j^2} + \mathrm{y_j^2}). \tag{10.3.1b}$$

The Dirac energy raising and lowering operators, $\mathrm{H_j}$ are, respectively,

$$\mathrm{b_j^\dagger} = \exp(-i\omega_\mathrm{j}\mathrm{t})(\mathrm{y_j} - \partial/\partial \mathrm{y_j}), \tag{10.3.1c}$$

$$\mathrm{b_j} = \exp(i\omega_\mathrm{j}\mathrm{t})(\mathrm{y_j} + \partial/\partial \mathrm{y_j}). \tag{10.3.1d}$$

Expressed in terms of them,

$$\mathrm{H_j} = (\omega_\mathrm{j}/2)(\mathrm{b_j^\dagger b_j} + \mathrm{b_j b_j^\dagger}). \tag{10.3.1e}$$

The Hamiltonian for the corresponding N-dimensional oscillator is

$$\mathrm{H} = \Sigma_1^\mathrm{N}\mathrm{H_j}. \tag{10.3.2}$$

The oscillator is isotropic if all the ω_j have the same value, ω. The ground state of such an oscillator has energy

$$\mathrm{E_0} = (\mathrm{N}/2)\omega. \tag{10.3.3}$$

In 1956, Baker[5] showed that because this H is invariant under operations of the unitary group, U(N), the n-th energy level of the oscillator, with energy

$$\mathrm{E}_n = \mathrm{E}_0 + n\omega, \tag{10.3.4}$$

has degeneracy

$$\mathrm{D} = \frac{(\mathrm{N} + n - 1)!}{n!(\mathrm{N} - 1)!}. \tag{10.3.5}$$

Fowler had previously determined D by working out the way in which n quanta can be assigned N degrees of freedom.[6] The operator that shifts a quantum of energy from degree of freedom j to degree of freedom k is

$$\mathrm{b}_\mathrm{k}^\dagger \mathrm{b}_\mathrm{j} = (1/2)(\mathrm{y}_\mathrm{k} - \partial/\partial \mathrm{y}_\mathrm{k})(\mathrm{y}_\mathrm{j} + \partial/\partial \mathrm{y}_\mathrm{j}). \tag{10.3.6}$$

Its adjoint, $\mathrm{b}_\mathrm{j}^\dagger \mathrm{b}_\mathrm{k}$, carries out the inverse transfer. These $\mathrm{N}(\mathrm{N}-1)$ operators together with the N operators H_j comprise a set of N^2 operators that generate the degeneracy group U(N).

Several different types of dynamical groups are known for N-dimensional isotropic harmonic oscillators. One of them, denoted $\mathrm{N}_{\mathrm{U(N)}}$,[7] is generated by these operators, together with the 2N Dirac shift operators and the identity operator of the Heisenberg group "N".

A dynamical group of three-dimensional isotropic harmonic oscillators that is particularly useful is the group Sp(6), whose Lie algebra contains the double-jump operators mentioned above. It may be used to interconvert states containing odd numbers of excitation quanta, or interconvert states containing even numbers of these quanta. For a more extensive discussion, the reader is directed to Chapter 20 of Wybourne's book,[13] and references therein.

The dynamical group of non-isotropic oscillators is the same as that of the isotropic ones, but the generators $\mathrm{b}_\mathrm{k}^\dagger \mathrm{b}_\mathrm{j}$ involving degrees of freedom with different ω's become time-dependent operators which alter energies, and cease being generators of the degeneracy group. However, if the ω's in two different degrees of freedom are so related that there are two integers I_a and I_b such that $\omega_\mathrm{a}/\omega_\mathrm{b} = (\mathrm{I}_\mathrm{a}/\mathrm{I}_\mathrm{b})$, then degeneracies will appear when $\mathrm{n}_\mathrm{a}/\mathrm{n}_\mathrm{b} = (\mathrm{I}_\mathrm{b}/\mathrm{I}_\mathrm{a})$.

The dynamical symmetries of harmonic oscillators have come to play a large role in nuclear shell theory. This begins with the assumption that nucleons move in approximate harmonic oscillator potential wells. It then develops the consequences of j–j coupling between spins of the nucleons. Appendix 10.B lists a number of references which will introduce the reader to this application of dynamical symmetries which is now very extensive.

10.4 Linearization of Energy Spectra by Time Dilatation; Spectrum-Generating Dynamical Group of Rigid Rotators

The spectrum of eigenvalues of a Cartan labeling operator N_j is linear:

$$\begin{aligned} n_j \to n_j' &= n_j + b_j, \\ n_j' \to n_j'' &= n_j' + b_j = n_j + 2b_j, \end{aligned} \tag{10.4.1}$$

and so forth. When the energy of a system is not a linear function of its quantum numbers, requiring that N_j satisfy

$$[H, N_j] = 0, \quad [i\partial/\partial t, N_j] = 0, \tag{10.4.2a,b}$$

does not ensure that it satisfies

$$[N_j, Q_{j\pm}] = db_j Q_{j\pm}, \quad d = 1, 2, \ldots. \tag{10.4.2c}$$

When one knows the dependence of E upon a quantum number n, methods developed by Kumei[8] enable one to circumvent this difficulty. They produce determining equations for Cartan labeling and shift operators analogous to those obtained for the harmonic oscillator in Sec. 9.8.

Given the Schrödinger equations

$$H\Psi = i\partial\Psi/\partial t, \quad H\Phi_n = E(n)\Phi_n, \tag{10.4.3a,b}$$

and let

$$\Psi = \Sigma c_n \Phi_n(q) \exp(-iE(n)t). \tag{10.4.3c}$$

One begins by making a transformation that converts $\exp(-iE(n)t)$ to $\exp(-int)$. To accomplish this, let $E = \lambda(n)$, and let N be an operator $N(q, p_{op})$ with eigenvalues n, so $H = \lambda(N)$, and N commutes with H. The dilatation operator

$$D = \exp(\alpha t\partial/\partial t), \quad \alpha = \ln(N/\lambda(N)), \tag{10.4.4a}$$

then converts

$$t \to t' = e^{\alpha} t = \frac{N}{\lambda(N)} t, \tag{10.4.4b}$$

and

$$\partial/\partial t \to \partial/\partial t' = e^{-\alpha}\, \partial/\partial t = \frac{\lambda(N)}{N} \partial/\partial t. \tag{10.4.4c}$$

In these expressions N must not, of course, be a function of t or $\partial/\partial t$. As N commutes with H, the operator (10.4.4a) converts

$$(H - i\partial/\partial t)\Sigma c_n \Phi_n(q) \exp(-iE(n)t)$$

to

$$\begin{aligned}
&\mathrm{D}(\mathrm{H} - \mathrm{i}\partial/\partial \mathrm{t})\, \Sigma \mathrm{c}_n \Phi_n(\mathrm{q}) \exp(-\mathrm{i}\mathrm{E}(n)\mathrm{t}) \\
&\quad = (\mathrm{H} - \mathrm{D}\,\mathrm{i}\partial/\partial \mathrm{t}\mathrm{D}^{-1})\mathrm{D}\Sigma \mathrm{c}_n \Phi_n(\mathrm{q}) \exp(-\mathrm{i}\mathrm{E}(n)\mathrm{t}) \\
&\quad = (\mathrm{H} - (\lambda(\mathrm{N})/\mathrm{N})\mathrm{i}\partial/\partial \mathrm{t})\Sigma \mathrm{c}_n \Phi_n(\mathrm{q}) \exp(-\mathrm{i}n\mathrm{t}).
\end{aligned} \tag{10.4.5}$$

As

$$\mathrm{N}\, \Phi_n(\mathrm{q}) \exp(-\mathrm{i}n\mathrm{t}) = \mathrm{i}\partial/\partial \mathrm{t}\, \Phi_n(\mathrm{q}) \exp(-\mathrm{i}n\mathrm{t}) = n\, \Phi_n(\mathrm{q}) \exp(-\mathrm{i}n\mathrm{t}), \tag{10.4.6a}$$

on writing

$$\Psi' = \Sigma \mathrm{c}_n \Phi_n(\mathrm{q}) \exp(-\mathrm{i}n\mathrm{t}), \tag{10.4.6b}$$

one obtains a transformed version of (10.4.3a,b), which has been converted to

$$(\mathrm{H} - \lambda(\mathrm{N}))\Psi' = 0, \tag{10.4.6c}$$

and, equivalently to

$$(\mathrm{H} - \lambda(\mathrm{i}\partial/\partial \mathrm{t}))\Psi' = 0. \tag{10.4.6d}$$

As both N and $\mathrm{i}\partial/\partial\mathrm{t}$ have linear spectra, both may be considered to be Cartan labeling operators. Consequently, on seeking operators $\mathrm{Q}(\mathrm{q}, \mathrm{t}, \mathrm{p}_{\mathrm{op}}, \mathrm{i}\partial/\partial \mathrm{t})$ that generate transformations which leave equation (10.4.6d) invariant, one expects to find Cartan shift operators $\mathrm{Q}_{\pm}$ for N and $\mathrm{i}\partial/\partial\mathrm{t}$.

The determination of the spectrum-generating group of a rigid rotor provides an example of the foregoing general discussion. Because, of the rotor's assumed rigidity, its Schrödinger wave function does not have a radial dependence, and the Schrödinger equations for a rotor with moment of inertia I, are

$$(\mathrm{L}^2/2\mathrm{I} - \mathrm{i}\partial/\partial \mathrm{t})\Psi(\mathrm{t}, \theta, \phi) = 0, \quad (\mathrm{L}^2/2\mathrm{I} - \mathrm{E}_l)\Psi_l(\mathrm{t}, \theta, \phi) = 0, \tag{10.4.7a,b}$$

with

$$\mathrm{E}_l = l(l+1)\hbar^2/2\,\mathrm{I}. \tag{10.4.7c}$$

Here, L^2 is the total angular momentum operator. The operator $\mathrm{L_z} = -\mathrm{i}\,\hbar\partial/\partial\phi$ commutes with L^2, and has eigenvalues $m\hbar$. The two operators, together have eigenstates Φ_{lm} $(\mathrm{t}, \theta, \phi)$, with m an integer constrained by the condition $|m| \leq l$. The $\Psi(\mathrm{t}, \theta, \phi)$ can be a linear combination of such eigenstates, and the general solution can be put in the form

$$\Psi(\mathrm{t},\theta,\phi) = \Sigma_l\Sigma_m \mathrm{c}_{lm}\Phi_{lm}(\mathrm{t},\theta,\phi) \tag{10.4.7d}$$

the c_{lm} being constants.

Multiplying Eqs. (10.4.7a,b,c) by $2\mathrm{I}/\hbar^2$ gives

$$(L^2 - \mathrm{i}\partial/\partial t)\Psi = 0, \quad (L^2 - E_l)\Psi = 0, \quad E_l = l(l+1),$$

with

$$L^2 = (\mathrm{L}^2/\hbar^2) = -\{(\sin(\theta)^{-1}\partial/\partial\theta(\sin(\theta)\partial/\partial\theta) + \sin(\theta^{-2}\partial^2/\partial\phi^2\},$$
$$t = (\hbar^2/2\mathrm{I})\mathrm{t}, \quad E_l = (2\mathrm{I}/\hbar^2)E_l. \tag{10.4.8a–c}$$

Expressed in terms of spherical harmonics, $\mathrm{Y}_{lm}(\theta,\phi)$, the general solution of (10.4.8a) is

$$\Psi = \Sigma_l\Sigma_m \mathrm{c}_{lm}\psi_{lm}(\mathrm{x},\phi,\mathrm{t}), \tag{10.4.9}$$

with

$$\psi_{lm}(\mathrm{x},\phi,\mathrm{t}) = \exp(-\mathrm{i}l(l+1)\mathrm{t})\mathrm{Y}_{lm}(\theta,\phi).$$

Given (10.4.9) one seeks operators Q that satisfy

$$(\mathrm{L}^2 - \mathrm{i}\partial/\partial t)\mathrm{Q}\Psi = 0. \tag{10.4.10}$$

The operator that depends only upon q, $\mathrm{p_{op}}$ and has eigenvalues l is

$$\gamma = (-1 + \mathrm{sqrt}(1 + 4L^2))/2. \tag{10.4.11a}$$

As $L^2 \to \gamma(\gamma+1)$, Eqs. (10.4.4) require that

$$\mathrm{D_t} = \exp(\ln(\gamma + l)t\,\partial/\partial t). \tag{10.4.11b}$$

$\mathrm{D_t}$ converts $\exp(-\mathrm{i}l(l+1)\mathrm{t})$ to $\exp(-\mathrm{i}l\mathrm{t})$ and carries out the transformation

$$\mathrm{i}\partial/\partial\mathrm{t} \to (\mathrm{i}\partial/\partial\mathrm{t})(\mathrm{i}\partial/\partial\mathrm{t} + 1). \tag{10.4.12}$$

It converts Eq. (10.4.10) to a determining equation for the Q':

$$\mathrm{AQ'f} = 0, \tag{10.4.13a}$$

with

$$\mathrm{A} = L^2 - (\mathrm{i}\partial/\partial t)(\mathrm{i}\partial/\partial t + 1), \tag{10.4.13b}$$

$$\mathrm{f} \equiv \mathrm{D_t}\Psi = \Sigma \mathrm{c}_{lm}\exp(-\mathrm{i}l\mathrm{t})\mathrm{Y}_{lm}(\theta,\phi), \tag{10.4.13c}$$

$$\mathrm{Q}' = \mathrm{D_t}\mathrm{Q}\mathrm{D_t^{-1}}. \tag{10.4.13d}$$

It is convenient to make the change of variables $\mathrm{x} = \cos(\theta)$, $\mathrm{y} = \sin(\theta)$ whence

$$L^2 = -\mathrm{y}^2\partial^2/\partial\mathrm{x}^2 + 2\mathrm{x}\partial/\partial\mathrm{x} - \mathrm{y}^{-2}\partial^2/\partial\phi^2, \quad \mathrm{y}^2 = 1 - \mathrm{x}^2. \tag{10.4.14}$$

Seeking Q' that contain derivatives no higher than the second, Kumei set

$$Q' = Q'^{\phi\phi}\partial^2/\partial\phi^2 + Q'^{x\phi}\partial^2/\partial\phi\partial x + Q'^{x}\partial/\partial x + Q'^{\phi}\partial/\partial\phi + Q'^{t}\partial/\partial t + Q'^{0}. \tag{10.4.15}$$

The derivatives in the determining Eq. (10.4.13) produce 13 linearly independent derivatives of f. This provides 14 linearly independent functions which one may choose to be:

$$f, f_t, f_\phi, f_x, f_{\phi\phi}, f_{x\phi}, f_{t\phi}, f_{tt}, f_{xt}, f_{\phi\phi\phi}, f_{x\phi\phi}, f_{tt\phi}, f_{xt\phi}, f_{t\phi\phi} \tag{10.4.16}$$

Inserting (10.4.13b) and (10.4.15) into (10.4.13a) and collecting the terms that multiply each of the functions in (10.4.16) he used the linear independence of the functions to obtain the following 14 equations:

$$\begin{gathered}
Q_t'^{\phi\phi} = 0, \quad Q_t'^{x\phi} = 0, \quad Q_x'^{\phi\phi} - y^{-4}Q_\phi'^{x\phi} = 0, \quad Q_x'^{x\phi} + xy^{-2}Q'^{x\phi} = 0,\\
Q_\phi'^{\phi\phi} - xy^{-2}Q'^{x\phi} = 0, \quad Q_x'^{t} - y^{-2}Q_t'^{x} = 0, \quad Q_t'^{t} - Q_x'^{x} - xy^{-2}Q'^{x} = 0,\\
Q_t'^{\phi} - y^{-2}Q_\phi'^{t} = 0, \quad AQ'^{x\phi} - 2Q'^{x\phi} - 2y^{-2}Q_\phi'^{x} - 2y^2Q_x'^{\phi} = 0,\\
AQ'^{\phi\phi} + 2y^{-2}Q_x'^{x} - 4x^2y^{-4}Q'^{x} - 2y^2Q_\phi'^{\phi} = 0,\\
AQ'^{x} - 2Q'^{x} - 4xQ_x'^{x} - 4x^2y^{-2}Q'^{x} - 2y^2Q_x'^{0} = 0,\\
AQ'^{t} + 2Q_t'^{0} + 2iQ_x'^{x} + 2ixy^{-2}Q'^{x} = 0,\\
AQ'^{\phi} - 2y^{-2}Q_\phi'^{0} = 0, \quad AQ'^{0} = 0.
\end{gathered} \tag{10.4.17}$$

Solving these equations yields 14 linearly independent operators for Q', three of which are products of the others. The remaining 11 operators may be chosen to be

$$\begin{gathered}
Q_0' = 1, \quad Q_{1,2}' = e^{\pm i\phi}(y\,\partial/\partial x - i(\pm x/y)\partial/\partial\phi),\\
Q_3' = \partial/\partial\phi, Q_4' = e^{-it}(iy^2\partial/\partial x + x\,\partial/\partial t - ix),\\
Q_5' = e^{it}(-iy^2\partial/\partial x + x\partial/\partial t), \quad Q_6' = \partial/\partial t,\\
Q_{7,8}' = e^{it}e^{\pm i\phi}(ixy\partial/\partial x \pm y^{-1}\partial/\partial\phi + y\partial/\partial t),\\
Q_{9,10}' = e^{-it}e^{\pm i\phi}(-ixy\partial/\partial x \pm (-y^{-1})\partial/\partial\phi + y\partial/\partial t - iy).
\end{gathered} \tag{10.4.18}$$

The corresponding operators that act on the functions Ψ of (10.4.9) are then

$$Q_j = D_t^{-1}Q_j'D_t. \tag{10.4.19}$$

The Q_j and Q_j', with $j = 1, \ldots, 10$, generate SO(3,2) groups.

10.5 The Angular Momentum Shift Algebra; Dynamical Group of the Laplace Equation

The results of the previous section may be used to construct shift and labeling operators for the angular momentum eigenfunctions, and skew- or self-adjoint operators of the SO(3, 2) group that acts upon these functions. We shall suppose that units have been chosen in which $\hbar$ has unit value. The simplest way to proceed is to makes use of the functions having the form given in (10.4.13c). The Schrödinger scalar product of two such functions, f and f', is

$$\langle \mathrm{f}, \mathrm{f}' \rangle = \int_0^{2\pi} \mathrm{d}\phi \int_0^{\pi} \sin(\theta)\mathrm{d}\theta\, \mathrm{f}^*\mathrm{f}'. \tag{10.5.1a}$$

Using the substitution $\mathrm{x} = \cos(\theta)$, $\sin(\theta)\mathrm{d}\theta = -\mathrm{d}(\cos(\theta)) = -\mathrm{dx}$, yields

$$\langle \mathrm{f}, \mathrm{f}' \rangle = \int_0^{2\pi} \mathrm{d}\phi \int_{-1}^{1} \mathrm{dx} \quad \mathrm{f}^*\mathrm{f}'. \tag{10.5.1b}$$

Thus $\partial/\partial \mathrm{x} = -(1/\sin(\theta))\partial/\partial\theta$ is skew-adjoint, while $\mathrm{i}\partial/\partial\mathrm{x}$ is self-adjoint:

$$\langle \mathrm{f}, (\mathrm{d}/\mathrm{dx})\mathrm{f}' \rangle = \langle (-\mathrm{d}/\mathrm{dx})\mathrm{f}, \mathrm{f}' \rangle = -\langle (\mathrm{d}/\mathrm{dx})\mathrm{f}, \mathrm{f}' \rangle,$$

and

$$\begin{aligned}\langle \mathrm{f}, (\mathrm{id}/\mathrm{dx})\mathrm{f}' \rangle &= \mathrm{i}\langle \mathrm{f}, (\mathrm{d}/\mathrm{dx})\mathrm{f}' \rangle \\ &= \langle (\mathrm{id}/\mathrm{dx})\mathrm{f}, \mathrm{f}' \rangle = (\mathrm{i}^*)\langle (-\mathrm{d}/\mathrm{dx})\mathrm{f}, \mathrm{f}' \rangle.\end{aligned} \tag{10.5.2}$$

We shall suppose that the spherical harmonics $\mathrm{Y}_{lm}(\theta, \phi)$ are normalized to unity and that they have Condon and Shortley phase factors, so that

$$\mathrm{Y}_{lm}(\theta, \phi)^* = (-1)^m \mathrm{Y}_{l-m}(\theta, \phi). \tag{10.5.3}$$

The m labeling and m shift operators for the functions $\exp(-\mathrm{i}lt)\mathrm{Y}_{lm}$ that correspond to the standard operators for the Y_{lm} are [9]:

$$\mathrm{M}_0 = -\mathrm{i}\mathrm{Q}_3' = -\mathrm{i}\partial/\partial\phi, \tag{10.5.4a}$$

$$\begin{aligned}\mathrm{M}_+ &= -\mathrm{Q}_2' = \mathrm{e}^{+\mathrm{i}\phi}(-\mathrm{y}\partial/\partial\mathrm{x} + \mathrm{i}\mathrm{x}\mathrm{y}^{-1}\partial/\partial\phi) \\ &= \mathrm{e}^{+\mathrm{i}\phi}(\partial/\partial\theta + \mathrm{i}\cot(\theta)\partial/\partial\phi),\end{aligned} \tag{10.5.4b}$$

$$\begin{aligned}\mathrm{M}_- &= \mathrm{Q}_3' = \mathrm{e}^{-\mathrm{i}\phi}(\mathrm{y}\partial/\partial\mathrm{x} + \mathrm{i}\mathrm{x}\mathrm{y}^{-1}\partial/\partial\phi) \\ &= \mathrm{e}^{-\mathrm{i}\phi}(-\partial/\partial\theta + \mathrm{i}\cot(\theta)\partial/\partial\phi).\end{aligned} \tag{10.5.4c}$$

They have long been known to satisfy the standard commutation relations of the A_2 Lie algebra of SO(3):

$$[\mathrm{M}_0, \mathrm{M}_+] = \mathrm{M}_+, \quad [\mathrm{M}_0, \mathrm{M}_-] = -\mathrm{M}_-, \quad [\mathrm{M}_+, \mathrm{M}_-] = 2\mathrm{M}_0. \tag{10.5.5}$$

Their actions on normalized functions ψ_{lm} are

$$\begin{aligned} \mathrm{M}_0\psi_{lm} &= \mathrm{m}\psi_{lm}, \\ \mathrm{M}_+\psi_{lm} &= ((l-m)(l+m+1))^{1/2}\psi_{lm+1}, \\ \mathrm{M}_-\psi_{lm} &= ((l+m)(l-m+1))^{1/2}\psi_{lm-1}. \end{aligned} \tag{10.5.6}$$

The $\mathrm{M}_\pm$ are skew-adjoint:

$$< \psi_{lm'}, \mathrm{M}_\pm\psi_{lm} >= - < \mathrm{M}_\pm\psi_{lm}, \psi_{lm'} > . \tag{10.5.7}$$

In terms of Cartesian coordinates,

$$\mathrm{z}_1 = \mathrm{r}\sin(\theta)\sin(\phi), \quad \mathrm{z}_2 = \mathrm{r}\cos(\theta)\sin(\phi), \quad \mathrm{z}_3 = \mathrm{r}\cos(\phi):$$

$$\begin{aligned} \mathrm{L}_{12} &= \mathrm{M}_0 = -\mathrm{i}(\mathrm{z}_1\partial/\partial\mathrm{z}_2 - \mathrm{z}_2\partial/\partial\mathrm{z}_1), \\ \mathrm{L}_{23} &= (\mathrm{M}_+ + \mathrm{M}_-)/2 = -\mathrm{i}(\mathrm{z}_2\partial/\partial\mathrm{z}_3 - \mathrm{z}_3\partial/\partial\mathrm{z}_2), \\ \mathrm{L}_{31} &= -\mathrm{i}(\mathrm{M}_+ + \mathrm{M}_-)/2 = -\mathrm{i}(\mathrm{z}_3\partial/\partial\mathrm{z}_1 - \mathrm{z}_1\partial/\partial\mathrm{z}_3). \end{aligned} \tag{10.5.8}$$

The $\mathrm{L}_{\mathrm{ij}} = -\mathrm{L}_{\mathrm{ji}}$ obey standard quantum mechanical SO(3) commutation relations

$$[\mathrm{L}_{\mathrm{ab}}, \mathrm{L}_{\mathrm{bc}}] = -\mathrm{i}\,\mathrm{L}_{\mathrm{ac}}. \tag{10.5.8}$$

The Casimir operator of the group is L^2, and for each value of l the set of functions ψ_{lm} provide the basis for a $(2l+1)$ dimensional UIR of the group SO(3) with the generators (10.5.8)

The determining equations for the Q's also yield an l labeling operator,

$$\mathrm{L}'_0 = \mathrm{i}\partial/\partial t$$

and l shift operators,

$$\mathrm{L}'_+ = \mathrm{i}\mathrm{Q}_4, \quad \text{and} \quad \mathrm{L}'_- = -\mathrm{i}\mathrm{Q}_5,$$

with

$$\begin{aligned} \mathrm{L}'_+ &= \mathrm{e}^{-\mathrm{it}}(-\mathrm{y}^2\partial/\partial\mathrm{x} + \mathrm{ix}\,\partial/\partial\mathrm{t} + \mathrm{x}) = \mathrm{e}^{-\mathrm{it}}(\sin(\theta)\partial/\partial\theta + \cos(\theta)(\mathrm{i}\partial/\partial\mathrm{t} + 1)), \\ \mathrm{L}'_- &= \mathrm{e}^{\mathrm{it}}(\mathrm{y}^2\partial/\partial\mathrm{x} + \mathrm{ix}\,\partial/\partial\mathrm{t}) = \mathrm{e}^{\mathrm{it}}(-\sin(\theta)\partial/\partial\theta + \cos(\theta)(\mathrm{i}\partial/\partial\mathrm{t})). \end{aligned} \tag{10.5.10a, b}$$

It is convenient to define the l labeling operator to be

$$\mathrm{L}'_0 = \mathrm{i}\partial/\partial\mathrm{t} + 1/2, \tag{10.5.10c}$$

The l shift operators have the following action on the $\psi_{lm} = \exp(-\mathrm{i}\,l\,\mathrm{t})\mathrm{Y}_{lm}$:

$$\mathrm{L}'_\pm\psi_{lm=\mathrm{c}_\pm(l,m)}\psi_{l\pm1m}$$

with

$$c_+(l,m) = \{(2\,l+3)/(2\,l+1)\}^{1/2}((l+1+m)(l+1-m))^{1/2},$$

and

$$c_-(l,m) = \{(2l+1)/(2l-1)\}^{1/2}((l+m)(l-m))^{1/2}. \qquad (10.5.11)$$

The L′s satisfy the commutation relations

$$[L'_0, L'_+] = L'_+, \quad [L'_0, L'_-] = -L'_-, \quad [L'_+, L'_-] = 2L'_0. \qquad (10.5.12)$$

The operators

$$L'_{45} = -L'_{54} = L'_0, \quad K'_{34} = K'_{43} = i(Q'_4 + Q'_5)/2,$$

and

$$K'_{35} = K'_{53} = (Q'_4 - Q'_5)/2. \qquad (10.5.13)$$

obey the quantum mechanical SO(2,1) commutation relations

$$[L'_{ab}, K'_{bc}] = -i\,K'_{ac}, \quad [K'_{ab}, K'_{bc}] = -i\,L'_{ac}. \qquad (10.5.14)$$

No linear combination of L'_+ and L'_- is self-adjoint under the scalar product (10.5.1a). The similarity transformation

$$L'_\pm \to \underline{L}'_\pm = L_0^{'1/2} L'_\pm L_0^{'-1/2} \qquad (10.5.15)$$

corrects this, and removes the factors in curly brackets from the expressions for the $c_\pm(l,m)$ in (10.5.11), so that

$$\underline{L}'_+ \psi_{lm} = ((l+1+m)(l+1-m))^{1/2} \psi_{l+1m} \qquad (10.5.16a)$$

and

$$\underline{L}'_- \psi_{lm} = ((l+m)(l-m))^{1/2} \psi_{l-1m}. \qquad (10.5.16b)$$

The transformation does not affect the other group generators L'. All may be subjected to the inverse of the spectrum-linearizing transformation, which does not change the relations (10.5.6) and (10.5.16).

When written in spherical polar coordinates, Laplace's equation,

$$\Delta\Psi = 0, \qquad (10.5.17a)$$

becomes

$$\{\partial^2/\partial r^2 + (2/r)\partial/\partial r - L^2/r^2\}\Psi(r,\theta,\phi) = 0. \qquad (10.5.17b)$$

Separating it yields solutions of the form

$$\Psi = \Sigma_l \Sigma_m c_{lm} r^l Y_{lm}(\theta, \phi). \tag{10.5.17c}$$

In this case, the spectrum may be linearized by the dilation operator $D_r = \exp(\ln(r\partial/\partial r))$. Now

$$\Delta r^l Y_{lm}(\theta, \phi) = 0, \tag{10.5.18}$$

consequently, D_r converts (10.5.15) to

$$\{l(l+1)/r^2 - L^2/r^2\}\Sigma_l \Sigma_m c_{lm} r^l Y_{lm}(\theta, \phi) = 0. \tag{10.5.19}$$

In this case it has been the radial variable r, rather than time, t, which has been used in the linearization of the spectrum of an operator.

In Cartesian coordinates, the invariance generators of the Laplace equation are:

the generators of the Euclidean group E(3):

$$T_j = \partial/\partial x_j, \quad R_{jk} = x_j\partial/\partial x_k - x_k\, \partial/\partial x_j; \tag{10.5.20a,b}$$

the generator of dilations

$$D = \Sigma\, x_j\partial/\partial x_j + 1/2, \tag{10.5.20c}$$

and the three generators of special conformal transformations:

$$S_j = 2x_j D - (x_1^2 + x_2^2 + x_3^2)\partial/\partial x_j. \tag{10.5.20d}$$

They satisfy commutation relations isomorphic to those of SO(4,1). This is most easily seen in quantum mechanical applications if one makes the identifications

$$L_{jk} = -i\, R_{jk}, \quad L_{j4} = -iS_j, \quad K_{45} = -iD, \quad K_{j5} = -iT_j, \;\; j = 1, 2, 3. \tag{10.5.21}$$

The operator $\exp(\boldsymbol{\alpha} \cdot \mathbf{S})$ has the following action on functions $f(\mathbf{x})$:

$$\exp(\boldsymbol{\alpha} \cdot \mathbf{S})f(\mathbf{x}) = (1 - 2\boldsymbol{\alpha} \cdot \mathbf{x} + \alpha^2 x^2)^{-1/2} f(\mathbf{x}'), \tag{10.5.22}$$

with

$$\mathbf{x}' = (\mathbf{x} - \boldsymbol{\alpha}\mathbf{x}^2)/(1 - 2\boldsymbol{\alpha}{\cdot}\mathbf{x}\mathbf{x}^2).$$

Solutions of the Laplace equation are also converted to solutions by the reflections $R_j x_k = -\delta_{jk} x_k$, *and by the inversion transformation*, U, defined by

$$\mathrm{Uf}(\mathbf{x}) = \sqrt{(\mathrm{x}^{-2})}\mathrm{f}(\mathbf{x}/\mathrm{x}^2). \qquad (10.5.23)$$

The reflections and the inversion are their own inverse, and

$$\mathrm{UT_jU^{-1}} = -\mathrm{S}, \quad \mathrm{UR_{jk}U^{-1}} = \mathrm{R_{jk}}, \quad \mathrm{UDU^{-1}} = -\mathrm{D}. \qquad (10.5.24)$$

On the space of solutions, Ψ, of the Laplace equation, because $\Delta\Psi = 0$, one has

$$\mathbf{T}\cdot\mathbf{T}\,\Psi = 0, \quad \mathbf{S}\cdot\mathbf{S}\,\Psi = 0, \quad \mathbf{T}\cdot\mathbf{R}\,\Psi = 0, \qquad (10.5.25\text{a–c})$$

and

$$(\mathbf{R}\cdot\mathbf{R}{+}\mathrm{D}^2)\Psi = 1/4\Psi. \qquad (10.5.25\text{d})$$

In Chapter 11 the reader will find that the invariance group of Laplace's equation and its generalization to Euclidean four-space becomes relevant to the theory of the hydrogen atom. In Chapter 14, the generalization of the group to Minkowski space provides invariance transformations of Maxwell's equations.

10.6 Dynamical Groups of Systems with Both Discrete and Continuous Spectra

The invariance groups of quantum mechanical systems with both discrete and continuous spectra are noncompact. These noncompact groups have at least a one-parameter compact subgroup, and also at least a one-parameter noncompact subgroup. Generators of the compact subgroup have discrete spectra, in contrast to those of the noncompact subgroup that have continuous spectra.

Time-dependent Schrödinger equations invariant under any of the locally isomorphic three-parameter groups, SO(2, 1), SU(1, 1), Sp(2, r), exemplify the general principles involved. When their three generators $\mathrm{R_k}$ are functions of real variables and derivatives, they obey the commutation relations

$$[\mathrm{R_1, R_2}] = -\mathrm{R_3}, \quad [\mathrm{R_2, R_3}] = \mathrm{R_1}, \quad [\mathrm{R_3, R_1}] = \mathrm{R_2}. \qquad (10.6.1\text{a})$$

The corresponding quantum mechanical generators $\mathrm{Q_k} = -\mathrm{i}\,\mathrm{R_k}$ satisfy

$$[\mathrm{Q_1, Q_2}] = -\mathrm{iQ_3}, \quad [\mathrm{Q_2, Q_3}] = \mathrm{iQ_1}, \quad [\mathrm{Q_3, Q_1}] = \mathrm{iQ_2}. \qquad (10.6.1\text{b})$$

The Casimir invariant can be chosen to be either

$$\mathrm{Q_3^2 - Q_1^2 - Q_2^2}, \qquad (10.6.2)$$

or its negative. Q_3 is the generator of the compact one-parameter subgroup of the groups; Q_1 and Q_2 are the generators of their noncompact subgroups.

The operators $Q_\pm = Q_1 \pm iQ_2$ obey the following commutation relations:

$$[Q_3, Q_\pm] = \pm Q_\pm, \quad [Q_+, Q_-] = 2Q_3. \tag{10.6.3}$$

The $Q_\pm$ thus shift the eigenvalues of Q_3 by one unit, so Q_3 has a discrete spectrum. An operator that is a linear combination of Q_3 and a generator of the noncompact subgroup, e.g., an operator of the form

$$Q = c_1\, Q_1 + c_3\, Q_3, \tag{10.6.4}$$

can be converted to a constant times Q_3 if $|c_3| > |c_1|$, and to a constant times Q_1 if $|c_1| > |\, c_3|$. This can be accomplished by the action of a finite transformation of the group. In this case one makes use of the relations

$$\exp(i\alpha Q_2) Q_1 \exp(-i\alpha Q_2) = Q_1 \cosh(\alpha) + Q_3 \sinh(\alpha), \tag{10.6.5a}$$

$$\exp(i\alpha Q_2) Q_3 \exp(-i\alpha Q_2) = Q_3 \cosh(\alpha) + Q_1 \sinh(\alpha), \tag{10.6.5b}$$

which have the consequence

$$\begin{aligned}&\exp(i\alpha Q_2) Q \exp(-i\alpha Q_2)\\ &\quad = Q_1(c_1 \cosh(\alpha) + c_3 \sinh(\alpha)) + Q_3(c_3 \cosh(\alpha) + c_1 \sinh(\alpha)).\end{aligned} \tag{10.6.5c}$$

If, for example, $c_3 > c_1$, one may require that α is such that

$$c_1 \cosh(\alpha) + c_3 \sinh(\alpha) = 0, \tag{10.6.5d}$$

whence

$$\tanh(\alpha) = -c_1/c_3, \quad \alpha = (1/2)\ln\{(c_3 - c_1)/(c_3 + c_1)\}. \tag{10.6.5e,f}$$

10.7 Dynamical Group of the Bound States of Morse Oscillators

Morse's potential,

$$V(r) = d(\exp(-2a(r - r_o)) - 2\exp(-a(r - r_o))), \tag{10.7.1a}$$

gives rise to Schrödinger equations that can be solved exactly for systems with no angular momentum. Illustrated in **Fig. 10.7.1**, it provides reasonable approximations to the potential governing vibrational motions in diatomic molecules. Assuming eigenstates ψ_n of energy E_n and the time-dependent Schrödinger equation

$$(H - i\partial/\partial t)\Psi = 0, \tag{10.7.1b}$$

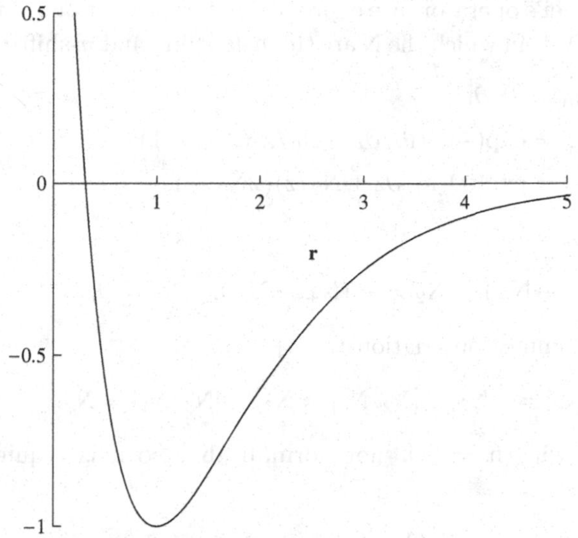

Fig. 10.7.1 Morse potential.

Morse showed that for the bound states of H,

$$E_n = -(a^2/2m)(n + 1/2 - k)^2, \quad k = (2md)^{1/2}/a. \tag{10.7.1c}$$

The operator

$$D = \exp(\ln((n_{op} + 1/2 - k)/H)t\partial/\partial t), \tag{10.7.2}$$

linearizes the spectrum of $-\, i\partial/\partial t$ in (10.7.1b) and converts $\exp(-iE_n t)$ to $\exp(-int)$. Kumei[8] set up and solved the determining equations for the generators Q' of the dynamical group that leaves invariant the resulting analog of (10.7.1), *viz.*

$$\begin{aligned}\{(-1/2m)(\partial^2/\partial r^2 + r^{-1}\partial/\partial r) + V(r)\\ -(a^2/2m)(\partial/\partial t)^2\}\, \Sigma c_n \exp(-int)\psi_n = 0.\end{aligned} \tag{10.7.3}$$

In solving his Schrödinger equation, Morse[10] introduced the change of variable

$$r \to z = 2d\exp(-a(r - r_o)). \tag{10.7.4}$$

When Kumei's operators are expressed in terms of z,t one obtains operators $Q = DND^{-1}$, in which the N are the n-labeling and n-shifting operators

$$N_0 = i\partial/\partial t, \tag{10.7.5a}$$
$$N_+ = \exp(-it)(i\partial/\partial z - N_0/z)(2N_0 + 1) + d, \tag{10.7.5b}$$
$$N_- = \exp(it)(i\partial/\partial z + N_0/z)(2N_0 - 1) - d. \tag{10.7.5c}$$

The operators

$$N_1 = (N_+ + N_-), \quad N_2 = -i(N_+ - N_-), \quad N_3 = \partial/\partial t, \tag{10.7.5d–f}$$

satisfy the commutation relations:

$$[N_1, N_2] = -N_3, \quad [N_2, N_3] = N_1, \quad [N_3, N_1] = N_2. \tag{10.7.6}$$

The function with $n = 0$ is not normalizable, so n is required to be a semi-integer:

$$n = 1/2, 3/2, \ldots, n_{max} \leq d - 1/2. \tag{10.7.7}$$

The dynamical group is SU(1,1).[11]

10.8 Dynamical Group of the Bound States of Hydrogen-Like Atoms

In spherical polar coordinates the time-independent Schrödinger equation with $H = p^2/2m - Z/r$ has solutions which we write as $\psi_{nlm}(p_{0n}r,\ \theta,\ \phi)$. The time-dependent equation can then be expressed as

$$(-1/2)\{\partial^2/\partial r^2 + r^{-1}\partial/\partial r - L^2/r^2 + 2Z/r - 2i\partial/\partial t\}\Sigma c_{nlm}\exp(-iE_n t)\psi_{nlm} = 0, \tag{10.8.1a}$$

where

$$p_{0n} = (-2E_n)^{1/2}, \quad E_n = (-1/2)Z^2/n^2, \tag{10.8.1b,c}$$

and, as in (10.4.14),

$$L^2 = -y^2\,\partial^2/\partial x^2 + 2x\,\partial/\partial x - y^{-2}\,\partial^2/\partial\phi^2. \tag{10.8.1d}$$

Defining

$$P_{0op} = (-2H)^{1/2}, \tag{10.8.2}$$

the operator that linearizes the spectrum,

$$D_t = \exp\{\ln 2Z\ (P_{0op})^{-3} t\partial/\partial t\}, \tag{10.8.3}$$

converts the operator in (10.8.1) to

$$(-1/2)\{\partial^2/\partial r^2 + r^{-1}\partial/\partial r - L^2/r^2 + 2Z/r - 2Z\,(P_{0op})^{-3}2i\partial/\partial t\}, \tag{10.8.4}$$

and converts the $\exp(-iE_n t)$ to $\exp(int)$, but does not alter $\psi_{nlm}(2p_{0n}r, \theta, \phi)$.

The operator identity $P_{0op} = iZ/(\partial/\partial t)$, which holds on the space of solutions of (10.8.4), can be used to convert (10.8.4) to

$$(-1/2)\{\partial^2/\partial r^2 + r^{-1}\partial/\partial r - L^2/r^2 + 2Z/r + Z^2(\partial/\partial t)^{-2}\}. \tag{10.8.5}$$

To eliminate the $(\partial/\partial t)^{-2}$ term Kumei[8] acted on (10.8.4), and the functions $\exp(int)\psi_{nlm}(p_{0n}r, \theta, \phi)$, with the operator

$$D_r = \exp\{\ln(ai\partial/\partial t) r\partial/\partial r\}, \tag{10.8.6}$$

in which $a = -1/2Z$. D_r following the action of D_t converts (10.8.1) to $(\partial^2/\partial t^2/8Z^2)$ times the standard equation[11]

$$\Omega f(r, \theta, \phi, t) = 0,$$

with

$$\Omega = \partial^2/\partial r^2 + r^{-1}\partial/\partial r - L^2/r^2 + 2i\partial/\partial t/r - 1/4 \tag{10.8.7}$$

and

$$f(r, \theta, \phi, t) = \Sigma c_{nlm} \exp(-int)\psi_{nlm}(r/2, \theta, \phi).$$

The ground state solution, ψ_{100}, is $\exp(-(1/2)r)\exp(it)$, and in all these ψ_{nlm} the radial functions are sturmians with the the same coefficient in the exponent, cf., Table 11.3.1 on page 351.

If Q is to represent generators of the Lie group that leaves (10.8.7) invariant, one must have

$$\Omega Q f(r, \theta, \phi, t) I = 0. \tag{10.8.8}$$

The generators that act on the solutions of the original time-dependent Schrödinger equation

$$\exp(-iE_n t)\psi_{nlm}(2p_{0n}r, \theta, \phi) = D_t^{-1}D_r^{-1}\exp(-int)\psi_{nlm}(r/2, \theta, \phi) \tag{10.8.9}$$

are then

$$Q' = D_t^{-1}D_r^{-1}Q\, D_r D_t. \tag{10.8.10}$$

Allowing Q to contain derivative operators of zero, first, and second order in all the variables, Kumei obtained and solved a set of 22 coupled determining equations. The 16 generators Q that are not products of others, are $Q_{16} = 1$, and $Q_1, \ldots, Q_{15}$, which satisfy SO(4,2) commutation relations.

Of these generators, Q, three are the standard angular momentum m labeling operator $M_0 = L_{12}$, and m shift operators $M_\pm = L_{23} \pm iL_{31}$. Another three represent the components of $A_z = L_{34}$, $A_x \pm iA_y = L_{14} \pm iL_{24}$, that act on the functions $\psi_{nlm}(r/2, \theta, \phi)$. These change the value of the angular momentum quantum number, l. One of the operators, $-i\partial/\partial t$, is the n labeling operator for the functions $\exp(-int)\psi_{nlm}(r/2,\ \theta, \phi)$. The remaining eight operators all change the value of n, and hence, E, and are consequently explicit functions of t. In the expressions for the generators given below, $x = \cos(\theta)$, $y = \sin(\theta)$. Generators of compact subgroups are labeled $L_{ab}(= -L_{ba})$. Generators of non-compact subgroups are labeled $K_{ab}(= +K_{ba})$. The generators satisfy the commutation relations

$$[L_{ab}, L_{bc}] = i\,L_{ac}, \quad [L_{ab}, K_{bc}] = i\,K_{ac}, \quad [K_{ab}, K_{bc}] = i\,L_{ac}. \qquad (10.8.11)$$

Using these commutation relations, one is able to obtain the explicit form of all the generators from the following:

$$
\begin{aligned}
&L_{12} = -i\partial/\partial\phi, && (10.8.12)\\
&M_\pm = \exp(\pm i\phi)(ixy^{-1}\partial/\partial\phi - \pm y\partial/\partial x) = (L_{23} \pm iL_{31}), &&\\
&L_{34} = 2\left\{-y^2\partial^2/\partial r\partial x + r^{-1}xy^2\partial^2/\partial x^2 + r^{-1}xy^{-2}\partial^2/\partial\phi^2 + x\partial/\partial r\right. &&\\
&\qquad\left. -2r^{-1}x^2\partial/\partial x - (i/2)x\partial/\partial t\right\}, && (10.8.12a–e)\\
&K_{45} = \cos(t)(-ir\partial/\partial r - i) - \sin(t)(i\partial/\partial t + r/2), && (10.8.13)\\
&L_{56} = i\partial/\partial t. &&
\end{aligned}
$$

The generators not listed above are L_{14}, L_{24}, K_{15}, K_{25}, K_{35}, K_{16}, K_{26}, K_{36}, and K_{46}. The generator K_{46} may, for example, be determined from the relation

$$[L_{56}, K_{45}] = -iK_{46}, \qquad (10.8.12f)$$

which implies that

$$K_{46} = -\sin(t)(-ir\partial/\partial r - i) - \cos(t)(i\partial/\partial t + r/2). \qquad (10.8.12g)$$

When acting on the functions in (10.8.7), $i\partial/\partial t$ multiplies them by n, so in the generators (10.8.12c–e,g) it may be replaced by $N_{op} = Z(P_{0op})^{-1}$. If one is primarily interested in the action of the generators on the position-space wave functions $\psi_{nlm}(r/2, \theta, \phi)$, one may set $t = 0$ in the operators that result from this replacement of $i\partial/\partial t$. This process carries out conversions such as

$$K_{45} \to K'_{45} = -ir\partial/\partial r - i, \quad K_{46} \to K'_{46} = N_{op} + r/2. \qquad (10.8.13a,b)$$

The generators are self-adjoint if one replaces the r^2dr in the Schrödinger scalar product by rdr. (cf. Secs. 10.8, 11.1 and 11.3).

The generators L'_{34} and K'_{45} have the following action on the functions $\psi_{nlm}(r, \theta, \phi)$:

$$L'_{34}\psi_{nlm}(r, \theta, \phi) = a^{+}_{nlm}\psi_{nl+1m}(r, \theta, \phi) + a^{-}_{nlm}\psi_{nl-1m}(r, \theta, \phi) \quad (10.8.14a)$$

with

$$a^{+}_{nlm} = \{(n^2 - (l+1)^2)((l+1)^2 - m^2)\}^{1/2}\{(2l+1)(2l-1)\}^{-1/2} \quad (10.8.14b)$$

$$a^{-}_{nlm} = \{(n^2 - l^2)(l^2 - m^2)\}^{1/2}\{(2l+3)(2l+1)\}^{-1/2} \quad (10.8.14c)$$

and

$$K'_{45}\psi_{nlm}(r, \theta, \phi) = d^{+}_{nlm}\psi_{n+1lm}(r, \theta, \phi) + d^{-}_{nlm}\psi_{n-1lm}(r, \theta, \phi) \quad (10.8.15a)$$

with

$$d^{\pm}_{nlm} = \pm\{n(n \pm 1) - (l(l+1)\}^{1/2} \quad (10.8.15b)$$

The action of the generators on the functions $\exp(-iE_n t)\ \psi_{nlm}(p_{0n}r, \theta, \phi)$ may be obtained by acting with the operator $D_t^{-1}D_r^{-1}$ on the functions $\exp(-in't)\ \psi_{n'l'm}(r, \theta, \phi)$, where in (10.8.15), $n' = n \pm 1$.

In Chapter 11 we will more extensively investigate the isomorphic SO(4, 2) dynamical group of the regularized Schrödinger hydrogen-like atoms.

10.9 Matrix Representations of Generators and Group Operators

Because Schrödinger's differential equations are linear, the differential operators in them may be represented by matrices that operate on vectors defined on the space of solutions of the equations. All operators that act on these wave functions may then be represented by matrices.

As an example, suppose that Ψ is a linear combination of p_{-1}, p_0, p_{+1} atomic wave functions

$$\Psi = c_{-1}\psi_{-1} + c_0\psi_0 + c_{+1}\psi_{+1}, \quad (10.9.1)$$

and that one seeks the matrix representation of the operator

$$L_0 = -i(x\partial/\partial y - y\partial/\partial x). \quad (10.9.2)$$

One has

$$\Psi' \equiv L_0\Psi = (-1)c_{-1}\psi_{-1} + (0)c_0\psi_0 + (+1)c_{+1}\psi_{+1}. \quad (10.9.3)$$

If one considers the ψ_m to be unit vectors, then (10.9.1) and (10.9.3) may be represented, respectively, by the vectors

$$\mathbf{C} = (c_{-1}, c_0, c_{+1}), \quad \mathbf{C}' = (- c_{-1}, 0, c_{+1}). \tag{10.9.4}$$

L_0 is then represented by a 3×3 matrix, $\mathbf{L}_0$, whose off-diagonal elements are zero and diagonal elements are -1, 0, $+1$. The functions ψ_m are said to *provide the basis* for the *matrix representation* of the operator.

The analogous matrix representations of $\mathbf{L}_+ = \mathbf{L}x + i\mathbf{L}y$, and $\mathbf{L}_- = \mathbf{L}x - i\,\mathbf{L}y$, are defined by their action on ψ_m, for which one finds

$$\begin{aligned} \mathbf{L}_+\psi_m &= \psi_{m+1}, \\ \mathbf{L}_-\psi_m &= \psi_{m-1}. \end{aligned} \tag{10.9.5}$$

The reader may verify that the resulting matrix representations, $\mathbf{L}_+$ and $\mathbf{L}_-$, satisfy the commutation relations

$$\begin{aligned} \mathbf{L}_0\mathbf{L}_+ - \mathbf{L}_+\mathbf{L}_0 &= \mathbf{L}_+, \\ \mathbf{L}_0\mathbf{L}_- - \mathbf{L}_-\mathbf{L}_0 &= -\mathbf{L}, \\ \mathbf{L}_+\mathbf{L}_- - \mathbf{L}_-\mathbf{L}_+ &= 2\mathbf{L}_0, \end{aligned} \tag{10.9.6}$$

which are *isomorphic* to the commutation relations obeyed by the corresponding differential operators. As there is a one-to-one correspondence between the matrices and the differential operators and a one to one correspondence between their Lie algebras, it is said that the basis provides a *faithful* representation of the Lie algebra. When such one-to-one correspondence does not exist, the commutation relations of the matrices and that of the differential operators are merely *homomorphic.*

The reader will find that all the matrices $\mathbf{M}$ in (10.9.6) are *unitary,* that is

$$\mathbf{M}^{-1} = \mathbf{M}^{*T}. \tag{10.9.7}$$

It can also be proven that these matrices cannot be reduced to the form

$$\begin{pmatrix} \mathbf{A} & \mathbf{B}' \\ \mathbf{B} & \mathbf{D} \end{pmatrix} \tag{10.9.8}$$

in which the elements of the submatrices $\mathbf{B}$ and/or $\mathbf{B}'$ are all zero. Matrices that cannot be put into this form are said to be *irreducible.* (When both $\mathbf{B}$ and $\mathbf{B}'$ are zero matrices, the matrix (10.9.8) is the product of the two matrices $\mathbf{A}$ and $\mathbf{D}$). All these conditions being fulfilled, the basis ψ_{-1}, ψ_0, ψ_{+1}, of the vector in (10.9.1) is said to be that of a *unitary irreducible representation,* abbreviated UIR.

The eigenfunctions of L^2 with quantum number l provide the bases for the $(2l+1)$-dimensional UIR of SO(3), and of a B_1 Lie algebra. They also provide the basis for infinite dimensional UIR of the SO(3,2) groups of Secs. 10.4 and 10.5.

A straightforward introductory treatment of the properties of group representations will be found in Chapter 2 of Wybourne's monograph, *Classical Groups for Physicists.*[13] An extensive treatment is provided in Barut and Raczka's *Theory of Group Representations and Applications.*[14] Work of Gelfand and Zeitlin provides a very useful scheme for describing and determining UIR of the orthogonal and unitary groups.[15]

We will summarize only a few basic facts here.

1. Compact groups and noncompact groups are not distinguishable if one allows the generators to have complex or imaginary eigenvalues, and/or the group parameters to be complex or imaginary. When the distinction is possible:

1a. The UIR of compact groups are finite dimensional.

1b. The UIR of noncompact groups are either one-dimensional or infinite dimensional. The latter may in some cases be labeled by quantum numbers that take on discrete values, and in other cases they must be labeled by quantum numbers that take on continuously variable values.

2. If **X** is a matrix representation of a group generator Q that is a real function of real variables and real derivative operators, (such as that of $R_z = x\partial/\partial y - y\partial/\partial x$ on the basis provided by x, y, z) and α is a real parameter, the expansion of $\exp(\alpha X)$ in powers of α can only be convergent only if the absolute values of the matrix elements in **X** have an upper bound. (In the example given in (10.9.2), one can replace R_z by L_z, and α by $i\alpha$.)

10.10 Invariant Scalar Products

In using the functions in (10.9.1) to define the vectors (10.9.4) of the previous section it was implicitly assumed that the functions ψ_m could be treated as unit vectors. This assumption presupposes that a scalar product is available so that one can define the meaning of *unit vector*. As the time-independent parts of the ψ_m in the example are the spherical harmonics, $Y_{lm}(\theta,\phi)$, the scalar product that defines their normalization, is

$$\langle f, g\rangle = \int_0^{2\pi} d\phi \int_0^{\pi} \sin(\theta) d\theta\, f^* g. \tag{10.10.1}$$

In this integral, the value of $\sin(\theta)d\theta d\phi$ is invariant under rotations. Because of this, the definition of the unit vectors in (10.9.1) is unambiguous. However, in other cases this is not necessarily so. A similar problem arises when one attempts to define the adjoint of an operator: the adjoint cannot be defined if the scalar product is not defined.

In short, when using a basis to associate matrices with differential operators and their adjoints it becomes necessary to determine what is an appropriate scalar product. Thus, for example in the case of the SO(4) group of the hydrogen atom, it is necessary to determine whether one's definition of Runge–Lenz operators **A** generates transformations that leave invariant the usual scalar product

$$\langle f, g\rangle = \int_0^\infty r^2 dr \int_0^{2\pi} d\phi \int_0^\pi \sin(\theta) d\theta\, f^* g. \tag{10.10.2}$$

In Fock's realization of the SO(4) group the *volume element* involved is the area element on a unit hypersphere in four-space. Fock's stereographic projection transforms it to a modified volume element in momentum three-space, and Fourier transformation converts it to a volume element in position three-space which is r^{-1} times the volume element in (10.10.2)[16]. Consequently, self-adjoint rotation operators on the Fock hypersphere, produce Runge–Lenz operators in momentum space which transform into position-space Runge–Lenz operators that are self-adjoint only if a r^{-1} factor is inserted into the Schrödinger scalar product. Another example is provided by the Morse oscillator. A number of ways of dealing with the invariant scalar product of its SO(2, 1) group will be found in Ref. 17.

In the general case one may use the extended generators of a group to determine the volume element, dv', that is left invariant by the group.

10.11 Direct-Products: SO(3)⊗SO(3) and the Coupling of Angular Momenta

The product of spherical harmonics $Y_{l_1 m_1}(\theta_1, \phi_1) Y_{l_2 m_2}(\theta_2, \phi_2)$ is an eigenstate of $Lz_1 + Lz_2$ with eigenvalue $M = m_1 + m_2$. From a set of such products, with the same values of M, one may build an eigenstate of $\mathbf{L}^2 = (\mathbf{L}_1 + \mathbf{L}_2)^2$ with eigenvalue $L(L+1)$. Because $\mathbf{L}^2$ commutes with $\mathbf{L}_1^2$ and $\mathbf{L}_2^2$, it can also be an eigenstate of $\mathbf{L}_1^2$with eigenvalue $l_1(l_1+1)$, and of $\mathbf{L}_2^2$ with eigenvalue, $l_2(l_2+1)$. If $\psi_{LM(l_1 l_2)}$ is such an eigenstate, one has

$$\psi_{LM(l_1 l_2)} = \Sigma C(l_1 l_2 LM; m_1 m_2) Y_{l_1 m_1}(\theta_1, \phi_1) Y_{l_2 m_2}(\theta_2, \phi_2), \tag{10.11.1}$$

where the sum is over the values of m_1 and m_2 whose sum is M. The coefficients $C(l_1 l_2 LM; m_1 m_2)$ are known as Clebsch–Gordon, or Wigner, coefficients. For normalized spherical harmonics Y_{lm} satisfying a given phase convention they may be completely determined using the commutation relations satisfied by the components of the vectors $\mathbf{L}_1$ and $\mathbf{L}_2$. A discussion of the process will be found in Edmond's monograph.[9]

If $\mathbf{L}_1$ and $\mathbf{L}_2$ represent the angular momentum vectors of particles 1 and 2, respectively, and their sum, $\mathbf{L}$, represents the total angular momentum vector of a two-particle system, then $\psi_{LM(l_1 l_2)}$ represents a state of the system with definite total angular momentum $|\mathbf{L}^2|^{1/2} = |L(L+1)|^{1/2}$. Because

$$\mathbf{L}^2 = \mathbf{L}_1^2 + \mathbf{L}_2^2 + 2\mathbf{L}_1 \cdot \mathbf{L}_2, \tag{10.11.2}$$

$\psi_{LM(l_1 l_2)}$ is also an eigenfunction of $\mathbf{L}_1 \cdot \mathbf{L}_2$:

$$\begin{gathered}(\mathbf{L}_1 \cdot \mathbf{L}_2)\psi_{LM(l_1 l_2)} = \mathrm{B}^2\psi_{LM(l_1 l_2)}, \\ \mathrm{B}^2 = \{L(L+1) - l_1(l_1+1) - l_2(l_2+1)\}/2.\end{gathered} \tag{10.11.3}$$

Physically speaking, one has *coupled* motions allowed to $\mathbf{L}_1$ and $\mathbf{L}_2$ by establishing a fixed angle, β, between them, such that

$$\cos(\beta) = |\mathbf{L}_1 \cdot \mathbf{L}_2|/|\mathbf{L}_1||\mathbf{L}_2| = \mathrm{B}^2/\{l_1(l_1+1)l_2(l_2+1)\}^{1/2}. \tag{10.11.4}$$

In Dirac notation, Eq. (10.11.1) may be expressed as

$$|LM(l_1 l_2)\rangle = \Sigma \mathrm{C}(l_1 l_2 LM; m_1 m_2)|l_1 m_1\rangle|l_2 m_2\rangle. \tag{10.11.5}$$

In this form it holds true for all eigenstates $|lm\rangle$, of $\mathbf{L}^2$, Lz that satisfy Condon and Shortley's phase conventions. As (10.11.1) is derived using only SO(3) commutation relations, and these are the same as those of SU(2), it also applies to the coupling of spin angular momenta. Thus (10.11.5) may be replaced by the more general equation

$$|JM(j_1 j_2)\rangle = \Sigma \mathrm{C}(j_1 j_2 JM; m_1 m_2)|j_1 m_1\rangle|j_2 m_2\rangle. \tag{10.11.6}$$

The coupling of the direct-product of representations generalizes to all Lie groups.

10.12 Degeneracy Groups of Non-interacting Systems; Completions of Direct-Products

We begin this section with an example. Suppose that the components of

$$\mathbf{L}_\mathrm{a} = (\mathrm{L}_{12}, \mathrm{L}_{23}, \mathrm{L}_{31}) \quad \text{and} \quad \mathbf{L}_\mathrm{b} = (\mathrm{L}_{45}, \mathrm{L}_{56}, \mathrm{L}_{64}) \tag{10.12.1}$$

are the generators of the two SO(3) subgroups of the direct-product group $\mathrm{SO(3)_a} \otimes \mathrm{SO(3)_b}$, that K is a constant and that the Hamiltonian of a system is

$$\mathrm{H} = \mathrm{K}(\mathbf{L}_\mathrm{a}^2 + \mathbf{L}_\mathrm{b}^2). \tag{10.12.2}$$

Then $SO(3)_a \otimes SO(3)_b$ is a degeneracy group of the system with Schrödinger equation

$$H\psi = E\psi, \quad E = E_a + E_b = K(l_a(l_a+1) + l_b(l_b+1)). \qquad (10.12.3a)$$

If

$$H\psi = i\partial\psi/\partial t,$$

then

$$\psi = \Sigma_{m_a,m_b} c(l_a l_b m_a m_m) Y_{l_a m_a}(\theta_a, \phi_a) Y_{l_b m_b}(\theta_b, \phi_b) \exp(-i(E_a + E_b)t). \qquad (10.12.3b)$$

Particular sets of solutions are obtained when

$$c(l_a l_b m_a m_b) = C(l_a l_b LM; m_a m_b), \qquad (10.12.3c)$$

and the sum is over m_a, m_b such that $m_a + m_b = M$. Then $\psi = \psi_{LM(l_a l_b)}$.

Now $SO(3)_a \otimes SO(3)_b$ is contained in an SO(6) group, whose Lie algebra contains the components of $\mathbf{L}_a$ and $\mathbf{L}_b$, and nine further rotation generators

$$L_{14}, L_{15}, L_{16}, L_{24}, L_{25}, L_{26}, L_{34}, L_{35}, L_{36}. \qquad (10.12.4)$$

If x_1,x_2,x_3 are Cartesian coordinates of particle a and x_4,x_5,x_6 are those of particle b, then the operators in (10.12.4) are those of generators of rotations acting on a basis which is provided by a product of single-particle states, regardless of whether it is coupled or uncoupled. As an example one might have a six-dimensional basis provided by products of s and p orbitals of two electrons. The rotation operators with generators (10.12.4) would then mix (hybridize) these orbitals in the manner of valence-bond theory. No matter what the choice of two-particle basis is, SO(6) provides operators that can mimic the effects of a much wider variety of interactions than the operators provided by $SO(3) \otimes SO(3)$.

One might term SO(6) the *completion* of the direct-product group $SO(3) \otimes SO(3)$. Though its Casimir operators commute with the generators of $SO(3) \otimes SO(3)$, it is not a degeneracy group of the original system. It can become a dynamical group of the modified system if all the generators L_{jk} are considered to be time-independent versions of invariance generators, Q, of an appropriate time-dependent Schrödinger equation. The discussion of Eqs. (10.8.12) and (10.8.13) provides an example of this relationship between time-independent generators and invariance generators of a time-dependent Schrödinger equations.

Completions of direct-product groups have general properties that arise from relationships between the Casimir operators of the groups. The case at hand illustrates some of these. If one defines $\mathbf{L}_{\rm ab}^2$ to be the sum of the squares of the generators in (10.12.4), then the quadratic Casimir operator of the group is

$$C_2 = \mathbf{L}_{\rm a}^2 + \mathbf{L}_{\rm b}^2 + \mathbf{L}_{\rm ab}^2. \tag{10.12.5}$$

As the components of $\mathbf{L}_{\rm a}$ *and* $\mathbf{L}_{\rm b}$ *commute with the Casimir operator,* C_2*, and with* $\mathbf{L}_{\rm a}^2$ *and* $\mathbf{L}_{\rm b}^2$*, they necessarily commute with* $\mathbf{L}_{\rm ab}^2$. The components of $\mathbf{L}_{\rm ab}^2$ of course commute with C_2, however, they do not commute with the components of $\mathbf{L}_{\rm a}$, $\mathbf{L}_{\rm b}$, nor with those of $\mathbf{L}_{\rm a} + \mathbf{L}_{\rm b}$. Properties such as these have several consequences:

a. Adding a constant times $\mathbf{L}_{\rm ab}^2$ to the Hamiltonian (10.12.2) leaves $\mathbf{L}_{\rm a}^2$ and $\mathbf{L}_{\rm b}^2$, and hence $(\mathbf{L}_{\rm a} + \mathbf{L}_{\rm b})^2$ as constants of motion, so it alters the energy of the unpertubed states by an amount determined by the eigenvalues of C_2.
b. As SO(6) is of rank 3, it also has two further Casimir operators. Adding these to the Hamiltonian (10.17.2) also leaves $l_{\rm a}$, $l_{\rm b}$ and L as labeling eigenvalues of constants of motion.

In both cases, adding the additional Casimir operators to the original Hamiltonian does not alter its eigenstates, it simply alters their energies.

In the general case, the completion of a direct-product group has two useful properties:

i) The Casimir operators of the group obtained by completion of a direct-product group necessarily commute with the generators of the direct-product group and so provide labeling operators whose physical relevance can be directly assessed.
ii) The generators of the enlarged group will have known matrix elements which can be used as a guide in developing approximate treatments of actual physical systems. In the next section, and in Chapter 12, we will discuss several other direct-product groups and their completions.

10.13 Dynamical Groups of Time-dependent Schrödinger Equations of Compound Systems; Many-Electron Atoms

Let $H_{\rm a} = H(\mathbf{r}_{\rm a})$ be the Hamiltonian of system $\bar{\rm a}$, and let $H_{\rm b} = H(\mathbf{r}_{\rm b})$ be that of system $\bar{\rm b}$, and suppose that both Hamiltonians have the same functional form. Let k, k' represent sets of quantum numbers or labels, and let

$$\phi_{a,k} = \phi_k(\mathbf{r}_a),\ \phi_{b,k'} = \phi_{k'}(\mathbf{r}_b),\ \psi_a = \psi(\mathbf{r}_a, t),\ \psi'_b = \psi'(\mathbf{r}_b, t). \tag{10.13.1a–d}$$

Require

$$H_a\phi_{a,k} = E_{a,k}\phi_{a,k}, \quad H_b\phi_{b,k'} = E_{b,k'}\phi_{b,k'}, \tag{10.13.1e,f}$$

and

$$H_a\psi_{a,} = i\partial\psi_{a,}/\partial t, \quad H_b\psi_{a,} = i\partial\psi_b/\partial t. \tag{10.13.1g,h}$$

The time-dependent functions ψ can be expressed as sums (or integrals) over the functions $c(k)\exp(-iE_k t)\phi_k$. We shall suppose that one has available a spectrum-generating algebra of operators

$$Q_a = Q(\mathbf{r}_a, \partial/\partial\mathbf{r}_a, \partial/\partial t), \quad Q_b = Q(\mathbf{r}_b, \partial/\partial\mathbf{r}_b, \partial/\partial t), \tag{10.13.2}$$

for Eqs. (10.13.1e,f), and that these Q satisfy the determining equations arising from the time-dependent Schrödinger equations (10.13.1g,h).

Next consider the compound system with

$$H = H_a + H_b, \tag{10.13.3a}$$

and let

$$\Phi_{a,b} = \Phi_{k,k'} = \phi_{a,k}\ \phi_{b,k'}, \quad \text{and} \quad \Psi_{a,b} = \Psi_{k,k'} = \psi_{a,k}\psi'_{b,k'}. \tag{10.13.3b,c}$$

Then

$$H\,\Phi_{k,k'} = (E_{a,k} + E_{b,k'})\Phi_{k,k'}, \tag{10.13.3d}$$

and

$$\begin{aligned} H\,\Psi_{a,b} &= \psi'_b H_a\psi_a + \psi_a H_b\psi'_b \\ &= i(\psi'_b\partial\psi_a/\partial t + \psi_a\,\psi_b/\partial t). \end{aligned} \tag{10.13.3e}$$

This last equation becomes separable if its right-hand side is replaced by

$$i(\psi'_b\partial\psi_a/\partial t_a + \psi_a\partial\psi_b/\partial t_b). \tag{10.13.3f}$$

To make this replacement one may set $t_a + t_b = t$, which implies that the physical time-evolution operator is

$$\partial/\partial t = \partial/\partial t_a + \partial/\partial t_b. \tag{10.13.4}$$

The dynamical group then becomes a *constrained* direct-product of the dynamical groups G_a and G_b. In it t_a and t_b, as well as $\partial/\partial t_a$ and $\partial/\partial t_b$, can be considered to be mathematically independent until one imposes the physical requirement in (10.13.4).

Completion of such direct-products yields groups which can be the dynamical groups of systems whose Hamiltonian is obtained from $H(\mathbf{r}_a) + H(\mathbf{r}_b)$ by adding interaction terms, or by introducing parameters that produce new potential and/or kinetic energy terms in it. For the group to be a dynamical group of the altered system, the changes in the Hamiltonian must be expressible by functions of the generators of the group, and the solutions Φ and Ψ, of the resulting Schrödinger equations, must be expressible as convergent linear combinations, $(\Sigma + \int)$, of the solutions of (10.13.3):

$$\Phi = \left(\Sigma + \int\right)(c(k,k')\Phi_{k,k'}), \tag{10.13.5a}$$

$$\Psi = \left(\Sigma + \int\right)(c(k,k')\Psi_{k,k'}). \tag{10.13.5b}$$

If these conditions are satisfied the completed group will be a dynamical group of the altered system.

In the case of two-electron atoms, one can begin with $H_a = p_a^2/2 - Z/r_a$, $H_b = p_b^2/2 - Z/r_b$, with eigenstates ψ_a, ψ_b. The basis states of the direct-product $SO(4,2)_a \otimes SO(4,2)_b$ and its completion, SO(8,4), can then be chosen to be $\psi_a\psi_b$. One may add the electron repulsion potential $1/r_{ab}$ to $H_a + H_b$, and use operations available within $SO(4,2)_a \otimes SO(4,2)_b$ or SO(8,4) to deal with the effects of the added potential, as well as those of other interactions of interest. For atoms with N electrons, the analogous group is SO(4N, 2N).

Kato investigated the question of the convergence of the series (10.13.5a) for two-electron atoms of nuclear charge Z when the added term is the electron repulsion potential.[18] He established that the series converges for Z > 4.1, but did not establish that this value is the lowest possible for Z. Even in cases where convergence of many-particle analogs of the series (10.13.5) is problematic, the groups can prove useful if they lead to discoveries of new approximate (or exact) degeneracies and an understanding of their physical consequences.

Chapter 12 contains a discussion of SO(4)⊗SO(4) degeneracies and symmetry breaking in several-electron atoms. It also contains brief discussions of the relevance of its SO(8,4) completion to studies of systems containing more than one electron.

Appendix A. References Providing Invariance Groups of Schrödinger Equations

L. Infeld and T. E. Hull, The Factorization Method, *Rev. Mod. Phys.*, **23** (1951) 21.

B. L. Dunlap and L. Armstrong Jr., SO(2,1) and the Hulthe'n Potential, *Phys. Rev. A* **6** (1972) 1370.

A. O. Barut and G. L. Bornzin, Algebraic solution of the Schrödinger equation for a class of velocity-dependent potentials, *Lett. Nuovo Cimento* **6** (1973) 177.

G. Bluman, S. Kumei and G. J. Reid, New classes of symmetries for partial differential equations, *J. Math. Phys.* **29** (1988) 806.

N. Ibragimov, *CRC Handbook of Lie Analysis of Differential Equations*, I, II, III, (CRC Press 1994–1996).

F. Cooper, A. Khare and U. Sukhatme, *Supersymmetry in Quantum Mechanics* (World Scientific, Sinagapore, 2001).

Appendix B. References to Work Dealing with Dynamical Symmetries in Nuclear Shell Theory

M. Moshinsky, *Group Theory and the Many-body Problem* (Gordon and Breach, N.Y., 1968).

K. Helmers, Symplectic invariants and Flowers' classification of shell model states, *Nucl. Phys.* **23** (1961) 594.

G. Racah, Group theory and spectroscopy, *Erg. D. Exakt. Naturwiss* **37** (1965) 30.

D. J. Rowe, Dynamical symmetries of nuclear collective models, *Prog. Part. Nucl. Phys.* **37** (1996).

Exercises

1a. Show that the quantum mechanical version, -iD, of the dilation operator D defined in (10.5.18), is not self-adjoint under the Schrödinger scalar product, but becomes self-adjoint if one replaces r^2dr by rdr.

1b. Construct symmetrized quantum-mechanical versions of all the other generators in (10.5.18) that are self-adjoint under this scalar product.

2a. Determine the commutation relations of the generators defined in (10.5.18).

2b. Prove that they are isomorphic to commutation relations of the generators of SO(4,1), and establish the relation of the generators in (10.5.18) to the generators $L_{ab} = -L_{ba}$ of the SO(4) subgroup of SO(4, 1), and those, $K_{ab} = K_{ba}$, of its noncompact subgroup.

3a. Completion of the direct-product group SO(2)⊗SO(2) produces SO(4). Consider the circles defined by the one-parameter groups generated by L_{12} and L_{34}, and suppose that these are contained in orthogonal planes, so that $(x_1, x_2)\cdot(x_3, x_4) = 0$. How are these circles affected by rotations generated by L_{23}, L_{14}?

3b. Determine the small rotations in the SO(4) group that leave invariant the scalar product $(x_1, x_2) \cdot (x_3, x_4)$.

References

[1] L. Infeld and T. E. Hull, *Rev. Mod. Phys.* **23** (1951) 21.
[2] Chap. IX Ref. 5.
[3] F. Cooper, A. Khare and U. Sukhatme, *Supersymmetry in Quantum Mechanics* (World Scientific Publishers, Sinagapore, 2001).
[4] R. L. Anderson, S. Kumei and C. E. Wulfman, *Rev. Mex. de Fisica* **21** (1972) 1–33.
[5] G. A. Baker Jr., *Phys. Rev.* **103** (1956) 1119.
[6] R. H. Fowler, *Statistical Mechanics* (Cambridge University Press, N. Y., 1955), Sec. 2.21.
[7] A. O. Barut and A. Bohm, *Phys. Rev. B* **139** (1965) 1107.
[8] S. Kumei, Group Theoretic Properties of Schrödinger Equations– Systematic Derivation, M.Sc. Thesis, Department of Physics, University of the Pacific (Stockton, CA, 1972).
[9] A. R. Edmonds, *Angular Momentum in Quantum Mechanics* (Princeton University Press, Princeton, 1957).
[10] P. M. Morse, *Phys. Rev.* **34** (1929) 57.
[11] P. Cordero and S. Hojman, *Lett. Nuovo Cimento* **4** (1970) 1123.
[12] A. Messiah, *Quantum Mechanics* (John Wiley, N.Y., 1962), Vol. I, Chap. XI.
[13] B. G. Wybourne, *Classical Groups for Physicists* (Wiley-Interscience, N.Y., 1974).
[14] A. O. Barut and R. Raczka, *Theory of Group Representations and Applications*, 2nd edn., (World Scientific, Singapore, 1986).
[15] I. M. Gelfand and M. L. Tsetlein, *Dokl. Akad. Nauk. SSSR* **71** (1950) 25; ibid. 1017.
[16] T. Shibuya, Fock's Representation of Molecular Orbitals, M.Sc. Thesis, Department of Physics, University of the Pacific (Stockton, CA, 1965); cf., T. Shibuya and C. E. Wulfman, *Proc. Roy. Soc.* A **286** (1965) 376.
[17] C. E. Wulfman, *Phys. Rev.* A **54** (1996) R987.
[18] T. Kato, *Perturbation Theory for Linear Operators* (Springer, N. Y. 1966), pp. 410–413.

CHAPTER 11

Dynamical Symmetry of Regularized Hydrogen-like Atoms

Introduction

This chapter relates the dynamical symmetries of one-electron atoms to the dynamical symmetries of regularized classical Keplerian systems. It first describes the quantum mechanical position-space formulation of the regularized system and the generators of its SO(4,2) group, then sets forth the corresponding momentum-space formulation and Fock's projection of its momentum space onto the hypersphere S(3). The chapter concludes with a description of the SU(2) $\otimes$ SU(2) structure of the SO(4) degeneracy group. To simplify the presentation in this and the subsequent chapter, we will use units in which the charge of the electron, e, has unit value, and as in the two previous chapters, will use units in which both the reduced mass of the electron and $\hbar = \mathrm{h}/(2\pi)$ also have unit value.

11.1 Position-space Realization of the Dynamical Symmetries

The self-adjoint Schrödinger analogs of classical PB operators are not, in general, those appropriate for the regularized system in position-space. The reason is that the regularized equation which defines classical Keplerian motion when the Hamiltonian is $\mathrm{p}^2/2 - 1/\mathrm{r}$ has, as its simplest quantum mechanical analog, the equation

$$(\mathrm{W_{op}} - 1)\psi = 0, \qquad (11.1.2\mathrm{a})$$

in which $\mathrm{W_{op}}$ is obtained from Schrödinger's time-independent equation by multiplying it from the left with r:

$$\mathrm{W_{op}} = \mathrm{r}(\mathbf{p}_{\mathrm{op}}^2 - 2\mathrm{E})/2, \quad \mathbf{p}_{\mathrm{op}}^2 = -\boldsymbol{\nabla}^2. \qquad (11.1.2\mathrm{b})$$

When one inserts $r(H_{op} - E)$ into the Schrödinger scalar product

$$\int \psi^* \psi' dv = \int \psi^* \psi' dxdydz = \int \psi^* \psi' r^2 dr \sin(\theta) d\theta d\phi, \tag{11.1.3a}$$

one obtains

$$\int \psi^* r(H_{op} - E)\psi' dv. \tag{11.1.3b}$$

For W_{op} to be self-adjoint it is necessary to alter the definition of the scalar product. If X is an operator, let

$$\langle \psi, X\psi' \rangle \equiv \int \psi^* (1/r) X \psi' dv. \tag{11.1.4a}$$

This implies

$$\langle \psi, r(H_{op} - E)\psi' \rangle = \int \psi^* (H_{op} - E)\psi' dv, \tag{11.1.4b}$$

and

$$\langle \psi, r\psi' \rangle = \int \psi^* \psi' dv. \tag{11.1.4c}$$

Though r and $-r\nabla^2$ are self-adjoint under the scalar product (11.1.4a), the operator $-i\partial/\partial x$ is not. However, operators, such as ordinary angular momentum generators, which commute with r and are self-adjoint under the Schrödinger scalar product, remain self-adjoint under (11.1.4a).

The dynamical symmetry of the classical Keplerian systems investigated in Chapter 8 can be enlarged to an SO(4,2) symmetry.[1–3] The simplest way to do this is to add to the SO(4,1) generators of Sec. 8.6, the five operators:

$$K_{j6} = \{rp_j, \cdot\} \equiv K_{6j}, \ \ j = 1, 2, 3; \quad K_{46} = \{(r/2)(p^2 - 1), \cdot\} \equiv K_{64},$$
$$J_{56} = \{-(r/2)(p^2 + 1), \cdot\} \equiv -J_{65}. \tag{11.1.1}$$

The Lie-algebraic extension of the correspondence principle ensures that this SO(4,2) dynamical symmetry is also a dynamical symmetry of hydrogen-like atoms.

We will use Bednar's form of the appropriate self-adjoint quantum mechanical position-space SO(4,2) generators.[4] These are the fifteen operators:

$$
\begin{aligned}
\mathrm{J_{jk}} &= -\mathrm{i}(\mathrm{q_j}\partial/\partial\mathrm{q_k} - \mathrm{q_k}\partial/\partial\mathrm{q_j}) = -\mathrm{J_{kj}}, \quad \mathrm{j} < \mathrm{k} = 1,2,3;\\
\mathrm{J_{k4}} &= -(\mathrm{x_k}\boldsymbol{\nabla}^2 - 2\partial/\partial\mathrm{x_k} - 2\mathbf{r}\cdot\boldsymbol{\nabla}\partial/\partial\mathrm{x_k} + \mathrm{x_k})/2 \equiv -\mathrm{J_{4k}}, \quad \mathrm{k} = 1,2,3;\\
\mathrm{J_{56}} &= -(\mathrm{r}/2)(\boldsymbol{\nabla}^2 - 1) \equiv -\mathrm{J_{65}}; \quad \mathrm{K_{45}} = -\mathrm{i}(1 + \mathbf{r}\cdot\boldsymbol{\nabla}) \equiv \mathrm{K_{54}};\\
\mathrm{K_{k5}} &= -(\mathrm{x_k}\boldsymbol{\nabla}^2 - 2\partial/\partial\mathrm{x_k} - 2\mathbf{r}\cdot\boldsymbol{\nabla}\partial/\partial\mathrm{x_k} - \mathrm{x_k})/2 \equiv \mathrm{K_{5k}}, \quad \mathrm{k} = 1,2,3;\\
\mathrm{K_{46}} &= -(\mathrm{r}/2)(\boldsymbol{\nabla}^2 + 1) \equiv \mathrm{K_{64}}; \quad \mathrm{K_{k6}} = -\mathrm{i}\mathrm{r}\partial/\partial\mathrm{x_k} \equiv \mathrm{K_{6k}}, \quad \mathrm{k} = 1,2,3.
\end{aligned}
\tag{11.1.5}
$$

The relations

$$[\mathrm{J_{ab}}, \mathrm{J_{bc}}] = \mathrm{iJ_{ac}}, \quad [\mathrm{J_{ab}}, \mathrm{K_{bc}}] = \mathrm{iK_{ac}}, \quad [\mathrm{K_{ab}}, \mathrm{K_{bc}}] = \mathrm{iJ_{ac}}, \tag{11.1.6}$$

determine all the nonzero commutators of the generators. It is convenient to collect twelve of the classical generators into the vectors

$$
\begin{aligned}
\mathrm{L} &= (\mathrm{J_{12}}, \mathrm{J_{23}}, \mathrm{J_{31}}) \sim \{(\mathbf{r}\times\mathbf{p}), \cdot\}, \quad \mathrm{A_-} = (\mathrm{J_{14}}, \mathrm{J_{24}}, \mathrm{J_{34}}) \sim \{\mathrm{A}(-1/2), \cdot\},\\
\mathrm{A_+} &= (\mathrm{K_{15}}, \mathrm{K_{25}}, \mathrm{K_{35}}) \sim \{\mathrm{A}(1/2), \cdot\}, \quad \boldsymbol{\Gamma} = (\mathrm{K_{16}}, \mathrm{K_{26}}, \mathrm{K_{36}}) \sim \{(\mathrm{r}\mathbf{p}), \cdot\}.
\end{aligned}
\tag{11.1.7a}
$$

The components of each vector are transformed amongst themselves by rotations generated by the components of L. We will use $\mathbf{L}, \mathbf{A}_-, \mathbf{A}_+$, and $\boldsymbol{\Gamma}$ to denote their quantum mechanical counterparts. The three further generators

$$\mathrm{K_{45}} \sim \{(\mathbf{r}\cdot\mathbf{p}), \cdot\}, \mathrm{K_{46}} \sim \{\mathrm{W}(1/2), \cdot\}, \mathrm{J_{56}} \sim \{\mathrm{W}(-1/2), \cdot\}, \tag{11.1.7b}$$

and their quantum-mechanical counterparts, $\mathrm{K_{45}}$, $\mathrm{K_{46}}$, $\mathrm{J_{56}}$, transform as scalars under rotations.

SO(4,2) possesses the semisimple subgroups listed in Table 11.1, together with their generators, and nonvanishing Casimir operators.

Table 11.1.

Subgroup	Generators	Casimir Operators, Eigenvalues
SO(3)	$\mathbf{L}$	$\mathbf{L}^2 \to l(l+1)$
SO(2,1)	$\mathrm{K_{45}}$, $\mathrm{K_{46}}$, $\mathrm{J_{56}}$	$(\mathrm{J_{56}})^2 - (\mathrm{K_{45}})^2 - (\mathrm{K_{46}})^2 \to l(l+1)$
SO(4)	$\mathbf{L}$, $\mathbf{A}_-$	$\mathbf{L}^2 + \mathbf{A}_-^2 \to n^2 - 1$; $(\mathbf{A}_- \cdot \mathbf{L})^2 \equiv 0$
SO(3,1)	$\mathbf{L}$, $\mathbf{A}_+$	$\mathbf{L}^2 - \mathbf{A}_+^2 \to \nu^2 - 1$; $(\mathbf{A}_+ \cdot \mathbf{L})^2 \equiv 0$
SO(4,1)	$\mathbf{L}$, $\mathbf{A}_-$, $\mathbf{A}_+$, $\mathrm{K_{45}}$	$\mathbf{A}_+^2 + (\mathrm{K_{45}})^2 - \mathbf{A}_-^2 - \mathbf{L}^2 \equiv 2$
SO(4,1)	$\mathbf{L}$, $\mathbf{A}_-$, $\boldsymbol{\Gamma}$, $\mathrm{K_{46}}$	$\boldsymbol{\Gamma}^2 + (\mathrm{K_{46}})^2 - \mathbf{A}_-^2 - \mathbf{L}^2 \equiv 2$

As SO(4,2) has rank three, it has three functionally independent Casimir operators. The quadratic Casimir operator

$$C^{[2]}_{SO(4,2)} = \mathbf{A}_-^2 + \mathbf{L}^2 + (J_{56})^2 - (\mathbf{A}_+^2 + \mathbf{\Gamma}^2 + (K_{45})^2 + (K_{46})^2). \quad (11.1.8)$$

has unit value. Both higher order SO(4,2) Casimir operators vanish. The SO(4,2) basis that is defined by the values of these Casimir operators is countably-infinite discrete. Because **A** is perpendicular to **L**, the quartic Casimir operator of SO(4) vanishes identically. As a result the SO(4) representations contained in the SO(4,2) have dimension n^2. The n^2-fold degeneracy of hydrogen atom energy levels is a physical expression of this group-theoretic property. The basis states of the SO(4) representations, with a given value of n, are most commonly chosen to be those of n different SO(3) representations, labeled by l, which takes on the values $0, 1, \ldots, n-1$, and by the integer m which takes on the $2l + 1$ values for which $|m| \leq$ l. The resulting basis for the representation of SO(4,2) is provided by the functions which satisfy the equations

$$\begin{aligned} C^{[2]}_{SO(4,2)}\chi_{nlm} &= \chi_{nlm}, J_{56}\chi_{nlm} = n\chi_{nlm}, \\ J_{12}\chi_{nlm} &= m\chi_{nlm}, \ \mathbf{L}^2\chi_{nlm} = l(l+1)\chi_{nlm}, \end{aligned} \quad (11.1.9)$$

An important alternative basis will be discussed in Sec. 11.4 below.

Utilizing the particular form of the generators given in (11.1.5), these abstract group theoretic considerations establish analytic relationships. One finds that the χ_{nlm} can be realized as functions, which when normalized under the scalar product (11.1.3a), are

$$\chi_{nlm}(r, \theta, \phi) = R_{nl}(r)Y_{lm}(\theta, \phi), \quad (11.1.10a)$$

with

$$R_{nl}(r) = N_{nl}(2r)^l \exp(-r)F(l + 1 - n; 2l + 2; 2r). \quad (11.1.10b)$$

Here F is the confluent hypergeometric function; the normalizing factor

$$N_{nl} = [(-i)^l(2^{3/2})/(2l + 1)!][(l + n)!/n(n - l - 1)!]^{1/2}, \quad (11.1.10c)$$

has been assigned the phase factor, $(-i)^l$.[5] The functions χ_{nlm}, now commonly called *Sturmians*, differ from those introduced in Sec. 10.8, and are not the same as the usual normalized bound-state electronic wave functions,

$\Psi_{nlm}(\mathrm{r}, \theta, \phi)$, which satisfy the Schrödinger equation

$$(-\nabla^2/2 - 1/\mathrm{r})\Psi_{nlm} = \mathrm{E}_n\Psi_{nlm}. \tag{11.1.11a}$$

If the phase choice is the same, the two types of wave functions are related by

$$\Psi_{nlm}(\mathrm{r}, \theta, \phi) = n^{1/2}\chi_{nlm}(\mathrm{r}/n, \theta, \phi). \tag{11.1.11b}$$

The basis provided by the χ_{nlm} is discrete, and *complete.* That is, any normalizable piecewise continuous function $\mathrm{f}(\mathrm{r}, \theta, \phi)$ can be expressed as a convergent series of these harmonics. The action of the group generators on this basis is given in Appendix A. The discreteness and completeness of the Sturmian basis stands in marked contrast to the usual complete hydrogenic bases, which contain a continuum of scattering states as well as a countable infinity of bound states. Because of this, working with the regularized system has the advantage that it is much simpler to expand a given bound-state function as a linear combination of Sturmian eigenfunctions than as a linear combination of hydrogenic eigenfunctions. Chapter 12 deals with the exploitation of these observations in studies of systems more complex than hydrogen-like atoms.

As in the classical system, SO(4) transformations interconvert bound-state solutions — wave functions here — of the same energy. When the energy is positive, SO(3,1) transformations interconvert degenerate solutions. The additional transformations available in the SO(4,1) and SO(4,2) groups superpose eigenfunctions of differing energies, but cannot superpose functions whose energies differ in sign. In the quantum mechanical systems the states of the discrete spectrum respond to perturbations differently than those of the continuous spectrum, and those of classical trajectories — only a few states of the discrete spectrum, the n^2 degenerate ones, mix appreciably when subjected to small perturbations. Consequences of this have been pointed out in Chapter 1, and they organize much of the discussion in Chapter 12.

The equation governing regularized hydrogen-like systems with nuclear charge of magnitude Z is

$$(\mathrm{W}_{\mathrm{op}} - \mathrm{Z})\psi = 0, \tag{11.1.12a}$$

$$\mathrm{W}_{\mathrm{op}} = \mathrm{r}(\mathbf{p}_{\mathrm{op}}^2 - 2\mathrm{E})/2, \tag{11.1.12b}$$

$$\mathbf{p}_{\mathrm{op}}^2 = -\nabla^2. \tag{11.1.12c}$$

It can be obtained by the action of the unitary dilatation operator $\exp(\mathrm{i}\alpha \mathrm{J}_{45})$ on Eq. (11.1.2). We will suppose that polar coordinates are being used, and set $\psi = \Psi(\mathrm{r}, \theta, \phi)$. One has

$$\exp(\mathrm{i}\alpha \mathrm{J}_{45})(\mathrm{r}(\mathbf{p}_{\mathrm{op}}^2 - 2\mathrm{E})/2 - 1)\Psi(\mathrm{r}, \theta, \phi) = 0, \tag{11.1.13a}$$

so

$$(\exp(\mathrm{i}\alpha \mathrm{J}_{45})(\mathrm{r}(\mathbf{p}_{\mathrm{op}}^2 - 2\mathrm{E})/2 - 1)\exp(-\mathrm{i}\alpha \mathrm{J}_{45}))\exp(\mathrm{i}\alpha \mathrm{J}_{45})\Psi(\mathrm{r}, \theta, \phi) = 0. \tag{11.1.13b}$$

Carrying out the dilations of r and **p** converts this equation to

$$((\mathrm{e}^{-\alpha}\mathrm{r}\mathbf{p}_{\mathrm{op}}^2 - 2\mathrm{E}\mathrm{e}^{\alpha}\mathrm{r})/2 - 1)\Psi'(\mathrm{r}, \theta, \phi) = 0, \tag{11.1.13c}$$

with

$$\Psi'(\mathrm{r}, \theta, \phi) = \mathrm{e}^{\alpha}\Psi(\mathrm{e}^{\alpha}\mathrm{r}, \theta, \phi). \tag{11.1.13d}$$

Thus

$$\mathrm{e}^{-\alpha}(\mathrm{r}(\mathbf{p}_{\mathrm{op}}^2 - 2\mathrm{E}\mathrm{e}^{2\alpha})/2 - \mathrm{e}^{\alpha})\Psi'(\mathrm{r}, \theta, \phi) = 0. \tag{11.1.13e}$$

If one multiplies this last equation by e^{α} and then sets $\mathrm{e}^{\alpha} = \mathrm{Z}$, it can be rewritten as

$$(\mathrm{r}(\mathbf{p}_{\mathrm{op}}^2 - 2\mathrm{E}')/2 - \mathrm{Z})\Psi'(\mathrm{r}, \theta, \phi) = 0, \tag{11.1.14a}$$

with

$$\mathrm{E}' = \mathrm{Z}^2\mathrm{E}, \quad \Psi'(\mathrm{r}, \theta, \phi) = \mathrm{Z}\Psi(\mathrm{Z}\mathrm{r}, \theta, \phi). \tag{11.1.14b,c}$$

We shall henceforth drop the prime on E, assume n, l, m are the appropriate quantum numbers, and express (11.1.14a) as

$$(\mathrm{r}(\mathbf{p}_{\mathrm{op}}^2 - 2\mathrm{E})/2 - \mathrm{Z})|znlm\rangle = 0, \tag{11.1.15a}$$

or as

$$(\mathrm{W}_{\mathrm{op}} - \mathrm{Z})|znlm\rangle = 0. \tag{11.1.15b}$$

Having established that the operator of a one-parameter subgroup of SO(4,2) can be used to directly change the nuclear charge, we will for the time being concentrate attention on the Z = 1 hydrogen atom system, for which

$$\chi_{nlm} \rightarrow |1nlm\rangle \equiv |nlm\rangle. \tag{11.1.15c}$$

Using the definitions of J_{56} and K_{46}, one finds that

$$\mathrm{r} = \mathrm{J}_{56} - \mathrm{K}_{46}, \quad \mathrm{r}\mathbf{p}_{\mathrm{op}}^2 = \mathrm{J}_{56} + \mathrm{K}_{46}. \tag{11.1.16}$$

Thus if $\mathrm{E} = -1/2$, $\mathrm{W}_{\mathrm{op}} = \mathrm{J}_{56}$, and if $\mathrm{E} = +1/2$, $\mathrm{W}_{\mathrm{op}} = \mathrm{K}_{46}$, and in general

$$\mathrm{W}_{\mathrm{op}} = (1/2 - \mathrm{E})\mathrm{J}_{56} + (1/2 + \mathrm{E})\mathrm{K}_{46}. \tag{11.1.17}$$

When E is negative, the coefficient of J_{56} is greater than that of K_{46}, and when E is positive the reverse is true. The unitary dilatation operators $\exp(\pm i\alpha J_{45})$ can also be used to relate a system with arbitrary E to the systems with the E values $\pm 1/2$. For $E < 0$ one uses the relation

$$\exp(-i\alpha J_{45}) J_{56} \exp(i\alpha J_{45}) = \cosh(\alpha) J_{56} + \sinh(\alpha) K_{46}. \qquad (11.1.18)$$

Letting k to be a constant, one may in (11.1.17) set

$$(1/2 - E) = k\cosh(\alpha), \quad (1/2 + E) = k\sinh(\alpha). \qquad (11.1.19a)$$

This requires

$$(1/2 - E)^2/k^2 - (1/2 + E)^2/k^2 = 1, \qquad (11.1.19b)$$

whence

$$k = (-2E)^{1/2}, \qquad (11.1.19c)$$

and

$$\cosh(\alpha) - \sinh(\alpha) = e^{-\alpha} = (-2E)^{1/2}. \qquad (11.1.19d)$$

It follows that, when $E < 0$, the equation

$$(W_{op} - 1)\psi = 0 \qquad (11.1.20a)$$

can be expressed as

$$(\exp(-i\alpha J_{45}) J_{56} \exp(i\alpha J_{45}) - 1)/(-2E)^{1/2})\psi = 0. \qquad (11.1.20b)$$

Multiplying this through by $\exp(i\alpha J_{45})$ gives

$$(J_{56} - 1/(-2E)^{1/2})\psi' = 0, \quad \psi' = \exp(i\alpha J_{45})\psi. \qquad (11.1.20c)$$

Thus ψ' is an eigenfunction of J_{56}, which has eigenvalues n, and the eigenfunctions χ_{nlm} defined Lie-algebraically in (11.1.9). Consequently,

$$n - 1/(-2E)^{1/2} = 0, \quad e^{-\alpha} = 1/n, \quad E = -1/(2n^2). \qquad (11.1.21)$$

One can thus write

$$\psi'_{nlm} = \exp(-i\alpha J_{45})\chi_{nlm}(r, \theta, \phi) = e^{-\alpha}\chi_{nlm}(e^{-\alpha}r, \theta, \phi), \qquad (11.1.22a)$$

and

$$\psi'_{nlm} = (1/n)\chi_{nlm}(r/n, \theta, \phi). \qquad (11.1.22b)$$

The Schrödinger eigenfunctions $\Psi_{nlm}(r, \theta, \phi)$ defined in (11.1.11) differ from these ψ'_{nlm} only because of the difference in normalization that arises from the difference in scalar products. That is,

$$\Psi_{nlm}(\mathrm{r}, \theta, \phi) = n^{-1/2}\psi'_{nlm}. \tag{11.1.22c}$$

When E is positive, one replaces (11.1.18) with

$$\exp(-\mathrm{i}\alpha \mathrm{J}_{45})\mathrm{K}_{46}\exp(\mathrm{i}\alpha \mathrm{J}_{45}) = \cosh(\alpha)\mathrm{K}_{46} + \sinh(\alpha)\mathrm{J}_{56}, \tag{11.1.23}$$

and relates the general solution of (11.1.20a) to the eigenfunctions and eigenvalues of the particular solution that is an eigenfunction of K_{46}. This can be obtained from the eigenfunction of J_{56} because the relation between these solutions can be established with the aid of a non-unitary dilatation. On setting $\alpha = -\mathrm{i}\pi/2$ in (11.1.18), one finds that

$$\exp(-(\pi/2)\mathrm{J}_{45})\mathrm{J}_{56}\exp((\pi/2)\mathrm{J}_{45}) = \cos(\pi/2)\mathrm{J}_{46} - \mathrm{i}\sin(\pi/2)\mathrm{K}_{46} = -\mathrm{i}\mathrm{K}_{46}.$$

Multiplying this through from the right by $\exp(-(\pi/2)\mathrm{J}_{45})$ yields

$$\mathrm{K}_{46}\exp(-(\pi/2)\mathrm{J}_{45}) = \mathrm{i}\exp(-(\pi/2)\mathrm{J}_{45})\mathrm{J}_{56}. \tag{11.1.24}$$

Thus

$$\begin{aligned}\mathrm{K}_{46}(\exp(-(\pi/2)\mathrm{J}_{45})\chi_{nlm}) &= \mathrm{i}\exp(-(\pi/2)\mathrm{J}_{45})\mathrm{J}_{56}\chi_{nlm}\\ &= \mathrm{i}n(\exp(-(\pi/2)\mathrm{J}_{45})\chi_{nlm}),\end{aligned}$$

and the functions $\exp(-(\pi/2)\mathrm{J}_{45})\chi_{nlm}$ are eigenfunctions of K_{46} with eigenvalues $\mathrm{i}n$. Let us set $\mathrm{i}n$ equal to ν, and denote these eigenfunctions by $\chi_{\nu lm}$. Using (11.1.22a) establishes then that

$$\chi_{\nu lm} = \mathrm{e}^{-\alpha}\chi_{nlm}(\mathrm{e}^{-\alpha}\mathrm{r}, \theta, \phi) = \mathrm{i}\chi_{nlm}(\mathrm{ir}, \theta, \phi). \tag{11.1.25}$$

11.2 The Momentum-space Representation

In the discussion of the SO(4) symmetry of classical Keplerian motions in Chapter 8 it was noted that the hodographs of all trajectories are circles. Then it was shown that the PB operators which generate the motions in phase space produce rotations in an invariant momentum subspace — a space in which the generators do not depend upon functions of the position variables. These properties of Keplerian motion have the consequence that in quantum mechanics the realization of the dynamical symmetry of hydrogen-like atoms is both simpler and more direct in momentum space than it is in position space. In this and the following section, we lay the mathematical groundwork required to exploit this fact.

The momentum-space analogs of Schrödinger functions $\psi(\mathbf{r})$ are their Fourier transforms

$$\phi(\mathbf{p}) = \int (2\pi)^{-3/2} \exp(-\mathrm{i}\mathbf{p}\cdot\mathbf{r})\psi(\mathbf{r})\mathrm{d}^3\mathrm{r}, \quad \mathrm{d}^3\mathrm{r} = \mathrm{dxdydz}. \qquad (11.2.1)$$

Dirac's introduction of his delta function, $\delta(\mathrm{x})$, greatly simplified the theory of Fourier transforms. It has the properties[6]

$$\int_{-\infty}^{\infty} \mathrm{f(x)}\delta(\mathrm{x}-\mathrm{x}')\mathrm{dx} = \mathrm{f}(\mathrm{x}'), \qquad (11.2.2a)$$

and

$$= (2\pi)^{-1}\int_{-\infty}^{\infty} \exp(\mathrm{ipx})\mathrm{dp}\delta(\mathrm{x}). \qquad (11.2.2b)$$

Letting $\delta(\mathbf{r})$ represent $\delta(\mathrm{x})\delta(\mathrm{y})\delta(\mathrm{z})$ one has

$$\delta(\mathbf{r}) = (2\pi)^{-3}\int \exp(\mathrm{i}\mathbf{p}\cdot\mathbf{r})\mathrm{d}^3\mathrm{p} \qquad (11.2.3a)$$

and

$$\int \mathrm{f}(\mathbf{r})\delta(\mathbf{r}-\mathbf{r}')\mathrm{d}^3\mathrm{r} = \mathrm{f}(\mathbf{r}'). \qquad (11.2.3b)$$

Using (11.2.3) one may derive that the inverse of transformation (11.2.1) is

$$\psi(\mathbf{r}) = \int (2\pi)^{-3/2} \exp(\mathrm{i}\mathbf{p}\cdot\mathbf{r})\phi(\mathbf{p})\mathrm{d}^3\mathrm{p}, \quad \mathrm{d}^3\mathrm{p} = \mathrm{dp_x dp_y dp_z}. \qquad (11.2.4)$$

The position-space analog of $\mathrm{i}\partial/\partial\mathrm{p_x}\ \phi(\mathbf{p})$ is

$$\mathrm{i}\partial/\partial\mathrm{p_x}\int (2\pi)^{-3/2} \exp(-\mathrm{i}\mathbf{p}\cdot\mathbf{r})\psi(\mathbf{r})\mathrm{d}^3\mathrm{r} = \int (2\pi)^{-3/2} \exp(-\mathrm{i}\mathbf{p}\cdot\mathbf{r})\ \mathrm{x_k}\psi(\mathbf{r})\mathrm{d}^3\mathrm{r}, \qquad (11.2.5a)$$

and the p-space analog of $-\mathrm{i}\partial/\partial\mathrm{x}\,\psi(\mathbf{r})$ is

$$-\mathrm{i}\partial/\partial\mathrm{x}\int (2\pi)^{-3/2} \exp(\mathrm{i}\mathbf{p}\cdot\mathbf{r})\phi(\mathbf{p})\mathrm{d}^3\mathrm{p} = \int (2\pi)^{-3/2} \exp(\mathrm{i}\mathbf{p}\cdot\mathbf{r})\mathrm{p_x}\phi(\mathbf{p})\mathrm{d}^3\mathrm{p}. \qquad (11.2.5b)$$

These connections are those expected from the correspondence principle.

However the product of the Fourier transforms of two functions is not the Fourier transform of the product of the functions. Symbolically

$$\mathrm{FT}(\mathrm{f}\cdot\mathrm{g}) \neq \mathrm{FT}(\mathrm{f})\cdot\mathrm{FT}(\mathrm{g}). \qquad (11.2.6a)$$

For two functions of $\mathbf{r}$ one has

$$\mathrm{FT}(\mathrm{f}(\mathbf{r})\cdot \mathrm{g}(\mathbf{r})) = \int (2\pi)^{-3/2}\exp(-\mathrm{i}\mathbf{p}\cdot\mathbf{r})(\mathrm{f}(\mathbf{r})\cdot \mathrm{g}(\mathbf{r}))\mathrm{d}^3\mathrm{r}. \qquad (11.2.6\mathrm{b})$$

By expressing $\mathrm{g}(\mathbf{r})$ as the Fourier transform of $\mathrm{G}(\mathbf{p}')$, and $\mathrm{f}(\mathbf{r})$ as that of $\mathrm{F}(\mathbf{p})$, the right-hand side can be written as

$$\int (2\pi)^{-3/2}\exp(-\mathrm{i}\mathbf{p}\cdot\mathbf{r})\mathrm{f}(\mathbf{r})\left\{\int (2\pi)^{-3/2}\exp(\mathrm{i}\mathbf{p}'\cdot\mathbf{r})\mathrm{G}(\mathbf{p}')\mathrm{d}^3\mathrm{p}'\right\}\mathrm{d}^3\mathrm{r}$$
$$= \int (2\pi)^{-3/2}\left\{\int (2\pi)^{-3/2}\exp(-\mathrm{i}(\mathbf{p}-\mathbf{p}')\cdot\mathbf{r})\mathrm{f}(\mathbf{p})\mathrm{G}(\mathbf{p}')\mathrm{d}^3\mathrm{p}'\right\}\mathrm{d}^3\mathrm{r}$$
$$= \int (2\pi)^{-3/2}\left\{\int (2\pi)^{-3/2}\exp(-\mathrm{i}(\mathbf{p}-\mathbf{p}')\cdot\mathbf{r})\mathrm{f}(\mathbf{p})\mathrm{d}^3\mathrm{r}\right\}\mathrm{G}(\mathbf{p}')\mathrm{d}^3\mathrm{p}'. \qquad (11.2.6\mathrm{c})$$

Thus

$$\mathrm{FT}(\mathrm{f}(\mathbf{r})\cdot \mathrm{g}(\mathbf{r})) = \int (2\pi)^{-3/2}\mathrm{F}(\mathbf{p}-\mathbf{p}')\mathrm{G}(\mathbf{p}')\mathrm{d}^3\mathrm{p}'.$$

An analogous string of connections establish the following relation between position-space and momentum-space scalar products:

$$\int \phi^*(\mathbf{p})\phi'(\mathbf{p})\mathrm{d}^3\mathrm{p} = \int \psi^*(\mathbf{r})\psi'(\mathbf{r})\mathrm{d}^3\mathrm{r}. \qquad (11.2.7\mathrm{a})$$

To prove this relation one may apply (11.2.1), to its left-hand side, and obtain

$$(2\pi)^{-3}\exp\int\left(\int \exp(-\mathrm{i}\mathbf{p}\cdot\mathbf{r})\Psi(\mathbf{r})\mathrm{d}^3\mathrm{r}\right)^*\left(\int \exp(-\mathrm{i}\mathbf{p}\cdot\mathbf{r}')\Psi'(\mathbf{r}')\mathrm{d}^3\mathrm{r}'\right)\mathrm{d}^3\mathrm{p}$$
$$= (2\pi)^{-3/2}\int\int (2\pi)^{-3/2}\int\exp(\mathrm{i}\mathbf{p}\cdot(\mathbf{r}-\mathbf{r}'))\ \mathrm{d}^3\mathrm{p}\ \Psi(\mathbf{r})^*\mathrm{d}^3\mathrm{r}\ \psi'(\mathbf{r}')\ \mathrm{d}^3\mathrm{r}'$$
$$= \int\int \delta(\mathbf{r}-\mathbf{r}')\psi(\mathbf{r})^*\mathrm{d}^3\mathrm{r}\psi'(\mathbf{r}')\mathrm{d}^3\mathrm{r}' = \int \psi^*(\mathbf{r})\psi'(\mathbf{r})\mathrm{d}^3\mathrm{r}, \qquad (11.2.7\mathrm{b})$$

and thus Eq. (11.2.7a) is established.

The momentum-space analog of the Schrödinger equation

$$(\mathrm{p}_{\mathrm{op}}^2/2 + \mathrm{v}(\mathrm{r}) - \mathrm{E})\phi = 0, \qquad (11.2.8)$$

is

$$(\mathrm{p}^2/2 + \mathrm{v}(\mathrm{r}_{\mathrm{op}}) - \mathrm{E})\phi = 0, \quad \mathrm{r}_{\mathrm{op}} = \sqrt{-\nabla^2}. \qquad (11.2.9)$$

Because working with the function of r_{op} is often very difficult, one proceeds differently. Taking the Fourier transform of the original Schrödinger equation, one has

$$(2\pi)^{-3/2}\int\{\exp(-i\mathbf{p}\cdot\mathbf{r})(p_{op}^2/2+v(r)-E)\psi\}d^3r=0, \qquad (11.2.10)$$

which becomes

$$(p^2/2-E)\phi(\mathbf{p}) = (2\pi)^{-3/2}\int \exp(-i\mathbf{p}\cdot\mathbf{r})v(r)\psi(\mathbf{r})d^3r$$

$$= \int \phi(\mathbf{p}')V(|\mathbf{p}'-\mathbf{p}|)d^3p', \qquad (11.2.11)$$

where V(p) is the Fourier transform of v(r). This equation provides an alternative to (11.2.9).

For hydrogen-like systems in which $v(r) = -Z/r$, the Fourier transform, V(p) is $(2/\pi)^{1/2}(-Z/|p|^2)$. Setting $E = -p_0^2/2$, Eq. (11.2.11) then becomes

$$(1/2)(p_0^2+p^2)\phi(\mathbf{p}) = (2/\pi)^{1/2}Z\int \phi(\mathbf{p}')/|\mathbf{p}'-\mathbf{p}|^2\, d^3p'. \qquad (11.2.12)$$

To solve this equation, Fock[7] began by using a stereographic projection to transform the integrand. His stereographic projection maps points with coordinates p_x, p_y, p_z in momentum-space to points with coordinates x_1, x_2, x_3, x_4 on a unit hypersphere. The transformation, defined by $\theta \to \theta$, $\phi \to \phi$, and

$$p/p_0 \to \tan(\alpha/2) = \sqrt{\frac{(1-\cos(\alpha))}{(1+\cos(\alpha))}}, \qquad (11.2.13)$$

sets $\mathbf{x} = (x_1, x_2, x_3, x_4)$, with

$$x_4 = (p_0^2-p^2)/(p_0^2+p^2) = \cos(\alpha), \qquad (11.2.14a)$$

$$x_3 = 2p_0p_z/(p^2+p_0^2) = \sin(\alpha)\cos(\theta), \qquad (11.2.14b)$$

$$x_2 = 2p_0p_y/(p^2+p_0^2) = \sin(\alpha)\sin(\theta)\sin(\phi), \qquad (11.2.14c)$$

$$x_1 = 2p_0p_x/(p^2+p_0^2) = \sin(\alpha)\sin(\theta)\cos(\phi). \qquad (11.2.14d)$$

It converts $1/(\mathbf{p}'-\mathbf{p})^2$ to

$$(2p_0^2/(p_0^2+p^2))(1/|\mathbf{x}-\mathbf{x}'|^2)(2p_0^2/(p_0^2+p'^2)), \qquad (11.2.15a)$$

and converts d^3p' to

$$((p_0^2+p'^2)/2p_0)^3\, d\Omega'. \qquad (11.2.15b)$$

Here,

$$d\Omega' = \sin^2(\alpha')\sin(\theta')\, d\alpha'\, d\theta'\, d\phi' \qquad (11.2.15c)$$

is the area element on the unit hypersphere defined by $|\mathbf{x}'|^2 = 1$. Both $\mathbf{x}$ and $\mathbf{x}'$ have the same origin and terminate on this hypersphere. Inserting these relations into the integrand, and multiplying both sides of (11.2.12) by $(p^2 + p_0^2)$, converts the equation to

$$(p^2 + p_0^2)^2 \phi(\mathbf{p}) = (Z/2p_0)(1/\pi)^2 \int (p_0^2 + p'^2)^2 \phi(\mathbf{p}')/|\mathbf{x} - \mathbf{x}'|^2 d\Omega'. \tag{11.2.16a}$$

On setting

$$(p^2 + p_0^2)^2 \phi(\mathbf{p}) = \Phi(\mathbf{x}), \tag{11.2.16b}$$

and multiplying (11.2.16a) through by p_0, it takes on the simpler form

$$p_0 \Phi(\mathbf{x}) = (Z/\pi^2) \int \Phi(\mathbf{x}')/|\mathbf{x} - \mathbf{x}'|^2 d\Omega'. \tag{11.2.17}$$

Fock recognized that this equation is invariant under the rotations of the group SO(4). To gain insight into how this invariance is expressed by the equation itself, we begin with the much used expansion of the Coulomb potential $1/|\mathbf{r} - \mathbf{r}'|$:

$$1/|\mathbf{r} - \mathbf{r}'| = \Sigma_j (r_<^j / r_>^{j+1})(2j+1)^{-1} Y_{jm}^*(\theta', \phi') Y_{jm}(\theta, \phi). \tag{11.2.18a}$$

When both $\mathbf{r}$ and $\mathbf{r}'$ terminate on a unit sphere, this becomes

$$1/|\mathbf{r} - \mathbf{r}'| = \Sigma_j (2j+1)^{-1} Y_{jm}^*(\theta', \phi') Y_{jm}(\theta, \phi). \tag{11.2.18b}$$

The analog of (11.2.18b) in Euclidean four-space is the equation[8]

$$1/|\mathbf{x} - \mathbf{x}'|^2 = \Sigma_k (k+1)^{-1} Y_{kjm}^*(\alpha', \theta', \phi') Y_{klm}(\alpha, \theta, \phi), \tag{11.2.18c}$$

in which the functions Y_{klm} are hyperspherical harmonics on S(3). Like the spherical harmonics on S(2), these hyperspherical harmonics are members of an orthornormal set. That is, one has

$$\int Y_{kjm}^*(\alpha, \theta, \phi) Y_{k'l'm'}(\alpha, \theta, \phi) d\Omega = \delta_{k,k'}, \delta_{l,l'} \delta_{m,m'}. \tag{11.2.19}$$

On inserting (11.2.18c) into (11.2.17) one obtains

$$\begin{aligned} p_0 \Phi(\mathbf{x}) &= (Z/\pi^2) \int \Phi(\mathbf{x}') \left\{ \Sigma_k (k+1)^{-1} Y_{kjm}^*(\alpha', \theta, \phi') Y_{klm}(\alpha, \theta, \phi) \right\} d\Omega' \\ &= (Z/\pi^2) \Sigma_k (k+1)^{-1} Y_{klm}(\alpha, \theta, \phi) \int \Phi(\mathbf{x}') Y_{kjm}^*(\alpha', \theta, \phi') d\Omega'. \end{aligned} \tag{11.2.20}$$

If one sets $\Phi(\mathbf{x}') = Y_{k'j'm'}(\mathbf{x}')$ in the integral

$$\int \Phi(\mathbf{x}') Y_{kjm}^*(\alpha', \theta, \phi') d\Omega', \tag{11.2.21}$$

the orthonormality of the Y's requires that it becomes unity if Φ is $Y_{kjm}(\mathbf{x})$. Equation (11.2.20) then states that

$$p_0\Phi(\mathbf{x}) = (Z/(k+1))Y_{kjm}(\mathbf{x}). \tag{11.2.22a}$$

If both $\Phi(\mathbf{x})$ and $Y_{k'j'm'}(\mathbf{x})$ are normalized, this relation requires that

$$p_0 = Z/(k+1), \quad \Phi(\mathbf{x}) = Y_{kjm}(\mathbf{x}). \tag{11.2.22b,c}$$

Identifying $k+1$ with the quantum number n, one obtains the energy, $E = -Z^2/2n^2$, and the associated wave function $Y_{k'j'm'}(\mathbf{x})$.

For a fixed value of $n = k+1$, there are $n^2 Y_{klm}$ with different values of l, m. One may insert any one of them, or any normalized linear combination of them, into (11.2.2a), without changing the relation $p_0 = Z/(k+1)$. Fock used this fact to explain the n^2-fold degeneracy of the energy spectrum of hydrogen atoms. Shortly thereafter Bargmann showed that Pauli's Runge–Lenz operators,[9] together with the angular momentum operators, obeyed SO(4) commutation relations.[10] This removed the possibility that Fock's remarkable result was an artifact produced by his stereographic projection. Bargmann's six operators are position-space realizations of the generators which act in the S(3) projective momentum space of $\mathbf{x}$. These are the six operators

$$J_{ij} = \sqrt{-1}(x_i\partial/\partial x_j - x_j\partial/\partial x_i) = \sqrt{-1}(\pi_{i<}\partial/\partial\pi_{j<} - \pi_{j<}\partial/\partial\pi_{i<}), \tag{11.2.23}$$

which are the quantum mechanical versions of the classical SO(4) generators introduced in Sec. 8.4.

11.3 The Hyperspherical Harmonics Y_{klm}

The relations

$$\begin{aligned}(p^2 + p_0^2)^2\phi(\mathbf{p}) &= \Phi(\mathbf{x}),\\ \Phi(\mathbf{x}) &= Y_{klm}(\mathbf{x}),\end{aligned} \tag{11.3.1}$$

of the previous section, together with the equations that define Fock's sterographic projection, enable one to establish the relation between the hyperspherical harmonics $Y_{n-1lm}(\alpha,\theta,\phi)$, and the momentum–space eigenfunctions $\phi_{nlm}\,(p,\theta,\phi)$ that satisfy (11.2.17). Tracing through the connections, the reader will find that the relation is

$$\phi_{nlm}(p/p_0,\theta,\phi) = N p_0^{-3/2}\varpi^{-2}Y_{klm}(\alpha,\theta,\phi), \tag{11.3.2a}$$

where

$$\varpi = (p^2 + p_0^2)/2p_0^2 = ((p/p_0)^2 + 1)/2,$$

and N is a normalization factor that may also contain a phase factor. Conversely,

$$Y_{klm}(\alpha, \theta, \phi) = N^{-1} p_0^{3/2} \varpi^2 \phi_{nlm}(p/p_0, \theta, \phi), \tag{11.3.2b}$$

The momentum-space functions ϕ_{nlm}, are of the form

$$\rho_{nl}(p/p_0) Y_{lm}(\theta, \phi). \tag{11.3.2c}$$

The hyperspherical harmonics are directly defined in terms of Gegenbauer polynomials C_ν^λ with $\nu = n - (l+1) = k - l$, and $\lambda = l + 1$. One has[11]

$$Y_{klm}(\alpha, \theta, \phi) = N_l C_\nu^\lambda(\cos(\alpha)) \sin^l(\alpha) Y_{lm}(\theta, \phi), \tag{11.3.3a}$$

with

$$N_l = (-i)^l (2\pi^2)^{-1/2}. \tag{11.3.3b}$$

The phase factor, $(-i)^l$, which makes the normalization factor l dependent, has been chosen so that the hyperspherical harmonics satisfy a standard phase convention[5]

$$Y_{klm}^* = (i)^{l-m} Y_{kl-m}. \tag{11.3.3c}$$

The term $\sin^l(\alpha) Y_{lm}(\theta, \phi)$ in (11.3.3), is a function of x_1, x_2, x_3 that is obtained from the usual $Y_{lm}(\theta, \phi)$ when one replaces x, y, z defined in momentum-space by the x_1, x_2, x_3 defined above. This leads us to define a unit vector **y** in p-space, analogous to the unit vector **x** in the projective space. Then one of the effects of Fock's projection is to convert $Y_{lm}(\mathbf{y})$ to $Y_{lm}(\mathbf{x})$.

Its other effect is to convert the radial function $\rho_{nl}(p/p_0)$ to $N_l p_0^{-3/2}$ times

$$\begin{aligned} ((p/p_0)^2 + 1)^{-2} C_\nu^\lambda\,(\cos(\alpha)) &= (1 + \cos(\alpha))^2 C_\nu^\lambda(\cos(\alpha)) = (1 + x_4)^2 C_\nu^\lambda(x_4) \\ &= (1 + x_4)^2 C_\nu^\lambda(x_4). \end{aligned} \tag{11.3.4}$$

Variations in the length of the momentum vector are mapped into variations of x_4, and the consequent variations in the radial function $\rho_{nl}(p/p_0)$ are expressed through variations in the Gegenbauer functions and the factor $(1+x_4)^2$. The color figures in Chapter 1 depict the stereographic projection of the planar analogs of 2s and 2p momentum-space functions onto an ordinary sphere.[12]

The C_ν^λ satisfy Gegenbauer's differential equation

$$\{(1-x^2)(\partial/\partial x)^2 + (2\lambda+1)x\partial/\partial x + \nu(\nu+2\lambda)\}C_\nu^\lambda(x) = 0. \qquad (11.3.5)$$

The Gegenbauer polynomials $C_\nu^\lambda(x)$ play the same role in the theory of the rotation groups SO(N), $N > 3$, that Legendre polynomials play in the theory of SO(3). The $C_\nu^\lambda(x)$ have a generating function, G, defined by

$$G \equiv (1-2hx+h^2)^{-\lambda} = \Sigma h^\nu C_\nu^\lambda(x), \qquad (11.3.6)$$

a relation that is easily exploited using computer mathematics systems. Two recent books by Avery provide clear and extensive treatments of the properties of Gegenbauer functions and hyperspherical harmonics in spaces of arbitrary dimension.[8,13] Operators that shift the ν in Gegenbauer functions are discussed in Appendix B. Table 11.3.1 displays a few of the Y_{klm}, along with their corresponding Sturmian position-space functions. A more extensive tabulation is contained in Judd's *Angular Momentum Theory for Diatomic Molecules*,[14] which deals with many of the topics in this chapter.

The volume element in the four-dimensional Euclidean space of the x's is

$$dx_1dx_2dx_3dx_4 = \rho^3 d\rho\, d\Omega, \quad \rho^2 = x_1^2 + x_2^2 + x_3^2 + x_4^2, \qquad (11.3.7)$$

the area element being

$$d\Omega = \sin^2(\alpha)\,\sin(\theta)\; d\alpha\, d\theta\, d\phi.$$

Table 11.3.1. Hyperspherical harmonics and sturmians.

Harmonic	$(2\pi^2)^{1/2}Y_{klm}(\alpha,\theta,\phi)$	$(2\pi^2)^{1/2}Y_{klm}(x)$	Radial Sturmian
Y_{000}	1	1	$2e^{-r}$
Y_{100}	$-2\cos(\alpha)$	$-2x_4$	$2\sqrt{2}(1-r)e^{-r}$
Y_{110}	$i\,2\sin(\alpha)\cos(\theta)$	$i\,2x_3$	$(2\sqrt{2}/\sqrt{3})re^{-r}$
Y_{111}	$i\,2\sin(\alpha)\sin(\theta)e^{i\phi}$	$-i\,2(x_1+ix_2)/2^{1/2}$	$(2\sqrt{2}/\sqrt{3})re^{-r}$
Y_{11-1}	$2\sin(\alpha)\sin(\theta)e^{-i\phi}$	$i2(x_1-ix_2)/2^{1/2}$	$(2\sqrt{2}/\sqrt{3})re^{-r}$
Y_{200}	$4\cos^2(\alpha)-1$	$4x_4^2-1$	$2\sqrt{3}(1-2r+2r^2/3)e^{-r}$
Y_{210}	$-i6^{1/2}\sin^2(\alpha)\cos(\theta)$	$-i6^{1/2}x_4x_3$	$(4\sqrt{2}/\sqrt{3})r(1-r/2)e^{-r}$
$Y_{21\pm1}$	$\pm(3)^{1/2}\sin(2\alpha)\sin(\theta)e^{\pm i\phi}$	$\pm i(3)^{1/2}2x_4(x_1\pm ix_2)$	$(4\sqrt{2}/\sqrt{3})r(1-r/2)e^{-r}$
Y_{220}	$-2^{1/2}\sin^2(\alpha)(3\cos^2(\theta)-1)$	$-2^{1/2}(3x_3^2+x_4^2-1)$	$(4/(\sqrt{30}))r^2e^{-r}$
$Y_{22\pm1}$	$\pm(3)^{1/2}\sin^2(\alpha)\sin(2\theta)e^{\pm i\phi}$	$\pm i(3)^{1/2}2x_3(x_1\pm ix_2)$	$(4/(\sqrt{30}))r^2e^{-r}$
$Y_{22\pm2}$	$-(3)^{1/2}\sin^2(\alpha)\sin^2(\theta)e^{\pm 2i\phi}$	$-(3)^{1/2}(x_1\pm ix_2)^2$	$(4/(\sqrt{30}))r^2e^{-r}$

The area element bears the following relation to the volume element d^3p in momentum space:

$$d\Omega = \varpi^{-3} p_0^{-3}\, d^3p. \tag{11.3.8}$$

The scalar product introduced in (11.2.19), that is,

$$\int_0^\pi d\alpha \int_0^\pi d\theta \int_0^{2\pi} d\phi \sin^2(\alpha)\sin(\theta)\, Y^*_{klm} Y_{k'l'm'} \tag{11.3.9}$$

will in the following be expressed as $\int Y^*_{klm} Y_{k'l'm'} d\Omega$, or $(Y_{klm}, Y_{k'l'm'})$. From the previous discussion it follows that

$$\int Y^*_{klm} Y_{k'l'm'} d\Omega = \int \phi^*_{nlm} \varpi \phi_{n'l'm'}\, d^3p. \tag{11.3.10}$$

The weight factor, ϖ, in the momentum-space scalar product corresponds to the factor $1/r$ in the position-space scalar product,[15] for one has

$$(p^2 + p_0^2)/2 = p^2/2 - E = 1/r. \tag{11.3.11}$$

Thus Fock's realization of hydrogen-like systems is the momentum–space analog of the realization of the regularized Schrödinger system that was discussed in Sec. 11.1. The two realizations provide isomorphic realizations of the dynamical symmetries of regularized hydrogen-like systems.

The radial Sturmians are normalized using the rdr scalar product. Normalization with the r^2dr *scalar product multiplies them by* $\sqrt{n}$, $n = k + 1$.

At this point, let us summarize some general relationships. The correspondence principle sets up an operator correspondence $X_q \to X_p$ and careful attention to phase and normalization factors has ensured that the position-space scalar products

$$\int \chi^*_{nlm}(1/r) X_q\, \chi_{n'l'm'} dv = \langle \chi_{nlm}, X_q\, \chi_{n'l'm'} \rangle, \tag{11.3.12a}$$

and their momentum-space counterparts,

$$\int d^3p\, \phi^*_{nlm} \varpi X_p\, \phi_{n'l'm'} \equiv \langle \phi_{nlm}, X_p \phi_{n'l'm'} \rangle, \tag{11.3.12b}$$

have the same values. To extend the correspondence to operators Ξ that act in the hyperspherical momentum-space, we required that the scalar products (11.3.12a,b) and

$$\int Y^*_{klm} \Xi\, Y_{k'l'm'} d\Omega \equiv (Y_{klm}, \Xi Y_{k'l'm'}), \tag{11.3.12c}$$

also have the same values. The phase factors have been so chosen that

$$\begin{aligned}\int d^3p\, p_0^{-3}\varpi^{-2}Y^*_{klm}\varpi\, X_p\, \varpi^{-2}\, Y_{k'l'm} \\ = \int d\Omega\, \varpi^{-3}\, \varpi^{-2}\, Y^*_{klm}\varpi\, X_p\, \varpi^{-2}\, Y_{k'l'm} \\ = \int d\Omega\, Y^*_{klm}\varpi^2\, X_p\, \varpi^{-2}\, Y_{k'l'm}. \qquad (11.3.24)\end{aligned}$$

Thus if

$$X_p \to \Xi = \varpi^2\, X_p\, \varpi^{-2}, \qquad (11.3.25a)$$

then

$$(Y_{klm}, \Xi\, Y_{k'l'm'}) = \langle \phi_{nlm},\; X_p\, \phi_{n'l'm'}\rangle. \qquad (11.3.25b)$$

The inverse of the similarity transformation defined by (11.3.25a), that is

$$\Xi \to X_p = \varpi^{-2}\, \Xi\, \varpi^2, \qquad (11.3.25c)$$

connects the SO(4) rotation generators J_{i4}, to momentum-space versions of Runge–Lenz operators.

In Fock's treatment, the dynamics of the atom is governed by the equation,

$$p_0\Phi(\mathbf{x}) = (Z/\pi^2)\int \Phi(\mathbf{x}')/|\mathbf{x}-\mathbf{x}'|^2 d\Omega', \qquad (11.3.26a)$$

which replaces the regularized Schrödinger equation

$$(W_{op} - Z)\psi = 0. \qquad (11.3.26b)$$

The equation

$$i\partial\Phi(\mathbf{x},\tau)/\partial\tau = (Z/\pi^2)\int \Phi(\mathbf{x}',\tau)/|\mathbf{x}-\mathbf{x}'|^2\, d\Omega', \qquad (11.3.27)$$

is an analog of Schrödinger's time-dependent equation in which, $i\partial/\partial t$ can be considered equivalent to $(-p_0/2)\; i\partial/\partial\tau$. The time-dependent equation associated with (11.3.26b) is the analog of the corresponding classical equation of Chapter 8.

There are several important aspects of regularized quantum mechanical treatments that make them particularly relevant to investigations of more complicated atomic, molecular, and solid state systems. At this point we wish only to call attention to a consequence of the fact that the hyperspherical harmonics Y_{klm} comprise a discrete and complete set of functions: any piecewise continuous function $\varphi(\mathbf{x})$ defined on the hypersphere can be expanded as series of these harmonics. When such a piecewise continuous function is stereographically projected onto the surface of a hypersphere, the projection is necessarily piecewise continuous, except, at the "South pole".

Finally, it should be noted that the reciprocal-space functions of solid state physics have a natural projection onto Fock's hypersphere, where they become superpositions of hyperspherical harmonics. Several depictions of molecular reciprocal-space functions will be found in Avery's book, *Hyperspherical Harmonics.*[13]

11.4 Bases Provided by Eigenfunctions of $\mathbf{J_{12}}$, $\mathbf{J_{34}}$, $\mathbf{J_{56}}$

The SO(4) group whose generators, $\mathrm{J_{ab}} = -\mathrm{J_{ba}}$, satisfy the classical commutation relations

$$[\mathrm{J_{ab}}, \mathrm{J_{bc}}] = \mathrm{J_{ac}}, \tag{11.4.1}$$

has two commuting generators, which may be choosen to be $\mathrm{J_{12}}$ and $\mathrm{J_{34}}$. The quantum mechanical position-space operators corresponding to the $\mathrm{J_{ab}}$ satisfy commutation relations in which the negative sign on the right-hand side is replaced by its square root, and the same may be said of the corresponding momentum-space operators. In the case of the SO(4) degeneracy group of hydrogen-like atoms, one may label degenerate eigenstates $|\rangle$ by the quantum number n, and by the eigenvalue m_{12} of $\mathrm{J_{12}}$ and the eigenvalue m_{34} of $\mathrm{J_{34}}$. Then one has the n^2 states defined by

$$\mathrm{J_{56}}|m_{12}, m_{34}, n\rangle = n|m_{12}, m_{34}, n\rangle, \tag{11.4.2a}$$

and

$$\mathrm{J_{12}}|m_{12}, m_{34}, n\rangle = m_{12}|m_{12}, m_{34}, n\rangle, \tag{11.4.2b}$$

$$\mathrm{J_{34}}|m_{12}, m_{34}, n\rangle = m_{34}|m_{12}, m_{34}, n\rangle. \tag{11.4.2c}$$

An alternative labeling turns out to be revealing.[16] Define

$$\begin{aligned} &\mathrm{M_1} = (\mathrm{J_{23}} + \mathrm{J_{14}})/2, \quad \mathrm{M_2} = (\mathrm{J_{31}} + \mathrm{J_{24}})/2, \quad \mathrm{M_3} = (\mathrm{J_{12}} + \mathrm{J_{34}})/2, \\ &\mathrm{N_1} = (\mathrm{J_{23}} - \mathrm{J_{14}})/2, \quad \mathrm{N_2} = (\mathrm{J_{31}} - \mathrm{J_{24}})/2, \quad \mathrm{N_3} = (\mathrm{J_{12}} - \mathrm{J_{34}})/2, \end{aligned} \tag{11.4.3}$$

one finds that all the M operators commute with all the N operators. It follows that SO(4) is the direct product of two three-parameter groups. The quantum mechanical version of the operators satisfy

$$[\mathrm{M_a}, \mathrm{M_b}] = \mathrm{i}\ \mathrm{e_{abc}M_c}, \quad [\mathrm{N_a}, \mathrm{N_b}] = \mathrm{i}\ \mathrm{e_{abc}M_c}, \quad [\mathrm{M_i}, \mathrm{N_j}] = 0. \tag{11.4.4}$$

One may choose M_3, N_3, and J_{56} to be a commuting set of operators defining the n^2 states that satisfy

$$\begin{aligned} J_{56}|m_3, n_3, n\rangle &= n|m_3, n_3, n\rangle, \\ M_3|m_3, n_3, n\rangle &= m_3|m_3, n_3, n\rangle, \\ N_3|m_3, n_3, n\rangle &= n_3|m_3, n_3, n\rangle. \end{aligned} \quad (11.4.5)$$

The labeling schemes (11.4.2) and (11.4.5) are connected by the relations

$$m_3 = (m_{12} + m_{34})/2, \quad n_3 = (m_{12} - m_{34})/2. \quad (11.4.6)$$

As m_3 and n_3 can have half-integer values, $SO(4) = SU(2) \otimes SU(2)$. As will be seen in the next chapter, this fact can be quite useful.

Appendix A. Matrix Elements of SO(4,2) Generators[3,16]

In the previous pages only the matrix elements of labeling operators have been presented. To represent the action of the remaining group generators and other important operators on the $|nlm\rangle = |1nlm\rangle$ basis, it is convenient to define the raising and lowering operators:

$$\begin{aligned} L^{\pm} = J_{23} \pm iJ_{31}, \quad N^{\pm} = K_{46} \pm iK_{45}, \quad \Gamma^{\pm} = K_{16} \pm iK_{26}, \\ A_{-}^{\pm} = J_{14} \pm iJ_{24}, \quad A_{+}^{\pm} = K_{15} \pm iK_{25}. \end{aligned} \quad (11.A.1)$$

It is also useful to define the functions

$$\begin{aligned} c_0(a, b) \equiv ((a^2 - b^2)/(4b^2 - 1))^{1/2}, \quad d_0(a, b) \equiv (a^2 - b^2)^{1/2}, \\ d_{\pm 1}(a, b) \equiv ((a \pm b)(a + 1 \pm b))^{1/2}. \end{aligned} \quad (11.A.2)$$

One has

$$N^{\pm}|nlm\rangle = (n(n+1) - l(l+1))^{1/2}|n \pm 1, l, m\rangle, \quad (11.A.3a)$$

$$\begin{aligned} J_{34}|nlm\rangle = c_0(n, l+1),\ d_0(l+1, m)|n, l+1, m\rangle + c_0(n, l), \\ d_0(l, m)|n, l-1, m\rangle \end{aligned} \quad (11.A.3b)$$

$$\begin{aligned} A_{-}^{\pm}|nlm\rangle = \pm\{-c_0(n, l+1, m)\ d_{\pm 1}(l, m)|n, l-1, m \pm 1\rangle \\ - c_0(n, l)\ d_{\pm 1}(l - m, -m)|n, l+1, m \pm 1\rangle\} \end{aligned} \quad (11.A.3c)$$

$$\begin{aligned} K_{35}|nlm\rangle = (1/2)c_0(n, l+1)\ d_0(l+1, m) \\ \{[(n+l+2)/(n-l-1)]^{1/2} \\ |n+1, l+1, m\rangle \\ + [(n-l-2)/(n+l+1)^{1/2}]|n-1, l+1, m\rangle\} \end{aligned} \quad (11.A.3d)$$

$$+(1/2)\,c_0(n,l)\,d_0(l,m)\{[(n-l+1)/(n+l)]^{1/2}|n{+}1,l-1,m\rangle$$
$$+\ [(n+l+2)/(n{-}l{-}1)]^{1/2}\ [(n+l-1)/(n-l)]^{1/2}$$
$$\times\ |n-1,l-1,m\rangle\}$$

$$K_{36}|nlm\rangle = c_0\ (n,l+1)\ d_0(l+1,m)\{(-1/2)[(n-l-2)/(n+l+1)]^{1/2}$$
$$\times\ |n-1,l+1,m\rangle + (1/2)[(n+l+2)/(n-l-1)]^{1/2}$$
$$\times\ |n+1,l+1,m\rangle\}$$
$$+\ c_0(n,l)\ d_0(l,m)\{(-1/2)[(n-l-2)/(n+l+1)]^{1/2}$$
$$\times\ |n-1,l-1,m\rangle$$
$$+\ (1/2)[(n+l+2)/(n-l-1)]^{1/2}|n+1,l-1,m\rangle\} \quad (11.A.3e)$$

Using the relations $r = (J_{56} - K_{46})$, and $z = (K_{35} - J_{34})$, one obtains:

$$r|nlm\rangle = (1/2)[(n-l-1)(n+l)]^{1/2}|n-1,l,m\rangle + n|nlm\rangle$$
$$+\,(1/2)[(n+l+1)(n-l)]^{1/2}|n+1,l,m\rangle.$$

and

$$z|nlm\rangle = c_0(n,l+1)d_0(l+,m)\{(1/2)[(n-l-1)(n+l+1)]^{1/2}$$
$$\times\,|n-1,l+1,m\rangle + |n,l+1,m\rangle$$
$$+\,(1/2)(n-l-1)^{1/2}|n+1,l+1,m\rangle\}$$
$$+\,c_0(n,l)\ d_0(l,m)\{(1/2)[(n-l-1)(n+l{+}1)]^{1/2}|n-1,l-1,m\rangle$$
$$+\,|n,l-1,m\rangle + (1/2)(n-l-1)^{1/2}|n+1,l-1,m\rangle\}$$

Appendix B. N-Shift Operators For the Hyperspherical Harmonics

Kumei[17] discovered that Gegenbauer's general differential equation,

$$\{(1-x_4^2)(\partial/\partial x_4)^2 + (2\lambda+1)x_4\partial/\partial x_4 + \nu(\nu+2\lambda)\}C_\nu^\lambda(x_4) = 0. \quad (11.B.1a)$$

has an SO(2,1) invariance group which contains ν-labeling and ν-shift operators.

In the special case of interest here, $\nu(\nu{+}2\lambda) = n^2 - l^2$, and the operators shift the princial quantum number n. Kumei linearized the spectrum of ν

by converting (11.B.1a) to

$$\{(1-x_4^2)(\partial/\partial x_4)^2 + (2\lambda+1)x_4\partial/\partial x_4 + i\partial/\partial\tau(i\partial/\partial\tau + 2\lambda)\} \qquad (11.4.1)$$
$$\times \exp(-i\nu\tau)C_\nu^\lambda(x_4) = 0. \qquad (11.B.1b)$$

The determining equations for the generators of the invariance group of this equation yield the required labeling and v-shift operators for the C_ν^λ when one allows Q's that correspond to first-order differential operators, U. The solutions of the determining equations are defined to within a multiplicative constant. He chose them to be

$$Q_1 = i\partial/\partial\tau, \quad Q_2 = 1, \qquad (11.B.2a)$$
$$Q_- = \exp(i\tau)\{(1-x_4^2)\partial/\partial x_4 + i\,x_4\partial/\partial\tau\}, \qquad (11.B.2b)$$
$$Q_+ = \exp(-i\tau)\{(1-x_4^2)\partial/\partial x_4 - i\,x_4\partial/\partial\tau - 2\lambda\,x_4\}. \qquad (11.B.2c)$$

Q_1 is the ν-labeling operator. The n-labeling operator of interest in the present context is

$$Q_0 = Q_1 + \lambda Q_2. \qquad (11.B.2d)$$

The operators Q_0, Q_+ and Q_- satisfy

$$[Q_+, Q_-] = 2Q_0, \quad [Q_\pm, Q_0] = \pm Q_\pm. \qquad (11.B.2e,f)$$

Consequently

$$Q_\pm \exp(-i\nu\tau)C_\nu^\lambda(x_4) = c_\pm(n,l)\exp(-i(\nu\pm1)\tau)C_{\nu\pm1}^\lambda(x_4). \qquad (11.B.2g)$$

They shift ν by ±1 without shifting l, and hence shift n by ±1. The Casimir operator is

$$C_2 = Q_0^2 + (Q_+Q_- + Q_-Q_+)/2. \qquad (11.B.2h)$$

The functions $\exp(-i\nu\tau)C_\nu^\lambda(x_4)$ are eigenfunctions of C_2 with eigenvalue $\lambda(\lambda-1)$, i.e., $l(l+1)$.

To obtain n-labeling and shift operators that act on the functions

$$\begin{aligned}\exp(-i\nu\tau)Y_{klm} &= \exp(-i\nu\tau)\ N_l\sin(\alpha)^l C_\nu^\lambda(x_4)\\ &= \exp(-i\nu\tau)\ N_l(1-x_4^2)^{l/2}C_\nu^\lambda(x_4),\end{aligned} \qquad (11.B.3)$$

we use a similarity transformation to exploit Kumei's general result. Let

$$Q\exp(-i\nu\tau)C_\nu^\lambda(x_4) \to N\exp(-i\nu\tau)(1-x_4^2)^{1/2}C_\nu^\lambda(x_4), \qquad (11.B.4a)$$

with

$$N = (1 - x_4^2)^{1/2}\ Q(1 - x_4^2)^{-1/2}. \tag{11.B.4b}$$

$Q_0 = i\partial/\partial\tau + \lambda \equiv N_0$ is not altered by the transformation. If we let

$$|nlm\rangle = \exp(-i\nu\tau)Y_{klm} \tag{11.B.5a}$$

then

$$N_0|nlm\rangle = n|nlm\rangle. \tag{11.B.5b}$$

Also let

$$\begin{aligned} N_\pm|nlm\rangle &= N_\pm \exp(-i\nu\tau)Y_{klm} = \{(1 - x_4^2)^{1/2}Q_\pm(1 - x_4^2)^{-1/2}\} \\ &\quad \times \exp(-i\nu\tau|N_l(1 - x_4^2)^{1/2}C_\nu^\lambda(x_4) \\ &= (1 - x_4^2)^{1/2}\ Q_\pm \exp(-i\nu\tau)\ N_l C_\nu^\lambda(x_4) \\ &= (1 - x_4^2)^{1/2}\ c_\pm(n, l)\{\exp(-i(\nu \pm 1)\tau)\ N_l C_{\nu\pm1}^\lambda(x_4)\} \\ &= c_\pm(n, l)|n \pm 1, l, m\rangle \end{aligned} \tag{11.B.5c}$$

One finds

$$N_- = \exp(+i\tau)\{(1 - x_4^2)\partial/\partial x_4 + x_4(N_0 - 1)\}, \tag{11.B.5d}$$

$$N_+ = \exp(+i\tau)\{(1 - x_4^2)\partial/\partial x_4 - x_4(N_0 + 1)\}. \tag{11.B.5e}$$

Because they have been obtained by similarity transformation, these Q' operators satisfy the same commutation relations as the original Q operators. The corresponding Casimir operator has eigenvalues $l(l + 1)$.

If $\mathbf{N}_+$ and $\mathbf{N}_-$ are matrices with elements $\langle n'lm, N_\pm|nlm\rangle$, then one finds that $\mathbf{N}_- = -\mathbf{N}_+^\dagger$: the operator N_- is the negative of the adjoint of N_+. In the general case,

$$'N_\pm|nlm\rangle = c_\pm(n, l)|n \pm 1, l, m\rangle \tag{11.B.6}$$

with

$$c_\pm(n, l) = \pm((l + 1 \pm n)(\pm n - l))^{1/2}.$$

The operators Q also yield operators $Q'' = \exp(i\nu\tau)Q\exp(-i\nu\tau)$ that can be used to produce operators which act directly on the Y_{klm}. After removing the time-dependent exponents from the Q'', the operator $Q_0'' \to N_0$, and one obtains operators $\Theta_\pm$ that convert Y_{klm} to $c_\pm(k + 1, l)\ Y_{k\pm1lm}$. These operators are

$$\begin{aligned} \Theta_- &= (1 - x_4^2)\partial/\partial x_4 + (k - l)x_4, \\ \Theta_+ &= (1 - x_4^2)\partial/\partial x_4 + (k - l)x_4 - 2(l + 1)x_4. \end{aligned} \tag{11.B.7}$$

References

[1] A. O. Barut and H. Kleinert, *Phys. Rev.* **156** (1967) 1541.
[2] C. Fronsdal, *Phys. Rev.* **156** (1967) 1665.
[3] Y. Nambu, *Phys. Rev.* **160** (1967) 1171.
[4] M. Bednar, *Ann. Phys. (N.Y.)* **75** (1973) 305.
[5] B. R. Judd, *Angular Momentum Theory for Diatomic Molecules* (Academic Press, N. Y., 1975), pp. 32–34.
[6] P. A. M. Dirac, *Quantum Mechanics*, 3rd edn. (Oxford, 1947), p. 53.
[7] V. Fock, *Z. Phys.* **98** (1935) 145.
[8] J. Avery, *Hyperspherical Harmonics and Generalized Sturmians* (Kluwer, Dordrecht, 2000), p. 63.
[9] W. Pauli, *Z. Phys.* **36** (1926) 336.
[10] V. Bargmann *Z. Phys.* **99** (1936) 576.
[11] J. Avery, *ibid.* (Ref. 8) p. 25.
[12] T. Shibuya and C. E. Wulfman, *Amer. J. Phys.* **33** (1965) 570.
[13] J. Avery, *Hyperspherical Harmonics* (Kluwer, Dordrecht, 1989).
[14] B. R. Judd, *ibid.* (Ref. 5) p. 222.
[15] T. Shibuya, C. E. Wulfman, *Proc. Roy. Soc. (London)* A **286** (1965) 376.
[16] L. C. Biedenharn, *J. Math. Phys.* **2** (1961) 433.
[17] S. Kumei, *M. Sc. Thesis* (University of the Pacific, 1972).

CHAPTER 12

Uncovering Approximate Dynamical Symmetries. Examples From Atomic and Molecular Physics

12.1 Introduction

In 1885, *Annalen der Physik* published a paper on the physical relevance of certain small integers.[1] In the paper, the Basel secondary-school teacher, J. J. Balmer, pointed out that the observed wavelengths, λ, of the H_α, H_β, H_γ, and H_δ lines in the sun's Fraunhofer spectrum[2] obeyed, with amazing accuracy, a relation

$$\lambda = \mathrm{G}n^2/(n^2 - 4), \tag{12.1.1}$$

in which G is a constant, and $n = 3, 4, 5, 6$. It was soon recognized that if one sets $n_1 = n$ and $n_2 = 2$, the corresponding frequencies, ν, of these *Balmer lines* satisfy the general equation

$$\nu = \mathrm{R}(1/n_2^2 - 1/n_1^2), \tag{12.1.2}$$

which is obeyed by subsequently related lines in the spectrum of hydrogen. In this equation, R, the Rydberg constant, is $109667.581\,\mathrm{cm}^{-1}$.

When Bohr introduced his theory of the atom,[3] the integers n became the principal quantum numbers of the Bohr and Bohr–Sommerfeld theory.[3,4] Ehrenfest then showed that n, and many other quantum numbers, defined closed areas of magnitude $n\,\mathrm{h}/2\pi$ in phase space. And then in 1935 Fock discovered that n defines an SO(4) symmetry.

The advent of quantum mechanics in 1926, together with extensive contributions from experimentalists, led to rapid strides in many branches of physics. The immediate consequences in atomic physics are evident in Condon and Shortley's 1935 monograph,[5] written shortly before Fock's discovery. By 1931 Wigner and Eckart had already produced, and demonstrated the utility of SO(3) coupling coefficients.[6] In the early 1940's this work was greatly extended by Racah, who also introduced the concept of seniority.[7] Many subsequent developments of the group theory relevant to atomic physics are expounded and reviewed in Edmond's monograph,[8] and in monographs by Biedenharn and Louck,[9] Wybourne,[10] and Judd.[11]

The development of methods for approximating the solutions of determining equations makes it possible to systematically investigate the dynamical groups of Schrödinger equations more complex than those considered in Chapter 11, but it does not obviate the need to use concepts, such as those of the theory of angular momenta, to simplify equations as complex as those governing many-particle systems, such as those arising in atomic and molecular physics. The development of relevant simplifications is the subject of the next few sections of this chapter, and much of Chapter 13.

In recent years it has been shown that the use of rather small Sturmian basis sets, or equivalent sets of hyperspherical harmonics, can lead to surprisingly accurate calculations of atomic energies.[12] Limited basis sets are able to produce much better approximate energy and potential energy curves than produced by LCAO or Gaussian bases of similar size. This approach will be outlined in Secs. 12.10 and 12.11. At this point we would note that in previous chapters it has become apparent that the generators of the dynamical groups SO(4), SO(4,1), and SO(4,2), are most simply expressed in these bases. This suggests that as the use of these bases increase, previously unrecognized dynamical symmetries, and their consequences, will be uncovered. The topics in this chapter can only hint at the possibilities.

12.2 The Stark Effect; One-Electron Diatomics

The regularization of the Hamiltonian for a hydrogen atom immersed in a uniform electric field of strength ε in the z direction gives

$$\mathrm{r(H-E) = r(p^2/2 - E} - \varepsilon \mathrm{z - 1/r) = W_0} - \varepsilon \mathrm{z\, r - 1}. \tag{12.2.1}$$

To understand how the field affects the degeneracy of the states $|nlm\rangle$ one need only examine the matrix elements of states of the same n;

$$\langle nlm, \mathrm{r\ z}\, nlm\rangle. \tag{12.2.2}$$

Now,

$$\mathrm{z = B_z - A_z} \tag{12.2.3}$$

and **B** only has nonzero matrix elements between states with different n values, so the removal of the degeneracy is due to $\mathrm{A_z = J_{34}}$. The $\mathrm{SU(2)} \otimes \mathrm{SU(2)}$ states $|m_{12}m_{34}n\rangle$ defined in Eq. (11.4.2) are eigenstates of $\mathrm{L_z}$ and $\mathrm{A_z}$

with eigenvalues m_{12} and m_{34}, respectively. These then become the zero-order eigenstates of the Stark effect. The $|nlm\rangle$ can be considered as being obtained from them by the coupling defined by

$$|nlm\rangle = \sum_{\mathrm{m_1 m_2}} (j_1 m_1 j_2 m_2 | j_1 j_2 lm) | j_1 m_1, j_2 m_2\rangle, \tag{12.2.4a}$$

for the special case in which $\mathrm{j_1 = j_2}$. As the coupling transformation

$$|j_1 m_1, j_2 m_2\rangle = \sum_{\mathrm{m}} (j_1 m_1 j_2 m_2 | j_1 j_2, lm) | nlm\rangle \tag{12.2.4b}$$

is unitary, and the energy shift defined by (12.2.2) may be calculated by evaluating the radial integral

$$\begin{aligned} &\int \chi_{nl'm'}(\mathbf{r})(1/\mathrm{r})\mathrm{r}\ \chi_{nlm}(\mathbf{r})\ \mathrm{r}^2 \mathrm{dr} \sin(\theta) d\theta\ d\phi \\ &= \int \chi_{nl'm'}(\mathbf{r})\ \mathrm{z}\ \chi_{nlm}(\mathbf{r})\ \mathrm{r}^2 \mathrm{dr} \sin(\theta) d\theta\ d\phi. \end{aligned} \tag{12.2.5}$$

The integral can be evaluated by relating the Sturmians to the ordinary hydrogen-atom functions Ψ_{nlm}. Referring back to (11.1.11b), one has

$$\Psi_{nlm}(\mathrm{r}, \theta, \phi) = \mathrm{i}^l\ n^{-3/2} \chi_{nlm}(\mathbf{r}/n), \tag{12.2.6a}$$

and

$$\exp(\mathrm{i}\alpha \mathrm{K}_{45})\ \chi_{nlm}(\mathbf{r}) = \chi_{nlm}(\mathbf{r}/n). \tag{12.2.6b}$$

Using these relations one can easily evaluate

$$\int \chi_{nl'm'}(\mathbf{r})\ \mathrm{z}\ \chi_{nlm}(\mathbf{r})\ \mathrm{r}^2 \mathrm{dr} \sin(\theta) d\theta\ d\phi. \tag{12.2.7}$$

A most enlightening discussion of the Stark effect is contained in Sommerfeld's *Atomic Structure and Spectral Lines.*[4]

In 1964 Redmond[13] found that the Stark effect Hamiltonian

$$\mathrm{p^2/2m - Ze^2/r + e}\mathbf{E}, \tag{12.2.8}$$

has an exact constant of motion, $\mathbf{C}$, that is obtained by adding a term $-(\mathbf{r}\,\mathbf{x}\,\mathbf{E})\,\mathbf{x}\,\mathbf{r}/2\mathrm{Ze}$ to the Runge–Lenz vector operator:

$$\mathbf{C} = \mathbf{r}/\mathrm{r} + (\mathbf{L}\,\mathbf{x}\,\mathbf{p} - \mathbf{p}\,\mathbf{x}\,\mathbf{L})/(2\mathrm{Ze^2m}) - (\mathbf{r}\,\mathbf{x}\,\mathbf{E})\,\mathbf{x}\,\mathbf{r}/2\mathrm{Ze}. \tag{12.2.9}$$

In 1949 Erikson and Hill[14] showed that one-electron diatomic molecule–ions possess a constant of motion in addition to the component, $\mathrm{m}\hbar$ of the electronic angular momentum on their axis. If the molecular axis is the z axis, and internuclear distance is 2a, the Hamiltonian, whose value is

E(a), is

$$H = p^2/2 - (Z_a/|\mathbf{q} - a\mathbf{k}| + Z_b/|\mathbf{q} + a\mathbf{k}|), \tag{12.2.10}$$

and the constant of motion can be expressed as

$$F = \mathbf{L}^2 + a^2p_z^2 + 2az(Z_a/|\mathbf{q}-a\mathbf{k}| - Z_b/|\mathbf{q} + a\mathbf{k}|). \tag{12.2.11}$$

The corresponding Poisson bracket operator is the generator of a one-parameter symmetry group in classical phase space. As F commutes with L_z, the degeneracy group is Abelian.

The non-crossing rule applies to sets of energy level curves such as E(a). Wigner and von Neumann proved that varying a single parameter cannot produce crossings of levels whose wave functions have the same symmetry; if they appear to do so, a symmetry has been overlooked.[15] One-electron diatomics provide one of the simplest such examples. Some levels of the same spatial symmetry cross because they have different dynamical symmetry. In diatomics with two or more electrons this symmetry is removed, and levels that previously crossed approach, but avoid, each other. The reader is referred to Judd's monograph[11] for discussions of a number of interesting aspects of the dynamical symmetry of one-electron diatomics. Included within it is a discussion of the simple solutions that arise for special sets of nuclear charges. In some later calculations on systems with $Z_a = 1$, and $Z_b = 3, \ldots, 8$, Ponomarev and Puzynia observed *pseudo-intersections.*[16] This raises a suspicion that as the bond between the proton and the other nucleus weakens, such systems may have a useful approximate dynamical symmetry larger than that of homonuclear systems.

12.3 Correlation Diagrams and Level Crossings: General Remarks

Whenever a change in a Hamiltonian produces a new degeneracy, new phenomena may develop. The symmetry involved can be revealing, and may have practical importance.

If the change in a Hamiltonian can be produced by changing a parameter, much can be learned from a simple plot of the energy level diagram arising from varying the parameter. Let λ be a parameter, and write

$$H = H_0 + \lambda H_1, \quad E = E_{m,0} + \lambda E_{m,1}. \tag{12.3.1}$$

As a simple example, suppose that H is the Hamiltonian of a molecule with one or more double bonds, such as the molecules depicted in **Figs. 12.3a–d**. In these figures, λ is the angle θ that measures the twist

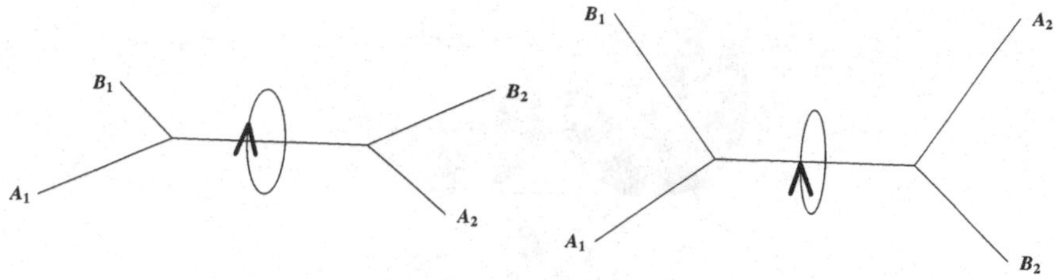

Fig. 12.3a Twisting of a substituted ethylene molecule.

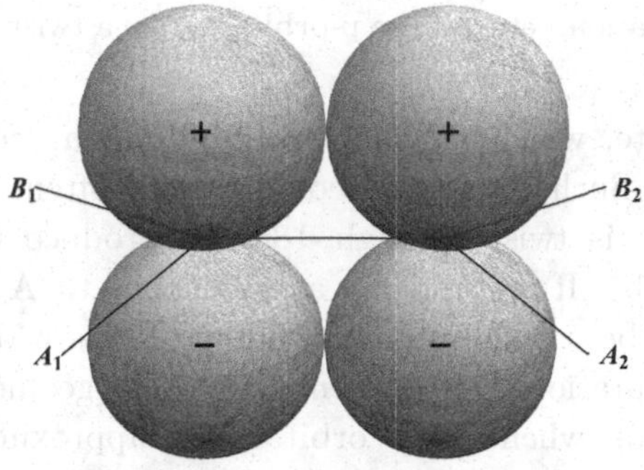

Fig. 12.3b The p-orbitals attached to atoms C_1 and C_2.

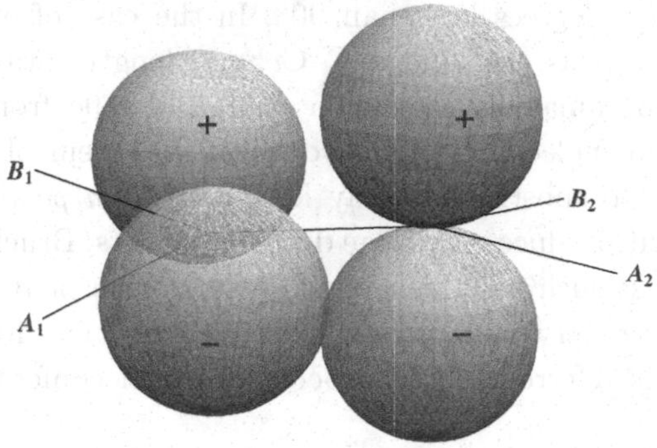

Fig. 12.3c Relation between the p-orbitals after a twist through 45°.

of the $A_2C_2B_2$ plane with respect to the $A_1C_1B_1$ plane. In (b) the two p orbitals that produce the π bond are depicted. As the twist increases to 90°, the overlap between them decreases to zero, and the bond is destroyed. Further increase of θ to 180° produces a state that has the same energy

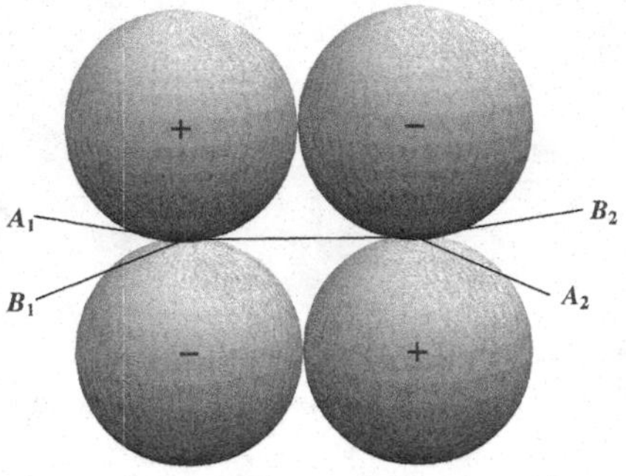

Fig. 12.3d Relation between the p-orbitals after a twist through 180°.

as the excited state, which may be produced when the molecule in (c) absorbs a photon. Such photo-excited molecules often undergo *cis-trans isomerism* — that is twist through 180° to produce the ground state of the molecule, (b). If A_1 and B_1 are identical to A_2 and B_2 respectively, there must be a level crossing that produces a degeneracy at the 90° twisted configuration.[17] In systems with less geometrical symmetry, degeneracies develop when the p orbitals are approximately perpendicular to each other. The resulting degenerate states are very sensitive to the presence of polarizing fields with components in the C_1–C_2 direction, and the molecules begin to develop large dipole moments when the twisting becomes a few degrees less than 90°. In the case of ethylene, a proton situated along this axis and a C–C bond length away produces the dipole moment obtained by moving a charge of 0.9e from one C atom to the other.[17] In molecules with a conjugated system of double bonds, asymmetries due to substituents may lead to *sudden polarizations* in the twisted states that produce very large dipole moments. Bruckner and Salem proposed that just such a sudden photo-polarization of retinal is responsible for initiating nerve impulses in the retina.[18] For further information the reader is referred to the proceedings of a conference on sudden polarization.[19]

When variation of a parameter changes the spatial point-group symmetry of a nonlinear molecular system, the theorem of Jahn and Teller can come into play.[20] It dictates that such a variation creates a degeneracy of energy levels, when in the region of the crossing the energies are linear functions of the parameter. It follows that if the degeneracy were to occur naturally, the molecule would undergo a distortion that removes the symmetry.

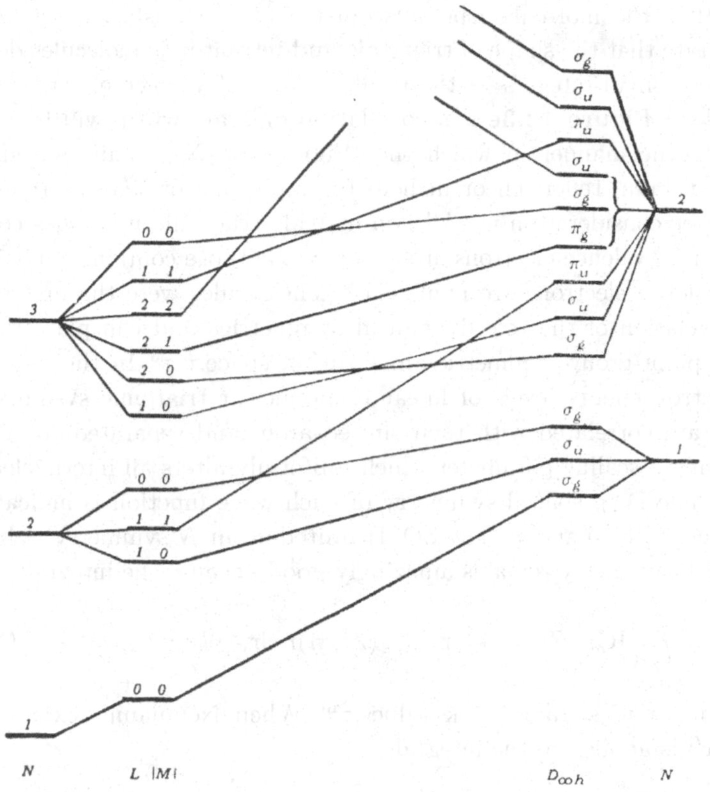

Fig. 12.3e Correlation between the united-atom and separated-atom energy levels of one-electron linear homonuclear triatomic molecules with identical bond lengths.

If variation of a parameter bends a linear system, an extension of the Jahn–Teller Theorem becomes relevant. In the linear system, rotational symmetry about the axis of linearity allows one to assign to each level a rotational quantum number m. As the linear system bends, the energies of all levels must depend quadratically upon the bending angle in the vicinity of the linearity. The bending produces a mixing of pairs of degenerate levels whose m values are the negative of one another. One of the new levels then exhibits a quadratically increasing energy, the other a quadratically decreasing energy. If the degeneracy develops naturally, the linear system bends on its own accord. A similar extension of the Jahn–Teller Theorem applies when planar systems are bent to produce nonplanar ones.

In 1953 the molecular spectroscopist Walsh published a set of rules which state that the shape of triatomic and tetraatomic molecules depends only upon small integers — the total number of valence electrons in the molecule.[21] **Figure 12.3e** is a correlation diagram which will be used to illustrate the manner in which the Wigner–von Neumann rule and the extended Jahn–Teller Theorem help to imply one of Walsh's rules. The rule under consideration is "All non-hydride triatomic molecules containing 15 or 16 valence electrons are linear, while those containing 17, 18, 19, or 20 valence electrons are nonlinear". These rules were the first to lead to a revelation of the role dynamical symmetries and can play in determining point-group symmetries in position space.[22,23] In the figure, the one-electron energy levels of linear homonuclear triatomic systems, such as N_3^-, are correlated with their united-atom and separated-atom limits by varying a scaling parameter which uniformly alters all internuclear distances. The $D_{\infty h}$ spatial symmetry of each wave function is indicated by the labels σ, π, g and u. The SO(4) united-atom N symmetry is broken, but the L symmetry remains amazingly good because the integrals

$$\int_0^\infty \mathrm{R}_{nl1}(\mathrm{Zr}/n)(\mathrm{r}^k/\mathrm{r})\mathrm{R}_{nl2}(\mathrm{Zr}/n)\mathrm{r}^2\mathrm{dr}, \quad l_2 \neq l_1, \qquad (12.3.2)$$

vanish for a wide range of k values.[24a] When Neumann's expansion of $1/|\mathbf{r} - \mathbf{r}'|$ is applied to the integrals

$$\int_0^\infty \Psi_{nl1m1}(\mathbf{Zr}/n)(1/|\mathbf{r} - \mathbf{R}|)\Psi_{nl2m2}(\mathbf{Zr}/n)\mathrm{dv}, \qquad (12.3.3)$$

some, as expected, vanish as a consequence of SO(3) symmetry. Others, however, vanish unexpectedly as a consequence of SO(4) symmetry.[22] Swamy, Kulkarni, and Biedenharn subsequently proved that whenever they vanish it is because of SO(4,2) symmetry.[24b]

In developing **Fig. 12.3e**, the center of nuclear charge of the system was assumed to be the origin of coordinates. This eliminates terms depending linearly on the nuclear displacements. The *forbidden crossings* that develop are those involving π_u orbitals. When electron-repulsion effects are taken into account, one finds that they take place in the united-atom. As a consequence, no matter what the size of the system is, the degenerate π_u orbitals always begin to be occupied when the system contain 17 valence electrons. This remains true when one mathematically shifts nuclei about to create non-hydride systems, such as NCO, that are no longer homonuclear. The extended Jahn–Teller Theorem then requires that molecules with 17, 18 or 19 valence electrons are nonlinear. As the systems are bent, the symmetry becomes C_{2v}. However, levels of the same C_{2v} symmetry, but with different l, cross, and the π_u orbitals correlate with orbitals of much lower energy. Systems with 15 or 16 valence electrons remain linear, but those with 17 to 19 valence electrons are bent. In 20 electron systems, bending increases the electron density at the center of nuclear charge, and this stabilizes the bent system. Extensive analyses have fully established the connection between all of Walsh's rules and the extended Jahn–Teller Theorem.[25]

The examples in this section have been chosen to illustrate a general strategy that is available whenever one is dealing with a system governed by differential equations. *If one begins to attack the problem by seeking the dynamical symmetry that, when broken, can produce the phenomenon of interest, several advantages result*:

a. The simplest form of perturbation theory (in quantum mechanics, degenerate perturbation theory) usually enables one to determine with certainty the physical interactions that are involved.
b. The extensive and well organized mathematics of the theory of Lie algebras and groups is available to guide and aid the investigation.

The strategy enables one to treat physical phenomena as displays of the consequences of symmetries of differential equations and their symmetry breaking.

For a Coulomb system of 3N degrees of freedom in position-space, we have seen that the relevant dynamical group may be as large as a constrained SO(4N,2N). However, in this chapter we can only adumbrate studies of the physical relevance of a few subgroups of the two-particle dynamical group SO(8,4). The goal is to gain a better understanding of the consequences of electronic repulsions. To apply the general strategy just outlined, several of the following sections are devoted to investigations of $SO(4) \otimes SO(4)$. An example of the direct physical relevance of the larger group is described in Sec. 12.8.

12.4 Coupling $SO(4)_1 \otimes SO(4)_2$ to Produce $SO(4)_{12}$

The mathematics described in this section applies to atoms containing two non-interacting electrons, each having associated Runge–Lenz and angular momentum vectors, $\mathbf{A}_j$ and $\mathbf{L}_j$. The components of $\mathbf{A} = \mathbf{A}_1+\mathbf{A}_2$ and the components of $\mathbf{L} = \mathbf{L}_1+\mathbf{L}_2$ close under commutation, and generate an

SO(4) group which is sometimes designated as $SO(4)_{12}$. In the absence of other interactions this becomes a degeneracy group of the two-electron system. As the generators of this group are the sum of the generators of two SO(4) groups denoted by $SO(4)_1$ and $SO(4)_2$, the group $SO(4)_{12}$ is their direct product, $SO(4)_1 \otimes SO(4)_2$. The situation is closely analogous to the $SO(3) \otimes SO(3)$ system discussed in Chapter 10.

Here we wish to produce functions that are eigenfunctions of the two-particle SO(4) Casimir operators $C_2 = \mathbf{A}^2 + \mathbf{L}^2$, and $C_2' = \mathbf{A} \cdot \mathbf{L}$ and also of $\mathbf{L}^2$ and L_z. The one-particle Casimir operators $\mathbf{A}_1^2 + \mathbf{L}_1^2$, $\mathbf{A}_2^2 + \mathbf{L}_2^2$, commute with C_2 and C_2', as of course do all 12 generators of $SO(4) \otimes SO(4)$. However, $\mathbf{A}_1$ does not commute with $\mathbf{L}_1^2$, and $\mathbf{A}_2$ does not commute with $\mathbf{L}_2^2$. It follows that the $SO(4)_{12}$ eigenfunctions which we seek must be expected to be linear combinations of products of one-particle states of differing l and m values. The eigenvalues of C_2 and C_4 are determined by two quantum numbers P and Q. These observations may be summarized with the statement that the desired $SO(4)_{12}$ eigenfunctions could be labeled $|PQLM(n_1, n_2)\rangle$. These functions are a linear combination of products, $|n_1 l_1 m_1\rangle |n_2 l_2 m_2\rangle$, of single particle functions:

$$|PQLM(n_1, n_2)\rangle = \sum_{l_1 l_2 m_1 m_2} B(n_1 l_1 m_1, n_2 l_2, m_2 | n_1 n_2, PQLM) |n_1 l_1 m_1\rangle |n_2 l_2 m_2\rangle. \quad (12.4.1)$$

The coefficients $B(n_1 l_1 m_1, n_2 l_2, m_2 | n_1 n_2, PQLM)$ are $SO(4) \otimes SO(4)$ coupling coefficients analogous to the $(l_1 m_1 l_2, m_2 | l_1 l_2, LM)$ coupling coefficients of $SO(3) \otimes SO(3)$. In 1961 Biedenharn determined their dependence upon the quantum numbers involved.[26] In doing so he imposed the phase convention of Condon and Shortley,[27] and the mathematics he used required him to make the assumption that **A**, like **L**, had the *derivative property* possessed by first-order derivative operators, D_{Op}, which is $D_{Op}(f^* g) = f^*(D_{Op}(g)) + (D_{Op}(f))^* g$. A determination of the $B(n_1 l_1 m_1, n_2 l_2, m_2 | n_1 n_2, PQLM)$ due to Wybourne does not require this last assumption.[28] Biedenharn also established that

$$C_2 |PQLM(n_1, n_2)\rangle = (P(P+2) + Q^2) |PQLM(n_1, n_2)\rangle, \quad (12.4.2)$$

$$\mathbf{A} \cdot \mathbf{L} |PQLM(n_1, n_2)\rangle = Q(P+1) |PQLM(n_1, n_2)\rangle. \quad (12.4.3)$$

Q may take on both positive and negative values. Because $\mathbf{A} \cdot \mathbf{L}$ is not invariant under the parity operation $\mathbf{x} \to -\mathbf{x}$, the SO(4) functions

$|PQLM(n_1, n_2)\rangle$ may not have definite parity when $n_1 \neq n_2$. Eigenfunctions of the O(4) Casimir operator $(\mathbf{A}\cdot\mathbf{L})^2$, with eigenvalue $Q^2(P+1)^2$, and definite parity, may be constructed by taking linear combinations of the functions obtained by interchanging the two quantum numbers.

The two-electron wave functions $|PQLM(n_1, n_2)\rangle$ may be thought of as obtained by a *recoupling* of the already coupled angular momentum eigenstates $|LM(l_1, l_2, n_1, n_2)\rangle$. In other words, the states $|PQLM(n_1, n_2)\rangle$ can be considered to be linear combinations of states $|LM(l_1, l_2, n_1, n_2)\rangle$ with differing l values:

$$|PQLM(n_1, n_2)\rangle = \sum_{l_1, l_2} \mathrm{C}(LM(l_1, l_2, n_1, n_2),\ PQLM)|LM(l_1, l_2, n_1, n_2)\rangle. \tag{12.4.4}$$

Multiplying these by the appropriate spin functions, $|\sigma\rangle$, produces states that are antisymmetric for particle exchange. When $n_1 = n_2 = 2$, for example, the two singlet $|LM(l_1, l_2, n_1, n_2)\rangle|^1\sigma\rangle$ states are

$$^1\mathrm{S}(2\mathrm{s}^2) = |00(0,0,2,2)\rangle|^1\sigma\rangle, \quad ^1\mathrm{S}(2\mathrm{p}^2) = |00(1,1,2,2)\rangle|^1\sigma\rangle. \tag{12.4.5}$$

The first state is simply a product of two 2s functions and a spin function, $|^1\sigma\rangle$, that is antisymmetric under exchange. In the second state, products of two 2p functions are coupled to produce a state of zero angular momentum that is symmetric under particle exchange. Multiplying this by $|^1\sigma\rangle$ produces the $^1\mathrm{S}(2\mathrm{p}^2)$ wave function. Together, these produce the $|PQLM(n_1, n_2)\rangle$ states[29]

$$\begin{aligned} |0000(2,2)\rangle|^1\sigma\rangle &= 0.524\ ^1\mathrm{S}(2\mathrm{s}^2) - 0.851\ ^1\mathrm{S}(2\mathrm{p}^2), \\ |2000(2,2)\rangle|^1\sigma\rangle &= 0.851\ ^1\mathrm{S}(2\mathrm{s}^2) + 0.524\ ^1\mathrm{S}(2\mathrm{p}^2). \end{aligned} \tag{12.4.6}$$

In this case, $Q = 0$ for both states, and $\mathbf{A}_1+\mathbf{A}_2$ is orthogonal to $\mathbf{L}$.

12.5 Coupling $\mathrm{SO}(4)_1 \otimes \mathrm{SO}(4)_2$ to Produce $\mathrm{SO}(4)_{1-2}$

The group SO(4) has a most surprising property[30]: two different $\mathrm{SO}(4)_{12}$ groups can be obtained by the coupling of $\mathrm{SO}(4)_1$ and $\mathrm{SO}(4)_2$. The alternative coupling is the subject of this section.

The components of $\mathbf{A}_- = \mathbf{A}_1 - \mathbf{A}_2$ and $\mathbf{L} = \mathbf{L}_1 + \mathbf{L}_2$ close under commutation and generate an SO(4) group which we designate $\mathrm{SO}(4)_{1-2}$.

The Casimir operators of the group are

$$C_2 = \mathbf{A}_-^2 + \mathbf{L}^2 \quad \text{and} \quad C_2' = \mathbf{A}_- \cdot \mathbf{L}. \tag{12.5.1}$$

The eigenstates of these operators and the operators

$$\mathbf{L}^2, L_z, \ \mathbf{A}_1^2 + \mathbf{L}_1^2, \ \mathbf{A}_2^2 + \mathbf{L}_2^2,$$

are of the form

$$|PQLM(n_1, n_2)\rangle = \sum_{l_1, l_2} D(LM(l_1, l_2, n_1, n_2), PQLM)|LM(l_1, l_2, n_1, n_2)\rangle. \tag{12.5.2a}$$

They are eigenfunctions of C_2 and C_2':

$$(\mathbf{A}_-^2 + \mathbf{L}^2)|PQLM(n_1, n_2)\rangle = (P(P+2) + Q^2)|PQLM(n_1, n_2)\rangle, \tag{12.5.2b}$$

$$(\mathbf{A}_- \cdot \mathbf{L})|PQLM(n_1, n_2)\rangle = Q(P+1)|PQLM(n_1, n_2)\rangle. \tag{12.5.2c}$$

When $n_1 \neq n_2$, O(4) eigenfunctions of $C_2'^2$ and definite parity may, as before, be obtained as linear combinations of the functions obtained by interchanging the two quantum numbers n_1, n_2.

The functions D are determined in Ref. 30. The $|PQLM(n_1, n_2)\rangle$ states analogous to the $|PQLM(n_1, n_2)\rangle$ states of Sec. 12.5 are

$$\begin{aligned} |0000(2,2)|^1\sigma\rangle &= -0.500\ {}^1S(2s^2) + 0.866\ {}^1S(2p^2), \\ |2000(2,2)|^1\sigma\rangle &= -0.866\ {}^1S(2s^2) - 0.500\ {}^1S(2p^2). \end{aligned} \tag{12.5.3}$$

The coupling process that produces these eigenfunctions of $\mathbf{A}_-^2 + \mathbf{L}^2$ and $(\mathbf{A}_- \cdot \mathbf{L})^2$, and the coupling process of Biedenharn which produces the eigenstates of $\mathbf{A}^2 + \mathbf{L}^2$ and $(\mathbf{A} \cdot \mathbf{L})^2$, begin with the same product of one-particle SU(2) $\otimes$ SU(2) states:

$$|j_{a1} m_{a1}, j_{b1} m_{b1}\rangle |j_{a2} m_{a2}, j_{b2} m_{b2}\rangle.$$

For them

$$\begin{aligned} \mathbf{j}_{a1} &= (\mathbf{L}_1 + \mathbf{A}_1)/2, \quad \mathbf{j}_{b1} = (\mathbf{L}_1 - \mathbf{A}_1)/2, \\ \mathbf{j}_{a2} &= (\mathbf{L}_2 + \mathbf{A}_2)/2, \quad \mathbf{j}_{b2} = (\mathbf{L}_2 - \mathbf{A}_2)/2, \end{aligned} \tag{12.5.4}$$

and

$$\mathbf{L}_1 = \mathbf{j}_{a1} + \mathbf{j}_{b1}, \quad \mathbf{A}_1 = \mathbf{j}_{a1} - \mathbf{j}_{b1}; \quad \mathbf{L}_2 = \mathbf{j}_{a2} + \mathbf{j}_{b2}, \quad \mathbf{A}_2 = \mathbf{j}_{a2} - \mathbf{j}_{b2}.$$

Biedenharn's result may be obtained by determining the couplings arising from:

i) adding $\mathbf{j}_{a1}$ and $\mathbf{j}_{a2}$ to obtain $\mathbf{J}_{a+} = (\mathbf{L}_1 + \mathbf{L}_2 + \mathbf{A}_1 + \mathbf{A}_2)/2$ and determining the eigenfunctions of $\mathbf{J}^2_{a+}$,
ii) adding $\mathbf{j}_{b1}$ and $\mathbf{j}_{b2}$ to obtain $\mathbf{J}_{b+} = (\mathbf{L}_1 + \mathbf{L}_2 - \mathbf{A}_1 - \mathbf{A}_2)/2$ and to determining the eigenfunctions of $\mathbf{J}^2_{b+}$,
iii) adding $\mathbf{J}_{a+}$ and $\mathbf{J}_{b+}$ to obtain $\mathbf{L} = \mathbf{L}_1 + \mathbf{L}_2$ and determining the eigenfunctions of $\mathbf{L}^2$.

$$P = J_{a+} + J_{b+} \text{ and } Q = J_{a+} - J_{b+}. \tag{12.5.5}$$

The result in (12.5.2) is obtained by determining the couplings arising from:

i) adding $\mathbf{j}_{a1}$ and $\mathbf{j}_{b2}$ to obtain $\mathbf{J}_{a-} = (\mathbf{L}_1 + \mathbf{L}_2 + \mathbf{A}_1 - \mathbf{A}_2)/2$ and determining the eigenfunctions of $\mathbf{J}^2_{a-}$,
ii) adding $\mathbf{j}_{a2}$ and $\mathbf{j}_{b1}$ to obtain $\mathbf{J}_{b-} = (\mathbf{L}_1 + \mathbf{L}_2 - \mathbf{A}_1 + \mathbf{A}_2)/2$ and determining the eigenfunctions of $\mathbf{J}^2_{b-}$,
iii) adding $\mathbf{J}_{a-}$ and $\mathbf{J}_{b-}$ to obtain $\mathbf{L} = \mathbf{L}_1 + \mathbf{L}_2$ and determining the eigenfunctions of $\mathbf{L}^2$.

$$\text{Then } P = J_{a-} + J_{b-} \text{ and } Q = J_{a-} - J_{b-}. \tag{12.5.6}$$

From the couplings (12.5.5) it follows that

$$2(\mathbf{J}^2_{a+} + \mathbf{J}^2_{b+}) = \mathbf{A}^2 + \mathbf{L}^2, \quad \mathbf{J}^2_{a+} - \mathbf{J}^2_{b+} = \mathbf{A} \cdot \mathbf{L}, \tag{12.5.7a,b}$$

and those of (12.5.6) imply

$$2(\mathbf{J}^2_{a-} + \mathbf{J}^2_{b-}) = \mathbf{A}^2_- + \mathbf{L}^2, \quad \mathbf{J}^2_{a-} - \mathbf{J}^2_{b-} = \mathbf{A}_- \cdot \mathbf{L}. \tag{12.5.7c,d}$$

The next several sections investigate the utility of these two couplings.

12.6 Configuration Mixing in Doubly Excited States of Helium-like Atoms

If one uses degenerate perturbation theory to determine the effect of the $1/r_{12}$ repulsions between electrons in the initially degenerate states of the previous two sections, one finds the configurations with definite l values, $^1S(2s^2)$ and $^1S(2p^2)$, mix to produce the following two ^1S wave functions:

$$\begin{aligned} \Psi_a &= -0.476\ {}^1S(2s^2) + 0.880\ {}^1S(2p^2), \\ \Psi_b &= -0.880\ {}^1S(2s^2) - 0.476\ {}^1S(2p^2). \end{aligned} \tag{12.6.1}$$

These are approximations to wave functions of an excited helium atom with two electrons in the $n = 2$ level. The overlap of these states with the corresponding $SO(4)_{12}$ states of Sec. 12.4 is, to three significant figures, 0.982.[29] To the same accuracy, their overlap with the $SO(4)_{1-2}$ states of Sec. 12.5 is 1.000.

The states given in (12.6.1) are defined by the eigenvectors of the interaction matrix

$$\begin{pmatrix} \langle S(2p^2), r_{12}^{-1} S(2p^2)\rangle & \langle S(2p^2), r_{12}^{-1} S(2s^2)\rangle \\ \langle S(2s^2), r_{12}^{-1} S(2p^2)\rangle & \langle S(2s^2), r_{12}^{-1} S(2s^2)\rangle \end{pmatrix} = \frac{1}{512}\begin{pmatrix} 111 & -15\sqrt{3} \\ -15\sqrt{3} & 77 \end{pmatrix}. \tag{12.6.2}$$

In the corresponding $SO(4)_{1-2}$ matrix that defines the states in (12.5.3), the lower right-hand entry, 77, is replaced by 81, with an increase of approximately 5%. The remarkable agreement of the $SO(4)_{1-2}$ configuration mixing of degenerate two-electron states and the mixing caused by the electron repulsion operator $1/r_{12}$ persists for a wide variety of two-electron n, n states.[30] It arises from the fact that in the regularized systems

$$\mathbf{r}_1 - \mathbf{r}_2 = (\mathbf{B}_1 - \mathbf{A}_1) - (\mathbf{B}_2 - \mathbf{A}_2). \tag{12.6.3}$$

This causes in-shell interactions to produce approximate eigenstates of $|\mathbf{A}_1 - \mathbf{A}_2|^2$.[31] To estimate energy changes produced by $1/r_{12}$ when one is using the regularized system defined by $W = 0$, with

$$W = r_1 r_2 (p_1^2/2 + p_2^2/2 - Z/r_1 - Z/r_2 + 1/r_{12} - 2E), \tag{12.6.4}$$

one may consider all wave functions to be built from Sturmian functions. To simplify notation, let us for the moment denote the collection of quantum numbers $LM\ l_1, l_2, n_1, n_2, \sigma$, by $\boldsymbol{k}$, and write $\Psi_{\boldsymbol{k}}(\mathbf{r}_1, \mathbf{r}_2)$ for $|LM(l_1, l_2, n_1, n_2)\rangle|\sigma\rangle$. The matrix elements of $1/r_{12}$ are then

$$\iint \Psi_{\boldsymbol{k}}^*(\mathbf{r}_1, \mathbf{r}_2)\ \Psi_{\boldsymbol{k}'}(\mathbf{r}_1, \mathbf{r}_2)\ (r_1 r_2 / r_{12}) dr_1 dr_2\ d\omega_1 d\omega_2. \tag{12.6.5}$$

These integrals vanish if $\boldsymbol{k}$ and $\boldsymbol{k}'$ do not have identical L, M, and σ. They are approximated in Ref. 31, and may be evaluated exactly with the aid of formulas such as those given in Ref. 32.

Sinanoglu and Herrick found that the $SO(4)_{1-2}$ coupling formula (12.5.2) quite accurately predicts configuration mixing when $n_1 \neq n_2$.[33] The $\mathbf{B}_j$ terms in (12.6.3) contribute to this mixing.

In 1980, Herrick and Kellman pointed out that the first two couplings outlined in (12.5.6) are couplings that determine internal motions which have not been coupled to produce rotational states of definite angular momentum. They made and developed the intriguing observation that these internal motions are analogous to the internal vibrations in triatomic molecules.[34]

12.7 Configuration Mixing Arising From Interactions Within Valence Shells of Second and Third Row Atoms

In the 2s-2p shell, the number of electrons, N, varies from 1 to 8. It turns out that in every case where the in-shell $1/r_{12}$ interactions act on initially degenerate $2s^D 2p^{N-D}$ configurations, the resulting states are very similar to eigenstates of $\mathbf{A}_-^2 \equiv \sum_{i<j} |\mathbf{A}_i - \mathbf{A}_j|^2$. In every case, the overlap between the corresponding states is 0.999.[35] As in the two-electron case, a slight difference in the configuration mixing arises because the actual 2s-2s diagonal element is slightly less than that required for eigenstates of $\mathbf{A}_-^2$. As the number of electrons increases, the difference decreases, becoming less than 1% when N reaches 8. The coupling (12.4.1) produces eigenstates of $\sum_{i<j} |\mathbf{A}_i + \mathbf{A}_j|^2$ which in all cases give poorer approximations to the actual configuration mixing.[29]

The effects of Coulomb repulsions within the 3s-3p shell are very similar to those within the 2s-2p shell. The overlap between the calculated states produced by configuration mixing and the eigenstates of $\mathbf{A}_-^2$ varies between 0.987 and 0.997.[36a]

In-shell configuration interaction between 3s, 3p, 3d electrons is well approximated by two-electron eigenstates of $\mathbf{A}_-^2$, and three-electron states if the orbital and spin angular momenta are fairly high, as in ^4S, ^4P, ^4F or ^2D states. However when d electrons are involved in ^2S,^2P or ^2D states, the overlap between the eigenstates of $\mathbf{A}_-^2$ and those of $1/r_{12}$ can become as small as 0.657.[36b]

12.8 Origin of the Period-Doubling Displayed in Periodic Charts

Contributions to a recent monograph, *The Mathematics of the Periodic Table*, well depict current understanding of the manner in which dynamical symmetries organize the properties of the chemical elements — and therefore much of the natural sciences.[37]

Barut was the first to use the dynamical group SO(4,2) for this purpose.[38] He pointed out that Madelung's rule, which describes the filling in of the periods, reflects a property of an SO(3,1) subgroup of SO(4,2).

The rule states that in the neutral atoms in a given period, orbitals of lowest $n + l$ are filled first, and that orbitals having the same $n + l$ value are filled in order of increasing n.[39] Unfortunately, as Ostrovsky notes, the SO(3,1) subgroups do not contain the orbitals required to deal with period doubling.[42]

Advances that follow Barut's are described in Ref. 37 in contributions from Novaro,[40] Kibler,[41] and Ostrovsky, who provides a very extensive list of references to the earlier literature.[42] These authors explore the relevance of the groups $SU(2) \otimes SU(2) \otimes SU(2), SO(4,2) \otimes SU(2)$, and $SO(4,2) \otimes SU(2) \otimes Z_2$. (In each of these direct-product groups the SU(2) subgroup is not due to electron spin).

Most periodic charts portray one period of length two, Period 1. Then, Periods 2 and 3 both have the same length, eight; Periods 4 and 5, both have length eighteen; Periods 6 and 7 both have length thirty-two.

Though SO(4) has unitary irreducible representations of the required dimensions, 1, 4, 9, and 16, each of these representations occurs only once in SO(4,2). Thus a definitive feature of the periodic chart can have no explanation within SO(4,2). A dynamical explanation of the observed order of filling of orbitals with nodal properties of hydrogen-like n, l, m orbitals must deal with this fundamental problem. Each of the authors mentioned above has approached it in a different way, but much remains to be done to develop a definitive deduction of period doubling.

The discussions in previous chapters make it evident that a dynamical explanation of period doubling will also provide a knowledge of the relevant subgroup(s) of the dynamical group of n-electron atoms.

In this connection, a key group-theoretic aspect of period doubling emerges from an investigation carried through by Kitigawara.[43] He established that the smallest special orthogonal groups with UIR of the required dimensions are the rank three group SO(7) and its noncompact generalizations SO(5,2), *etc.* No known single-particle system appears to have the required dynamical symmetry.

Fortunately, Kitigawara's analysis provides further insight into what is required to derive period doubling from the dynamics — and hence dynamical symmetries — encoded in many-electron Schrödinger equations. The length of each period is a reflection of the fact that the energy gaps between two successive periods is greater than the step-by-step energy changes that develop as each periods fills. One might conjecture that in the simplest of approximations the dynamical symmetry responsible for this could be a degeneracy group which is a subgroup of SO(8,4). It could be that of systems with a single outer electron moving in an approximate self-consistent field potential provided by a nucleus surrounded by inner-shell electrons. The inner shells must then have the effect of shifting the energy of an n−d state to that of an n+1 s or p state, and must have the effect of shifting the energy of an n − f state to that of an n + 2 s or p state. If η is the quantum number required to give a common label to all the states of a period, then states whose n, l values assign the same value to the expression $n + l + \delta_{l,0} - 1$ must have the same value of η. The appropriate UIR's of the group must contain only the wave functions whose n, l quantum numbers produce the same value of η. The dimensions of these UIR's must be η^2, so that when spin is taken into account, the number of initially degenerate outer-electron states with quantum number η will be $2\eta^2$. The quantum number η must, in all the periods, play much the same role as the principal quantum number n does in Periods 1, 2 and 3. It should be able to label the Period, and one should be able to express one-electron energies in the outer shell of neutral atoms by

$$\mathrm{E} = -\zeta^2/2\eta^2, \tag{12.8.1}$$

where the effective nuclear charge, ζ, is a function of η. Further, quantum numbers must be able to label orbitals in such a way that those filled in each period have the correct number of nodes. As the period fills, ζ can be expected to become a function of these additional quantum numbers.

To summarize:

Though well developed inductions of possible symmetries responsible for period doubling are available, there is not yet available a definitive deduction of period doubling from the relevant Schrödinger equations.

12.9 Molecular Orbitals in Momentum-Space; The Hyperspherical Basis

To make contact with current literature, in this and subsequent sections of the chapter we will replace the quantum number k of Chapter 11, by p, and use $p = n - 1$ to replace the label n in both position-space and momentum-space hydrogen-like wave functions. In the units we have been using the operator of translations that carries $\mathbf{r}$ to $\mathbf{r} + \mathbf{R}$, that is, $\exp(\mathbf{R} \cdot \boldsymbol{\nabla}_{\mathrm{q}})$, can be expressed as $\exp(\mathrm{i}\mathbf{R} \cdot \mathbf{p}_{\mathrm{op}})$. The corresponding operator that acts on momentum space functions and their associated hyperspherical harmonics is the operator of multiplication $\exp(\mathrm{i}\mathbf{p} \cdot \mathbf{R})$. If, for example, $\mathrm{Y}_{plm}(\Omega)$, $p =$

n − 1, are hyperspherical harmonics corresponding to Sturmian position-space functions, $\chi_{nlm}(\mathbf{r})$, and Schrödinger functions $\Psi_{nlm}(\mathbf{r})$, then

$$\exp(\mathrm{i}\mathbf{p}\cdot\mathbf{R})\mathrm{Y}_{nlm}(\Omega) \tag{12.9.1a}$$

corresponds to

$$\exp(-\mathrm{i}\mathbf{R}\cdot\mathbf{p}_{\mathrm{op}})\chi_{nlm}(\mathbf{r}) = \chi_{nlm}(\mathbf{r}-\mathbf{R}), \tag{12.9.1b}$$

and

$$\exp(-\mathrm{i}\mathbf{R}\cdot\mathbf{p}_{\mathrm{op}})\Psi_{nlm}(\mathbf{r}) = \Psi_{nlm}(\mathbf{r}-\mathbf{R}). \tag{12.9.1c}$$

To determine the effect of the translation one may expand the product $\exp(\mathrm{i}\mathbf{p}\cdot\mathbf{R})\mathrm{Y}_{plm}(\Omega)$ as a sum of hyperspherical harmonics, writing,[44]

$$\exp(\mathrm{i}\mathbf{p}\cdot\mathbf{R})\mathrm{Y}_{plm}(\Omega) = \sum \mathrm{S}^{nlm}_{n'l'm'}(\mathrm{p}_0\mathbf{R})\ \mathrm{Y}_{p'l'm'}(\Omega). \tag{12.9.2a}$$

Then

$$\mathrm{S}^{nlm}_{n'l'm'}(\mathrm{p}_0\mathbf{R}) = \int \mathrm{Y}_{p'l'm'}(\Omega)^*\exp(\mathrm{i}\mathbf{p}\cdot\mathbf{R})\mathrm{Y}_{nlm}(\Omega)\mathrm{d}\Omega. \tag{12.9.2b}$$

If $\mathbf{X}$ is a vector from the origin to a point on Fock's unit hypersphere, then his equation governing the motion of an electron in the field of a nucleus of charge Z is

$$\mathrm{p}_0\Psi(\Omega) = \int (\mathrm{Z}/|\mathbf{X}-\mathbf{X}'|^2)\ \Psi(\Omega')\ \mathrm{d}\Omega'. \tag{12.9.3}$$

Translations convert this into an equation governing the momentum of an electron moving in the electric field due to a set of nuclei of charge $\mathrm{Z_j}$ at positions $\mathbf{R}_\mathrm{j}$. The resulting generalization of Fock's equation is[44]

$$\mathrm{p}_0\Psi(\Omega) = \int \sum_{\mathrm{j}} \{\mathrm{Z_j}|\mathbf{X}-\mathbf{X}'|^{-2}\exp(-\mathrm{i}(\mathbf{p}-\mathbf{p}')\cdot\mathbf{R}_\mathrm{j}\}\ \Psi(\Omega')\ \mathrm{d}\Omega'. \tag{12.9.4a}$$

It defines the equation

$$(\mathbf{P}-\mathrm{p}_0\mathbf{I})\mathbf{C} = \mathbf{0}, \tag{12.9.4b}$$

in which $\mathbf{C}$ is the vector of coefficients in the expansion

$$\Psi(\Omega) = \sum \mathrm{C}_{plm}\mathrm{Y}_{plm}(\Omega). \tag{12.9.4c}$$

The matrix

$$\mathbf{P} = \sum \mathrm{Z_j}\mathbf{W}(\mathrm{p}_0\mathbf{R}_\mathrm{j}) = \sum \mathrm{Z_j}\mathbf{S}(\mathrm{p}_0\mathbf{R}_\mathrm{j})^{-1}\mathbf{\Pi}\mathbf{S}(\mathrm{p}_0\mathbf{R}_\mathrm{j}). \tag{12.9.4d}$$

The elements of the matrix $\mathbf{S}(\mathrm{p_0}\mathbf{R}_\mathrm{j})$ are $\mathrm{S}^{nlm}_{n'l'm'}(\mathrm{p_0}\mathbf{R}_\mathrm{j})$, and

$$\mathbf{S}(\mathrm{p_0}\mathbf{R}_\mathrm{j})^{-1} = \mathbf{S}(-\mathrm{p_0}\mathbf{R}_\mathrm{j}). \tag{12.9.4e}$$

The matrix $\mathbf{\Pi}$ is diagonal with elements $(1/\mathrm{n})\delta_{nlm,n'l'm'}$. Its inverse, $\boldsymbol{N}$, has elements $n\ \delta_{nlm,n'l'm'}$, and

$$\boldsymbol{N}_\mathrm{j} \equiv \mathbf{S}(\mathrm{p_0}\mathbf{R}_\mathrm{j})\boldsymbol{N}\mathbf{S}(\mathrm{p_0}\mathbf{R}_\mathrm{j})^{-1}. \tag{12.9.5a}$$

Thus

$$\mathbf{W}(\mathrm{p_0}\mathbf{R}_\mathrm{j}) = \boldsymbol{N}_\mathrm{j}^{-1} = (\mathbf{S}(\mathrm{p_0}\mathbf{R}_\mathrm{j})(\boldsymbol{N}^2)\mathbf{S}(\mathrm{p_0}\mathbf{R}_\mathrm{j})^{-1})^{1/2}. \tag{12.9.5b}$$

The analog of (12.9.4b) that corresponds to Schrödinger's time-dependent equation is

$$(\mathbf{P} - \mathbf{I}\ \mathrm{i}\partial/\partial\tau)\mathbf{C} = \mathbf{0}. \tag{12.9.6}$$

The expansions in the past few paragraphs can be translated into corresponding expansions in a Sturmian basis. Here we wish to call attention to two properties that are common to both expansions:

(i) The operator $\mathbf{P}$ becomes that of a united-atom as the $\mathbf{R}_\mathrm{j}$ approach zero and its eigenvalues approach n. As the $\mathbf{R}_\mathrm{j}$ increase from zero the degeneracies associated with each value of n are broken, and thereafter each wave function and energy undergoes adiabatic change until any further crossings develop. One can think of these altered n values as effective quantum numbers, and they enter into the Sturmian functions as such.

(ii) In Eq. (12.9.5), the independently variable quantities in the matrix elements of $\mathbf{P}$ are the components of the variables $\mathbf{s}_\mathrm{j} = \mathrm{p_0}\mathbf{R}_\mathrm{j}$. For each set of chosen values of the $\mathbf{s}_\mathrm{j}$, the equation determines the eigenvalue $\mathrm{p_0}$; then one can determine the $\mathbf{R}_\mathrm{j}$. Though at first sight a nuisance, this property has several important consequences:

   (a) It introduces a scale factor that multiplies the radial argument r of Sturmian basis functions in a manner which ensures that united-atom and separated-atom hydrogenic levels are properly correlated.
   (b) This helps enable limited basis sets to produce much better approximate energies and potential energy curves than are produced by small LCAO or Gaussian basis sets.[45,46]
   (c) Near nuclear configurations that make $\mathrm{p_0}$ a degenerate eigenvalue, the rank of the matrix $\mathbf{P}$ becomes unstable to small displacements. This provides a method for uncovering crossings and degeneracies at which new chemical and physical phenomena may develop.

12.10 The Sturmian Ansatz of Avery, Aquilanti and Goscinski

If one neglects center of mass motions and regularizes the Schrödinger equation for a helium atom by multiplying it by the radial coordinates r_1, r_2, of electrons number 1 and 2, one obtains an equation which may be written as

$$r_2\{r_1(p_1^2/2 - E_1) - Z\} + r_1\{r_2(p_2^2/2 - E_2) - Z\} + r_1 r_2/r_{12} = 0, \qquad (12.10.1)$$

with

$$E_1 + E_2 = E.$$

If the $r_1 r_2/r_{12}$ term were not present, the solutions of the Schrödinger equation would be a product of the scaled Sturmians that are solutions of the single-particle Schrödinger equations whose regularized Hamiltonians are the operators contained in the curly brackets. The Sturmian ansatz of Avery, Aquilanti and Goscinski may be thought of as a method, a very clever one, for parceling out portions of the $r_1 r_2/r_{12}$ term between the curly brackets. It is now a well developed system for approximating solutions to the regularized Schrödinger equations of multi-electron systems of atomic and molecular physics.

The following outline of the ansatz closely follows that of Avery and uses his symbolism.[45,46] He begins with the general Schrödinger equation

$$(-\Delta/2 + V(\mathbf{x}) - E)\psi = 0, \qquad (12.10.2a,b)$$

in which

$$\Delta = \sum_{1}^{d} \partial^2/\partial x_j^2, \quad \mathbf{x} = (x_1, x_2, \ldots, x_d), \qquad (12.10.2c)$$

and the total potential $V(\mathbf{x})$ is the sum of the attractive potentials, $V_0(\mathbf{x})$, the total electrostatic repulsion potentials, and any other relevant potentials.

Let us for the moment ignore the Pauli principle and spin, and expand ψ as a linear combination of many-particle Sturmian functions

$$\psi = \sum B_\nu \, \Phi_\nu(\mathbf{x}), \qquad (12.10.3)$$

in which the Φ_ν are required to satisfy the equations

$$(-\Delta/2 + \beta_\nu V_0(\mathbf{x}) - E)\Phi_\nu(\mathbf{x}) = 0. \qquad (12.10.4)$$

Here β_ν is chosen in such a way that the energy E in these equation does not depend on ν. The members of the set of solutions to Eq. (12.10.4) all correspond to the same energy. This energy is chosen according to the state that is being represented. It can be shown that the set of isoenergetic solutions to (12.10.4) obeys a potential-weighted orthogonality relation. It is convenient to normalize the configurations in such a way that these orthogonality relations take the form

$$\int \mathrm{dx}\, \Phi^*_{\nu'}(\mathbf{x})\, \mathrm{V}_0(\mathbf{x})\, \Phi_\nu(\mathbf{x}) = (2\mathrm{E}/\beta_\nu)\delta_{\nu,\nu'}, \tag{12.10.5}$$

where dx represents the d-dimensional volume element of the system. Equations (12.10.4) and (12.10.5) assign a portion of the attractive potential to each $\phi_\nu(\mathbf{x})$ in a manner which, when supplemented with side conditions given below, produces energies and wave functions that closely approximate self-consistent field results. Taken together, (12.10.4) and (12.10.5) require that

$$\sum_\nu(-\Delta/2 + \mathrm{V}(\mathbf{x}) - \mathrm{E})\, \mathrm{B}_\nu\Phi_\nu(\mathbf{x}) = \sum_\nu(-\beta_\nu \mathrm{V}_0(\mathbf{x}) + \mathrm{V}(\mathbf{X}))\, \mathrm{B}_\nu\Phi_\nu(\mathrm{x}) = 0. \tag{12.10.6}$$

Multiplying these relations from the left by $\Phi^*_{\nu'}(\mathbf{x})$ and using (12.10.5) one obtains the simple set of equations

$$\sum_\nu \left[\int \mathrm{dx}\, \Phi^*_{\nu'}(\mathbf{x})\, \mathrm{V}(\mathbf{x})\Phi_\nu(\mathbf{x}) - 2\mathrm{E}\delta_{\nu'\nu},\right] \mathrm{B}_\nu = 0. \tag{12.10.7}$$

Next, suppose that the functions $\Phi_\nu(\mathbf{x})$ are products of single particle Sturmians $\phi_\eta(\mathbf{x}), \eta = nlm$, and spin functions $\sigma_\pm$, and then write

$$\chi_\mu(\mathbf{x}_1) = \phi_\eta(\mathbf{x})\sigma_\pm, \quad \mu = \eta \pm 1/2. \tag{12.10.8a}$$

Then let

$$\phi_\nu(\mathbf{x}) = \chi_\mu(\mathbf{x}_1)\, \chi_{\mu'}(\mathbf{x}_1)\, \chi_{\mu''}(\mathbf{x}_1)\ldots \tag{12.10.8b}$$

and require:

(A) The functions $\chi_\mu(\mathbf{x}_\mathrm{j})$ are solutions of the equations

$$(-\Delta_\mathrm{j} + k_\mathrm{u}^2 - 2nk_\mathrm{u}/\mathrm{r_j})\, \chi_\mu(\mathbf{x}_\mathrm{j}) = 0 \tag{12.10.9a}$$

and

(B)

$$\int \mathrm{d}^3\mathrm{x_j}\, \chi_{\mu'}(\mathbf{x}_\mathrm{j})^*(1/\mathrm{r_j})\, \chi_\mu(\mathbf{x}_\mathrm{j}) = (k_\mu/n)\delta_{\mu,\mu'}, \tag{12.10.9b}$$

(C)

$$\int \mathrm{d}^3\mathrm{x_j}|\chi_\mu(\mathbf{x}_\mathrm{j})|^2 = 1. \tag{12.10.9c}$$

The k_u are required to be related by the constraint

(D)

$$k_\mu^2 + k_{\mu'}^2 + k_{\mu''}^2 + \cdots = p_0^2 = -2E \tag{12.10.9d}$$

and a further constraint, which for an atom with nuclear charge Z, is

(E)

$$n_\mu k_\mu = n_{\mu'} k_{\mu'}, = \cdots = Z\beta_\nu. \tag{12.10.9e}$$

Equations (A) and (D) taken together assign to the argument of each $\chi_\mu(\mathbf{x}_j)$ a scale factor k_μ, and hence eigenvalues

$$p_{o\mu} = k_\mu = (-2E_\mu)^{1/2}. \tag{12.10.9f}$$

Then equation E requires

$$k_\mu = Z\beta_\nu / n_\mu. \tag{12.10.9g}$$

Thus $Z\beta_\nu$ is an effective nuclear charge that becomes assigned to the Sturmian with single-particle energy

$$-(1/2)k_\mu^2 = -(1/2)(Z\beta_\nu)^2/n_\mu^2. \tag{12.10.10}$$

With the substitution given by (12.10.9g) the orbitals χ_μ become solutions to the one-electron hydrogen-like wave equation for an atom with effective nuclear charge Z β_ν. All the orbitals belonging to the configuration Φ_ν have the same effective nuclear charge.

Equation C defines a normalization of the scaled Sturmian functions that is obtained by multiplying the Sturmians of Table 11.3.1 by a factor of $k_\mu^{3/2}$ after their radial argument has been multiplied by k_μ. Thus, for example, the normalized radial function for a 2p function is the Sturmian $k_\mu^{3/2}\ 2(k_\mu r)\exp(-k_\mu r)$.

Now let us return to Eq. (12.10.7), that is to the equations

$$\sum_\nu \left[\int dx \Phi_{\nu'}^*(\mathbf{x})\, V(\mathbf{x})\, \Phi_\nu(\mathbf{x}) - 2E\delta_{\nu'\nu} \right] B_\nu = 0.$$

Let

$$V(\mathbf{x}) = V^0(\mathbf{x}) + V^1(\mathbf{x}), \tag{12.10.11a}$$

with $V^0(\mathbf{x})$ being the sum of nuclear–electronic attractive potentials, and

$$V^1(\mathbf{x}) = \sum_{i<j} 1/r_{ij}. \tag{12.10.11b}$$

Next define

$$\int dx\, \Phi^*_{\nu'}(\mathbf{x})\, V^0(\mathbf{x})\, \Phi_\nu(\mathbf{x}) = -p_0 T^0_{\nu'\nu}, \tag{12.10.11c}$$

$$\int dx\, \Phi^*_{\nu'}(\mathbf{x})\, V^1(\mathbf{x})\, \Phi_\nu(\mathbf{x}) = -p_0 T^1_{\nu'\nu}. \tag{12.10.11d}$$

The $T^1_{\nu'\nu}$ are negative dimensionless quantities and are independent of p_0. The $T^0_{\nu'\nu}$ are positive, also dimensionless, and in the atomic case, independent of p_0.

In the atomic case, conditions (12.10.9a–d) imply that

$$\begin{aligned} T^0_{\nu'\nu} &= -(1/p_0) \int dx\, \Phi^*_{\nu'}(\mathbf{x})\, V^0(\mathbf{x})\, \Phi_\nu(\mathbf{x}) = (-2E/p_0\beta_\nu)\delta_{\nu'\nu} \\ &= (p_0/\beta_\nu)\delta_{\nu'\nu} = (1/\beta_\nu)(\textstyle\sum k_u^2)^{1/2}\delta_{\nu'\nu}. \end{aligned} \tag{12.10.12}$$

If there are N_j electrons with principle quantum number n_j, it becomes convenient to define

$$R_\nu = \left(\sum N_j/n_j^2\right)^{1/2}, \tag{12.10.13}$$

a function whose value only changes when orbital occupancies change. Then (12.10.9e) implies

$$T^0_{\nu'\nu} = ZR_\nu\delta_{\nu'\nu}. \tag{12.10.14}$$

Inserting (12.10.11b) into (12.10.7), and dividing through by p_0 produces a set of secular equations. If one lets $\mathbf{T}^0 = \mathbf{T}^0\mathbf{I}$ be the diagonal matrix with elements ZR_ν, and $\mathbf{T}^1$ be the matrix with elements $T^1_{\nu'\nu}$, then the secular equations take on the matrix-vector form:

$$(\mathbf{T}^1 - (p_0 - \mathbf{T}^0)\mathbf{I})\mathbf{B} = \mathbf{0}. \tag{12.10.15}$$

Some of the key consequences of Eqs. (12.10.9) are revealed when the methodology is applied to neutral several-electron atoms while configuration interactions are neglected. If the many-electron states obtained by solving (12.10.15) are labeled by an index κ, and their momenta p_0 are labeled $p_{0\kappa}$, then

$$p_{0\kappa} = ZR_\nu - T^1_{\kappa\kappa}, \tag{12.10.16a}$$

and

$$E_\kappa = -(1/2)(ZR_\nu - T^1_{\kappa\kappa})^2. \tag{12.10.16b}$$

For the ground states of neutral atoms with 2 to 9 electrons, as Z runs from 2 through 9, $T^1_{\kappa\kappa}$ changes from 0.441942 to 2.41491, an average increase of ~0.281/electron. For Z > 3, the calculated energies are very close to the Hartree–Fock self-consistent field values.[45,46] However, with marked contrast to SCF calculations, the orbitals produced have a simpler form: they are all one-electron Sturmians. Avery and Avery have used this fact and a 1/Z expansion to neatly isolate and study the large configuration mixing effects that arise from the interaction of degenerate one-electron functions the same principal quantum number.[47]

When configuration interactions are included, the resulting many- particle functions can still be considered to be Sturmians whose momentum-space analogs correspond to functions defined on hyperspheres in spaces of dimension greater than 4. Avery, Aquilanti and others have extensively developed the connection between hyperspherical bases defined on n-dimensional hyperspheres, and corresponding multiparticle Sturmians.[12,46] Correlations between particle motions are introduced by both the radial and angular functions that multiply these terms. In the hyperspherical bases, these correlations are all expressed as angular correlations.

The theory has been developed and extensively applied when $V(\mathbf{x})$ is the potential function appropriate to many-electron atoms.[12] It has been used to turn the treatment in Sec. 12.9 into that of many-electron molecules.[46] It has also been extended to relativistic systems.[12]

In concluding this section, we would call attention to the fact that Eq. (12.10.16), and the previous deductions, require the development of an entirely new analysis of effective nuclear charges for regularized systems.

Exercises

1. Draw a united-atom versus separated-atom correlation diagram for the levels of H_2^+ that includes united-atom levels through $n = 3$. Identify the "forbidden crossings".
2. Express the integral (12.2.7) as a function of n, and use the result to determine the dependency of (12.2.2) on all relevant quantum numbers. Using these, determine the dependency of the wavenumber of Stark effect lines of hydrogen upon these quantum numbers (cf. Ref. 4).
3. Use Sturmians for the ground state of He to exemplify Eq. (12.10.2) through (12.10.14). Set $\Lambda = 0.4419$ and calculate the contribution of $\langle 1/r_{12} \rangle$ to p_0 and E.
4. Use (12.9.1) to express the hyperspherical harmonic analog of the bonding molecular orbital of H_2^+. Do the same for the corresponding antibonding orbital.

5. It has long been known that, in solutions, benzoquinone can exhibit a dipole moment — despite the fact that its molecules have a center of symmetry. Use *extended Huckel* or similar calculations to develop a correlation diagram for the energy levels of its hydrocarbon analog as the molecule undergoes a B_3u out-of-plane bending. (The corresponding mode in benzoquinone has a very low frequency, $\sim 100\,\mathrm{cm}^{-1}$). Identify crossings involving the highest occupied π level, if any, and estimate the direction of polarization that might result. Carry out analogous calculations using an SCF or a better program.

References

[1] J. J. Balmer, *Ann. Phys.* **25** (1885) 80.
[2] J. V. Fraunhofer, *Encylopedia Britannica*, 11th edn. (Cambridge, 1910).
[3] N. Bohr, *Phil. Mag.* **26** (1913) 1.
[4] Cf. A. Sommerfeld, *Atomic Structure and Spectral Lines*, 3rd German edn., translated by H. L. Brose (Methuen, London, 1923).
[5] E. U. Condon and G. H. Shortley, *The Theory of Atomic Spectra* (Cambridge University Press, 1935, 1951–1967).
[6] (a) C. Eckart, *Rev. Mod. Phys.* **2** (1930) 305;
(b) E. Wigner, *Gruppentheorie* (Viewig, 1931);
(c) E. Wigner, *Group Theory and its Application to the Quantum Mechanics of Atomic Spectra*, (Academic Press, N.Y., 1959).
[7] G. Racah, *Phys. Rev.* **61** (1942) 186; *Phys. Rev.* **62** (1942) 438; *Phys. Rev.* **63** (1943) 367; *Phys. Rev.* **76** (1949) 1352.
[8] A. R. Edmonds, *Angular Momentum in Quantum Mechanics* (Princeton, 1957).
[9] (a) L. C. Biedenharn and J. D. Louck, *Angular Momentum in Quantum Physics* (Addison Wesley, Reading, Mass, 1981);
(b) L. C. Biedenharn and J. D. Louck, *The Racah-Wigner Algebra in Quantum Theory* (Addison Wesley, Reading, Mass, 1981).
[10] B. G. Wybourne, *Symmetry Principles and Atomic Spectroscopy* (Wiley-Interscience, N.Y., 1970).
[11] B. R. Judd, Atomic Shell Theory Recast, *Phys. Rev.* **162** (1967) 28.
[12] J. Avery, *Generalized Sturmians and Atomic Spectra*, (World Scientific, Singapore, 2006).
[13] P. J. Redmond, *Phys. Rev.* **133** (1964) B1352.
[14] H. A. Erikson and E. L. Hill, *Phys. Rev.* **75** (1949) 29.
[15] J. von Neumann and E. Wigner, *Z. Phys.* **30** (1929) 467.
[16] L. I. Ponomarev and T. P. Puzynina, *Soviet Physics JETP* **25** (1967) 846.
[17] C. E. Wulfman and S. Kumei, *Science* **172** (1971) 1061.
[18] L. Salem, *Isr. J. Chem.* **12** (1979) 8.

[19] *Proc. Indian Acad. Sci.* (*Chem. Sci.*) **107**(6) (1995).
[20] H. A. Jahn and E. Teller, *Proc. Roy. Soc. (London)* **A161** (1937) 220.
[21] A. D. Walsh, *Disc. Faraday Soc.*, Series 2, (1947) 18; *J. Chem. Soc.* (1953), 2260, 2266, 2278, 2296, 2300, 2306, 2318, 2325, 2330.
[22] C. E. Wulfman, *J. Chem. Phys.* **31** (1959) 381.
[23] C. Wulfman, *Memoria do Primeiro Congresso Latinamericano De Fisica*, Vol. 23 (Estadio pr la Sociedad Mexicana de Fisica Mexico, D. F, 1969), p. 105.
[24] (a) S. Pasternack and R. M. Sternheimer, *J. Math. Phys.* **3** (1962) 1280;
(b) N. V. V. J. Swamy, R. G. Kulkarni and L. C. Biedenharn, *J. Math. Phys.* **11** (1970) 1165.
[25] (a) R. F. W. Bader, *Can. J. Chem.* **40** (1962) 1164;
(b) R. G. Pearson, *J. Am. Chem. Soc.* **91** (1969) 1252; **91** (1969) 4947;
(c) R. G. Pearson, *J. Chem. Phys.* **52** (1970) 2167.
[26] L. C. Biedenharn, *J. Math. Phys.* **2** (1961) 433.
[27] E. U. Condon and G. H. Shortley, *The Theory of Atomic Spectra* (Cambridge University Press, 1967), p. 48.
[28] B. G. Wybourne, *Lie Groups for Physicists* (Wiley-Interscience, New York, 1974), pp. 236–240.
[29] E. Chacon, *et al.*, *Phys. Rev. A* **3** (1971) 166.
[30] C. Wulfman, *Chem. Phys. Lett.* **23** (1973) 370.
[31] C. Wulfman and S. Kumei, *Chem. Phys. Lett.* **23** (1973) 367.
[32] J. Avery, *Hyperspherical Harmonics and Generalized Sturmians* (Kluwer, Dordrecht, 2000), p. 14.
[33] (a) O. Sinanoglu and D. Herrick, *J. Chem. Phys.* **62** (1975) 886;
(b) O. Sinanoglu and D. Herrick, *J. Chem. Phys.* **65** (1976) 850.
[34] D. Herrick and M. Kellman, *Phys. Rev. A* **21** (1980) 418.
[35] C. Wulfman, *Phys. Rev. Lett.* **51** (1983) 1159.
[36] (a) Y. Ho and C. Wulfman, *Chem. Phys. Lett.* **103** (1983) 35.
(b) Y. Ho and M.Sc. Thesis, Dept. of Phys., University of the Pacific, Stockton, CA. (1985).
[37] D. H. Rouvray and R. B. King, eds., *The Mathematics of the Periodic Table* (Nova Science Pub., N. Y., 2006).
[38] A. O. Barut, in *The Structure of Matter, Proceedings of the Rutherford Centennial Symposium*, edn. B. G. Wybourne (Christchurch, N. Z., 1972), p. 126.
[39] E. Madelung, *Math. Hilfsmittel des Physikers*, 6th edn. (Springer, Berlin, 1950), p. 611.
[40] O. Novaro, in *The Mathematics of the Periodic Table* (2006), p. 217.
[41] M. R. Kibler, *ibid.*, p. 237.
[42] V. N. Ostrovsky, *ibid.*, p. 265.
[43] Y. Kitigawara, M.Sc. Thesis, Dept. of Phys., University of the Pacific, Stockton, CA (1977); J. C. Donini (ed.), *Recent Advances in Group Theory and Their Application to Spectroscopy*, (Plenum Press, New York, 1978).
[44] T. Shibuya and C. E. Wulfman, *Proc. Roy. Soc. A* **286** (1965) 376.

[45] J. Avery, *Hyperspherical Harmonics and Generalized Sturmians* (Kluwer, Dordrecht, 2000), Ch. 1.
[46] Cf. J. Avery, *Hyperspherical Harmonics* (Kluwer, Dordrecht, 1989), and the extensive list of references contained in it, and in Ref. 45.
[47] J. Avery and J. Avery, *Generalized Sturmians and Atomic Spectra* (World Scientific, Singapore, 2006).

CHAPTER 13

Rovibronic Systems

13.1 Introduction

Lie-algebraic methods were introduced into the study of the spectra of rotating and vibrating molecules by physicists who had contibuted to the development of the liquid drop model of nuclei and its connections to models composed of discrete nucleons. Their early success in nuclear physics led to several thoroughly developed mathematical models of rovibrational motion: the harmonic oscillator model, the symplectic model and the vibron model. For information about the first two, the reader is referred to reviews by Moshinsky and Rowe.[1,2] A U(6) vibron model was introduced in nuclear physics by Arima and Iachello,[3] and Iachello introduced a U(4) vibron model into molecular physics.[4] They have been applied to diatomics,[5] triatomics,[6] tetra-atomics,[7] and to molecules at least as complex as benzene.[8] Several monographs have been devoted to applications of the model in nuclear physics and molecular physics.[9–11]

The vibron models are semi-empirical ones, and a major criticism leveled against their application to molecular physics has been that there does not appear to be a definitive connection between their Lie-algebraic structure and that associated with Schrödinger equations with definite potentials. Nevertheless, they represent an outstanding example of the inductive use of the mathematics of Lie algebras in the sciences. We will concentrate attention on the connection they establish between molecular spectroscopy and Lie algebras.

13.2 Algebraic Treatment of Anharmonic Oscillators With a Finite Number of Bound States

The Morse oscillator was one of the systems with both a discrete and a continuous spectrum investigated in Chapter 10. It was found to have a non-compact three-parameter dynamical group, SU(1,1).

Here we wish to deal only with bound states of anharmonic oscillators. The simplest way is to *compactify* the noncompact groups that are the dynamical groups of these systems with both continuous and discrete spectra. Compactifying the Lie algebras of the dynamical groups, SO(2,1) of SU(1,1), of one-dimensional systems, produces the semi-simple A_1 Lie algebra of the groups SU(2) and SO(3). It provides a useful spectrum-generating algebra for the Morse oscillator.[12] Its potential energy function can be expressed as

$$V = D_e(1 - \exp(-a(r - r_e))^2, \tag{13.2.1}$$

where r_e is the value of the internuclear distance, r, at which the potential takes on its minimum value, D_e. This function is plotted in **Fig. 13.2.1a**. The corresponding total energy surface, E(r,p), is depicted in **Fig. 13.2.1b.**

Let three Lie generators S_k, of SU(2) and SO(3), obey the commutation relations

$$[S_1, S_2] = iS_3, \quad [S_2, S_3] = iS_1, \quad [S_3, S_1] = iS_2. \tag{13.2.2}$$

In this notation, the shift operators for the labeling operator S_3 are

$$S_+ = S_1 + iS_2, \quad S_- = S_1 - iS_2. \tag{13.2.3}$$

The mathematical eigenstates of S_3 may be denoted $|jm\rangle$. They satisfy the relations

$$\begin{aligned} S_+|jm\rangle &= (j(j+1) - m(m+1))^{1/2}|jm+1\rangle, \\ S_-|jm\rangle &= (j(j+1) - m(m-1)))^{1/2}|jm-1\rangle. \end{aligned} \tag{13.2.4}$$

When the resulting Lie algebra and Lie group are those of the bound states of an oscillator, it is assumed that only the absolute value of m is a physical observable.

As the mathematical states $|jm\rangle$ and $|j-m\rangle$ correspond to the same physical state, it is the value of $|m|$ that is physically relevant. As

$$S_-|j,-j\rangle = 0, \quad S_+|j,j\rangle = 0, \tag{13.2.5a}$$

one has

$$|m| = 1/2, 3/2, \ldots, j \tag{13.2.5b}$$

or

$$|m| = 0, 1, 2, 3, \ldots, j. \tag{13.2.5c}$$

In the first case the oscillator has $2j-1$ energy levels and the group is SU(2). In the second case it has $j+1$ levels and the group is SO(3). However, as

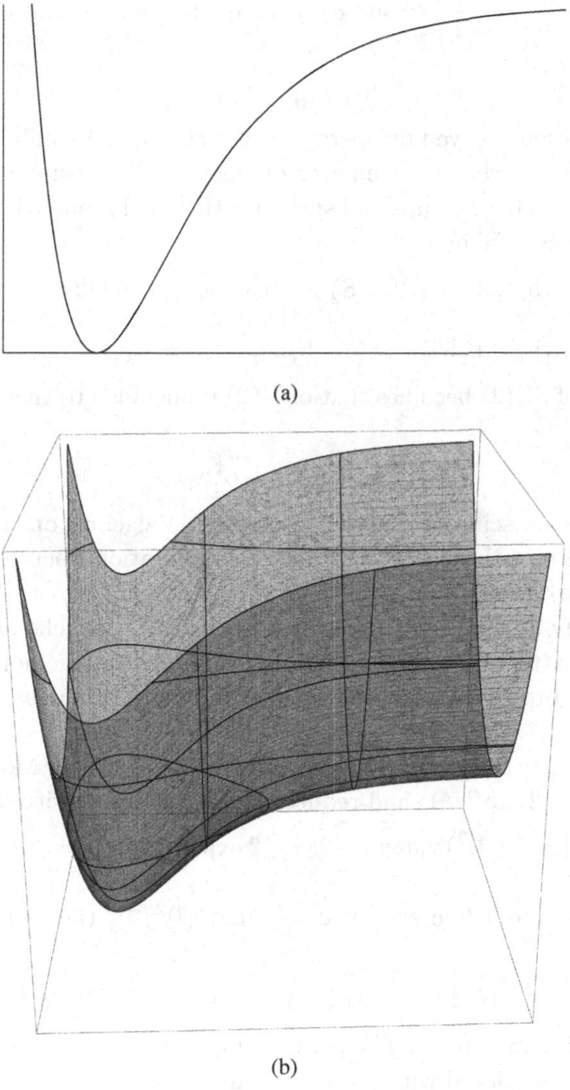

Fig. 13.2.1 (a) Morse potential function; (b) Morse energy surface.

seen in Sec. 10.7, for the Morse oscillator, the $m = 0$ case is excluded, so in the second case there are j energy levels.

To make contact with the vibron model which originated in nuclear shell theory, one expresses the group generators as binary products of *boson* creation and annihilation operators $\mathrm{b}_1^\dagger \mathrm{b}_1$, $\mathrm{b}_2^\dagger \mathrm{b}_2$, $\mathrm{b}_1^\dagger \mathrm{b}_2$, and $\mathrm{b}_2^\dagger \mathrm{b}_1$.

These operators are mathematically characterized by the commutation relations

$$[\mathrm{b}_\alpha, \mathrm{b}_\beta^\dagger] = \delta_{\alpha\beta}. \tag{13.2.6}$$

These are relations obeyed by operators that create and annihilate particles that have the interchange symmetry of bosons. The compound operators $\mathrm{b}_j^\dagger \mathrm{b}_k$ generate unitary groups and special orthogonal groups. For SU(2) and SO(3), letting $\mathrm{S} \to \mathrm{S}'$ one has

$$\mathrm{S}_1' = (\mathrm{b}_1^\dagger \mathrm{b}_2 + \mathrm{b}_2^\dagger \mathrm{b}_1)/2, \quad \mathrm{S}_2' = -i(\mathrm{b}_1^\dagger \mathrm{b}_2 - \mathrm{b}_2^\dagger \mathrm{b}_1)/2, \tag{13.2.7a,b}$$

$$\mathrm{S}_3' = (\mathrm{b}_1^\dagger \mathrm{b}_1 - \mathrm{b}_2^\dagger \mathrm{b}_2), \quad \mathrm{S}_+' = \mathrm{b}_1^\dagger \mathrm{b}_2, \quad \mathrm{S}_- = \mathrm{b}_2^\dagger \mathrm{b}_1 \tag{13.2.7c–e}$$

The algebra of SU(2) becomes that of U(2) if one adds to these the further generator

$$\mathrm{N_{op}} = \mathrm{b}_1^\dagger \mathrm{b}_1 + \mathrm{b}_2^\dagger \mathrm{b}_2. \tag{13.2.7f}$$

In the Morse oscillator realization of the Lie algebra, one can establish the dependence of the boson creation and anhillation operators upon the dynamical variables q,t and the operators $-\mathrm{i}\partial/\partial\mathrm{q}$, $i\partial/\partial\mathrm{t}$. To do so we slightly modify a mathematical development due to Iachello and Levine.[13] To simplify matters for the reader we first observe that the notation of Sec. 10.7, is related to that of Iachello and Levine by the changes $d \to \mathrm{V}_0, a \to \beta, m \to \mu$.

If one sets $\xi = \beta(r - r_0)$, the Schrödinger equation with Morse potential $\mathrm{V}_0(\exp(-2\xi) - 2\exp(-\xi))$ and reduced mass μ can be written as

$$\{-\partial^2/\partial\xi^2 + (2\mu/(\hbar\beta)^2)\mathrm{V}_0(\exp(-2\xi) - 2\exp(-\xi))\}\Psi_\mathrm{E} = (2\mu/(\hbar\beta)^2)\mathrm{E}\Psi_\mathrm{E}. \tag{13.2.8a}$$

If one defines a unit-free energy, $\varepsilon = (2\mu/(\hbar\beta)^2)^{-1}$, the results of Morse require that

$$\mathrm{E} = -\varepsilon(M/2)^2, \; M/2 = n + 1/2 - \mathrm{k}, \quad \mathrm{k} = (\hbar\beta)^{-1}(2\mu\mathrm{V}_0)^{1/2}, \tag{13.2.8b}$$

with $M/2$ being an integer. The relationship of this expression to observed spectral lines associated with photons of frequencies ω is supplied by comparing it to the standard Dunham expansion,[14] which can be put in the form

$$\omega = \omega_e(\nu + 1/2) - \omega_e\chi_e(\nu + 1/2)^2 + \ldots. \tag{13.2.8c}$$

After a time dilation similar to that introduced in Sec. 10.7, one can write

$$\Psi_\mathrm{E} = \Phi_\mathrm{E}(\xi)\exp(\mathrm{i}M\mathrm{t}). \tag{13.2.8d}$$

Letting N be an eigenvalue of $\mathrm{N_{op}}$, and setting

$$(2\mu/(\hbar\beta)^2)\mathrm{V_0})^{1/2} = (N+1)/2, \quad \text{and} \quad \Phi_\mathrm{E}(\xi) = R_{N,M}(\xi), \tag{13.2.9a}$$

converts (13.2.8a) to the following equation:

$$\begin{aligned}&\{-\partial^2/\partial\xi^2 + ((N+1)/2)^2(\exp(-2\xi) - 2\exp(-\xi)) + (M/2)^2\}\\ &\quad\times R_{N,M}(\xi)\exp(\mathrm{i}M\mathrm{t}) = 0.\end{aligned} \tag{13.2.9b}$$

To express this in terms of the boson operators one may define

$$\mathrm{x}' = \rho\cos(\mathrm{t}), \quad \mathrm{y}' = \rho\sin(\mathrm{t}) \tag{13.2.9c}$$

and, a la Dirac, define the shift operators

$$\begin{aligned}\mathrm{b}_1 &= (\mathrm{x}' + \partial/\partial\mathrm{x}')/\sqrt{2}, \quad \mathrm{b}_1^\dagger = (\mathrm{x}' - \partial/\partial\mathrm{x}')/\sqrt{2},\\ \mathrm{b}_2 &= (\mathrm{y}' + \partial/\partial\mathrm{y}')/\sqrt{2}, \quad \mathrm{b}_2^\dagger = (\mathrm{y}' - \partial/\partial\mathrm{y}')/\sqrt{2}.\end{aligned} \tag{13.2.9d–g}$$

On setting

$$\exp(-\xi) = \rho^2/(N+1), \tag{13.2.9h}$$

and letting

$$\mathrm{R}_{N,M}(\xi)\exp(\mathrm{i}M\mathrm{t}) \rightarrow |N, M\rangle, \tag{13.2.9i}$$

the result of Iachello and Levine establishes that the Morse Schrödinger equation is equivalent to the relations

$$\begin{aligned}\mathrm{N_{op}}|N, M\rangle &= N|N, M\rangle,\\ \mathrm{S}_3^2|N, M\rangle &= M^2|N, M\rangle.\end{aligned} \tag{13.2.10a,b}$$

These two last equations may be considered to be the defining equations of an abstract one-dimensional *vibron* — an oscillator with U(2) dynamical symmetry that is vibrating in one spatial dimension and possesses only a finite number of bound states. In this particular case the relations (13.2.8) and (13.2.9) define the vibron to be a Morse oscillator.

An unlimited number of realizations of the vibron algebra with spectrum (13.2.4) can be obtained by unitary transformations of the Morse generators and Hamiltonian, yielding operators $\mathrm{S}' = \mathrm{TST}^{-1}$. These will not alter the abstract group and they will not alter the spectrum of the oscillator, but can alter transition probabilities. The spectrum itself can be most easily altered by changing the parameters that determine N and $\mathrm{D_0}$, and by adding functions of S_3' to the Hamiltonian. A vibron Hamiltonian $\mathrm{H} = \mathrm{F}(\mathrm{S}_3')$ will have eigenstates $\mathrm{T}\ |jm\rangle$ with energy $\mathrm{E} = \mathrm{F}(\mathrm{m})$.

13.3 U(2) ⊗ U(2) Model of Vibron Coupling

In this model, a pair of non-rotating, initially independent, vibron oscillators are coupled.[15] The pair of uncoupled oscillators has as its spectrum generating group $U(2)_a \otimes U(2)_b$. The U(2) groups are as defined in the previous section. This two-vibron group contains several subgroup chains, of which we list the following three:

$$\begin{aligned}
&\text{(i)}\ \ \mathrm{U(1)_{ab}} \otimes \mathrm{SU}(2)_a \otimes \mathrm{SU}(2)_b \supset \mathrm{U(1)_{ab}} \otimes \mathrm{SU(2)_{ab}} \supset \mathrm{SO(2)_{ab}},\\
&\text{(ii)}\ \ \mathrm{U(2)_{ab}} \supset \mathrm{U(2)_{ab}} \supset \mathrm{SU(2)_{ab}} \supset \mathrm{SO(2)_{ab}},\\
&\text{(iii)}\ \ \mathrm{SO(2)_a} \otimes \mathrm{SO(2)_b} \supset \mathrm{SO(2)_{ab}}.
\end{aligned} \tag{13.3.1}$$

The generator of $\mathrm{U(1)_{ab}}$ is

$$\mathrm{N_{op}} \equiv \mathrm{N_{opa}} + \mathrm{N_{opb}} = \mathrm{b^\dagger_{1a} b_{1a}} + \mathrm{b^\dagger_{2a} b_{2a}} + \mathrm{b^\dagger_{1b} b_{1b}} + \mathrm{b^\dagger_{2b} b_{2b}}, \tag{13.3.2a}$$

and the generators of $\mathrm{SU}(2)_{ab}$ are the three corresponding operators

$$\mathrm{S_{jab}} = \mathrm{S_{ja}} + \mathrm{S_{jb}}. \tag{13.3.2b}$$

The group $\mathrm{SO(2)_{ab}}$ is usually chosen to have $\mathrm{S_{jab}}$ as its generator.

The coupling of the direct product $\mathrm{SU(2)_a} \otimes \mathrm{SU(2)_b}$ produces the second-order Casimir invariant

$$\mathrm{C_{2ab}} = (\mathbf{S}_\mathrm{a} + \mathbf{S}_\mathrm{b}) \cdot (\mathbf{S}_\mathrm{a} + \mathbf{S}_\mathrm{b}) = \mathrm{S_a^2} + \mathrm{S_b^2} + 2\mathbf{S}_\mathrm{a} \cdot \mathbf{S}_\mathrm{b}. \tag{13.3.3}$$

It can be used to impose a correlation of the motions of the two vibrons. The coupled states are those defined by (10.11.6), *viz.*

$$|JM(j_a, j_b)\rangle = \Sigma \mathrm{C}(j_a j_b JM; m_a m_b)|j_a m_a\rangle|j_b m_b\rangle. \tag{13.3.4}$$

They are eigenstates of $\mathbf{S}_\mathrm{a} \cdot \mathbf{S}_\mathrm{b}$ with eigenvalue $\{J(J+1) - j_a(j_a+1) - j_a(j_a+1)\}/2$. In the vibron model, $\mathbf{S}_\mathrm{a} \cdot \mathbf{S}_\mathrm{b}$ couples the normal modes of two vibrons. Using the notation of the previous section, all two-vibron Hamiltonians of the form

$$\mathrm{H} = \mathrm{F}(\mathrm{C'_{1,a}}, \mathrm{C'_{1,b}}, \mathrm{C'_{2,a}}, \mathrm{C'_{2,b}}, \mathrm{C'_{2ab}}, \mathrm{J'_{ab}}, \mathrm{S'_{3a}}, \mathrm{S'_{3b}}), \tag{13.3.5a}$$

will have eigenstates that can be linear combinations of $T(|j_a m_a\rangle|j_b m_b\rangle)$ and $\mathrm{T}|JM(j_a, j_b)\rangle$, with eigenvalues

$$\mathrm{E} = \mathrm{F}(j_a, j_b, j_a(j_a+1), j_b(j_b+1), J(J+1), M, m_a, m_b). \tag{13.3.5b}$$

Strong couplings of normal modes take place when the modes are degenerate, or very nearly so. As this specificity exerts the greatest strain on coupled U(2) ⊗ U(2) vibron models, it is worth considering them in some detail. A seminal study of the bond vibrations of Y–X–Y molecules by van Roosmalen, Benjamin and Levine did just this.[16] They found the most relevant group chains to be (i) and (iii) The states of chain (i) are the $T|JM(j_a, j_b)\rangle$ and those of chain (iii) are local mode states $T(|j_a m_a\rangle|j_b m_b\rangle)$. For their most general vibron Hamiltonian with $SO(2)_{ab}$ symmetry the authors choose one with three parameters, A, B and λ, and of the form

$$H = E_0 + A(S_{3a}^2 + S_{3b}^2) + B(S_{3a} + S_{3b})^2 + \lambda(N_a N_b/2 - 2\mathbf{S}_a \cdot \mathbf{S}_b). \quad (13.3.6a)$$

Its eigenvalues are

$$E_0 + A(m_a^2 + m_b^2) + B(m_a + m_b)^2 + \lambda(j_a(j_a+1) + j_b(j_b+1) + 2j_a j_b - J(J+1)). \quad (13.3.6b)$$

H commutes with an operator, Π, which is the sum of the two operators

$$\Pi_J = (b_{1j}^\dagger b_{1j}^\dagger - b_{2j}^\dagger b_{2j}^\dagger)(b_{1j} b_{1j} - b_{2j} b_{2j}). \quad (13.3.7)$$

Here, the parameters are determined from the vibrational spectra of the system.

The model was tested using least-squares fits to the vibrational spectra of H_2O, O_3 and SO_2 to determine the values of A, B and λ for each molecule. The integer, *N*, was set to yield the correct number of bound vibrational states of the molecule. The root mean square deviations from vibron energies were found to be 3.99 cm^{-1}, 23.7 cm^{-1} and 1.23 cm^{-1}, respectively. The model was also applied to the C–H and C–D bond stretching vibrations of acetylene and deuteroacetylene, for which root mean square deviations were found to be 3.91 cm^{-1} and 4.85 cm^{-1}. These results show that the model is able to predict vibrational energies with accuracies better than 1%, even in molecules with almost degenerate vibrational modes.

The comparisons with experimental data noted above turn out to be similar to those obtained in a number of subsequent investigations of n-vibron models of vibrations in polyatomic molecules with n modes of vibration. It has even proved possible to include terms in the vibron Hamiltonian that deal properly with Fermi resonances[17] and overtone vibrations in linear[18] and bent[19] triatomics. To obtain results of such accuracy the model usually requires fewer parameters than are required in the previously standard Dunham expansions. This implies that the vibron models express group structure that must be very similar to that of the dynamical groups of the physical systems.

13.4 Spectrum Generating Groups of Rigid Body Rotations

Edmond's description and notation for Euler's angular coordinates α, β and γ will be used to describe the orientation of the principal axes of inertia of a rigid body with respect to a space-fixed axes.[20] Setting $\mathrm{x} = \cos(\beta)$, the Hamiltonian of a symmetric top is

$$(-1/2I_1)\{(1-\mathrm{x}^2)\partial^2/\partial\mathrm{x}^2 - 2x\partial/\partial\mathrm{x} + (I_1/I_3 + \mathrm{x}^2(1-\mathrm{x}^2)^{-1})\partial^2/\partial\gamma^2 \\ - 2\mathrm{x}(1-\mathrm{x}^2)^{-1}\partial^2/\partial\alpha\partial\gamma\}.$$

The solutions of

$$(\mathrm{H} - \mathrm{i}\partial/\partial\mathrm{t})\Psi = 0 \quad (13.4.1)$$

may be expressed in terms of Wigner's D functions.[20] We set

$$\Psi = \sum \mathrm{C}_{jmn}\Psi_{jmn}(\alpha,\beta,\gamma)\exp(-\mathrm{i}\mathrm{E}_{jn}\mathrm{t}) \quad (13.4.2a)$$

with

$$\Psi_{jmn}(\alpha,\beta,\gamma) = (2j+1)^{1/2}\mathrm{D}^j_{mn}(\alpha,\beta,\gamma)/4\pi, \quad (13.4.2b)$$

and

$$\mathrm{E}_{jn} = j(j+1)/2\mathrm{I}_1 + n^2(\mathrm{I}_1 - \mathrm{I}_3)/2\mathrm{I}_1\mathrm{I}_3, \quad (13.4.2c)$$

with

$$j = 0, 1/2, 1, 3/2, \ldots, \ |m| < j, \ |n| < j. \quad (13.4.2d)$$

The $\psi_{jmn}(\alpha,\beta,\gamma)$ are normalized D functions, and are orthonormal under the scalar product

$$(\mathrm{f},\mathrm{g}) = \int_{-2\pi}^{2\pi} \mathrm{d}\alpha \int_0^{2\pi} \sin(\beta)\mathrm{d}\beta \int_0^{2\pi} \mathrm{d}\gamma\ \mathrm{f}^*\mathrm{g}. \quad (13.4.3)$$

Bopp and Haag have proposed that the D functions with half-integer j values could provide spatial realizations of the wave functions of particles with half-integer spin.[21]

To determine the dynamical group of (13.4.1), Kumei[22a] began by using

$$\mathrm{D}_{\mathrm{op}} = \exp\{\mathrm{t}\partial/\partial\mathrm{t}\ln((\mathrm{J}_{\mathrm{op}} + 1/2)/\mathrm{H})\} \quad (13.4.4a)$$

to linearize the spectrum of j. In (13.4.4a),

$$\mathrm{J}_{\mathrm{op}} + 1/2 \equiv (1/2)\{1 + 8\mathrm{I}_1[\mathrm{H} + ((\mathrm{I}_1 - \mathrm{I}_3)/2\mathrm{I}_1\mathrm{I}_3)\partial^2/\partial\gamma^2]\}^{1/2}. \quad (13.4.4b)$$

The dilatation operator D_{op} converts (13.4.1) to

$$\{(1-\mathrm{x}^2)\partial^2/\partial\mathrm{x}^2 - 2\mathrm{x}\partial/\partial\mathrm{x} + (1-\mathrm{x}^2)^{-1}(\partial^2/\partial\alpha^2 + \partial^2/\partial\gamma^2) \\ -2\mathrm{x}(1-\mathrm{x}^2)^{-1})\partial^2/\partial\alpha\partial\gamma + (\mathrm{i}\partial/\partial\mathrm{t})^2 - 1/4\}\Phi = 0.$$

with

$$\Phi = \sum C_{jmn}\phi_{jmn}, \quad \phi_{jmn} = \Psi_{jmn}(\alpha,\beta,\gamma)\exp(-i(j+1/2)t). \tag{13.4.5a,b}$$

Setting up and solving the 14 determining equations required to obtain first- and second-order differential operators yields 16 generators Q' of the invariance group of (13.4.5). One of these is the generator $i\Phi\partial/\partial\Phi$ of the group that can change the phase of Φ. The other 15 were denoted $Q_j'^k$. They obey the SL(4,R) commutation relations

$$[Q'^{i}_{j}, Q'^{k}_{m}] = \delta^{i}_{m}Q'^{k}_{j} - \delta^{k}_{j}Q'^{i}_{m}. \tag{13.4.6}$$

Here we summarize some of the general properties of this dynamical group. For a fuller discussion the reader is referred to the original paper.[22b]

Converting the Q'^{k}_{j} to skew-adjoint or self-adjoint generators of invariance transformations of the original Schrödinger equation produces the generators of its dynamical group. The action of the resulting labeling operators $-i\partial/\partial\alpha, -i\partial/\partial\gamma$ and $i\partial/\partial t$, on the eigenfunctions

$$\begin{aligned}\phi_{jmn} &= \Psi_{jmn}(\alpha,\beta,\gamma)\exp(-i(j+1/2)t)\\ &= ((2j+1)^{1/2}/4\pi)D^{j}_{mn}(\alpha,\beta,\gamma)\exp(-i(j+1/2)\end{aligned} \tag{13.4.7}$$

is given by

$$\begin{aligned}&-i\partial/\partial\alpha\ \phi_{jmn} = m\phi_{jmn}, \quad -i\partial/\partial\gamma\ \phi_{jmn} = n\ \phi_{jmn},\\ &i\partial/\partial t\ \phi_{jmn} = (j+1/2)\phi_{jmn}.\end{aligned}$$

Kumei found that all of the generators which shift j shift it by 1/2 unit. They must consequently shift both m and n by 1/2 unit. The Lie algebra is consequently that of the group SU(2,2). The group contains operators that mix states with integral and half-integral values of j. The physical interpretation of these mixed states present problems that have been addressed by Bohm and Teese.[23] Double-jump operators for j may be constructed from products of the single-jump operators. However they do not close under commutation with the other SL(4,R) generators.

We next consider the degeneracy groups of the systems. When all three moments of inertia are equal, the resulting spherical top has $SU(2)\otimes SU(2)$ as its degeneracy group. In the notation of Ref. 22 the shift and labeling operators of the group are

$$\begin{aligned}Q_9 = Q^{\dagger}_{10} &= \exp(i\alpha)\{(1-x^2)^{1/2}\partial/\partial x - ix(1-x^2)^{-1/2}\partial/\partial\alpha\\ &\quad + i(1-x^2)^{-1/2}\partial/\partial\gamma\},\end{aligned} \tag{13.4.8a}$$

$$\begin{aligned}Q_{11} = Q^{\dagger}_{12} &= \exp(i\gamma)\{(1-x^2)^{1/2}\partial/\partial x - ix(1-x^2)^{-1/2}\partial/\partial\gamma\\ &\quad + i(1-x^2)^{-1/2}\partial/\partial\alpha\}.\end{aligned} \tag{13.4.8b}$$

These generators have the following action on the ϕ_{jmn}:

$$\begin{aligned}
Q_9\, \phi_{jmn} &= (j(j+1) - m(m+1))^{1/2}\phi_{jm+1n},\\
Q_{10}\, \phi_{jmn} &= -(j(j+1) - m(m-1))^{1/2}\phi_{jm-1n},\\
Q_{11}\, \phi_{jmn} &= -(j(j+1) - n(n+1))^{1/2}\phi_{jmn+1},\\
Q_{12}\, \phi_{jmn} &= (j(j+1) - n(n-1))^{1/2}\phi_{jmn-1}.
\end{aligned} \qquad (13.4.9a–d)$$

For the symmetric top with $I_1 = I_2 \neq I_3$ the generators Q_{11} and Q_{12} become time-dependent, but remain invariance generators of Eq. (13.4.1). The degeneracy group is reduced to $\mathrm{SU}(2) \otimes \mathrm{U}(1)$.

This change in the degeneracy group is an example of a general phenomenon with interesting consequences.[24] Suppose that $\Psi_a(\mathrm{q,t})$ is a general solution of the time-dependent Schrödinger equation with Hamiltonian $\mathrm{H_a}$, and that $\Psi_\mathrm{b}(\mathrm{q,t})$ is a general solution of the time-dependent Schrödinger equation with Hamiltonian $\mathrm{H_b}$. Next, expand these $\Psi(\mathrm{q,t})$ as a sum of solutions of their time-dependent eigenstates, $\Psi_{\mathrm{an}}(\mathrm{q})\exp(-\mathrm{iE_{an}t})$ and $\Psi_{\mathrm{bn'}}(\mathrm{q})\exp(-\mathrm{iE_{bn}}t)$, respectively. Now, suppose also that all the functions $\Psi_{\mathrm{an}}(\mathrm{q})$ and $\Psi_{\mathrm{bn'}}(\mathrm{q})$ are contained in the same Hilbert space of square-integrable functions f(q). Then, if $\mathrm{Q_a}$ is an operator satisfying the equation

$$[(\mathrm{H_a} - \mathrm{i}\partial/\partial\mathrm{t}), \mathrm{Q_a}]\Psi_\mathrm{a}(\mathrm{q,t})| = 0, \qquad (13.4.10a)$$

and if

$$\mathrm{S_{ab}} = \exp(-\mathrm{i}(\mathrm{H_b} - \mathrm{H_a})\mathrm{t}), \qquad (13.4.10b)$$

is well defined,[24] then

$$\mathrm{Q_{ab}} = \mathrm{S_{ab}Q_aS_{ab}^{-1}}, \qquad (13.4.10c)$$

will satisfy

$$[(\mathrm{H_b} - \mathrm{i}\partial/\partial\mathrm{t}), \mathrm{Q_{ab}}]\Psi_\mathrm{b}(\mathrm{q,t})| = 0. \qquad (13.4.10d)$$

When $\mathrm{Q_a}$ is a time-independent constant of motion of $\mathrm{H_a}$, the operator Q_{ab} will be a time-dependent constant of motion if $\mathrm{Q_a}$ does not commute with $\mathrm{H_b}$. This is the case when a spherical top is converted to a symmetric top. One then has

$$\mathrm{H_b} - \mathrm{H_a} = (-1/2)(\mathrm{I_2} - \mathrm{I_1})\partial^2/\partial\gamma^2. \qquad (13.4.11a)$$

Thus

$$\mathrm{S_{ab}} = \exp(\mathrm{i}(\mathrm{t}/2))(\mathrm{I_2} - \mathrm{I_1})\partial^2/\partial\gamma^2 \equiv \exp(\mathrm{at}\partial^2/\partial\gamma^2). \qquad (13.4.11b)$$

It acts on the term $\exp(\pm i\gamma)$ in Q_{11} and Q_{12}, converting both generators to functions of time.

The transformation from the generators Q_a to Q_{ab} is a similarity transformation. Application to all the generators, Q_a, of the Lie algebra associated with Hamiltonian H_a does not change their commutation relations: the corresponding set of generators Q_{ab} are those of an isomorphic Lie algebra. Its generators are the quantum mechanical representatives of time-dependent constants of motion of system b. An analogous change occurs when the symmetry of the top is reduced to that of the asymmetric top whose degeneracy group is simply SO(3).

The inverse of the transformation S_{ab} is the transformation S_{ab}^{-1}, which can convert time-dependent generators of the dynamical group of the symmetric top, to the generators Q_{11} and Q_{12} of the degeneracy group of the spherical top. Usually one expects perturbations to shrink degeneracy groups. Here one has a perturbation that enlarges a degeneracy group, and does so in a most obvious manner. As pointed out in Chapter 1, if one does not know the dynamical group of a system, the perturbations that produce increases or changes in observed symmetries are very difficult to detect.

In 1984 Harter and Patterson introduced the concept of rotational energy surface. Such surfaces depict the relation between the energy of a rotating object and the orientation of its angular momentum vector in a body fixed system. The angular momentum vector is assumed to have a given direction and magnitude in a space-fixed system. For a classical top with angular momentum of magnitude L^2 and principal moments of inertia $I_x = 1/2A, I_y = 1/2B, I_z = 1/2C$,

$$H = A\,L_x^2 + B\,L_y^2 + C\,L_z^2. \tag{13.4.12}$$

If the Euler coordinates of **L** in the body-fixed frame are $-\alpha, -\beta, -\gamma$, then the energy, E, of the top is given by the expression[25]

$$\begin{aligned} E = J^2\{&(1/3)(A+B+C) + (1/6)(2C-A-B)(3\cos^2(\beta)-1) \\ &+ (1/2)(A-B)\sin^2(\beta)\cos(2\gamma)\}. \end{aligned} \tag{13.4.13}$$

Figures 13.4 illustrate the resulting constant energy surfaces for prolate, oblate and spherical tops. They are obtained by setting β and γ equal to the polar coordinates $-\theta$ and $-\phi$, respectively, then considering $E = r(\theta, \phi)$. Because the classical and quantum mechanical dynamical groups of these

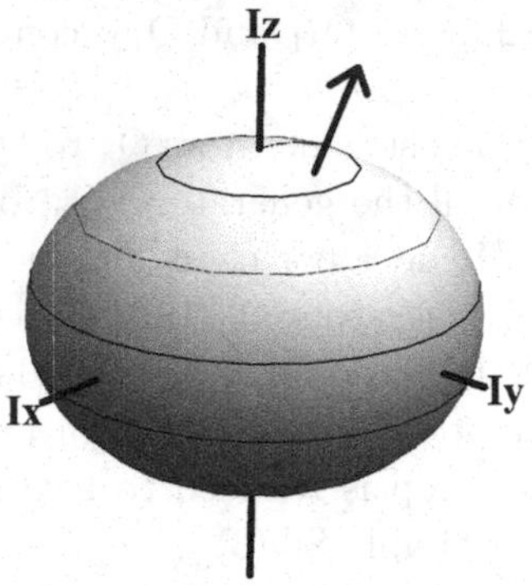

Fig. 13.4.1 Momental ellipsoid of a prolate symmetric top in the body-fixed system. The angular momentum vector is stationary in the space-fixed system. $I_z > I_x = I_y$.

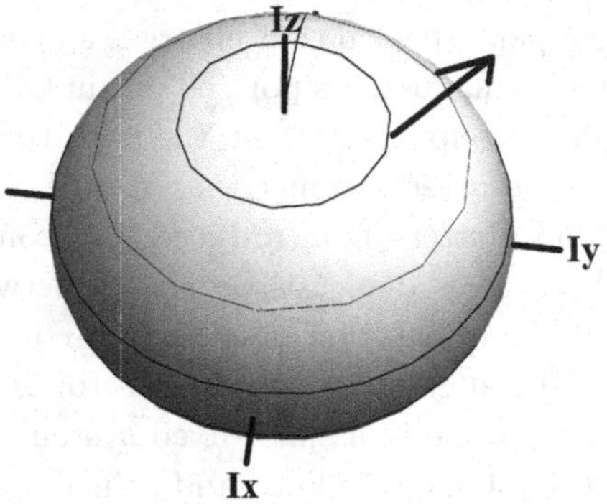

Fig. 13.4.2 Momental ellipsoid of an oblate symmetric top in the body-fixed system. $I_z > I_x = I_y$.

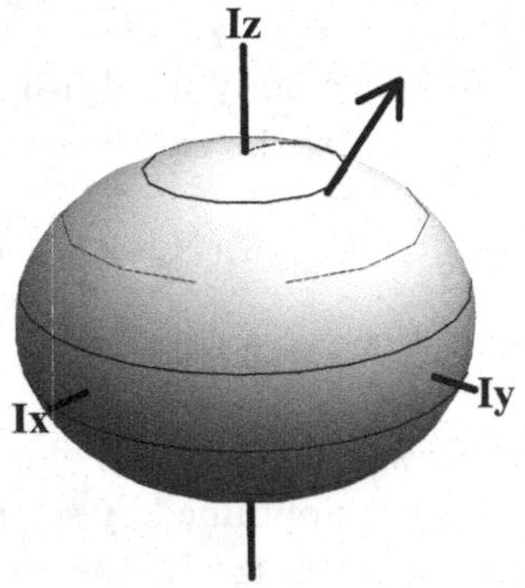

Fig. 13.4.3 Momental ellipsoid of a spherical top in the body-fixed system. $I_z = I_x = I_y$.

systems are locally isomorphic, the relation between classical trajectories of constant E on these surfaces can be used to graphically illustrate their dynamical symmetries. The energy surfaces of the prolate and oblate symmetric tops express an SO(2) rotational symmetry about an axis; that of the asymmetric top has no such symmetry. The reader is referred to Harter's article for a discussion of the classical trajectories of constant energy on the surface of these tops.

13.5 The U(4) Vibron Model of Rotating Vibrating Diatomics

If there were no coupling between the rotational motion and vibrational motion of a diatomic molecule with a spectrum generating U(2) or SO(3) group, one would expect the dynamical group of the vibrating and rotating system to be U(2) $\otimes$ SO(3) or SO(3) $\otimes$ SO(3). However, one knows that the moment of inertia of the molecule will increase when centrifugal forces stretch the bond, so one anticipates interaction between the two groups. To deal with this interaction Iachello introduced a U(4) model.[4] Its Hamiltonian must be invariant under rotations of a coordinate system with origin at the center of mass of the molecule. There are two group chains starting with U(4) that meet this requirement:

$$\begin{aligned}(a) &: U(4) \supset U(3) \supset U(2) \supset O(2),\\ (b) &: U(4) \supset O(4) \supset O(3) \supset O(2).\end{aligned} \qquad (13.5.1a,b)$$

For chain (a), Iachello set

$$H = H^a = E_0 + \alpha^1 C_1(U(3)) + \alpha^2 C_2(U(2)) + \lambda \mathbf{L}^2, \qquad (13.5.2a)$$

and for chain (b), set

$$H = H^b = E_0 + A\ C_2(SO(4)) + B\mathbf{L}^2. \qquad (13.5.2b)$$

The variables E_0, α^1, α^2, λ, A, B, are parameters whose value is determined from experimental data.

The eigenvalue, N, of the U(4) number operator $\sum_i^4 b_j^\dagger b_j$, fixes the dimension of the U(4) UIR, and thus determines the total number of bound states the system can support. Let n be the eigenvalue of the U(3) number operator, $\sum_i^3 b_j^\dagger b_j$, and let l, m be standard SO(3), SO(2) quantum numbers. Then the rovibrational state $|Nnlm\rangle$ will have

$$\begin{aligned} \mathrm{E}^{\mathrm{a}} &= \mathrm{E}_0 + \alpha^1 n + \alpha^2 n(n+3) + \beta l(l+1), \\ n &= N, N-1, \ldots, 0, \ l = n, n-2, \ldots, 1 \text{ or } 0. \end{aligned} \tag{13.5.3}$$

If w is the SO(4) quantum number with values 1, 2,..., then the rovibrational state $|Nwlm\rangle$ will have

$$\begin{aligned} \mathrm{E}^{\mathrm{b}} &= \mathrm{E}_0 + \mathrm{A}w(w+2) + \mathrm{B}l(l+1), \\ w &= N, N-2, \ldots, 1 \text{ or } 0, \quad l = w, w-1, \ldots, 0. \end{aligned} \tag{13.5.4}$$

To identify the $l = 0$ case of H^{a} with a Morse Hamiltonian one must have

$$\begin{aligned} \mathrm{E}^{\mathrm{a}} &= \mathrm{E}_0 + \alpha^1 n + \alpha^2 n(n+3), \\ \mathrm{E}_0 = -(1/4)(\beta^2/2\mu), &\quad \alpha^1 = (5/2)(\beta^2/2\mu), \quad \alpha^2 = -(\beta^2/2\mu). \end{aligned} \tag{13.5.5}$$

If one wishes to generalize to the $l \neq 0$ case, one must also take into account the fact that n is an integer which jumps by unity from 0 up to N, while $|m|$ is half integer and jumps from 1/2 to j. This can only be arranged by setting $|m| = n + 1/2$, and $N = j + 1/2$. This, however, requires l to be shifted by 2 units, so the change in $l\,\hbar$ is that of a two-photon transition. This indicates that model (a) should be interpreted as applying to systems other than a Morse oscillator. For further discussion of this algebraic Hamiltonian the reader is referred to the review by Oss.[26]

To identify H^b with a Morse Hamilton when $l = 0$, one must take into account that w is an integer which jumps by two units from 0 to N. This requires that $|m| = w/2 + \mathrm{k}$, where $\mathrm{k} = 0$ if N is odd and $\mathrm{k} = 1/2$ if N is even. One then finds that $\mathrm{A} = -(1/4)(\beta^2/2\mu) = -(\mathrm{k}/2)(\beta^2/2\mu)$, so k must be chosen as 1/2. It follows that N must be an even number. Also, E_0 must have the same value it has under case (a). Consequently

$$\mathrm{E}^{\mathrm{b}} = E_0 + (1/4)(\beta^2/2\mu)(w+1)^2 + \mathrm{B}l(l+1). \tag{13.5.6}$$

As in the atomic case considered in previous chapters, the SO(4) Lie algebra contains operators that shift l by one unit.

In both cases (a) and (b), the energy expression is the sum of a kinetic energy of rotation and an internal vibrational energy. As the internal energy of the vibration motion increases, the number of rotational states allowed to the bound system must decrease. If ν is the vibrational quantum number, as it increases n and w must decrease. This is accommodated in model (a) by setting $n = j - \nu - 1/2$. In model (b) one sets $(w + 1)/2 = j - \nu$.

The generators of U(4) and SO(4) that couple U(2) $\otimes$ U(2) relate the vibrational and rotational motions, while converting product wave functions or the form $\mathrm{R(r)Y}_{lm}(\theta, \phi)$ to functions $\mathrm{Z}(\mathrm{r}, \theta, \phi)$ that could be expressed as a sum of functions of the general form $\mathrm{R}_l(\mathrm{r})\mathrm{Y}_{lm}(\theta, \phi)$. The existing work on the U(4) model is examined in the monograph of Iachello and Levine, who conclude that chain (b) "provides a reasonably good description of the spectra of rigid diatomics", but "quite often the acual situation deviates somewhat from that of a simple Morse oscillator." They find that some general expansions which are polynomial functions of the Casimir operators of model (b) converge more rapidly than standard Dunham expansions.

13.6 The U(4) $\otimes$ U(4) Model of Rotating Vibrating Triatomics

Conceptually, mathematically and physically, the study of the coupling of rotational and vibrational motions in systems with more than a few degrees of freedom is nontrivial. For molecular physics, the correlation diagrams of Kellman, Amar and Berry that connect the energy levels of rigid and floppy triatomics are of significant assistance.[27] From them, and the previous discussion of tops, one would expect that the coupling of vibrational and rotational motions might be modeled as a coupling of the generators of the vibron model with those generators of the dynamical groups of the tops that affect the relation between the space-fixed total angular momentum vector and the principle axes of inertia.

In their treatment of the rovibronic motion of triatomics, van Roosmalen, Iachello, Levine and Diepernik have introduced a model in which a U(4) algebra is used to model each *vectorial* degree of freedom in the molecule.[29] In a triatomic molecule, A–B–C, there is one vectorial degree of freedom along bond A–B, and another along B–C. The total number of internal degrees of freedom is then six, of which three describe the internal configuration of the molecule, and three describe its orientation in an inertial system. With this as a starting point, they investigate the consequences of assuming that the Hamiltonian can be expressed in terms of generators of $U(4)_1 \otimes U(4)_2$.

Each of the U(4) groups can be reduced using the chains (a), or (b) above. Couplings between the chains can also be introduced, starting with the coupling of $U(4)_1 \otimes U(4)_2$ to $U(4)_{12}$, or at subsequent points along the chains. They established that $U(4)_1 \otimes U(4)_2$ can be reduced to SO(3) by ten different intermediate chains, and determined those that are physically relevant. These chains and their rough interpetations are:

$$
\begin{array}{lll}
\text{Ia: } U(3)_1 \otimes U(3)_2 \supset O(3)_1 \otimes O(3)_2 & \text{One rigid bond} & \\
\text{Ib: } U(3)_1 \otimes O(4)_2 \supset O(3)_1 \otimes O(3)_2 & \text{One rigid bond} & \\
\text{Ic: } O(4)_1 \otimes O(4)_2 \supset O(3)_1 \otimes O(3)_2 & \text{One rigid bond} & \\
\text{IIa: } U(3)_1 \otimes U(3)_2 \supset U(3)_{12} & \text{No rigid bond} & (13.6.1a\text{–}g) \\
\text{IIb: } U(4)_{12} \supset U(3)_{12} & \text{No rigid bond} & \\
\text{IIIa: } O(4)_1 \otimes O(4)_2 \supset O(4)_{12} & \text{Two rigid bonds} & \\
\text{IIIb: } U(4)_{12} \supset O(4)_{12} & \text{Two rigid bonds} &
\end{array}
$$

There are three further chains in which, according to the notation of Chapter 12.5, reductions such as $O(4)_1 \otimes O(4)_2 \supset O(4)_{1-2}$ are used. These are

$$
\begin{array}{lll}
\text{IIIc: } U(4)_{1-2} \supset O(4)_{1-2} & \text{Two rigid bonds} & \\
\text{IVa: } U(3)_1 \otimes U(3)_2 \supset (3)_{1-2} & \text{No rigid bond} & (13.6.1h\text{–}j) \\
\text{IVb: } U(4)_{1-2} \supset U(3)_{1-2} & \text{No rigid bond} &
\end{array}
$$

The authors first point out that there is not enough group structure in the type I cases to substantially reduce the size of the matrices that must be diagonalized when calculating the eigenvalues and eigenvectors of a Hamiltonian with the assumed group structure. They then turn to type II models. Quantum numbers associated with each group in the chain are used to lable eigenstates of a Hamiltonian that is a linear combination of the Casimir operators of each group. This gives an expression for the rovibrational energy as a function of the quantum numbers, which can be compared with a Dunham expansion. It was found that chain IIIa is the most useful. In this case the Lie-algebraic Hamiltonian is

$$
\begin{aligned}
H = E_0 + A_1\, C_2(SO(4)_1) + A_2\, C_2(SO(4)_2) + B\, C_2(SO(4)_{12}) \\
+ B'\, C_2'(|SO(4)_{12}|) + C\, C_2(SO(3)_{12}).
\end{aligned} \tag{13.6.2}
$$

Here $C_2'(|SO(4)_{12}|)$ is the square root of the quartic Casimir operator of $SO(4)_{12}$. The eigenvalues of the Casimir operators and their relationship to standard spectroscopic vibrational *normal-mode* quantum numbers are as follows:

$$
\begin{aligned}
C_2(SO(3)_{12})&: J(J{+}1), \text{ the measure of the total angular momentum of the molecule}\\
C_2(SO(4)_1)&: w_1(w_1+2), w_1 = N_1 - 2v_1\\
C_2(SO(4)_2)&: w_2(w_2+2), w_2 = N_2 - 2v_2\\
C_2(SO(4)_{12})&: \tau_1(\tau_1+2) + (\tau_2)^2\\
C_2'(|SO(4)_{12}|)&: \tau_2(\tau_2+1)
\end{aligned} \tag{13.6.3a–e}
$$

The quantum numbers v_j are those of Morse-like normal vibrational modes.

For bent molecules,

$$\tau_1 = N_1 + N_2 - 2(v_1 + v_2 + v_3) - K, \quad K = \tau_2. \tag{13.6.3f}$$

For linear molecules,

$$\tau_1 = N_1 + N_2 - 2(v_1 + v_2) - K_2, \quad K_2 = \tau_2. \tag{13.6.3g}$$

The quantum number K is that of a Poeschl–Teller oscillator[28] that has potential function

$$V = -d_0\, p(p+1)/\cosh(a\theta)^2,$$

with p an integer. Its energy levels are

$$E(\nu) = -d_0(p-\nu)^2, \quad \nu = 0, 1, 2, \ldots, p.$$

K and K_2 are the projection of $\mathbf{J}$ onto the corresponding principal axis of the body-fixed system.

The coefficient B' is zero for linear molecules. When $B' = 0$, the H of (13.6.2) has four fewer empirical parameters than the corresponding Dunham expansion. For this case, van Roosmalen, Iachello, Levine, and Diepernik[29] use a least-squares method to fit the parameters in this H to the values required to best fit the spectra of HCN and CO_2, which run from 667 to 18,377 cm^{-1}. They obtained fits with root-mean-square deviation of 9 to 12 cm^{-1}. Clearly, the dynamical symmetries of the model express real and exploitable dynamical symmetries of molecular motions. It is particularly noteworthy that the model deals well with known Fermi resonances in CO_2 .

The authors attempted an analogous fit to calculated spectra of the nonlinear ion H_3^+. This uncovered the inability of the Hamitonian to deal with degeneracies removed by Darling–Dennison couplings.[30]

In a more recent work Iachello and Oss used a vibron Hamiltonian closely related to that in (13.4.12) to deal with a class of Darling–Dennison couplings in bent XY_2 molecules. In it they consider $C_2(SO(4)_1)$ and $C_2(SO(4)_2)$ to refer to local vibrational modes rather than normal modes.

They then add to their Hamiltonian the coupling term in the quadratic Casimir operator of $U(4)_{12}$. This removes the degeneracies of the local-modes that it requires only minor degenerate perturbation calculations to accomodate. This is because the term only couples states with identical $SO(4)_{12}$ quantum numbers, i.e., with $\tau_1 = \tau_2$. They find that with this sophistication, the algebraic Hamiltonian of chain IIIa determines overtone frequencies of the nonlinear molecules H_2O^{16}, H_2O^{18}, D_2O^{16}, H_2S^{32} and $S^{32}O_2^{16}$, with root mean square deviations of 1 to 5 cm^{-1}.[16c] A similar, but slightly more elaborate treatment of the spectra of N_2O, $C^{12}O_2$, $C^{13}O_2$, OCS and HCN, produced results of the same accuracy.[16d]

Harter and his coworkers developed extensions to semi-rigid rotors of the energy surfaces of rigid rotors noted above. This produced amazing insights that led to a number of useful simplifications in treatments of the rotation of systems at least as complex as SF_6.[3]

13.7 Concluding Remarks

The remarkable successes of vibron models strongly suggest that for a wide range of molecules, they and their classical counterparts express real dynamical symmetries defined by differential operators as yet unknown. The ability of some of the models to deal with the rotational stretching of bonds suggests that even in the case of diatomics they may reflect dynamical symmetries that exist when the Born–Oppenheimer approximation is not imposed. If this guess is validated, these models may also find application in technologically important studies of vibrational–electronic coupling in solid state systems.

Exercises

1. Hold a claw hammer by its handle, with its head–claw axis in a horizontal plane, and note the direction in which the head points. Throw the hammer up in the air, giving it a rotation about its intermediate axis of inertia. Catch it by its handle as it descends after one revolution about this axis. In what direction does the head point? The figures on page 327 of Harter's article, Ref. 25, can help in developing an explanation of your observations.

References

[1] M. Moshinsky, *Group Theory and the Many-Body Problem* (Gordon & Breach, New York, 1967).

[2] D. J. Rowe, in *Dynamical Groups and Spectrum Generating Algebras*, Vol. I, eds. A. Bohm, Y. Ne'eman and A. O. Barut (World Scientific, Singapore, 1988).
[3] A. Arima and F. Iachello, *Phys. Rev. Lett.* **35** (1975) 1069.
[4] F. Iachello, *Chem. Phys. Lett.* **78** (1981) 581.
[5] F. Iachello and R. D. Levine, *J. Chem. Phys.* **77** (1982) 3046; also cf. A. Frank and R. Lemus, *J. Chem. Phys.* **84** (1986) 2698.
[6] O. S. van Roosmalen, F. Iachello, R. D. Levine and A. E. L. Dieperink, *J. Chem. Phys.* **79** (1983) 2515.
[7] (a) F. Iachello, S. Oss and R. Lemus, *J. Mol. Spect.* **149** (1991) 132;
(b) F. Iachello, N. Maniniu and S. Oss, *J. Mol. Spect.* **156** (1992) 190;
(c) T. A. Holme and R. D. Levine, *J. Chem. Phys.* **131** (1989) 169;
[8] (a) F. Iachello and S. Oss, *Chem. Phys. Lett.* **187** (1991) 500;
(b) F. Iachello and S. Oss, *J. Mol. Spect.* **153** (1993) 225;
(c) F. Iachello and S. Oss, *J. Chem. Phys.* **99** (1993) 7337;
(d) F. Iachello and S. Oss, *Chem. Phys. Lett.* **205** (1993) 285.
[9] F. Iachello and A. Arima, *The Interacting Boson Model* (Cambridge University Press, Cambridge, 1991).
[10] F. Iachello and R. D. Levine, *Algebraic Theory of Molecules* (Oxford University Press, Oxford, 1994).
[11] A. Frank and P. Van Isaker, *Algebraic Methods in Molecular and Nuclear Structure Physics* (Wiley, New York, 1994).
[12] R. D. Levine and C. E. Wulfman, *Chem. Phys. Lett.* **13** (1979) 372.
[13] F. Iachello and R. D. Levine, *Algebraic Theory of Molecules*, pp. 32–34.
[14] G. Herzberg, *Molecular Spectra and Molecular Structure I. Spectra of Diatomic Molecules* (D. Van Nostrand, Princeton, 1950) pp. 106–109.
[15] (a) F. Iachello and R. D. Levine, *J. Chem. Phys.* **77** (1982) 3046;
(b) I. Benjamin and R. D. Levine, *J. Mol. Spect.* **126** (1987) 486;
(c) F. Iachello and S. Oss, *Phys. Rev. Lett.* **66** (1991) 2776;
(d) F. Iachello and S. Oss, *J. Mol. Spect.* **153** (1992) 225;
(e) A. Frank and R. Lemus, *Phys. Rev. Lett.* **68** (1992) 413.
(f) A. Mengoni and T. Shirai, *J. Mol. Spect.* **162** (1993) 246.
[16] O. S. van Roosmalen, I. Benjamin and R. D. Levine, *J. Chem. Phys.* **81** (1984) 5986.
[17] F. Iachello and R. D. Levine, *Algebraic Theory of Molecules*, pp. 96–98.
[18] F. Iachello and R. D. Levine, *Algebraic Theory of Molecules*, pp. 104–106.
[19] F. Iachello and R. D. Levine, *Algebraic Theory of Molecules*, pp. 106-108.
[20] A. R. Edmonds, *Angular Momentum in Quantum Mechanics* (Princeton University Press, Princeton, 1957), pp. 64–67.
[21] F. Bopp, R. Haag and Z. Naturforsch **5a** (1950) 644.
[22] (a) S. Kumei, M. Sc. Thesis (University of the Pacific, 1972);
(b) Robert L. Anderson, S. Kumei and C. E. Wulfman, *J. Math. Phys.* **14** (1972) 1527.
[23] A. Bohm and R. B. Teese, *J. Math. Phys.* **17** (1976) 94.
[24] R. L. Anderson, T. Shibuya and C. E. Wulfman, *Revista Mexicana de Fisica* **23** (1974) 257.

[25] W. G. Harter, *Comp. Phys. Rep.* **8** (1958) 321.
[26] S. Oss, *Algebraic Models in Molecular Spectroscopy*, *Adv. Chem. Phys.* XCIII, eds. S. Prigogine, A. Rice (Wiley, New York, 1996).
[27] M. E. Kellman, F. Amar and R. S. Berry, *J. Chem. Phys.* **73** (1980) 2387.
[28] G. Poeschl and E. Teller, *Z. Phys.* **83** (1933) 143.
[29] O. S. van Roosmalen, F. Iachello, R. D. Levine and A. E. L. Diepernik, *J. Chem. Phys.* **79** (1983) 2515.
[30] B. T. Darling and D. M. Dennison, *Phys. Rev.* **15** (1940) 128.
[31] W. G. Harter, *ibid.* (Ref. 25), Section 3.

CHAPTER 14

Dynamical Symmetry of Maxwell's Equations

In his 1905 paper introducing the theory of relativity, Einstein prefaced his arguments with the observation, "... the unsuccesful attempts to discover any motion of the earth relative to the (aether), suggest that the phenomena of electrodynamics as well as mechanics possess no property corresponding to the idea of absolute rest."[1] He then made four key assumptions in developing his theory of special relativity. These are:

i. Sets of rigid rods and clocks may be used to define sets of local coordinate systems that assign identical space and time intervals to successive events.
ii. The rods and clocks are not affected by motions at constant velocity with respect to one another.
iii. Measurements made using these comoving coordinate systems are compared via electromagnetic radiation.
iv. Electromagnetic radiation moves with a constant velocity, c, with respect to all observers.

Einstein established that these assumptions imply the transformation equations of Lorentz,[2] which relate measurements of time intervals and spatial displacements made by observers in relative motion with a constant velocity.

Four years later Cunningham and Bateman[3a–c] showed that Maxwell's equations are invariant under Lorentz transformations, and that the validity of Einstein's special theory is a necessary consequence of this invariance.

However, Bateman and Cunnningham also proved that Maxwell's equations are invariant under the transformations of a larger group, which is a conformal group. This preserves the angle between spacetime vectors in Minkowski space, but contains inversion transformations that interconvert small and large spacetime separations. To quote Cunningham, "The question arises whether the theory of relativity also holds for the types of motion of an electromagnetic system derived from one another by such a transformation."

The following pages briefly describe the transformations of Maxwell's equations carried out by their Poincare invariance group, and their physical interpretation. The action of the Bateman–Cunningham inversion transformation is then described, and questions associated with its physical interpretation are considered. The primary concern throughout this chapter is the physical interpretation of invariance transformations of Maxwell's equations — none of which can change the velocity of light in vacuo.

14.1 The Poincare Symmetry of Maxwell's Equations

Maxwell, in his treatise *Electricity and Magnetism*,[4] provides an extensive and enlightening discussion of the experimental measurement of electrical and magnetic forces and fields. In discussing these it is necessary to use variables whose units must be carefully defined. For our purposes one may consider the concept of electrical charge, the measurement of the forces between charges, and the measurement of forces between current carrying wires, as fundamentals. Experiments beginning with Coulomb[5a] found that charge, like mass, can be quantified, and thereafter established what became known as Coulomb's law. In it the Newtonian concept of point mass, M, is replaced by that of point charge, Q. The magnitude of the forces between two stationary point charges a and b, measured experimentally, is found to be related to the distance r_{ab} between their centers by the law

$$F_{el} = \kappa Q_a Q_b / r_{ab}^2.$$

If Q is measured in coulombs C, and force is measured in newtons, i.e., $kgms^{-2}$, then κ is measured in newton $m^2/C^2 = kg\,m^3 s^{-2} C^{-2}$. It turns out that setting $\kappa = 1/4\pi\varepsilon_o$ is quite revealing. Here $\varepsilon_o \cong 10^{-9}/36\pi\,kg^{-1}\,m^{-3}\,s^2 C^2$, is termed the electrical permitivity of the space. Electrical current is so defined that a charge of one coulomb flowing by a point in one second generates a current of one ampere. Experiments culminating with those of Ampere[5b] established that the magnetic force between two current-carrying parallel wires with centers r meters apart in empty space, has magnitude

$$F_m = \mu_o I_a I_b / 2\pi r.$$

If the current measured in wire a, I_a, and that in wire b, I_b, are both measured in amperes, then the constant, μ_o, the magnetic permeability of empty space, is found experimentally to have the value $4\pi\ 10^{-7}\,kg\,m/C^2$.

Readers of Maxwell's collected works[6] can gain a good deal of understanding of his thought processes, and of the way in which he utilized Faraday's graphic idea of lines of force in guiding the development of his mathematical conceptions of electromagnetic fields. Maxwell wrote down his equations shortly before Gibbs' concept of a vector became known, but quickly became a supporter of Gibbs.[7] In a vector notation[8] now commonly used, his equations interrelate the following quantities.

Symbol	Name	Units	
$\mathbf{E}(\mathbf{r},t)$,	electric field strength,	volts/meter ($ms^{-2}kg/C$),	
$\mathbf{H}(\mathbf{r},t)$,	magnetic field strength,	amperes/meter ($m^{-1}s^{-1}C$),	
$\mathbf{B}(\mathbf{r},t)$,	magnetic flux density,	webers/meter^2 ($s^{-1}kg/C$),	(14.1.1)
$\mathbf{D}(\mathbf{r},t)$,	electric displacement,	coulombs/meter^2 ($m^{-2}C$),	
$\mathbf{J}(\mathbf{r},t)$,	electric current density,	amperes/meter^2 ($m^{-2}s^{-1}C$),	
$\rho(\mathbf{r},t)$,	electric charge density,	coulombs/meter^3 ($m^{-3}C$).	

Maxwell's original set of 12 equations can be written as

$$\begin{aligned}\nabla \times \mathbf{E}(\mathbf{r},t) + \partial\mathbf{B}(\mathbf{r},t)/\partial t &= 0,\\ \nabla \times \mathbf{H}(\mathbf{r},t) - \partial\mathbf{D}(\mathbf{r},t)/\partial t &= \mathbf{J}(\mathbf{r},t),\\ \nabla \cdot \mathbf{B}(\mathbf{r},t) &= 0,\\ \nabla \cdot \mathbf{D}(\mathbf{r},t) &= \rho(\mathbf{r},t).\end{aligned} \tag{14.1.2a–d}$$

In regions of space in which there is no electric charge, there may, however, be a displacement current, so $\mathbf{J}(\mathbf{r},t)$ need not vanish. In vacuo, $\mathbf{B} = \mu_o\mathbf{H}$, and $\mathbf{D} = \varepsilon_o\mathbf{E}$. Equations (14.1.2) then become

$$\begin{aligned}\nabla \times \mathbf{E}(\mathbf{r},t) + \mu_o\partial\mathbf{H}(\mathbf{r},t)/\partial t &= 0,\\ \nabla \times \mathbf{H}(\mathbf{r},t) - \varepsilon_o\partial\mathbf{E}(\mathbf{r},t)/\partial t &= 0,\\ \nabla \cdot \mathbf{H}(\mathbf{r},t) &= 0,\\ \nabla \cdot \mathbf{E}(\mathbf{r},t) &= 0.\end{aligned} \tag{14.1.3a–d}$$

This set of first-order PDEs implies the second-order PDE

$$\nabla^2\mathbf{E}(\mathbf{r},t) = \mu_o\varepsilon_o\partial^2\mathbf{E}(\mathbf{r},t)/\partial t^2, \tag{14.1.4a}$$

which governs electric fields. Observing that the velocity of propagation of the resulting electric field from $\mathbf{r}$ to $\mathbf{r}'$ is $1/(\mu_o\varepsilon_o)^{1/2}$, and knowing the results of measurements of μ_o and ε_o, Maxwell established that this is c, the velocity of light.[9] Thus in vacuo, electromagnetic radiation obeys the equation

$$\nabla^2\mathbf{E}(\mathbf{r},\mathrm{t}) - \partial^2\mathbf{E}(\mathbf{r},\mathrm{t})/\partial(\mathrm{ct})^2 = 0. \tag{14.1.4b}$$

Also,

$$\mathbf{H}\cdot\mathbf{E} = 0. \tag{14.1.4c}$$

The density of energy in the field is, in units of Jm^{-3}, given by

$$(\varepsilon_o\mathrm{E}^2 + \mu_o\mathrm{H}^2)/2. \tag{14.1.5a}$$

Formulae are simplified if one absorbs $\sqrt{\varepsilon_o}$ into $\mathbf{E}$, and $\sqrt{\mu_o}$ into $\mathbf{H}$. We shall do this and denote the resulting field vectors by $\bar{\mathbf{E}}$ and $\bar{\mathbf{H}}$. Then, for example, (14.1.5a) becomes

$$(\bar{\mathrm{E}}^2 + \bar{\mathrm{H}}^2)/2. \tag{14.1.5b}$$

Defining $\mathrm{x} = \mathrm{x}_1$, $\mathrm{y} = \mathrm{x}_2$, $\mathrm{z} = \mathrm{x}_3$, and $\mathrm{ct} = \mathrm{x}_4$, the generators of the Lorentz invariance group of Maxwell's equations are

$$\mathrm{L_{ij}} = \mathrm{x_i}\partial/\partial\mathrm{x_j} - \mathrm{x_j}\partial/\partial\mathrm{x_i} = -\mathrm{L_{ji}}, \quad \mathrm{i} < \mathrm{j} = 1,2,3; \tag{14.1.6a}$$

$$\mathrm{K_{j4}} = \mathrm{x_j}\partial/\partial\mathrm{x_4} + \mathrm{x_4}\partial/\partial\mathrm{x_j} = \mathrm{K_{4j}}, \quad \mathrm{j} = 1,2,3. \tag{14.1.6b}$$

Their Poincare invariance group is generated by these, and the further operators,

$$\mathrm{Tp_j} = \partial/\partial\mathrm{x_j}, \quad \mathrm{j} = 1,\ldots,4. \tag{14.1.6c}$$

The commutation relations satisfied by these generators are

$$\begin{aligned} &[\mathrm{L_{ab}},\mathrm{L_{bc}}] = \mathrm{L_{ac}}, \quad [\mathrm{L_{ab}},\mathrm{K_{b4}}] = \mathrm{K_{a4}}, \quad [\mathrm{K_{a4}},\mathrm{K_{b4}}] = \mathrm{L_{ab}}, \\ &[\mathrm{L_{ab}},\mathrm{Tp_b}] = \mathrm{Tp_a}, \quad [\mathrm{K_{a4}},\mathrm{Tp_b}] = \mathrm{Tp_a}, \quad [\mathrm{Tp_a},\mathrm{Tp_b}] = 0. \end{aligned} \tag{14.1.7}$$

The physical interpretation of the transformations of the group, dealt with in Chapter 2, is summarized in the next paragraph.

Before beginning to deal with the action of Lorentz transformations on electromagnetic fields we must define their action on points in spacetime. Let observer b have velocity $\mathbf{V}_\mathrm{t}$ with respect to observer a. Let x, y, z and t be the spacetime coordinates of an event as measured in the coordinate system of observer a, and let x',y',z' and t' be the coordinates that observer b assigns to the same event. If $\beta \equiv (1-(\mathrm{v/c})^2)^{-1/2}$, then the observations of b are related to those of a by the Lorentz transformations

$$x' = \beta(x - vt), \quad y' = y, \quad z' = z, \quad t' = \beta(t - vx/c^2),$$

i.e. by (14.1.8)

$$x' = \beta(x - (v/c)ct), \quad y' = y, \quad z' = z, \quad t' = \beta(ct - (v/c)x)/c.$$

The operators $\exp(\alpha_k T p_k)$ translate points in spacetime, and leave $\Delta x, \Delta y, \Delta z$ and Δt, unchanged.

Now suppose that observer a determines the components of **E**, **H** at a point x, y, z and t in spacetime to be E_x, E_y, E_z and H_x, H_y, H_z, and that both observers are using right-handed coordinate systems. Then the analysis of Cunningham and Bateman establishes that if observer b is moving at velocity $v = v_x$ with respect to a, then observer b will find these components to be

$$\begin{aligned} E'_x &= E_x, \quad E'_y = \beta(E_y - (v/c)H_z), \quad E'_z = \beta(E_z + (v/c)H_y), \\ H'_x &= H_x, \quad H'_y = \beta(H_y + (v/c)E_z), \quad H'_z = \beta(H_z - (v/c)E_y). \end{aligned} \tag{14.1.9}$$

Cunningham and Bateman showed that Eqs. (14.1.4) and (14.1.5) are invariant under Lorentz transformations, and that the same is true of Eq. (14.1.2). **Figures. 14.1.1a,b** depict, in the frame of a, the electric field $E = E_y\mathbf{k}$, between the plates of a capacitor, and the magnetic field **H** generated by a current running through a wire on the x axis a. **Figures. 14.1.1c,d** depict the fields $\mathbf{E}'=$ and $\mathbf{H}'$ that are observed by b, moving at velocity $\mathbf{v} = 0.7c\mathbf{i}$ with respect to a.

14.2 The Conformal and Inversion Symmetries of Maxwell's Equations; Their Physical Interpretation

In the late 1860's and early 1870's Lie investigated the continuous groups of point transformations that leave metrics invariant, and he established connections between the continuous group of conformal transformations that leaves angles invariant in euclidean spaces and conformal transformations of the *Grupe der reciproken Radien* that may alter their metrics.[10] In Sec. 14.4 of Bateman's 1909 paper,[3b] he refers to this work, and using Minkowski's transformation to spacetime coordinates, he and Cunningham generalize Lie's results. Cunningham[3a] establishes that if k is a real constant, the

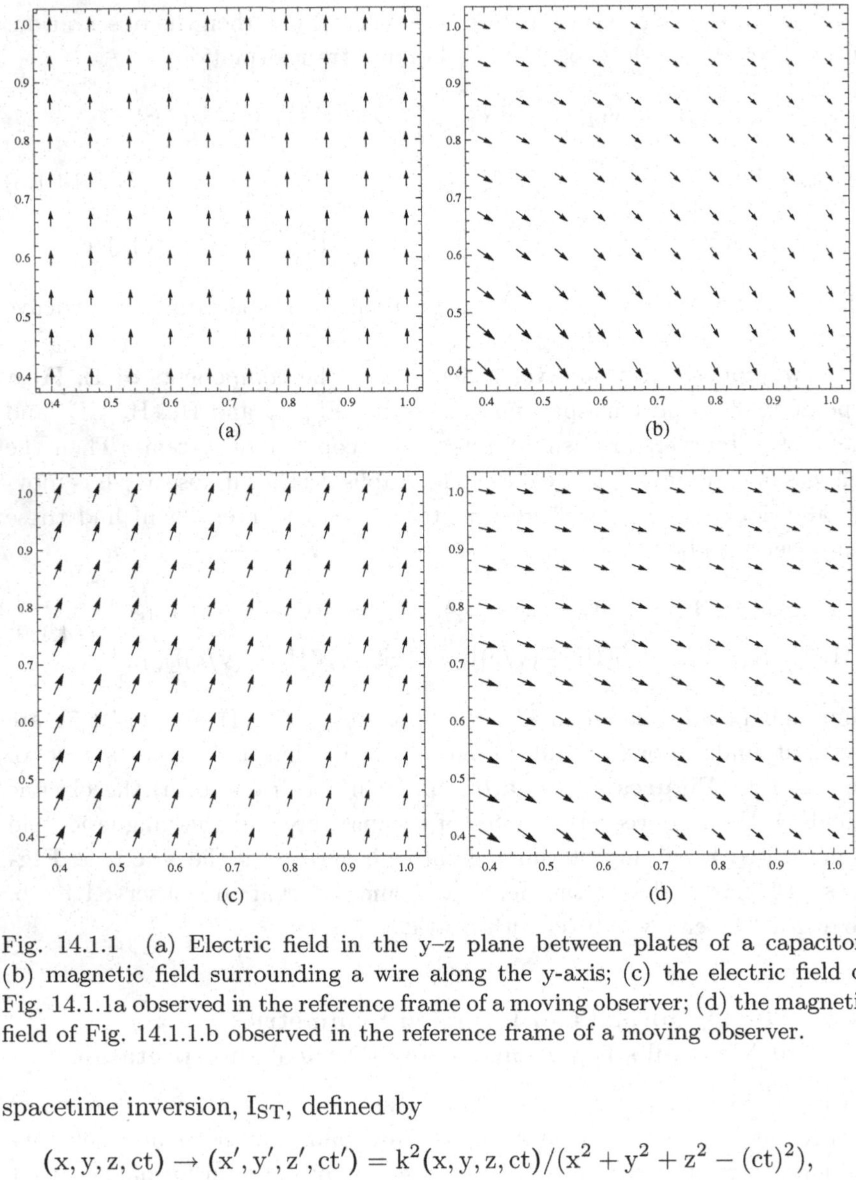

Fig. 14.1.1 (a) Electric field in the y–z plane between plates of a capacitor; (b) magnetic field surrounding a wire along the y-axis; (c) the electric field of Fig. 14.1.1a observed in the reference frame of a moving observer; (d) the magnetic field of Fig. 14.1.1.b observed in the reference frame of a moving observer.

spacetime inversion, I_{ST}, defined by

$$(x, y, z, ct) \rightarrow (x', y', z', ct') = k^2(x, y, z, ct)/(x^2 + y^2 + z^2 - (ct)^2), \tag{14.2.1a}$$

converts solutions of Maxwell's equations into solutions. For all real values of k, one may define the dimensionless four-vector

$$s = (x, y, z, ct)/k. \tag{14.2.1b}$$

Then k can be considered to be a scale parameter with the dimension of length. To further simplify notation, we define the scalar product of **s** with itself by

$$\mathbf{s}\cdot\mathbf{s} = \Sigma_j g_{jj} s_j^2 = s_1^2 + s_2^2 + s_3^2 - s_4^2 \equiv s^2. \qquad (14.2.1c)$$

With this change to dimensionless variables, (14.2.1a) becomes

$$\mathbf{s} \to \mathbf{s}' = \mathbf{s}/s^2. \qquad (14.2.1d)$$

It then follows that

$$s'^2 = s^2/(s^2)^2, \quad s'^2 s^2 = 1, \qquad (14.2.2a,b)$$

and that I_{ST}^{-1} is defined by

$$\mathbf{s} = \mathbf{s}'/s'^2 \qquad (14.2.2c)$$

Thus, I_{ST} is its own inverse.

In discussing the effect of the inversion (14.2.1) on electromagnetic fields, it is useful to define

$$\boldsymbol{\sigma} = (s_1, s_2, s_3) = (x, y, z)/k = \mathbf{r}/k, \qquad (14.2.3a)$$

$$\tau = s_4, \qquad (14.2.3b)$$

$$u^2 = s_1^2 + s_2^2 + s_3^2 + s_4^2 = \sigma^2 + \tau^2. \qquad (14.2.3c)$$

Cunningham[3a] showed that the inversion converts solutions **E** and **H** of Maxwell's equations, to solutions **E′** and **H′**, which may be defined by

$$\mathbf{E}' = (-s^2)\{u^2\mathbf{E} - 2(\boldsymbol{\sigma}\cdot\mathbf{E})\boldsymbol{\sigma} + 2\tau(\boldsymbol{\sigma}\times\mathbf{H})\}, \qquad (14.2.3d)$$

and

$$\mathbf{H}' = (s^2)\{u^2\mathbf{H} - 2(\boldsymbol{\sigma}\cdot\mathbf{H})\boldsymbol{\sigma} - 2\tau(\boldsymbol{\sigma}\times\mathbf{E})\}. \qquad (14.2.3e)$$

This result holds true for all choices of the origin of the spacetime coordinates for which **s** does not vanish.

Figures 14.2.1 illustrate the effect of inversions on the fields **E** and **H** of **Figs. 14.1.1a,b**. **Figures 14.2.2** depict the corresponding solutions, **E′** and **H′**, of Maxwell's equations produced by the relations (14.2.3).

In dealing with the effect of the Cunningham–Bateman inversion on the Lorentz metric

$$dx^2 + dy^2 + dz^2 - d(ct)^2 = dx_1^2 + dx_2^2 + dx_3^2 - dx_4^2, \qquad (14.2.4a)$$

we will usually use the dimension free metric

$$ds^2 = (dx^2 + dy^2 + dz^2 - d(ct)^2)/k^2. \qquad (14.2.4b)$$

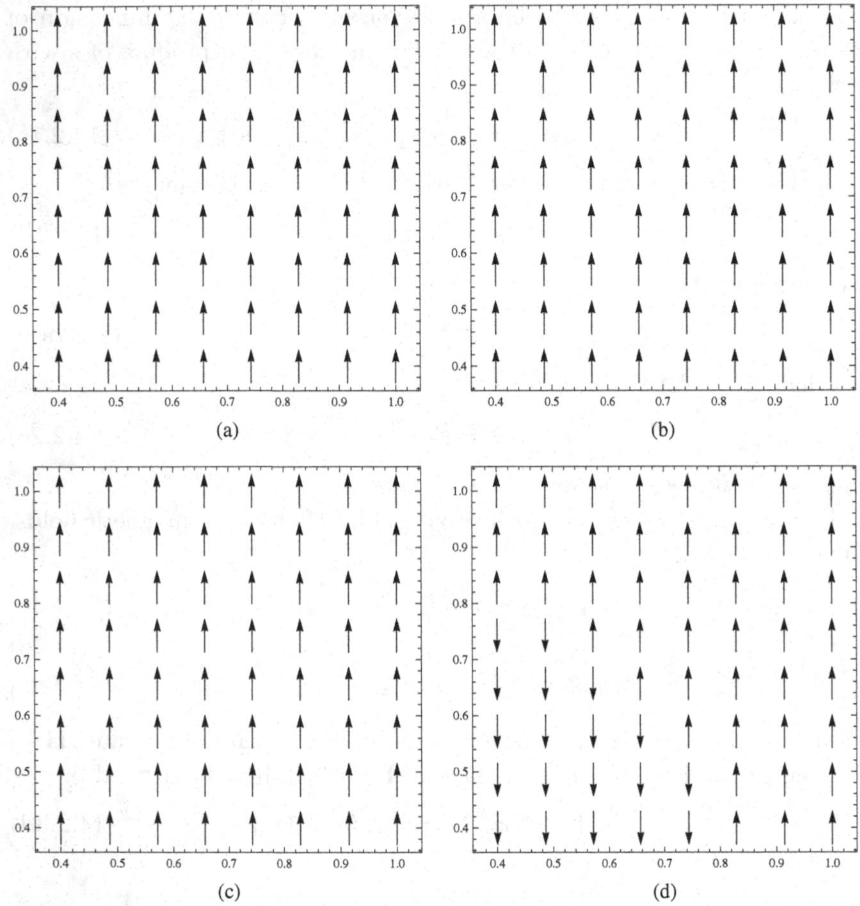

Fig. 14.2.1 (a) Electric field in the y–z plane between plates of a capacitor; (b) effect of the inversion on the electric field at t = 0; (c) effect of the inversion on the electric field at t = 0.5c; (d) effect of the inverson on the electric field at t = 0.9c.

Inversion converts (14.2.4b) to

$$\mathrm{ds}'^2 = (1/\mathrm{s}^2)^2 \mathrm{ds}^2. \tag{14.2.4c}$$

As both s^2 and ds^2 are Lorentz invariant, for all nonvanising $\mathbf{s}$, ds'^2 is also invariant under the operations of the Lorentz group. *However, the inversion does not leave invariant the Lorentz invariant metrics* (14.2.4a) or (14.2.4b) if $\mathrm{ds}^2 \neq 0$. *And, though ds^2 is invariant under Poincare transformations of* $\mathbf{s}$, ds'^2 *is not when* $\mathrm{ds}^2 \neq 0$. *This has the consequence that Maxwell's equations*

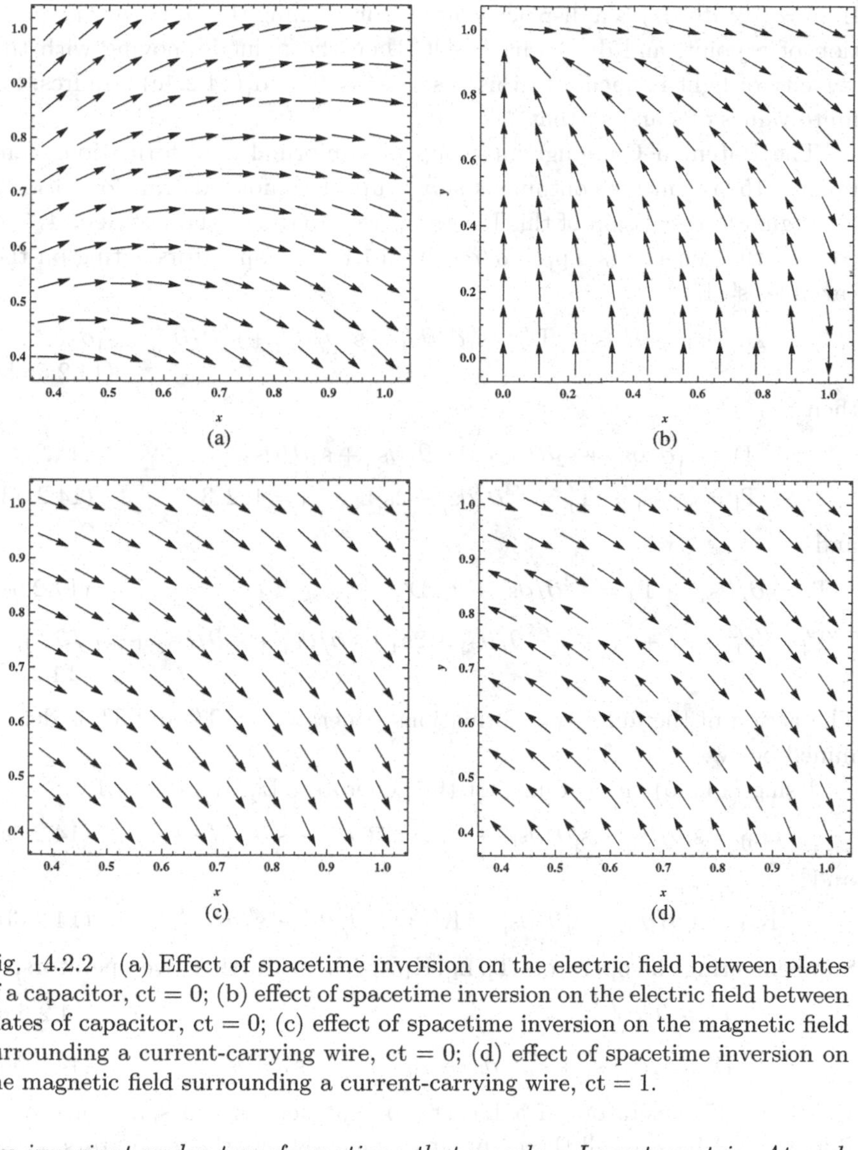

Fig. 14.2.2 (a) Effect of spacetime inversion on the electric field between plates of a capacitor, ct = 0; (b) effect of spacetime inversion on the electric field between plates of capacitor, ct = 0; (c) effect of spacetime inversion on the magnetic field surrounding a current-carrying wire, ct = 0; (d) effect of spacetime inversion on the magnetic field surrounding a current-carrying wire, ct = 1.

are invariant under transformations that rescale a Lorentz metric. At each origin the metric takes on a radial position dependence.

Using (14.2.2b), relation (14.2.4c) may be expressed in the more symmetrical form,

$$\mathrm{ds'^2/s'^2 = ds^2/s^2}. \tag{14.2.4d}$$

If $\mathbf{dr} = (dx, dy, dz)$ is a displacement vector defining the time-varying position of a point, and $dr \equiv |\mathbf{dr}| = cdt$, then the point is moving with the velocity of light in vacuo. In this case, $ds^2 = 0$, and (14.2.4c) requires, for finite values of s and s′, that $ds'^2 = 0$.

The Bateman–Cunningham group of conformal transformations contains a 15 parameter continuous subgroup of conformal transformations. To begin our discussion of this Lie group we use the method of **Sec. 4.7** to convert the generators $Tp_j = \partial/\partial x_j$ in (14.1.7) to operators acting on the variables s'_j. If

$$T_j \equiv k\partial/\partial x_j = \partial/\partial s_j, \quad D' \equiv s'_1\partial/\partial s'_1 + s'_2\partial/\partial s'_2 + s'_3\partial/\partial s'_3 + s'_4\partial/\partial s'_4, \tag{14.2.5a,b}$$

then

$$D \equiv s_1\partial/\partial s_1 + s_2\partial/\partial s_2 + s_3\partial/\partial s_3 + s_4\partial/\partial s_4, \tag{14.2.5c}$$

$$T_j = \partial/\partial s_j \to T'_j = s'^2\partial/\partial s'_j - 2s'_jD', \quad j = 1, 2, 3, \tag{14.2.5d}$$

and

$$T_4 = \partial/\partial s_4 \to T'_4 = s'^2\partial/\partial s'_4 + 2s'_4D', \tag{14.2.5e}$$

$$T'_4 = (s'^2_1 + s'^2_2 + s'^2_3 + s'^2_4)\,\partial/\partial s'_4 + 2s'_4\,(s'_1\partial/\partial s'_1 + s'_2\partial/\partial s'_2 + s'_3\partial/\partial s'_3). \tag{14.2.5f}$$

The action of the finite transformations generated by T'_1 and T'_4 is determined below.

Using (14.2.5) one obtains, for the generators L_{ij}, K_{j4} of (14.1.6),

$$L_{ij} = s_i\partial/\partial s_j - s_j\partial/\partial s_i \to L'_{ij} = s'_i\partial/\partial s'_j - s'_j\partial/\partial s'_i, \tag{14.2.6a}$$

and

$$K_{j4} = s_j\partial/\partial s_4 + s_4\partial/\partial s_j \to K'_{j4} = s'_j\partial/\partial s'_4 + s'_4\partial/\partial s'_j. \tag{14.2.6b}$$

Supplementing the operators T_j, L_{ij}, K_{ij}, with the further five operators,

$$P'_j \equiv \partial/\partial s'_j, \tag{14.2.6.c}$$

$$D' \equiv s'_1\partial/\partial s'_1 + s'_2\partial/\partial s'_2 + s'_3\partial/\partial s'_3 + s'_4\partial/\partial s'_4 \tag{14.2.6d}$$

yields the 15 generators of a Lie group that acts in the space of the s'_j. Replacing s'_j by s_j in all these operators, one obtains the generators of the isomorphic conformal group acting in the space of the s_j. These are

$$K_{j4}, \quad L_{ij},\ P_j = \partial/\partial s_j \quad \text{and} \quad D = \Sigma_1^4\ s_j\partial/\partial s_j, \tag{14.2.7a–d}$$

and

$$C_j \equiv s^2\partial/\partial s_j - 2s_jD, \quad j = 1, 2, 3. \tag{14.2.7e}$$

They satisfy the commutation relations of the Poincare group, cf. (14.1.7), and the following commutation relations:

$$\begin{aligned}
&[\mathrm{D}, \mathrm{L_{ab}}] = 0,\ [\mathrm{D}, \mathrm{K_{bc}}] = 0,\ [\mathrm{D}, \mathrm{P_a}] = -\mathrm{P_a}, && [\mathrm{D}, \mathrm{C_a}] = \mathrm{C_a},\\
&[\mathrm{L_{ab}}, \mathrm{C_b}] = \mathrm{C_a},\ \ \mathrm{b} = 1,2,3; && [\mathrm{L_{ab}}, \mathrm{C_4}] = 0,\\
&[\mathrm{K_{a4}}, \mathrm{P_b}] = -\delta_{\mathrm{a,b}}\, \mathrm{P_4},\ \ \mathrm{b} = 1,2,3; && [\mathrm{K_{a4}}, \mathrm{P_4}] = -\mathrm{P_a},\\
&[\mathrm{K_{a4}}, \mathrm{C_b}] = -\delta_{\mathrm{a,b}}\, \mathrm{C_4},\ \ \mathrm{b} = 1,2,3; && [\mathrm{K_{a4}}, \mathrm{C_4}] = -\mathrm{C_a},\\
&[\mathrm{P_a}, \mathrm{C_b}] = 2(\mathrm{L_{ab}} - \delta_{\mathrm{a,b}}\, \mathrm{D}),\ \ \mathrm{b} = 1,2,3; && [\mathrm{P_a}, \mathrm{C_4}] = 2\mathrm{K_{a4}},\ \ \mathrm{a} = 1,2,3;\\
&[\mathrm{P_4}, \mathrm{C_4}] = 2\mathrm{D}, && [\mathrm{C_a}, \mathrm{C_b}] = 0.
\end{aligned} \tag{14.2.8}$$

The $\mathrm{C_j}$ are known as generators of the special conformal group.[11] They, and D, inherit from the inversion transformation the ability to change Lorentz metrics.

The action of the group operators $\exp(\alpha_i \mathrm{C_i})$, can, *a la* Lie, be obtained from the translation operators, $\exp(\alpha_i \mathrm{P_i})$, by inversion. Noting that

$$\exp(\alpha_i \mathrm{C_i}) = (\mathrm{I_{st}})^{-1} \exp(\alpha_i \mathrm{P_i}) \mathrm{I_{st}}, \tag{14.2.9a}$$

one finds

$$\exp(\alpha_i \mathrm{C_i})\ \mathrm{s_j} = (\mathrm{s_j} + \delta_{\mathrm{i,j}} \alpha_i \bar{\mathrm{s}}^2)/(1 + 2\mathrm{g_{ii}} \mathrm{a_i} \mathrm{s_i} + \mathrm{a_i^2} \bar{\mathrm{s}}^2). \tag{14.2.9b}$$

With this result in hand, one can easily obtain all the finite transformations of the full 15 parameter conformal group. *The operations of this continuous group do not, in general, leave the Lorentz metric invariant.* The group is (see the exercises) locally isomorphic to a SO(4,2) group.

Cunningham argued that, just as in the case of the Poincare transformations of special relativity, the inversion and special conformal transformations of Maxwell's equations could not make it possible "for an observer to discriminate between the sequence of electromagnetic phenomena as he knows them and the sequence obtained by this transformation." Cunningham arrived at this conclusion before the advent of quantum mechanics. A fundamental contribution by Barut and Haugen[12] implies that it can only continue to be correct if one replaces the usual concept of (Poincare invariant) mass by a concept of mass that is invariant under the subgroup of special conformal transformations. Barut and Raczka point out that the operations of this subgroup of the Bateman–Cunningham invariance group of Maxwell's equations also leave invariant relativistic wave equations for massless particles of spin 0 and 1/2.[12] However, wave equations for massive particles are only invariant under its actions if the mass m is also transformed by a dilation operator that also carries out the transformation

$$m^2 \to m'^2 = e^{2\beta}m^2, \quad (14.2.10a)$$

and the special modified conformal transformations must convert

$$m^2 \to m'^2 = (1 + 2\boldsymbol{\alpha} \cdot \mathbf{s} + \alpha^2 s^2)m^2, \quad (14.2.10b)$$

and so yield a mass that evolves in spacetime in a manner that could be difficult to detect if the group parameters are small.[13]

To carry out these transformations, the authors added an additional term to the expressions for the group generators C_j and D. Several other attempts to use conformal transformations to extend special relativity will be found in papers by Fulton, Rohrlich and Witten,[13a] and by Hill[13b–d].

Taking a nonrelativistic limit of the generators of their conformal group, and converting them to Schrödinger operators, Barut and Raczka obtain the time-independent and time-dependent constants of motion of the Schrödinger equation of a free-particle, and of the Schrödinger equation of an isotropic oscillator. The author has shown how a related limiting process can be used to convert generators of the special conformal group to the Runge–Lenz operators of bound and unbound states of regularized hydrogen-like atoms, and produce the conjugate Fock variables.[14] The transformation converts relativistic free-particle motion to classical Kepler motion. Many other applications of conformal groups are discussed in an engaging and extensive historical review by Kastrup which also contains a critical analysis of Ref. 13.[15]

14.3 Alteration of Wavelengths and Frequencies by a Special Conformal Transformation: Interpretation of Doppler Shifts in Stellar Spectra

In this section we will consider further macroscopic consequences of the invariance of Maxwell's equations under the Bateman–Cunningham conformal group. As will be shown below, transformations of the special conformal group alter the frequencies and wavelengths of electromagnetic waves without altering their velocity. This can produce effects similar to the Doppler effect, and as will appear below, these may have been detected in radar signals received from spacecraft.

When a source emitting radiation with wavelength λ has velocity V' with respect to a spectrometer, the wavelength, λ', measured by a spectrometer or interferometer is given by the equation[16]

$$\lambda'/\lambda = \left[\frac{1 + V'/c}{1 - (V'/c)}\right]^{1/2}. \tag{14.3.1}$$

Expanding this in powers of V'/c and keeping only the first two terms, one obtains the non-relativistic Doppler equation

$$\lambda' = (1 + V'/c)\lambda. \tag{14.3.2a}$$

Rearranging this produces the usual Doppler relation

$$V' = c(\lambda'/\lambda - 1). \tag{14.3.2b}$$

The special conformal transformations that alter λ to λ' can produce the same effect as a relative velocity, V', of a source with respect to an observer.

Because the velocities of distant heavenly bodies are estimated using measured Doppler shifts of wavelengths in the spectra of radiation emitted by them, special conformal transformations can, in principle, introduce ambiguities into astronomy, astrophysics, and cosmology. Beginning in the 1920's Hubble made an extensive series of astronomical observations of Cepheid variable stars and the Doppler shifts of spectral lines in light received from them.[17] He concluded that his observations were consistent with the supposition that in a galaxy the stars at a distance R from the earth are receding from earth with a mean radial velocity V' proportional to R. This relation may be expressed as

$$V' = H_0 R. \tag{14.3.3a}$$

If V represents the radial velocity of star with respect to this mean, this becomes Hubble's equation

$$V' = H_0 R + V. \tag{14.3.3b}$$

In the mid 1940's, it was estimated that Hubble's constant, H_0 had a value approximately equal to $1.8 \times 10^{-17} s^{-1}$. The currently accepted value of H_0 is $(2.19 \pm 0.56) \times 10^{-18} s^{-1}$.[18a] The estimated values of V, R, and H_0 depend in part upon observations and empirical relations that supplement Doppler measurements — most notably, relations between the period and brightness of Cepheid variables.[18b] The analyses and correlations of observations that support Hubble's theory and equation (14.3.3b) are now very extensive. They have led to the conclusion that throughout the universe objects not gravitationally bound to each other are uniformly receding from each other.

Thus distances between most heavenly objects appear to slowly increase in much the way distance between points on the surface of a balloon increase as the balloon is inflated.

To take this inflation into account, cosmologists often alter Minkowski's metric. It is supposed that in regions free of strong local gravitational effects the metric governing heavenly motions is a Robertson–Walker metric, a metric of the form[18c]

$$\mathrm{ds}^2 = \mathrm{dx}_4^2 - \mathrm{a}(\mathrm{x}_4)^2\{\mathrm{dr}^2/(1-(\mathrm{r}/\rho)^2) + \mathrm{r}^2(\mathrm{d}\theta^2 + \sin^2(\theta)\mathrm{d}\phi^2)\}. \quad (14.3.4)$$

In this expression, ρ is a constant with the dimension of length, and the function $\mathrm{a}(\mathrm{x}_4)$ expresses the time dependence of the radius $\rho\mathrm{a}(\mathrm{x}_4)$ of the universe.

In 1945, Hill[19] noticed an unexpected property of the special conformal transformations generated by

$$\mathrm{C}_4 = (\mathbf{r}^2 + \mathrm{x}_4^2)\partial/\partial\mathrm{x}_4 + 2\mathrm{x}_4\mathbf{r}\partial/\partial\mathbf{r}. \quad (14.3.5)$$

The group operator $\exp(\beta\mathrm{C}_4)$ carries out the transformations

$$\mathbf{r} \to \mathbf{r}' = \gamma\mathbf{r}, \quad \mathrm{x}_4 \to \mathrm{x}_4' = \gamma\{\mathrm{x}_4 - \beta\mathrm{s}^2\}, \quad (14.3.6\mathrm{a,b})$$

in which

$$\mathrm{s}^2 = (\mathrm{x}_4^2 - \mathbf{r}^2),\ \gamma = \gamma(\beta, \mathrm{x}_4, \mathbf{r}) = (1 - 2\beta\mathrm{x}_4 + \beta^2\mathrm{s}^2)^{-1} \quad (14.3.6\mathrm{c,d})$$

To the first-order in the group parameter:

$$\mathbf{r}' = (1 + 2\beta\mathrm{x}_4)\mathbf{r}, \quad \mathrm{x}_4' = \mathrm{x}_4' + \beta(\mathrm{r}^2 + \mathrm{x}_4^2)/2. \quad (14.3.7\mathrm{a,b})$$

Hill wrote these equations as

$$\mathbf{r}' = (1 + \alpha\mathrm{t})\mathbf{r}, \quad \mathrm{t}' = \mathrm{t} + \alpha(\mathbf{r}^2/\mathrm{c}^2 + \mathrm{t}^2)/2, \quad (14.3.7\mathrm{c,d})$$

with $\alpha = 2\beta\mathrm{c}$. From them he derived the change in velocity from $\mathrm{v} = \mathrm{dr}/\mathrm{dt}$, to $\mathrm{v}' = \mathrm{dr}'/\mathrm{dt}'$:

$$\mathbf{v}' = \mathbf{v} + \alpha(\mathbf{r} - \mathbf{v}(\mathbf{r}\cdot\mathbf{v})/\mathrm{c}^2). \quad (14.3.8)$$

On neglecting terms of order $\mathbf{v}^2/\mathrm{c}^2$ this gives

$$\mathbf{v}' = \mathbf{v} + \alpha\mathbf{r}. \quad (14.3.9)$$

Hill observed that this equation could be interpreted as Hubble's relation if one sets α equal to H_0. However, in startling contrast to Hubble's relation, Hill's relation arises solely from a change in spacetime coordinate systems. Its implications have been overlooked for nearly 65 years. Here, we will concentrate our attention on one of the questions it raises, the question:

"What is the spacetime metric governing the motion of electromagnetic waves?"[20]

To address this question it is necessary to develop additional consequences of the transformation generated by C_4. Using (14.3.6) one finds that the transformation carries

$$\mathrm{dr} \rightarrow \mathrm{dr}' = \gamma^2(\mathrm{A\ dr} + \mathrm{B\ dx_4}), \tag{14.3.10a}$$

and

$$\mathrm{dx_4} \rightarrow \mathrm{dx}_4' = \gamma^2(\mathrm{B\ dr} + \mathrm{Adx_4}), \tag{14.3.10b}$$

with

$$\mathrm{A} = 1 - 2\beta \mathrm{x_4} + \beta^2(\mathrm{r}^2 + \mathrm{x}_4^2), \quad \mathrm{B} = 2\beta \mathrm{r}(1 - \beta \mathrm{x_4}). \tag{14.3.10c,d}$$

Thus

$$\mathrm{dr}'/\mathrm{dx}_4' = (\mathrm{A\ dr/dx_4} + \mathrm{B})/(\mathrm{B\ dr/dx_4} + \mathrm{A}). \tag{14.3.10e}$$

Using (14.3.10) one finds that the Minkowski metric

$$\mathrm{ds}^2 = |(\mathrm{dr})^2 - (\mathrm{dx_4})^2| \tag{14.3.11}$$

is transformed to the conformal metric

$$\mathrm{ds}'^2 = |(\mathrm{dr}')^2 - (\mathrm{dx}_4')^2| = \gamma^2|(\mathrm{dr})^2 - (\mathrm{dx_4})^2|. \tag{14.3.12a}$$

Thus

$$\mathrm{ds}'^2 = \gamma(\alpha/2\mathrm{c}, \mathrm{x_4}, \mathrm{r})^2\mathrm{ds}^2, \quad \mathrm{ds}^2 = \gamma(\beta, \mathrm{x}_4', \mathrm{r}')^2\mathrm{ds}'^2. \tag{14.3.12b,c}$$

Note that the conformal metric smoothly reduces to the Minkowski metric as the group parameter α approaches zero. This metric is specific to C_4. As it is the only special conformal metric we will deal with in the remainder of the chapter, we will sometimes, simply call it *the conformal metric.*

The function γ introduces a relation between coordinates that destroys the Poincare invariance of spacetime with metric ds^2. In this respect the conformal metric is qualitatively similar to the Robertson–Walker metric. However, the conformal metric produces a local multiplicative scaling of both dr and $\mathrm{dx_4}$, while the Robertson–Walker metric only applies such a rescaling to dr. Most importantly, in the conformal metric the choice of spacetime origin is arbitrary. However because the origin $\mathrm{r} = 0, \mathrm{t} = 0$, is an invariant point of the conformal transformation, there one also has the special property $\mathrm{r}' = 0, \mathrm{t}' = 0$.

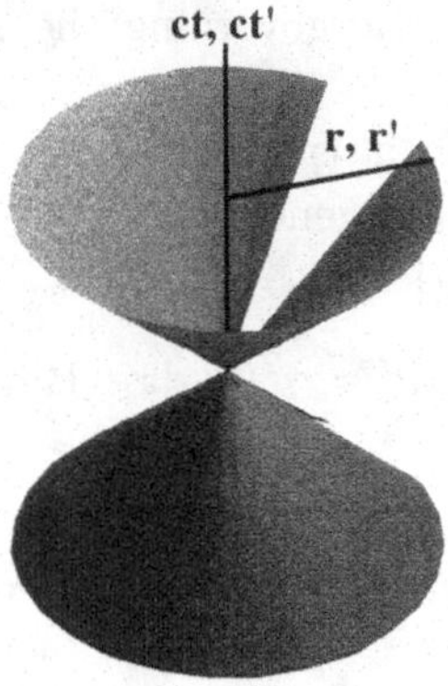

Fig. 14.3.1 Radial and axial coordinates of points on a light cone.

Though the conformal metric is not Poincare-invariant, $\exp(\beta_4 C_4)$ defines similarity transformations that convert relations between group generators that subsist in Minkowski spacetime into isomorphic relations in the conformal spacetime. Thus the Poincare group reappears with a different realization as an invariance group of spacetime with conformal metric ds'^2.

We will henceforth consider that Eqs. (14.3.6), together with Eqs. (14.3.10), define *mappings* that relate the two sets of coordinates, $(r, t, v = dr/dt)$ and $(r', t', v' = dr'/dt')$, of any point, P. As will perhaps become evident in the following paragraphs, thinking in terms of mappings can help one avoid the insertion of unrecognized presuppositions into descriptions of physical phenomena.

Let a point P on a trigonometric wave have ordinary coordinates (r,t), and conformal coordinates (r', t'). If it is an electromagnetic wave, the point will have radial velocities

$$dr/dt = v_{EM} = dr'/dt', \quad |v_{EM}| = c, \tag{14.3.13}$$

the same in both coordinate systems. Suppose $P_1 = P+dP$ and $P_2 = P-dP$ are two such points on an electromagnetic wave, and suppose they are separated by one wavelength. Denote this wavelength "at P," as measured in the two systems, by λ and λ', and denote the corresponding frequencies by ν and ν'. (As a consequence of Eq. (14.3.12), one has $\lambda\nu = c = \lambda'\nu'$.) The frequency of the wave in the region centered at P, as measured in the two coordinate systems, is the inverse of the time interval that it takes for this one wavelength to pass through a point with fixed position coordinates r, r'. Let δt and $\delta t'$ be the interval as measured in the two systems. Equation

(14.3.10b) then implies that

$$\delta t'/\delta t = \nu/\nu' = \frac{1 - 2\beta x_4 + \beta^2(s^2 + 2r^2)^2}{(1 - 2\beta x_4 + \beta^2 s^2)}. \tag{14.3.14}$$

Thus as a point P on the wave propagates with it, (14.3.14) requires that the *conformal wavelength*, λ', of the wave at P must vary with the coordinates (r, x_4) according to the law

$$\lambda'/\lambda = \{1 - 2\beta x_4 + \beta^2(s^2 + 2r^2)\}/(1 - 2\beta x_4 + \beta^2 s^2)^2. \tag{14.3.15a}$$

Consequently,

$$\lambda'/\lambda = 1 + 2\beta x_4 + 3\beta^2(x_4^2 + r^2) + O(\beta^3). \tag{14.3.15b}$$

Because frequencies and wavelengths measured in Minkowski space do not vary with position or time, λ may be considered constant in this equation. It follows that the wavelength in the conformal coordinate system increases or decreases with t as βct is positive or negative. The two wavelengths are equal at the common origin of coordinates, (0, 0). As there is no evidence for supposing that λ' differs from λ in regions near a wave source, we shall choose the origin (0, 0) to be at the source of interest.

When a wave with wavelengths λ, λ' at (r, x_4), (r', x_4'), travels from its source to the observer, the change in r must equal the change in x_4. If the source is fixed at the origin, r remains equal to x_4, and substituting $\beta = \alpha/2c$ into (14.3.15b) yields

$$\lambda'/\lambda = 1 + \alpha r/c + 3/2(\alpha r/c)^2 + O((\alpha/c)^3). \tag{14.3.16}$$

This implies that when r reaches R,

$$c(\lambda'/\lambda - 1) = c\{\alpha R/c + 3/2(\alpha R/c)^2 + O((\alpha/c)^3). \tag{14.3.17}$$

Interpreting the left-hand side of this equation as a Doppler shift, one obtains to $O(\alpha R/c)$,

$$c(\lambda'/\lambda - 1) = V' = \alpha R. \tag{14.3.18}$$

This Doppler shift will not be observed if the propagation of the wave has taken place in a space with Minkowski metric: it may be considered to have arisen because the conformal wavelength, λ', changes as the wave propagates.

When the observer has velocity $V = dR/dT$ relative to the source, then the wavelength λ in the (R,T) system undergoes the additional Doppler shift given by the usual formula

$$\Delta\lambda/\lambda = \left[\frac{1+V/c}{1-V/c}\right]^{1/2} - 1 = V/c + O(V^2/c). \qquad (14.3.19)$$

If the sign of V/c is positive when the source is receding from the observer, a reshift is produced. On neglecting terms of $O((\alpha R)^2/c)$ and $O(V^2/c)$, (14.3.18) then yields Hill's relation

$$V' = \alpha R + V. \qquad (14.3.20)$$

For values of α as small as H_0, the group parameter α could, in principle, be determined by experimental measurements of the radial positions and velocities of distant sources producing the measured Doppler shifts.

For sources as distant as the nearest star, only Doppler shifts can be directly measured. However, the Pioneer spacecraft program established that it is possible to measure, with an accuracy of one part in 10^{12}, Doppler shifts of S band radar frequencies that are developed over distances up to 70 AU.[21] This precision is sufficient to measure α values to the order of 10^{-18}sec^{-1} with an accuracy better than 1%.

Suppose, for simplicity of thought, that an observer on earth arranges to have a clock and interferometer with no accelerations. Let these have coordinates $R = R' = 0$. Suppose that at $T = T' = 0$, the observer sends a radar signal to a receding spacecraft. The radar signal will take a time T, T' to reach the spacecraft, and if it is immediately returned, when it reaches the observer the clock times will be $2T, 2T'$ respectively. The observer will conclude that when the signal reached the spacecraft the distance from it was $R = cT, R' = cT'$. Equations (14.3.7c,d) imply that to the first-order in α,

$$\begin{aligned} R' &= (1+\alpha T)R, \quad R = (1-\alpha T')R', \\ T' &= (1+\alpha T)T, \quad T = (1-\alpha T')T'. \end{aligned} \qquad (14.3.21a\text{–}d)$$

If V, V' is the radial velocity of the spacecraft with respect to the origin, then to the first order in α,

$$V' = V + \alpha R, \quad V = V' - \alpha R'. \qquad (14.3.21e,f)$$

In the Pioneer experiments systematic differences were found between the observed Doppler shifts and the shifts that were expected. These anomalous results have since been denoted the "Pioneer Anomaly."

To understand the origin of this anomaly it is necessary to rather carefully investigate how it arose. This first become possible in early 2010, when an extensive report on the Pioneer program was put on the internet[22]. The following discussion benefits from additional information kindly supplied by Toth, one of its authors.[23] In the Pioneer program, pulses of radar waves of fixed frequency were sent from sources on earth to the receding Pioneer 10 and 11 spacecraft, which transmitted back to earth an amplified signal that had been phase locked at the spacecraft to the incoming signal it had received. When these wave pulses arrived back at their source they were mixed with a wave of fixed frequency. The beat frequency this produced was the frequency shift developed in the wave during its journey.

At the time the experiments were carried out it was not possible to directly measure the distance to the spacecraft at the moment it receives the pulse sent back to earth. The time and distance had to be estimated by comparing the measured Doppler frequency with that produced by a sophisticated dynamical model solving the equations of motion for the spacecraft, given the forces acting upon it. The model was first supplied with initial values, $\mathrm{t_0, r_0, v_0}$. It then predicted the positions $\mathbf{r}(\mathrm{t})$ and velocities $\mathbf{v}(\mathrm{t})$ of the spacecraft, and the Doppler shift that would result at time t. The observed frequency shift was carefully matched with the predicted Doppler frequency to obtain the best estimate to the modeled value of v, and hence t, and r. The new values of $\mathbf{r}(\mathrm{t})$, $\mathbf{v}(\mathrm{t})$ were then evolved by the equations of motion, which were thereby expected to predict values of subsequent Doppler shifts that would better match the next observed Doppler shift. This matching and revision was repeated until the spacecraft reached a distance of about 70 AU and the signals became too weak to use. It was found that, as the distance, r, to the spacecraft increased beyond 20 AU, the Doppler shift corrections required by the matching process were small frequency shifts $\Delta\nu$. These frequency shifts appeared to correspond to the blueshift that one would obtain from anomalous velocity contributions

$$\mathrm{v} = -(2.80 \pm 0.42) \times 10^{-18}\mathrm{sec}^{-1}\mathrm{r}. \qquad (14.3.22)$$

However, in developing the dynamical model, it was presumed that when strong gravitational fields were not present, both the spacecraft and the radar waves moved in a space with Minkowski metric, a space in which the coordinates r, t, v are those that apply. The Doppler matching procedure used to modify the model, while enabling it to continue supposing that the measurements were made in a Minkowski space, transferred to the model the functional dependence on the $\mathrm{r', t', v'}$ coordinates that arises if the metric in which the waves move is the special conformal metric of (14.3.12). To understand the consequences it is helpful to reorganize Hill's approximate equation, which for $V \ll c$, can be written as

$$V'_C - V_M = \alpha R. \tag{14.3.23a}$$

As the distance R′ traversed by the radar waves was twice the distance r in (14.3.22), in Eq. (14.3.23c) one has $-2\alpha r = -(2.8 \pm 0.42) \times 10^{-18}$ sec^{-1} r, so $\alpha = (1.4 \pm 0.42) \times 10^{-18}$ sec^{-1}. This converts (14.3.23a) to

$$V'_M - V'_C = \alpha R', \text{i.e. } V'_C - V_M = -\alpha R'. \tag{14.3.23b,c}$$

As R′ differs from R only by terms of $O(\alpha^2)$ in these relations R′ may be replaced by R. This has the consequence that sets $-\alpha$ to have the anomalous reported value $-(2.80 \pm 0.42) \times 10^{-18}$sec^{-1}. Thus *the anomalous Pioneer blueshift corrections arise from a physical redshift of radar wavelengths. This shift has the magnitude, and the dependence on* R *that develop if the waves move in a space with special conformal metric that has the group parameter* α *approximately equal to 2/3 Hubble's constant* H_0. This explanation of the Pioneer anomaly was anticipated by Tomilchik before a detailed description of the modeling process was available.[22] However, he concluded that the group parameter is actually Hubble's constant.

Though the value of α deduced from the Pioneer observations is currently close to 2/3 the value $(2.19 \pm 0.56) \times 10^{-18}$sec^{-1} assigned to Hubble's constant, this relation between α and H_0 should not be assumed because the Pioneer matching algorithms have altered several parameters in the dynamical model, and at least one fixed parameter may have been assigned an incorrect value.[23] Moreover, in the actual matching process other parameters in the dynamical model were also altered. The whole process that uncovered the Pioneer anomaly is currently the subject of a series of further investigations which could have the effect of significantly altering the numerical value of α.[23,24]

It appears that spacecraft similar to Pioneer 10 and Pioneer 11 could provide quite accurate values of all the variables required to directly determine the group parameter α, and so determine the spacetime metric in whatever distant regions they might visit.[20] To do so, the spacecraft should, like the Pioneers, be equipped with a system to receive radar pulses from ground stations and return them as amplified pulses that have been phase-locked to the received pulses. If both the Doppler shifts, and time-delays between emission and receipt of successive pulses are measured, the necessary data are provided. Repeaters installed on the outer planets and their moons might also be able to provide useful restrictions on the value of α, and hence the metric.

If it is established that α has a nonzero value, it will be necessary to reckon with the fact that measurements of wavelengths are measurements of special conformal wavelengths λ', and that the currently accepted metrics must be changed to special conformal ones. A few of its consequences are suggested by considering the way in which it would affect the interpretations of Hubble's equation (14.3.3b). In doing so it is helpful to replace V by V$'$, R by R$'$, and set $\mathrm{H_0} = \mathrm{h_0} + \alpha$. The value of $\mathrm{h_0}$ will then determine the universal rate of expansion previously determined by $\mathrm{H_0}$. V$'$ and R$'$ will become the special conformal analogs of V and R.

Conclusion

Few physical implications of Lie's discoveries were noticed prior to the discoveries of Bateman and Cunningham. Hill seems to have been one of the first to investigate possible physical consequences of their conformal group. The increasing interest in the role of group theory in quantum mechanics that began in the 1960's led to an increasing recognition of possible consequences of the conformal symmetry of Maxwell's equations at the submicroscopic level. Some applications of the conformal group in quantum theory abandon the requirement of Lorentz invariance on the submicroscopic scale. Hill's work and the Pioneer results suggest that Poincare invariance of electromagnetic phenomena on the supermacroscopic scale should not be assumed. Though no conclusions as startling as that of Hill have emerged, a number of workers have investigated relativistic and nonrelativistic consequences of conformal invariance.[25] If further determinations of the conformal group parameter α show that it is nonzero, much physics will have to be revised. If it is found that α has a value of the same order of magnitude as Hubble's constant, the consequences for physics and cosmology will be even more extensive.

Over a century has passed since Bateman and Cunningham established the conformal invariance group of Maxwell's equations. However, as has become evident in this chapter, many fundamental physical implications of this invariance have yet to be established.

Exercises

1. Use the extended generators of the Poincare group to show that the group leaves the Lorentz metric (14.2.4a) invariant.

2. Determine whether or not the Lie subgroup of the Bateman–Cunningham conformal group leaves invariant the metric (14.2.4c).
3. Use the methods of Chapter 4 to convert the generators $u_j = \partial/\partial x_j$, of translations in spacetime, to Lie generators $U_j = \Sigma_k \xi_{jk}(X)\partial/\partial X_k$ that act on functions of the inverted variables X.
4. Do the same for the six generators of Lorentz transformations.
5. Show that the conformal Lie group with generators (14.2.5) is locally isomorphic to SO(4,2).
6. Determine the effect of finite special conformal transformations on straight-lines in spacetime, and suppose that the lines represent trajectories of a particle. Compare the physical interpretations of your results when you have independent evidence that a force is acting, and when no such evidence exists.
7. Integrate the infinitesimal transformation of $x_4 = ct$ and r generated by Hill's X_{14}, and determine its invariant function. Use Hill's physical interpretation of the transformation to give a physical interpretation to this function.
8. In a series of articles[26] Hill developed a generalization of special relativity by considering that coordinate systems moving at constant relative velocity are special cases of coordinate systems moving at constant relative acceleration. Can you detect any errors in Hill's analysis in Ref. 26a? What conclusions can be drawn about interpretations of Newton's laws? Cf. Kastrup's review.[15]
9. The action of $\exp(\beta C_4)$ on a Hamiltonian H converts it to
$$H' = H + 2(\alpha/c)[C_4, H] + O((\alpha/c)^2).$$
   (a) Show that to the first-order in α, a radial potential r^{-n} in H undergoes the dilatation
$$r^{-n} \to (1 - 4(\alpha/c)n)r^{-n}.$$
   (b) Show that to the first-order in α, the Laplacian operator Δ undergoes the conversion
$$\Delta \to (1 - 8\alpha t)\Delta - (4\alpha/c^2)(3 + 2r\partial/\partial r)\partial/\partial t,$$
   and that
$$\partial/\partial t \to \partial/\partial t - 4\alpha(t\partial/\partial t + r\partial/\partial r).$$
10. Determine, to the first-order in β, the effect of $\exp(\beta C_4)$ on the energies, transition frequencies of H atoms, and the evolution of its wave functions.

11. Generalize C_4 to a C'_4 that contains an additional term which acts on the mass M of the proton in a hydrogen atom such that $\exp(\beta C'_4)$ will leave invariant the Schrödinger equations governing the motion of an electron with reduced mass μ.

References

[1] A. Einstein, *Ann. Phys.*, **17** (1905) 891.
[2] H. Lorentz, *Proc. Amstrdam Acad.* **6** (1904) 809.
[3] (a) E. Cunningham, *Proc. London Math Soc.* **8**(2) (1909) 77;
(b) H. Bateman, *ibid.* 223;
(c) H. Bateman, *Proc. London Math Soc.* **7**(2) (1909) 70.
[4] J. C. Maxwell, *Electricity and Magnetism* (Cambridge University Press, Cambridge, 1873).
[5] a) W. F. Magie, *Source Book in Physics* (McGraw-Hill, N.Y., 1935), p. 408;
(b) E. Whittaker, *A History of the Theories of Aether and Electricity* (Thomas Nelson and Sons, Edinburgh, 1951), p. 83.
[6] W. D. Niven, ed., *The Scientific Papers of James Clerk Maxwell* (Cambridge University Press, Cambridge, 1880); photographic reprint (J. Hermann, Paris), (Steichert, N.Y., 1930).
[7] L. P. Wheeler, *Josiah Willard Gibbs* (Yale University Press, New Haven, 1952).
[8] J. A. Kong, *Electromagnetic Wave Theory*, 2nd edn. (Wiley Interscience, New York, 1990).
[9] J. C. Maxwell, *Trans. Royal. Soc. (London)* **CLV**, (1864), Collected works, XXV, A dynamical theory of the electromagnetic field, Part I, Sec. (20).
[10] S. Lie, *Transformations Grupen, III* (Leipzig, 1893), p. 351; reprinted (Chelsea, New York, N.Y., 1970).
[11] B. G. Wybourne, *Classical Groups for Physicists*, (Wiley, New York, 1974), pp. 345–7.
[12] (a) A. O. Barut and R. Raczka, *Theory of Group Representations and Applications*, 2nd revised edn. (World Scientific, Singapore, 1986);
(b) A. O. Barut and R. B. Haugen, *Ann. Phys.* **71** (1972) 519.
[13] T. Fulton, F. Rohrlich and L. Witten, *Rev. Mod. Phys.* **34** (1962) 442.
[14] C. Wulfman, *J. Phys. A: Math. Theor.* **42** (2009) 185301.
[15] H. A. Kastrup, *Ann. Phys.* **17** (2008) 631.
[16] M. Born, *Einstein's Theory of Relativity*, translated by H. Brose (Methuen, London, 1924), p. 240.
[17] E. Hubble, *The Realm of the Nebulae* (Yale University Press, New Haven, 1936).
[18] (a) P. J. E. Peebles, *Principles of Physical Cosmology* (Princeton, 1993), pp. 24, 71, 82;
(b) *ibid.*, pp. 20, 106–7;
(c) *ibid.*, pp. 73, 4.
[19] E. L. Hill, *Phys. Rev.* **68** (1945), 235.
[20] C. Wulfman, [arXiv: 1003.0507]; [arXiv: 1010.2139] (2010).

[21] J. D. Anderson et al., *Phys. Rev. D* **65** (2005) 082004.
[22] L. M. Tomilchik, *AIP Conf. Proc.* **1205** (2010) 177.
[23] S. G. Turyshev and V. T. Toth, [arXiv:1001.3686v].
[24] V. T. Toth, private communication.
[25] (a) F. Hoyle, G. Burbidge and N. Narlikar, *A Different Approach to Cosmology* (Cambridge University Press, Cambridge, 2000).
(b) E. F. Bunn and D. W. Hogg, *Am. J. Phys.* **77** (2009).
[26] (a) E. L. Hill, *Phys. Rev.* **67** (1945) 358;
(b) E. L. Hill, *Phys. Rev.* **72** (1947) 143;
(c) E. L. Hill, *Phys. Rev.* **94** (1951) 1165.

Index

Printed in the United States
By Bookmasters